Pharmacological Perspectives of Toxic Chemicals and Their Antidotes

Pharmacological Perspectives of Toxic Chemicals and Their Antidotes

Editors

S J S Flora
James A Romano
S I Baskin
K Sekhar

Springer-Verlag

Narosa Publishing House
New Delhi Chennai Mumbai Kolkata

Editors
S J S Flora
Division of Pharmacology and Toxicology
Defence Research and Development Establishment
Gwalior, India

James A Romano
U.S. Army Medical Research Institute of Chemical Defense
Aberdeen Proving Ground
Maryland, U.S.A.

Steven I Baskin
U. S. Army Medical Research Institute of Chemical Defense
Aberdeen Proving Ground
Maryland, U.S.A

K Sekhar
Defence Research and Development Establishment
Gwalior, India

Springer-Verlag Berlin Heidelberg New York
A member of Springer Science + Business Meda GnbH
Http://www.sprinngeronline.com

ISBN 3-540-22378-9 Springer-Verlag Berlin Heidelberg New York
ISBN 0-387-22378-9 Springer-Verlag New York Berlin Heidelberg
ISBN 81-7319-548-X Narosa Publishing House, New Delhi

Printed in India

Dedication

On 29th Oct. 2002 Dr. Satyanarayan Motilal "Satu" Somani, a renowned pharmacologist and toxicologist at Southern Illinois University, School of Medicine, Springfield, IL and senior member of Association of Scientist of Indian origin in America (ASIOA), passed away of an abdominal aortic aneurysm. Dr. Somani was a member of many professional scientific societies in the field of his research. He was an author of more than 120 research articles in various areas of pharmacology, toxicology and exercise pharmacology. He published numerous review articles and book chapters. He has edited four books: Environmental toxicology: Principles and policies (1981); Chemical warfare agents, Academic press (1992); Pharmacology in Exercise and Sports, CRC Press (1996); Chemical warfare agents: toxicity at low level, CRC Press (2001). He was a featured lecturer at many national and international scientific meetings. Dr. Somani was instrumental in organizing the international participation of the 35th Annual Meeting of the Indian Pharmacological Society. In particular he organized the sessions upon which the current book is based. Dr. Somani was a significant player in evaluating and conducting key research in the area of medical chemical defense. His stature was so significant that he was invited by the US Congress to testify on "Gulf War Syndrome" in April 1997. He received numerous research grants from US Army, American Heart association, EPA, and other agencies. Dr. Somani was also a supporter of Ayurvedic Medicine and lectured on Ayurveda as a complement to allopathic medicine, and helped the university put together a Complementary and Alternative Medicine task force. As a well-respected colleague, academician, researcher, scientist and friend, he has contributed immensely to the advancement of science. His scientific impact will be remembered and felt by scientific community for many years to come.

Foreword

It is a distinct honor to have been invited by my colleagues in the Chemical and Biological Defense arena to write the foreword to this very articulate and scientifically state-of-the-art book entitled "Pharmacological Perspectives of Toxic Chemicals and Their Antidotes". In addition it provides me the opportunity to express my personal tribute to the memory of a good friend and colleague Dr. Satu Somani.

Dr. Somani was internationally recognized and respected as a scholar, and educator, as well as a pharmacologist and toxicologist. He was admired and an inspiration to his colleagues and students. I recall his excitement at a Society of Toxicology meeting when he discussed his plans with me for the 35th Annual Conference of the Indian Pharmacology Society on "Chemical and Biological Warfare" (CBW). This conference was held in 2002 at Gwalior, India, and this volume was an outgrowth of the presentations that were compiled and edited by his colleagues and good friends, Drs. Flora, Romano, Baskin and Sekhar. The effort and expertise that Dr. Somani invested in both of his books on Chemical Warfare Agents, his publications in this field, as well as his orchestration of this symposium, makes the dedication of this book to his memory a well deserved and fitting tribute.

Dr. Satu Somani, Professor of Pharmacology and Toxicology at Southern Illinois University (SIU) since 1974, was dedicated to research and teaching, as well as to his students and his native India. As he had done on numerous occasions, he was planning to meet several of his SIU medical students in India following this conference to work with them to inform medical students in India about problem-based learning (PBL). This concept, although new to most of India's medical schools, is the core philosophy at SIU medical school. In a great part due to Dr. Somani's influence and the workshops he has held in India, some medical schools in India have implemented problem-based learning. His passion for science, India and its culture, as well as for his students and colleagues was demonstrable in his daily life.

Dr. Satu Somani left us for his laboratory in the sky on October 29, 2002, and is sorely missed by all.

History documents the use of chemical and biological agents, as well as weapons of mass destruction, against troops in warfare and against civilians collaterally or deliberately by terrorists. The actual use, or threat of their use, can cause not only physical, physiological and pathological effects, but also psychological effects. The U.S. abandoned its offensive chemical program in 1969, and the Chemical Weapons Convention (CWC) prohibits their use. However, there are many rogue states, such as those included in speeches by President George Bush designated as the "Axis of Evil" and by Undersecretary of State John Bolton designated as "Beyond the Axis of Evil", that are believed to still be pursuing weapons of mass destruction and supporting international terrorists. Thus, it is critical that the United States and other civilized democracies continue to conduct research and development in chemical and biological defense and medical countermeasures. During these times when we live with daily elevated threats of terrorists' use of weapons of mass destruction, we can be reassured by the advances being made by government and academic scientists to cope with these threats. The results of these efforts were presented at the conference and are captured by the chapters in this book.

The authors of the contributed chapters in this volume are from the United States, Germany, and India, and are internationally recognized as subject matter experts in their respective fields. In this book they provide the state of knowledge on mechanisms of action, prophylaxis, and therapy.

As detailed in the preface, the editors have organized the presentations into three sections. In Section A, entitled Chemical and Biological Weapons, the pharmacology and toxicology are presented along with the psychological and neuropsychological sequelae of chemical terrorism. In this section there are also chapters on detection, treatment, decontamination, and detoxification. In Section B, entitled Toxic Chemicals and Their Mode of Action, chemicals other than cholinesterase inhibitors are included. Capsaicin, a riot control agent is discussed, as is arsenic, a heavy metal water contaminant causing international health concerns. Chapters on the role of free radicals generated in response to noxious agent exposure and their metabolism by cytochrome P-450 are also in this section. The final section, Pharmacology of CBW Antidotes, contains chapters on clinical treatments of selected toxins, organophosphates, cyanide and metal overdosing. Also in this section, the implications of antioxidant treatment of Alzheimer's disease and cerebral ischemia are presented.

The editors of this book, Drs. S.J.S. Flora, James Romano, Steven Baskin, and K. Sekhar are to be commended for their skill in the organization of this book and for making the current state of the knowledge in chemical and biological defense and medical countermeasures available to the international community. Their skills are reflected in their scientific accomplishments. They are all involved in their respective governments' defense establishments. Dr. Flora is with the Defense Research & Development Establishment (DRDE) of India, and is internationally recognized expert in heavy metals, while Dr. Baskin is with the United States Medical Research Institute of Chemical Defense (USMRICD), and a world's authority on cyanide. Dr. Jim Romano, formerly the Commander of the USMRICD, is currently the Deputy Commander of the United States Army Medical Research and Material Command. Dr. Romano is a recognized authority in chemical and biological warfare agents, with special emphasis on their psychological effects and impacts.

This book provides a valuable source of information for scientists in the military and civilian sectors, as well as those involved in homeland security and defense.

Harry Salem

Chief Scientist Life Sciences,

U.S. Army Edgewood Research Development and Engineering Center

Aberdeen Proving Ground, MD, U.S.A.

Editor-in-Chief, Journal of Applied Toxicology

Preface

In order to create an inter-disciplinary forum for mutual exchange of information on protection against chemical warfare agents, the organizer of the 35th Annual Conference of Indian Pharmacology Society and Prof. Satu M. Somani jointly invited many U.S. and Indian defense scientists to a special symposium on "Chemical and Biological Warfare" at Gwalior, India. The response to the invitation was very positive and nearly 20 defense experts agreed to participate in the symposium. During the preparation for the symposium, we lost our motivator, Prof. Satu M. Somani, just hours before he was to board the flight to India. It was a shattering moment for us but, with the active cooperation of all Indian and U.S. participants, the symposium became a huge success. Dr. S.J.S. Flora and Colonel James A. Romano then agreed to publish a book including not only the topics covered in the symposium but a few others which are of utmost interest to the toxicologists these days. This was thought to be a fitting tribute to the memory of Prof. Satu Somani who was the main architect of this symposium and the book on Toxicology.

This book covers many interdisciplinary and wider topics of utmost current interest to pharmacologists and toxicologists ranging from toxic effects of chemical and biological warfare agents, cyanobacterial toxins, organophosphorus (OP) toxicants, metal or metalloid toxicity including their mode of action and possible control measures. The book has been divided into three sections; the first section includes chapters on Chemical and Biological Warfare written by experts from Indian and U.S. defense establishments. The book also covers information on some highly potent toxic chemicals that are drawing the attention of toxicologists and pharmacologists. Among these toxic chemicals are capsaicin (used as an incapacitating agent) and arsenic. Another major highlight of this book is a section on mechanistic approach to various toxicants that are of current interest. Lastly, there are a few chapters on the pharmacology of antidotes pertaining to chemical warfare agents, cyanide, and other toxic chemicals. The breadth of this book reflects the evolution of toxicology as a science applied to the protection of human health from highly toxic agents.

Among the highly toxic chemicals are chemical warfare agents that have direct toxic effects on humans, animals, and plants. These chemicals have attracted much media and public attention due to the number of terrorists' activities around the globe and people becoming aware of their potential risk and danger. Chemical warfare agents are horrifying because their effects can spread very fast. These agents have given substantial and psychological edge to a number of militant developing countries. Section A of the book includes chapters written principally by defense scientists which comprehensively discuss the many aspects of chemical warfare agents and provide a current source of knowledge about the chemicals which annihilate human, animal, and plant lives. In the first chapter of this section, psychological and neuropsychological sequelae of chemical terrorism have been briefly discussed by Romano and King. Lukey and his group in Chapter 2 focused upon skin decontaminating agents with a special emphasis on chemical warfare agents. Vijayaraghavan and his team provide evidence about the efficacy of amifostine and DRDE-07, an S- (ω-aminoalkylamino) alkylaryl sulfide against sulphur mustard (SM) administered by various routes. The specific protection of amifostine and DRDE-07 when SM is administered by percutaneous route and the more toxic effect of SM when administered by percutaneous route are reported. In Chapter 4, Gordon and his group describe their work on developing enzyme-immobilized polyurethanes configured as (1) biosensors for OPs or (2) as sponges to soak up and inactivate the OPs. They suggest that these immobilized enzyme biosensors and sponges by virtue of their high capacity for enzymes, stability, specificity, sensitivity, and resistance to harsh environmental conditions could be used under diverse conditions encountered by troops and civilian first responders. In Chapter 5, Doctor and Saxena describe their interesting work on an ideal bioscavenger for organophosphate chemical warfare agent toxicity. They suggest that at the present time butyrylcholinesterase (BChE) is a better bioscavenger for OP poisoning than Hu AchE and suggest the scientific basis for that conclusion. Chapter 6, by Raza and Jaiswal, stresses the need for developing reliable analytical methods for the retrospective determination of SM in biological samples collected from the subjects allegedly exposed to it. They provide experimental evidence for various monitoring methods in order to detect SM, its metabolite(s), or its *in-vivo* reaction product(s) within the body. They further suggest that the reaction of SM with the nucleophilic sites in macromolecules results in covalent reaction products (adducts), which could be responsible for the toxicity produced by it, and are termed as biomarkers. Measurement of such biomarkers is considered to be a suitable monitor of SM exposure. The presence or absence of biomarkers indicates whether exposure to SM has taken place or not and will, therefore, be quite useful in verification activities also. Cheng and Rastogi's paper (Chapter 7) summarized the data available on the potential applications of organophosphorus hydrolase (OPH) and organophosphorus anhydrolase A (OPAA) enzymes for personnel/ casualty decontamination. Reddy et al. (Chapter 8) suggest that SM undergoes hydrolysis and forms products such as thiodiglycol (TG), 1,4-oxathiane, 1,4-dithiane, 2-vinylthionethanol and mustard chlorohydrin in biological or environmental systems. Thiodiglycol (TG) has been detected as a contaminant of soil and water at certain Army installations. These authors provide a summary of the toxicity information on TG that they developed as well as a proposed oral Reference Dose (RfD) for the U.S. Environmental Protection Agency (USEPA) for the inclusion in the Integrated Risk Information System (IRIS). Chapter 9 by Baskin and his group is one of the two papers in this book which provides information on various strategies for the therapies required for treating cyanide intoxication. Baskin et al. describe possible alternate approaches to develop therapies against cyanide. They claim that several of these methods appear promising. Hypoxic inducible factor – 1 (HIF-1), and 4-amino thiazoline 2-carboxylic acid (ATCA) antagonists may provide other approaches towards the treatment of cyanide poisoning. Administration of ATCA, the cystine metabolite of cyanide, or cyanide results in unconsciousness in a dose-related fashion. They stressed the need to conduct more studies to establish if these alternative treatments can provide safe, effective therapy for cyanide intoxication. There are few chapters on biological weapons describing principally the effects of anthrax, plague Bacilli, brevetoxin and cyanobacterial toxin. Cyanobacteria, which are broadly classified as oxygenic phototrophs containing chlorophyll-a and accessory pigments, are among the oldest life forms on earth. Two interesting chapters, one by Batra and his colleagues and one by Nancy Khardori, provide indepth information about anthrax and plague bacilli. These chapters described toxicity in various animal species, predicted lethality in humans, mechanisms of toxicity, and various treatment modalities (pre and post treatment). Chapter 12 by Rao provides information about these cyanobacterias, their toxicity, human health effects, risk assessment, mode of action, and various environmental conditions that may trigger toxic algal blooms and possible control measures, etc. Deshpande and Singh (Chapter 13), on the other hand, summarized information about brevetoxins that periodically and fatally contaminate the water supplies of wild, domestic, and marine animals and can harm humans leading to serious health problems. They suggest that brevetoxins can also be potential biological warfare agents.

Section B of the book provides information about some selected toxicants and, in particular, various mechanisms of their toxic action. There are a number of chapters which cover areas of current interest. Chapter 14 briefly reviews the findings illustrating the wide array of additional macromolecular targets for anticholinesterases. Pepper spray containing Capsaicin is used by the police as a weapon and has also been licensed for private use. Chapter 15 by Anna Barbara Fischer evaluates the possible health risks entailed by such use. Premkumar and Van Buren (Chapter 16) provide information on inflammatory mediators—their production, transduction, and role in processing of the pain sensation. Chronic arsenic poisoning due to contaminated drinking water in many areas of Indo-Bangladesh has become a major international environmental health issue of major concern. This is the world's most recent and perhaps worst arsenic contamination of drinking water affecting approximately 38 million. Chapter 17 by Flora and Sekhar provides detailed information about the toxicity of arsenic, biochemical and pharmacological mechanisms, and possible therapeutic measures. Another interesting chapter by Sharma and Dixit describes the mechanisms involved in the regulation of free radical generation from neutrophils. The chapter provides important details about the potential pathways for the stimulation of neutrophil apoptosis, which could be critical in protecting tissue from damage and exacerbation of the inflammatory response. Parmar and his group in their Chapter 19 discuss and provide experimental evidence suggesting that the modulation of the levels of P450s significantly influences the neurobehavioral toxicity of many pesticides and insecticides like deltamethrin and lindane. They suggest that differences in the induction of P450s amongst the different brain regions could play a role in regulating the response of the brain to environmental chemicals. P450s may modulate chemical concentrations per se or modulate the levels of their active metabolites at the target site(s).

Section C of the book is principally on the pharmacology of antidotes. Among the various chapters selected in this section include current treatment modalities about botulium neurotoxin (Adler et al., Chapter 20). These authors discuss various problems associated with the subject, like binding affinity, toxicity of compounds, access of drugs to the nerve terminal, and targeting of drugs to the appropriate neurons to minimize side effects. Chapter 21 by Saxena et al. provides convincing evidence about Human Butyrylcholinesterase (Hu BchE) that, in their opinion, is a safe and effective bioscavenger that should be assessed for future inclusion into the protective regimen against chemical warfare nerve agents. Chapter 22 by Gupta et al. highlights the incidence of poisoning due to drugs particularly in Indian context. The drugs commonly consumed include antihistamines, barbiturates, benzodiazepines, carbamazepine, lithium, opioids, paracetamol and tricyclic antidepressants. Chapter 23 by Bhattacharya provides experimental evidence that combined administration of a-Ketoglutarate and sodium thiosulphate could be a better treatment option where sodium nitrite is contraindicated. An augmented protection could be anticipated by the adjunction of α-KG. Gupta and Sharma in their chapter (number 24) discuss the role of free radicals and oxidative stress in the pathogenesis and pathophysiology of Alzheimer's disease and cerebral ischemia and related toxicities. They recommend that administration of antioxidants may be useful in prevention and treatment of such diseases. In the chapter 25 current status about the treatment of chronic arsenic poisoning is being discussed. Experimental details about the role of succimer and some of their new analogues in the treatment of chronic arsenic poisoning are being discussed. Finally, in the last chapter 26 an in silico approach is presented suggesting how available data and models can be used in the development of new therapies for cyanide caused cardiac toxicity.

This volume attempts to provide a valuable source of information on many aspects of highly toxic chemicals (including chemical and biological weapons), their mode of action, and current treatment modalities. Military personnel, undergraduate and postgraduate students, scientific professionals in toxicology, pharmacology, biochemistry, and environmental sciences will find this volume useful. It may also be of interest to medical planners who would like a glimpse of future response capabilities to counter toxic agents in the environment.

This volume was the last dream project of Professor Satu M. Somani that he planned with us. We do not know if we can live up to his expectations. However, by reaching for the goal set by him, we hope to partially succeed in paying him a befitting tribute and homage to his memory. This book is thus dedicated to the memory of that great scientist.

We sincerely thank all contributors for submitting their chapters in a timely manner and working diligently with the editorial staff. We would also like to thank Mr. Tapan Das for his secretarial assistance. Our special thank to Dr. Ashish Mehta for all his untiring efforts and co-operation during the preparation of this volume

EDITORS

Contents

Section A

Chemical and Biological Weapons

Pharmacological Perspectives of Toxic Chemicals and Their Antidotes
Editors: S J S Flora, J A Romano, S I Baskin and K Sekhar

Psychological and Neuropsychological Sequelae of Chemical Terrorism

[a]James A. Romano Jr., Ph.D and [b]James M. King Ph.D.
[a]Colonel, Medical Service Corps
US Army Medical Research Institute of Chemical Defense
3100 Ricketts Point Rd., Aberdeen Proving Ground Maryland 21010-5400, USA
[b]Deputy Director, Chemical and Biological Defense Information Analysis Center
Building E3330, Aberdeen Proving Ground
Maryland 21010-5423, USA

INTRODUCTION

The Chemical Warfare Agents (CWAs) and Their Use

We begin by listing the principal classes of chemical warfare agents (CWAs)—nerve, blister, blood, and choking—and significant examples of each type of agent. The simplistic military chemical nomenclature adopted over 50 years ago captures the principal target organ of each class of agent—central nervous system (CNS), skin or other epithelial tissue, the oxygen carrying capability of blood, and/or the pulmonary system, respectively (Headquarters, Department of the Army, 1966). These CWAs were designed to be rapidly lethal (nerve, blood) or incapacitating (blister, choking) at relatively low concentrations. "G" nerve agents include tabun (GA), sarin (GB), soman (GD), or cyclosarin (GF). Another prominent nerve agent is VX, weaponized by both Cold War superpowers. The most prominent example of a blister agent is sulfur mustard (HD). Of course, the most well-known and widely employed and studied blood and choking agents are cyanide and phosgene, respectively. The latter two chemicals are both traditional CWAs and "toxic" industrial chemicals (TICs), i.e., chemicals produced in large amounts for commerce. The combination of toxicity plus availability leads to concern about their use in chemical terrorism. The rapid and toxic action of these agents requires prompt medical response. The major instances of each type of CWA and their principal physiologic effects are represented in Table 1.

Table 1 Morbidity of Four Classes of Chemical Warfare Agents and Their Principal Targets and Medical Countermeasures (Estimates Derived from U.S. Published Sources)

Class of CW Agent	*Historic Lethality/ Morbidity in Warfare*	*Principal Target Tissue*	*Physiological/ Performance Effects*
Choking (e.g., phosgene)	1%[1]	Deep lung compartment such as pulmonary capillary	Pulmonary edema,
Blood (e.g., cyanide)	Unknown[2]	Cellular respiratory enzymes	Depression of cortical function, unconsciousness, convulsions
Blister (e.g., "mustard gas")	2-4%[3]	Skin, airway, eyes, GI tract, bone marrow	Loss of function due to skin, lung, ocular lesions, recovery over time
Nerve (e.g., sarin)	Unknown[4]	CNS, neuromuscular junction; cholinergic synapse	GI tract, miosis, nausea, weakness, loss of consciousness, convulsions

[1] WWI figures for the U.S. are estimates because phosgene was often mixed with chlorine; however, a total of 6834 injured (average hospitalization = 49 days have been directly attributed to phosgene with 66 fatalities). [2] No data from wartime use; however, wartime experiences suggest difficulty in achieving militarily effective concentrations unless confined to closed spaces. [3] WWI, 2% with 27,711 U.S. injured; Iran-Iraq War, 4% with 45,000 estimated injured. [4] No data from wartime use; however, on 20 March 1995, using a primitive method of dispersal, Sarin was released on Tokyo subways with 5,500 people seeking medical care; approximately 1500 had defined symptoms of exposure, and 12 casualties died. Less well known is the fact that on June 27, 1999, sarin was released in Matsumoto, Japan, with estimates of 471 subjects exposed to sarin and 7 deaths.

Unfortunately, CWAs have been used or are thought to have been used in a number of major conflicts in the 20th century. The military effectiveness of these weapons is questionable~ their psychological value appears significant (Spies, 1986). Instances of CWA use include: (1) extensive use in WWI; (2) use by Italy against Ethiopia; (3) use in the Sino-Japanese War; (4) relatively well-studied use in the Iran/Iraq conflict; and (5) the Matsumoto/Tokyo terrorist incident. We briefly discuss the major implications of each of the above instances.

In WWI, HD was the most significant generator of CWA casualties (80%). It produced 125,000 casualties for the United Kingdom alone, and for the United States (U.S.), approximately 28,000 casualties, 2% of whom died. The average hospitalization of U.S. casualties was 42 days (Sidell, 1997). For each U.S. soldier exposed to HD, two sought medical care without evidence of exposure. A significant instance was the operation of the 3rd Division at Aisne-Marne. During that operation, the division reported 500 casualties over an 8-day period with such symptoms as fatigue, chest pain, dyspnea, coughing, throat pain (ranging from tingling to burning), and indefinite eye symptoms. Nevertheless, there was no clinical evidence of gas inhalation or burning (Medical History of 3rd Div, 1959). The 3rd Division experience is often used to highlight the psychological impact of these types of weapons.

During the Italian-Abysinnian conflict, the extensive use of HD seems to have had significant effect on performance. A New York Times report stated "[HD] burned their shoulders and feet, blinded them, and burned the mouths of their pack animals.. .." Historians conclude that "gas seemed to have impaired Abysinnian morale" and contributed to the speed of the Italian victory (Spies, 1986). Fifty years later, the pernicious nature of mustard was reinforced in the mid-1980s in the Iran-Iraq War where it produced an estimated additional 45,000 chemical casualties (Carus, 1988). Significant skin, eye, and pulmonary lesions were seen and required extensive hospitalization, often up to 10 weeks (Willems, 1989). Reports of neuropsychiatric effects, such as severe apathy, impaired concentration, and diminished libido have appeared in the clinical literature (Balili-Mood et al., 1986). Thus, many of the lessons we have learned in the 20th century are derived from military history, research, training exercises, and analyses. These have yielded valuable insights into the nature of the psychological reaction to the problem. We will probe the lessons of these experiences and combine them with the sociopsychological and neuropsychological analyses of disasters and/or chemical disasters.

Recently, the terrorist use of sarin (GB) in two separate instances in Japan has yielded some important new insights into the problem that forms the basis for this symposium. For example, in Tokyo, 5,500 patients were processed at nearby hospitals, only 1,500 of which required treatment of any type; there were 12 deaths (Yokoyama et al., 1998). Chronic health effects from this acute exposure, including CNS or behavioral changes that were inextricably linked to Post Traumatic Stress Disorder (PTSD) were observed (Yokoyama et al., 1998). As we have analyzed these events, a number of significant intrinsic differences in the problem of chemical terrorism vis-a-vis chemical warfare are revealed. These differences and the parameters of response are seen in Table 2.

TERRORISM AND ITS OUTCOMES

Terrorist acts involving use of CWAs evoke all of the outcomes of terrorist or catastrophic events, the most prominent of which is PTSD. We have observed this following the events of Matsumoto and Tokyo, and we can learn about this generalized response from many experiences. In the U.S., the most prominent recent example would be the World Trade Center. The most complete longitudinal study appears to have been done in France, which demonstrated a clear relationship between degree of injury and degree of PTSD and a high instance of PTSD in terrorist incidents (Abenheim et al., 1992).

Table 2 Chemical Warfare and Chemical Terrorism: Critical Differences

	Military	*Civilian*
Population level of training	Trained, protected troops	Generally untrained
Population age, gender	Ages 18-45, predominantly male[1]	Ages 0-75, male
Reported health	Generally good health, fitness	Wide range of health status
Availability of medical protection	Medical protection available, pretreated perhaps	Post-exposure treatment only
Psychological effects	2:1, limited misuse of equipment and antidotes	5:1, some misuse of equipment, antidotes[2]
Leadership	Military command and control	Law enforcement and public safety
Response	Retaliate with overwhelming force	Forensic, judicial response

[1]In January 2003, the proportion of males in the active U.S. forces was 85%.
[2]Experiences from Persian Gulf War, discussed below in text.

Terrorist acts involving use of CWA evoke all of the psychological outcomes of catastrophic events involving chemicals, the most notable of which revolve around: (1) the insidious and indiscriminate nature of chemicals, (2) the process of decontamination, and (3) the wearing of PPE. Each of these experiences brings its own unique fears and reactions to the situation. With respect to PPE, for example, there have been valuable insights gained through the study of training exercises, field studies, and simulation (King and Frelin, 1984). Many of the psychological effects of the use or threat of use of CW As stem from the wearing of the Mission-Oriented Protective Posture (MOPP) suit. The "classic triad" of symptoms includes anxiety, panic, and claustrophobia (Singh, 1992). The problem of field management of such casualties is complicated by the known CNS symptoms of mild intoxication by nerve agents. These include ataxia, confusion, slowing or loss of reflexes, slurred speech, coma, and paralysis, all of which have been seen in human studies (Sidell, 1997) and effects on learned behaviors, analgesia, and cardiac effects in animal studies (Karczmar, 1984). Premedications may have apparent behavioral (if not CNS-mediated) effects in man. Chemical warfare agents, particularly those that have high and rapid toxicity in CNS tissue, require medical countermeasures that may produce significant performance effects, especially in a stressed individual.

When the stressful event overwhelms ego defenses then some further negative outcomes may ensue. These include (a) conversion disorders, (b) mistaking normal physiological stress symptoms for exposure to CWA (despite significant efforts to train soldiers in proper recognition of signs of poisoning), (c) mistaking or magnifying the symptoms of minor illnesses and

(d) deliberate faking or malingering. One might add the possibility of an additional type of self-inflicted wound to this list (i.e., the inadvertent or misguided use of antidotal compounds, e.g., atropine and diazepam). Self-administration of two nerve agent antidote autoinjectors can produce headache, restlessness, and fatigue, symptoms that can be aggravated in a tired, dehydrated, or stressed individual (Romano and King, 2001).

DIRECT CNS EFFECTS OF CWAs

Acute Exposure

There have been a number of investigations as to the visible long-term consequences of an acute exposure to OP nerve agents, to include several studies of the effects of sarin-exposure victims from Japan. In this short paper, we will only cite two of the most pertinent papers. Yokoyama et al. (1998) evaluated 18 victims of the Tokyo subway incident 6 to 8 months after exposure. All but three of these victims have plasma ChE values below normal values on the day of exposure. Sarin-exposed individuals scored significantly lower than controls on a digit symbol substitution test (see Figure 1); they scored significantly higher than controls on a general health questionnaire (GHQ; psychiatric symptoms) and a profile of mood states (POMs; fatigue).

0	1	2	3	4	5	6	7	8	9
*	#	Π	λ	÷	$	!	?	Σ	Δ

*	#	÷	!	Σ	Δ	?	$	Π	#	*	#	÷	$ Σ
Δ	Σ	$	÷	#	*	*	λ	Π	!	?	Δ	*	$?

Fig. 1 An example of a digit symbol substitution test. Typical tests require a subject to substitute up to 90 symbols using the key directly above the test table. The key performance measures are speed and accuracy. Norms are based on sex and age.

Additionally, they had elevated scores on a post-traumatic stress disorder (PTSD) checklist; they had significantly longer P300 latencies on event-related brain-evoked potentials and longer PI00 latencies on brain visual-evoked potentials; and female exposed cases had significantly greater indexes of postural sway. The elevated scores on the GHQ and POMS were positively related to the increased PTSD scores and were considered to be due to PTSD. Nakajima et al. (1998) performed a cohort study of victims of the Matsumoto City sarin exposure one and three years following the incident. At 1 year following the exposure, they reported that 20 victims still felt some symptoms (fatigue, asthenopia, blurred vision, asthenia, shoulder stiffness, and husky voice). They had all lived close to the sarin release, and they had lower erythrocyte ChE activity than those who did not have symptoms. (Note: Not all the symptoms seen at one year have been related to nerve agent exposure historically.) At three years, some victims still complained of experiencing these symptoms, although with a reduced degree and frequency. Additionally, there have been two brief case reports of severely poisoned nerve agent victims (one sarin, one VX) in Japan who experienced retrograde amnesia, possibly due to prolonged periods of seizures and/or hypoxia. Additionally, one of the Matsumoto victims who experienced prolonged seizure activity was followed for at least 1 year and was found to have sporadic, sharp-wave complexes in the EEG during sleep and frequent premature ventricular contractions on Holter monitoring of the electrocardiogram (Sekijima et al., 1998).

CHRONIC OR REPEATED EXPOSURE

Chronic or repeated subclinical exposures to organophosphorus (OP) agents produce less consistent health effects than those observed following acute exposures. For example, the report of Burchfiel and Duffy (1976) about the effects of repeated low doses of sarin to rhesus monkeys producing a long-term increase in relative power in the electroencephalogram (EEG) beta frequency bands is the most cited study in support for a long-term health effect. Further reports in the literature of animal studies show that nerve agents can be administered repeatedly with minimal overt neurobehavioral effects if care is taken in choosing the dose and the time between doses (Russell et al., 1986). Blood and brain acetylcholinesterase (AChE) levels can be reduced to >20% of normal with no observable signs of toxicity with appropriate dosing schedules (Sterri et al., 1980). The most notable effect of such chronic dosing is the development of tolerance to the disruptive effects of each acute exposure on certain behaviors. In this respect, nerve agents act much like other OP compounds, and the possibility and mechanisms of tolerance development have been addressed in several studies (Russell et al., 1986). As for the human experience, workers exposed to small amounts of nerve agents that produced mild, non-threatening medical signs of exposure, reported CNS effects such as headache, insomnia, excessive dreaming, restlessness, drowsiness, and weakness. Clearly, use of nerve CWA increases the differential diagnosis of stress reactions from manifestations of organic brain syndromes that might be readily ameliorated, at least temporarily, by antidotal therapy.

Table 3 summarizes the types of psychological sequelae to CW terrorism that can be derived from disaster research, chemical spills, military training experiences, and the like.

Table 3 Sources of Information About Psychological/Behavioral Effects ofCW Terrorism

Sources of Information	*Component of the Catastrophic Event*	*Significant Psychological Outcome*
Natural disaster, terrorism, etc.	Stress	PTSD'
Chemical spills	Fear of chemicals, decon, etc.	Anxiety, depression
Military training	Wearing of PPE	"Mask phobia"[2]
History of chemical warfare	Direct and indirect effects of CWAs	PTSD, loss of libido,[3] neurasthenia
	Direct CNS effects	PTSD, insomnia, irritability, restlessness[4]

[1]Frequency of neuropsychological effects or PTSD directly related to extent of injury; see Abenhaim et al., 1992.

[2]Performance degradation improved by training; see King and Frelin, 1984.

[3]Iranian casualties demonstrated various neuropsychological sequelae; see Balali-Mood and Navaeian, 1986.

[4]Effects of nerve agents linger, perhaps for months; see Yokoyama et al., 1998.

SUMMARY AND CONCLUSIONS

Historical review of American forces' activity in WWI showed that initial experiences with chemical attacks produced large numbers of patients who thought they had been exposed to chemical agents, e.g., the experience of the 3rd Division at Rhine-Marne. As training and wartime experiences accumulated, however, this proportion sharply decreased. A similar trend was seen in Israel during the Gulf War. Historically, experience suggests that training and education of the populace should minimize the numbers of panicked "worried well" in a future incident.

Both the threat of use and the actual use of CWA will have important psychological and physiological consequences. The threat of use of CWA can produce significant stress effects in affected populations. If CWA are actually used, the psychological impacts of both CNS and peripheral effects of the CWAs and medical countermeasures must be considered. In the case of true exposure to nerve agents, chronic neurological sequelae can be expected in some cases, as can long-term PTSD-like effects. A variety of toxic industrial chemicals and materials also represent a "chemical" threat and must be considered in planning. Recent experiences with terrorist use of CWAs in Tokyo, as well as modeling efforts, suggest that psychological CWA casualties will outnumber physical CWA casualties by approximately four to one. Threatened or actual use of CWAs will require substantial mental health support to health care personnel and to the populations they service, and will have major impacts on the health care system.

References

Abenhaim L., Dab W. and Salmi L.R., Study of civilian victims of terrorist attacks (France 1982-1987). J. Clin. Epidemiol. **45:** 103-109, 1992.

Balali-Mood M. and Navaeian A. Clinical and paraclinical findings in 233 patients with sulfur mustard poisoning in Proceedings of the 2nd World Congress on New Compounds in Biological and Chemical Warfare: Toxicological Evaluation, Industrial Chemical Disasters, Civil Protection and Treatment, pp 464-473. Ghent, Belgium. 24-27 August 1986.

Bleich A., Dycian A., Koslowsky M., Solomon Z. and Weiner M. Psychiatric implications of missile attacks on a civilian population. Israel lessons from the Persian Gulf War. JAMA. **268(5):** 613, 1992.

Burchfiel J. and Duffy F. Persistent effect of sarin and dieldrin upon the primate electroencephalogram, Toxicol. Appl. Pharmacol., **35:** 365-379, 1976.

Carus W.S. Chemical weapons in the Middle East, the Washington Institute for Near East policy, Research Memorandum No.9, 1988.

General Accounting Office, Combating Terrorism: Need for Comprehensive Threat and Risk Assessments of Chemical and Biological Attacks. Report No. GAO/NSIAD 99-163, DTIC #A369111, Washington, DC Government Printing Office, 2000.

Headquarters U.S. Department of the Army; Employment of chemical agents, Field Manual 3-10, Washington, DC, March 1966.

Karczmar A., Acute and long-lasting central actions of organophosphorus agents. Fund. Appl. Toxicol. **4(2):** SI-17, 1984.

King J. and Frelin A.J. The impact of chemical protective clothing on the performance of basic medical tasks. Military Medicine. **199:** 496-501, 1984.

Medical History of 3rd Div, Quoted in Cochran RC The 3rd Division at Chateau Thierry July 1918, in US Army Chemical Corps Historical Studies: Gas Warfare in World War I, Washington, DC Office of the Chief Chemical Officer, US Army Chemical Corps Historical Office, 1959; 91, Study 14; 117.

Nakajima T., Ohta S., Morita H., Midorikawa Y., Mimura S. and Yanagisawa N. Epidemiological study of sarin poisoning in Matsumoto City, Japan. J. Epidemiol. **8:** 33, 1998.

Sekijima Y., Morita H., Shindo M., Okudera H., Shibita T. and Yanagiswa N. A case of severe poisoning in the sarin attack at Matsumoto—one year follow-up on the clinical findings and the laboratory data. Rinso Shinejaker (Clin. Neurol.). **35:** 12-42, 1995.

Sidell F., Urbanetti J., Smith W. and Hurst C.G., Vesicants in Medical Aspects of Chemical and Biological Warfare; 197-228. Washington, DC F Sidell, E Takafuji, D Franz, 1997.

Romano J.A. and King J.M., Psychological casualties resulting from chemical and biological weapons. Military Medicine. **166:** S2-21, 2001.

Russell R.W., Booth R.A., Lauretz S.D., Smith C.A. and Jenden D.J., Behavioral, neurochemical, and physiological effects of repeated exposure to sub symptomatic levels of the anticholinesterase, soman. Neurobehav. Toxicol. Teratol., **8:** 675-685, 1986.

Sidell F., Nerve agents in Medical Aspects of Chemical and Biological Warfare; 129-180. Washington DC, Sidell F., Takafuji E. and Franz D. 1997.

Singh V., Sociopsychological effects of chemical warfare agents in Chemical Warfare Agents, pp 1-12, San Diego, Somani, S. 1992.

Spies E.M., Chemical Warfare. Chicago, IL, University of Illinois Press, 1986.

Sterri S.N., Lyngaas S. and Fonnum F. Toxicity of so man after repetitive injection of sublethal doses in rats. Acta Pharmacol. et. Toxicol. **46:** 1, 1980.

Willems J.L. Clinical management of mustard gas casualties. Ann. Med. Milit. Belg., **35:** 1-61, 1989.

Yokoyama K., Araki S., Kaysuyutci M., Nishihitani M., Okumura T., Ishimatsu S., Takasu N. and White R.F. Chronic neurobehavioral effects of Tokyo subway sarin poisoning. J. Physiol. (paris) **92:** 317, 1998.

Pharmacological Perspectives of Toxic Chemicals and Their Antidotes
Editors: S J S Flora, J A Romano, S I Baskin and K Sekhar

CHAPTER

Six Current or Potential Skin Decontaminants for Chemical Warfare Agent Exposure—A Literature Review

Brian J Lukey[a], C. Gary Hurst[a], Richard K. Gordon[b], Bupendra P. Doctor[b], Edward Clarkson IV[a], Harry F. Slife[a]

[a]US Army Medical Research Institute of Chemical Defense
3100 Rickett's Point Road, Aberdeen Proving Ground
MD 21010, USA

[b]Walter Reed Army Institute of Research
503 Robert Grant Avenue, Silver Spring
MD 20910-7500, USA

INTRODUCTION

In light of the recent increase in terrorism and the rapid advancement of rogue nations developing and stockpiling chemical and biological warfare agents, the United States has recognized the imminent threat of chemical/biological warfare. We have established defensive measures as a number one priority in dealing with this threat. Consequently, the ability to optimally decontaminate casualties has become an utmost concern. This paper focuses upon decontaminating skin with a special emphasis on chemical warfare agents.

The United States Army Medical Research and Materiel Command (USAMRMC) has been the lead agency in the Department of Defense for chemical and biological medical defense research. The Command established a directed research effort to develop skin decontaminants for biological and chemical warfare agents. Basic research in this area is being conducted at the US Army Medical Research Institute of Chemical Defense, the Walter Reed Army Institute of Research and the US Army Medical Research Institute of Infectious Diseases. In addition, USAMRMC established contractual agreements with non-government research laboratories nationally and internationally. Also, the US Department of Defense established an Advance

Concept Technology Demonstration called Joint Service Family of Decontamination Systems, in which advanced products could be rapidly tested and fielded.

The first step was to identify the characteristics of an optimal decontaminant. By definition, decontamination can involve the removal or neutralization of the injurious agent. Of the two, the most important process for the exposed individual is to rapidly remove the agent from the skin. A decontaminant that destroys the agent in the process prevents the potential for another to be exposed. In the heat of the battle, the warfighter will not likely have time to identify the specific agent on his skin but will expect the decontaminant to universally remove all chemical and biological warfare agents. This decontaminant should obviously not facilitate the penetration of the agent into the skin but instead wick any penetrated agent out of the skin.

In addition to efficacy, the complete acquisition process must be evaluated. The entire life cycle should be relatively inexpensive, beginning with research and development; moving through manufacturing, logistical support, and fielding; and ending with disposal. An optimal candidate would move through the acquisition process quickly, and preferably be already commercially available for another use and easily approved by the Federal Drug Administration (FDA) for this intended use. The manufacturer must be able to scale up production to make sufficient quantities for all users, and do so under good manufacturing practice such that all lots are equally efficacious. The resulting product should have a long shelf-life and be stable under extreme temperature changes to support arctic and desert operations. Also, a single product deliverable from several devices for both mass casualty and individual use would be a logistical plus.

Finally, the warfighter's needs must be considered. Complicated decontamination procedures are unacceptable, since they increase the probability of error during the stressful exposure scenario and require excessive training. The warfighter must be able to employ the device quickly to minimize transdermal penetration. The individual device should be lightweight so as not to burden the carrier and should be generally acceptable for use. That is, it should not have a repulsive odor, leave an intolerable residue or form dust that irritates the eyes or is offensive to breathe. More importantly, the product must be medically safe, in that it is non-irritating and non-allergenic. Ideally, the product would also be environmentally safe to use by itself and render the chemical/biological agent environmentally safe.

Obviously the ideal candidate does not exist, and we must compromise on some of these ideal characteristics to get a fieldable product. VanHooidonk et al. (1983) evaluated a variety of household products as decontaminants for VX, soman and sulfur mustard. His paper provides an excellent comparison of these readily available decontaminants to include flour, soapy water, talcum powder, and tissue paper. This paper has selected several other candidate decontaminants that are emerging in advanced development or are currently fielded. These include the currently fielded US Department of Defense skin decontaminants (M291 resin kit and 0.5% hypochlorite), the Reactive Skin Decontaminant Lotion, Diphoterine, Sandia Laboratory's Decontamination Foam, and the Reactive Sponge.

HYPOCHLORITE

Current US doctrine specifies the use of a 0.5% sodium or calcium hypochlorite solution for decontaminating skin. This solution can be made by diluting a standard bleach solution (5% hypochlorite) ten-fold with water. The theory for its effectiveness involves both physical removal and neutralization of the agent. Bleach preparations as a decontamination solution first appeared in WWI in which they were initially used quite successfully as a field expedient antiseptic known as Dakin's solution (0.5-2% bleach) (Vedder 1925). Due to its oxidative chlorination properties, the solution was grandfathered as an acceptable decontamination preparation. However, Vedder (1925) preferred organic solvents such as kerosene over bleach because he believed the absorbed sulfur mustard could be pulled out of the skin by repeated application, thus greatly diminishing subsequent injury.

Full strength bleach has been used successfully for years as a universal chemical warfare agent decontaminant for equipment and facilities. The general rule has been to use copious amounts of bleach decontamination since some agents require more bleach than others. For example, VX requires up to 20 moles of active chlorine per mole of VX due to the formation of reactive intermediates if not enough chlorine is used (Yang 1999).

Unfortunately, undiluted bleach is itself toxic to the skin and may create more damage than no decontaminant. Zvirblis and Kondritzer (1953) noted that rabbits receiving 20 cubic mm of sarin (GB)/kg had no convulsions or deaths without decontamination but had a 100% incidence of convulsions and 75% incidence of lethality when decontaminated with 5% bleach. The bleach itself at a 5% hypochlorite concentration may destroy the protective barrier of the skin and facilitate GB transdermal absorption. Consequently, bleach has been diluted to a concentration (0.5%) that produces no appreciable irritation to human skin (Hobson and Snider, 1992).

The effectiveness of diluted bleach has been tested in a few studies. Dolzine and Logan (1991) conducted an in vitro test in which they determined the percentage of agent that was detoxified after a five-minute exposure to various concentrations of bleach. They found that 67% of HD, 69% of cyclosarin (GF) and 85% of GB were destroyed in 0.5% bleach. Undiluted bleach resulted in 79% of HD, 91 % of GF and 99% of GB destroyed in 5 minutes. Thus, even with undiluted bleach, approximately 20% of HD would be active after 5 minutes of decontamination.

Van Hooidonk et al. (1983) conducted in vitro penetration studies using a diffusion cell with guinea pig abdominal skin. They found that diluted bleach (concentration not defined) removed 96% and 97% of VX and GD respectively. However, they also found that tissue paper wetted with water removed 92% of VX and 91% of GD. In the in vivo experiment, dilute bleach decontamination resulted in a 100% survival rate when VX (0.25 mg) or GD (5.0 mg) was dermally administered to guinea pigs. However, soapy water and wet tissue paper each produced the same 100% survival rate for the same amount of agent decontaminated. They concluded that removing the chemical agent from the skin quickly is more important than which decontaminant is used.

Wormser et al. (2002) compared the effectiveness of 0.5% hypochlorite with water in decontaminating HD on guinea pigs. They found the skin HD content was reduced 64% and 68% by water and hypochlorite respectively, indicating that both are equally effective in removing HD. However, they found the gauze pads soaked with the decontaminant contained microgram quantities of HD when water was used but no detectable HD levels when 0.5% bleach was used. Thus, the most significant portion of the neutralizing effect of 0.5% bleach seems to occur after agent is removed from the skin.

Two studies looked at the effect of bleach decontamination on damaged skin exposed to CW agents. Gold et al. (1994) evaluated the effects of water or diluted bleach (0.5%) as a wound decontaminant two minutes after the hairless guinea pig was exposed to HD. The study found that 0.5% bleach and even water soaking five minutes in a wound contaminated with HD (20 mg/kg) cause greater necrosis than when no decontaminant was added. This does not mean that the wound should not be decontaminated but rather that bleach soaking in the wound is not the route to decontaminate.

Hobson and Snider (1992) evaluated the effectiveness of hypochlorite solutions in decontaminating rabbit intact skin and wounds exposed to VX or HD. When the intact skin was decontaminated with bleach at 5% or 0.5% hypochlorite concentrations one minute after HD exposure, lesion areas were reduced by 4.6- and 4.3-fold respectively. For VX-contaminated intact skin, 5% and 0.5% sodium hypochlorite increased the median lethal dose of VX by 19- and 16-fold respectively. The results indicate that 0.5% bleach is as effective as 5% in decontaminating HD and VX on intact skin. However, when VX was applied to a wounded site, the 0.5% bleach was not effective in increasing survival rate, whereas 5% bleach increased the median lethal dose two-fold.

The major mechanism of bleach skin decontamination is physical removal of the agent from the skin. Because bleach and water are nearly equally effective in physically removing the agent and because the primary goal in skin decontamination is to quickly remove the harmful agent from the skin, the much more abundant water decontaminant may be the better alternative in many cases.

M291 Skin Decontamination Kit

The M291 skin decontamination kit is the currently fielded US Department of Defense kit that individual warfighters carry for personal skin decontamination. The kit consists of a wallet-like carrying pouch, containing six individually packaged decontamination pads, intended for spot decontamination of the skin. Each pad has a loop that fits over the fingers so that the user can easily scrub it over the contaminated skin. Consequently, each pad provides the individual with a single step, nontoxic, nonirritating, decontamination application, which can be used on the skin, including the face and around wounds.

Each of these nonwoven, fiber-filled, laminated pads is impregnated with the decontamination compound Ambergard XE-555 resin, which is a black, free-flowing powder. This dry powder is a carbonaceous adsorbent, a polystyrene polymer and an ion-exchange resin. These characteristics afford both an adsorbent powder, removing the agent from the skin, and a reactive detoxifying substance, neutralizing the agent by ion exchange.

The resin itself was first tested for efficacy in rabbits by decontaminating shaved skin exposed to CW agents. For nerve agent studies, the treatment animals were decontaminated 2 minutes after agent exposure, and the lethal dose (LD_{50}s) for both the control and treated animals determined. The LD_{50}s for animals decontaminated with the resin were 2.0-, 10.4- and 22.8-fold higher for thickened soman (TGD), GD, and VX respectively, when compared with those groups not decontaminated (Joiner et al. 1988).

The manufactured M291kit was then shown to be effective against HD, Lewisite (L), TGD, and VX in rabbit studies. In these studies, rabbits were percutaneously exposed to a fixed volume of neat agent on their shaved dorsal skin and decontaminated at variable times ranging from 30 seconds to 5 minutes. The lesion areas were compared for HD and L, whereas the % blood actelycholinesterase inhibitions were compared for the nerve agents. With a one-minute decontamination time, the lesion areas were reduced 21-fold for HD and 22-fold for L relative to no decontamination. With a two-minute decontamination time, the dose of nerve agent required to produce 50% inhibition of AChE activity was increased 1.8-fold for TGD and 18-fold for VX (Hobson et al. 1993).

A follow-on study investigated the effectiveness of a topical skin protectant applied before agent dosing in combination with M291 kit decontamination after agent exposure. The amount of thicken GD required to produce 50% inhibition of AChE activity was increased 2.3-fold when decontaminated with M291 kit alone and 37-fold when the topical skin protectant pretreatment was combined with M291 kit decontamination. The lesion area from HD injury was reduced 8.5-fold for the pretreated topical skin protectant group that was decontaminated with M291 decontamination relative to the control group that received no pretreatment (Dill et al. 1991).

One of the drawbacks of the M291 kit is that the powder forms a dust cloud that could be inhaled or can get small particles in the eye. Obviously, this problem should not occur if the user is properly masked. Another is that the kit leaves a black powder on the skin that may deter its use, unless the soldier is absolutely convinced of contamination.

REACTIVE SKIN DECONTAMINATION LOTION (RSDL)

The Canadian Defence Research Establishment at Ottawa and Suffield developed and patented a liquid skin decontaminant for chemical warfare agent exposure called Reactive Skin Decontamination Lotion (RSDL). O'Dell Engineering Limited (Ontario, Canada) has the exclusive rights for sales.

Several improvements in the formulation have occurred up to the current formulation developed in 1995. The lotion is now a solution of 1.25 molar potassium 2,3-butanedione monoximate in polyethylene glycol mono ethyl ethers of average molecular weight 550 daltons (MPEG550) with 10% w/w water (pH 10.6) (Sabourin et al. 2001).

Also, several improvements occurred in the packaging. During the 1991 Gulf War, it was packaged in 125 mL suntan bottles. They then impregnated RSDL into cotton towelettes and packaged in heat sealed pouches. However, the towelettes were inefficient since only 5 ml of the total 45 ml packaged could be squeezed out with the hands. They settled upon developing a sponge-like plastic foam, Opcell@ foam, which is lightweight and easy to store.

Efficacy of the current formulation and packaging of RSDL was tested on the skin of hairless guinea pigs exposed to nerve and vesicant agents. Measurements were taken for lesion areas, necrosis scores, Draize scores and agent adducts in blood or skin. RSDL was compared in similar tests with three other decontaminants, Dutch Powder (Huid Ontsmettings Poeder), Fuller's Earth pad, and M291 Skin Decontamination Kit. RSDL ranked the most efficient in decontaminating L, thicken L, thickened HD, VX and thickened VX. However, it ranked last in decontaminating HD (van Hooidonk et al1996).

The US FDA has recently approved RSDL as a medical device. The decontaminant leaves an oily residue on the skin that makes one's hands slippery and makes handling items more difficult.

SANDIA FOAM

Sandia is a government-owned/contractor-operated facility in Albuquerque, New Mexico. Lockheed Martin manages Sandia for the US Department of Energy's National Nuclear Security Administration. They developed a decontamination foam derived from common chemicals (often found in household cleaners and medical products) to neutralize chemical and biological warfare agents.

Sandia National Laboratories has licensed its decontaminating foam to two companies: Modec Inc. (Denver, Colorado) and Envirofoam Technologies Inc. (Draper, Utah). One of these companies, Modec Inc. (Denver, Colorado), supplied our laboratory with its MDF-100 product in October 2001. MDF-100 is stored in two separate containers, one holding a solution of 6.6%-N, N, N, N', N'-Penta-methyl, -N'-Tallow Alkyl1, 3-Propanamine Diammonium; 2.6% -n-Tallow Pentamethyl Propane Quaternary Ammonium Compounds; Benzyl-C12-18 Alkyl Dimethyl; 1 %-Isopropyl Alcohol, and the other holding a solution of 8% hydrogen peroxide. When the two solutions are mixed and sprayed, they generate foam, which settles into a liquid in under 30 minutes.

We tested the ability of the MDF-100 foam to protect against chemical warfare agent exposure by clipping the fur from one side of a guinea pig, weighing the animal, sedating the animal, placing the animal in a fume hood and cutaneously exposing the animal to varying doses of either GD or VX. One minute after exposure, the organophosphonates were removed by wiping sterile gauze soaked in MDF-100 across the exposure site, in the direction of the fur. The soaked gauze was swiped across the area once, then rotated (so that a clean portion of gauze was now facing the animal) and three additional passes were made. The exposure area was then dried with a second piece of gauze in the same manner, so that the exposure site was wiped a total of eight times. The animals were placed in cages inside the fume hood, observed and euthanized 24 hours after exposure.

MDF-100 did an excellent job of protecting against GD exposure. In untreated animals, the LD_{50} for cutaneous GD exposure is 11.3 mg/kg. In animals decontaminated with MDF-100, the LD_{50} is 400 mg/kg, or a 35-fold protective ratio, which is by far the best protective ratio against GD seen to date in our laboratory. MDF-100 provides an even higher protective ratio against VX. In untreated animals, the LD_{50} for cutaneous VX exposure is 0.14 mg/kg. In animals decontaminated with MDF-100, the LD_{50} is 10.1 mg/kg, or a 72-fold protective ratio. Although the protective ratio for VX is high, other decontaminating solutions are more effective, providing > 100 fold protective ratio against VX.

MDF-100 has been shown to neutralize anthrax spores. In October 2001 the Environmental Protection Agency granted a waiver to allow the use of

MDF-100 to decontaminate rooms exposed to anthrax spores. This wavier was cancelled in March 2002, in part because of concerns that MDF-100 neutralized anthrax spores with insufficient speed. To make the foam faster acting and to make it less corrosive, Sandia developed another foam formulation, DF-200. DF-200 has a lower concentration of peroxide and surfactants, and contains an activator to increase the speed with which it breaks down agents. Our laboratory is in the process of evaluating this new decontaminant formulation.

DIPHOTERINE

Diphoterine® is a French product, advertised as a decontaminating solution for chemical spatters on the eye and skin. Prevor Laboratory, Vlamondois, France, manufactures it as an odorless, colorless liquid (pH of 7.4) dispensed as an eye wash or skin decontamination spray. It is classified as a slight dermal irritant (Jones et al. 1997) but a non-irritant in the eye (Clouzeau et al. 1990), as tested in the rabbit model.

Its effectiveness involves three modes of action. As a liquid, it rinses off surface contaminants. As an amphoteric and chelating molecule, it binds acids, bases, oxidizing agents, reducing agents, solvents, irritants, alkylating agents and radionuclides. As a hypertonic solution, its osmotic pressure pulls chemical contaminants out of tissue that the chemical has already penetrated (Laboratoir Prevor, 2002).

Most of the relevant studies tested the ability of Diphoterine to decontaminate the eye and are published in French. Hall et al. (2002) translated much of the work into English and wrote an excellent review article. In this review, they described Josset et al.'s work in which Diphoterine's effectiveness in decontaminating an alkali agent in rabbit eyes was studied. Josset et al. exposed the eye to filter paper soaked with concentrated sodium hydroxide for 1 minute and irrigated for 3 minutes with running water, an isotonic tears solution or Diphoterine. Although all lavage solutions did not protect against the destruction of the corneal epithelial surface, Diphoterine reduced the amount of stromal edema compared with water. Also, the endothelial cells were completely destroyed with water lavage, partially destroyed with the isotonic tear solution and slightly destroyed with Diphoterine.

Several case studies describe the effectiveness of Diphoterine in treating patients accidentally exposed to chemicals. Simon (2000) compared water and Diphoterine rinsing of 375 chemical splashes during a 7-year period at a chemical factory in France. This retrospective analysis identified 205 cases where water was used as the rinsing agent, resulting in 68 cases (33%) with no after effects, whereas 170 cases that used Diphoterine resulted in 88 cases (52%) with no after effects. Also, Hall et al. (2002) translated an unpublished report from Komad et al. Which described 45 occupational accidents at Bergheim, Germany, involving sodium hydroxide or other strong bases. The finding was a significant reduction in lost work time when Diphoterine was used relative to water. Also, Diphoterine-treated patients did not require further significant medical treatment, whereas the water-treated patients did.

Gerasimo et al. (2000) compared the ability of soapy water, physiological saline and Diphoterine to decontaminate HD. They exposed human skin obtained from elective abdominoplasty to C14 labeled HD in vitro for 5 minutes. They added the lavage to the test tube and removed the skin after 3 minutes, 10 minutes or 3 successive 10 minute-washes. In each case, Diphoterine significantly removed more HD than the other two treatments. For the three successive washes, Diphoterine removed 50% of the applied agent compared with 37% for soapy water and 32% for physiological saline.

Many industrial facilities in Germany and France currently use Diphoterine as a decontamination solution for chemical accidents. Also, special units of the French gendarmerie forces carry Diphoterine as a chemical/biological warfare agent decontaminant (VIALA 1999).

REACTIVE SPONGE

The reactive sponge is an emerging decontaminant with much promise. Currently, it is in the research and development phase but is worth discussing in this paper. The concept involves immobilizing enzymes such as cholinesterases to polyurethane foam such that chemical warfare agents would be absorbed by the sponge and destroyed by the enzymes.

A number of studies have shown that a variety of enzymes exhibit enhanced mechanical and chemical stability when immobilized on a solid support, producing a biocatalyst. The study of degradation of organophosphonates (OP) by immobilized enzymes dates back to Munnecke (1979), who attempted to immobilize a pesticide detoxification extract from bacteria by absorption on glass beads. The absorbed extract retained activity for a full day. Wood and coworkers (1982), - using isocyanate-based polyurethane foams (HypolÒ), found that a number of enzymes unrelated to OP hydrolysis could be covalently bound to this polymer; after that Havens and Rase (1993) successfully immobilized parathion hydrolase.

More recently, the enzyme bioscavenger approach (Carranto et al., 1994; Maxwell et al., 1992) has been shown to be effective against a variety of OP compounds in vitro and in vivo; pretreatment of rhesus monkeys with fetal bovine serum (FBS) acetylcholinesterase (AChE) or equine serum butyrylcholinesterase (BChE) protected them against a challenge of up to 5 LD_{50} of GD. While the use of cholinesterase (ChE) as a single pretreatment drug for OP toxicity provided complete protection, a stoichiometric amount of enzyme was required to neutralize the OP in vivo.

To increase the OP/enzyme stoichiometry, enzyme pretreatment was combined with oximes such as HI-6 so that the catalytic activity of OP-inhibited AChE is rapidly and continuously restored before irreversible aging of the enzyme-OP complex can occur. Thus, the OP is continuously detoxified.

Based on the two above observations, (a) that polyurethane foams are excellent adsorption materials for OPs such as pesticide vapors, and (b) that soluble ChEs and oxime together have the ability to detoxify OP compounds, Gordon et al. (1999) combined these features in a porous polyurethane foam formed in situ from water-miscible hydrophilic urethane prepolymers and the enzymes. Thus, they envision a reusable immobilized enzyme sponge of cholinesterases and oximes for OP decontamination.

Gordon et al. (1999) have demonstrated the rapid copolymerization of ChEs and other chemical warfare agent decontaminating enzymes at room temperature. The ChE-sponges exhibit high activity and stability, making them suitable for a wide variety of decontamination tasks. They have evaluated several additives to aid in the removal of OPs and HD from the skin. Once removed, OPs can then be detoxified in the enzyme-sponge pad with an oxime reactivating the enzyme, and in so doing prevents secondary contamination.

Also, in the presence of additional oxime, the enzyme-sponge would be reusable. In addition to decontamination of skin, wounds, and personnel, the enzyme-sponges can be utilized for preventing cross-contamination of medical and clinical personnel. Still more uses for these formulations could include decontamination foams as masks and in garments, thereby replacing carbon filters that absorb OPs without inactivating them. The sponge could be used in chemical-biological sensors and incorporated into the telemedicine initiative as electrochemical OP probes.

Chemical warfare agents in the environment could be contained and decontaminated if the sponge components were incorporated into fire fighting foams. The enzyme-foams could be used to decontaminate sensitive equipment without posing new environmental disposal problems, since the final products are rendered inert. Indeed, the sponge should be suitable for a variety of detoxification and decontamination schemes for both chemical weapons and civilians exposed to pesticides or highly toxic OPs such as sarin. The sponges should be suitable for a variety of biological surface detoxification and decontamination schemes for both chemical weapons and pesticides (Gordon et al. 1999).

Conclusion

Several decontaminants are readily available for chemical warfare exposure as described above. Since the events of 11 September 01, numerous individuals have proposed other candidate products to be effective. The true measure of effectiveness requires that all products to be tested identically and compared with each other, to include soapy water as van Hooidonk et al. (1983) did so well for common household products. Until such studies are completed, the effectiveness of the above products must be evaluated independently using the studies to date. The most important guiding principle in evaluating them is that the best decontaminant rapidly remove threat agents from the skin. This principle predicates the availability of a decontaminant immediately (within minutes) after chemical warfare agent exposure. The immediate use of soapy water would offer far better results than the best possible decontaminant that was delayed more than 30 minutes in its use. Future evaluation of potential decontaminants should consider all of the discussed factors to determine the best product.

References

Caranto G.R., Waibel K.II., Asher J.M., Larrison R.W., Brecht K.M., Schutz M.B., Raveh L., Ashani Y., Wolfe A.D., Maxwell D.M. and Doctor B.P. Amplification of the effectiveness of acetylcholinesterase for detoxification of organophosphorus compounds by bis-quaternary oximes. Biochem. Pharmacol. **47:** 347-357, 1994.

Clouzeau J. and Read M.H. Evaluation of ocular irritation in the rabbit, Study performed at the Centre International de Toxicologie, Evereux, France, Technical report. July, 1990.

Dill G.S., Blank J .A. and Menton R.G. The effect of a candidate topical skin protectant (ICD No. 1511) on the efficacy of the M291 skin decontamination kit against TGD and HD in the rabbit, USAMRICD, Edgewood Area, Aberdeen Proving Ground, MD, MREF Task 91-154, Final Report. August, 1991.

Dolzine T.W. and Logan T. Proceedings of the 1998 Medical Defense Bioscience Review, US Army Medical Research and Development Command, Bioscience 98. AD MOO 1167.

Gerasimo P., Blomet J., Mathieu L. and Hall A. Diphoterine decontamination OfC14–sulfur mustard contaminated human skin fragments in vitro. The Toxicologist. **54 (1):** 152, 2000.

Gold M.B., Bongiovanni R., Scharf B.A., Gresham V.C. and Woodard C.L. Hypochlorite solution as a decontaminant in sulfur mustard contaminated skin defects in the euthymic hairless guinea pig. Drug and Chemical Toxicology. **17(4):** 499-527, 1994.

Gordon R.K. Feaster S.R., Russell A.J., Lejeune K.E., Maxwell D.M., Lenz D.E., Ross M. and Doctor B.P. Organophosphate skin decontamination using immobilized enzymes. Chemico-Biolog.lnterac. **119-120:** 463-470,1999.

Hall A.H., Blomet J. and Mathieu L. Diphoterine from emergent eye/skin chemical splash decontamination: a review. Veterinary and Human Toxicology. 44 (4): 228-231, 2002.

Havens P.L. and Rase H.F. Reusable immobilized enzyme/polyurethane sponge for removal and detoxification of localized organophosphate pesticide spills. Ind. Eng. Chern. Res. **32:** 2254-2258, 1993.

Hobson D.W., Blank J.A. and Menton R.G. Test up to 12 Candidate Skin Decontaminants USAMRICD, Edgewood Area, Aberdeen Proving Ground, MD, MREF Task 89-11, Final Report. April, 1993. AD #ADB173995.

Hobson D. W. and Snider, T .H. Evaluation of the effects of hypochlorite solutions in the decontamination of wounds exposed to either the organophosphonate chemical surety materiel VX or the vesicant chemical surety materiel HD. USAMRICD, Edgewood Area, Aberdeen Proving Ground, MD, MREF Task 89-11, Final Report. June, 1992. AD #ADB165708.

Joiner R.L., Keys Jr., W.B., Harroff Jr. H.H. and Snider T.H. Evaluation of the effectiveness of two Rohm & Haas candidate decontamination systems against percutaneous application of undiluted TGD, DG, VX, HD, AND L on the laboratory albino rabbit. USAMRICD, Edgewood Area, Aberdeen Proving Ground, MD, MREF Task 86-25, Final Report. February, 1988. AD #ADBI20368.

Jones J.R., Guest R.L., and Warner P .A. Diphoterine Brevete: Official Record concerning skin irritation: Test done in rabbits. Project Number 133/3, Study performed at SafepharmLaboratories, Ltd. Derby UK, Technical report. October 1987.

Josset P., Pellosse B. and Saraux H. [Interest of an isotonic amphoteric solution in the early treatment of corneo-conjunctival base chemical bums] [French]. Bull. Soc. Opht. France. **6-7:** 765-769, 1986.

Laboratoir Prevor: Prevor Health, Chemical Bums, Diphoterine, http:// www.~revor.com/ Prevor-us/SITE/pages/Olsante/brulure chimiaue/diphoterine/OO diDhoterine.htm. Accessed 22 Oct 2002.

Maxwell D.M., Castro C.A., De La Hoz D.M., Gentry M.K., Gold M.B. Solana B.P., Wolfe, A.D. and Doctor B.P. Protection of rhesus monkeys against soman and prevention of performance decrement by pretreatment with acetylcholinesterase. Toxicol. Appl. Pharmacol. **115:** 44-49, 1992.

Munnecke D.M. Chemical physical, and biological methods for the disposal and detoxification of pesticides. Residue Rev. **70:** 1-26, 1979.

Sabourin C.L., Hayes T.L. and Snider T.H. A Medical Research and Evaluation Facility Study on Canadian Reactive Skin Decontamination Lotion. USAMRICD, Edgewood Area, Aberdeen Proving Ground, MD, MREF Task 0008, Final Report. February, 2001.

Simon F. Comparison water/Diphoterine: Rinsing of more than 600 chemical splashes during 7 years in the factory ATOCHEM SAINT-A VOLD. Abstract presented at the SFETB Congress. June 2000.

van Hooidonk C., van der Weil H.J. and Langerberg J.P., Prins Marutis Laboratory. Comparison of the efficacy of skin decontaminations II. In vivo test—Final Report. TNO Report PML 1996-A77.

van Hooidonk C., Ceulen B.I., Bock J. and van Genderen J. CW agents and the skin. Penetration and decontamination. Proc. Int. Symp. Stockholm, Sweden; 153-160, 1983.

Vedder E.B. Vesicants. In: Vedder EB, ed. The Medical Aspects of Chemical Warfare. Baltimore, Md: Williams & Wilkins. 125-166, 1925.

Viala B. Medical Counselor to the Director of Defense and Civil Security, France, Report on decontamination of sulfur mustard {Yperite} with Diphoterine. Sent to the French Republic, Liberty, Equality, Fraternity, and Ministry of the Interior, 1999.

Wood L.L., Hardegen F.J. and Hahn P.A. Enzymes bound to polyurethane. U.S. Patent 4, 342, 834, 1982.

Wormser U., Brodsky B. and Sintov A. Skin toxicokinetics of mustard gas in the guinea pig: effect of hypochlorite and safety aspects. Arch Toxicol **76 (9):** 517-22, 2002.

Yang Y.C. Chemical detoxification of nerve agent VX. Acc. Chern Res. **32:** 109-115, 1999.

Pharmacological Perspectives of Toxic Chemicals and Their Antidotes
Editors: S J S Flora, J A Romano, S I Baskin and K Sekhar

Prophylactic Efficacy of Amifostine and DRDE-07 Against Sulfur Mustard Administered by Various Routes

R. Vijayaraghavan, A. Kulkarni, Pravin Kumar, P.V. Lakshmana Rao, S.C. Pant, U. Pathak, S.K. Raza and D.K. Jaiswal
Defence Research and Development Establishment
Jhansi Road, Gwalior 474002, India

INTRODUCTION

Sulfur mustard (SM), chemically bis [2-chloroethyl] sulfide and commonly known as mustard gas is an alkylating agent that causes serious blisters upon contact with human skin. SM is frequently used as a chemical warfare agent (Wormser, 1991; Smith and Dunn, 1991; Eisenmenger et al., 1991; Momeni et al., 1992). Due to the simple method of preparation, SM being used clandestinely during war or by terrorist groups remains a threat, inspite of the successful implication of the Chemical Weapons Convention (Krutzsch and Trapp, 1994). SM forms sulfonium ion in the body and alkylates DNA leading to DNA strand breaks and cell death (Papirmeister et al., 1991; Lakshmana Rao et al., 1999). Due to high electrophilic property of the sulfonium ion, SM binds to a variety of cellular macromolecules and fatality may occur due to multi organ failure (Somani and Babu, 1989; Dacre and Goldman, 1996). Eyes, skin and the respiratory tract are the principal target organs of SM toxicity (Papirmeister et al., 1991; Pechura and Rall, 1993; Vijayaraghavan, 1997).

Several antidotes have been reported for the systemic toxicity of SM in experimental animals (Callaway and Pearce, 1958; Vojvodic et al., 1985; Vijayaraghavan et al., 1991; Dacre and Goldman, 1996; Kumar et al., 2000). All the antidotes so far screened have given only limited protection and decontamination of SM, immediately after contact is recommended as the best protection (Somani and Babu, 1989). SM is highly lipophilic and is absorbed very quickly after contact with skin. Hence, the decontamination has to be accomplished quickly to limit the absorption. The most commonly used decontaminant is Fuller's earth (a native form of aluminium silicate) that removes SM by adsorption, thereby reducing the toxicity (Marrs et al., 1996). Few chemical decontaminants for human use have also shown very good efficacy (Shih et al., 1999; Vijayaraghavan et al., 2002).

An effective prophylactic agent against SM is the requirement of the day especially for personnel engaged in the destruction of SM and during inspection by the Organisation for Prohibition of Chemical Weapons. We earlier reported that amifostine, an organophosphorothioate and DRDE-07, an S-(ω-aminoalkylamino) alkylaryl sulfide gave very good protection as a prophylactic agent against SM (Vijayaraghavan et al., 2001; Bhattacharya et al., 2001; Joshi et al., 2002). When given orally DRDE-07 was more efficacious than amifostine as a prophylactic agent against SM (Vijayaraghavan et al., 2001; Kumar et al., 2002).

We studied the efficacy of amifostine and DRDE-07 against SM administered by various routes. The specific protection of amifostine and DRDE-07 when SM is administered by percutaneous route, and the more toxic effect of SM when administered by percutaneous route are reported here.

MATERIALS AND METHODS

Chemicals

Amifostine and DRDE-07 were synthesised in the chemistry laboratory of DRDE (Defence Research and Development Establishment). The compounds were characterised by elemental analysis, IR, ^{1}H NMR and mass spectral analysis. The purity was checked by thin layer chromatography. SM was also synthesised and was found to be above 99 % pure by gas chromatographic analysis. SM was diluted in PEG 300 (Fluka, USA), fresh daily.

Amifostine : $NH_2 - CH_2 - CH_2 - CH_2 - NH - CH_2 - CH_2 - S - PO_3H_2$

DRDE-07 : $NH_2 - CH_2 - CH_2 - NH - CH_2 - CH_2 - S - C_6H_5$

Animals

Randomly bred Swiss female mice (25 to 30 g body weight) and Wistar male rats (200-250 g) maintained in the Establishment's animal house were used for the study. The animals were housed in polypropylene cages on dust free rice husk as the bedding material. The animals were provided with pellet diet (Amrut Ltd., India) and water *ad libitum*. The care and maintenance of the animals were as per the approved guidelines of the Committee for the Purpose of Control and Supervision of Experiments on Animals (CPCSEA, India). This study has the approval of the Establishment's Ethical Committee.

A day before dermal application (percutaneous administration) of SM, the hair on the back of mice and rats was closely clipped using a pair of scissors. The mice and rats were fasted for about two hours before oral feeding of the prophylactic agents.

LD_{50} DETERMINATION

LD_{50} of SM was determined by percutaneous, subcutaneous and oral routes. For this SM was diluted in polyethylene glycol (PEG-300) and about 0.1 ml for a mouse and 0.2 ml for a rat was administered. The animals were weighed daily and observed for mortality for a period of 14 days. For all LD_{50} determinations, 3 to 4 groups were used with each group consisting of four mice or four rats.

LC_{50} AND RD_{50} DETERMINATIONS

Mice were exposed head-only to SM vapor in a cylindrical inhalation exposure chamber of 50 cm long and 10 cm diameter made of polytetrafluoroethylene (PTFE), positioned horizontally. Four mice were exposed at a time in individual body plethysmographs made of glass. The generation of SM vapors and the analysis are given in detail elsewhere (Vijayaraghavan, 1997). Briefly, SM was diluted in 10 ml of acetone and the solution was vaporized in a compressed air nebulizer. The vapors were directed into the exposure chamber, and the outgoing air from the exposure chamber was decontaminated and then exhausted out in a fume hood. The chamber air was analyzed gas chromatographically. During inhalation exposure, they were monitored for respiratory modification. A differential pressure transducer was attached to the body plethysmograph for the measurement of respiratory flow. The flow signals from the individual transducers were amplified using an universal amplifier (Gould, Cincinnati, USA). The amplified signals were digitized using an analog to digital converter (Metrabyte, Taunton, Mass, USA), integrated as tidal volume (VT), stored and analyzed using a personal computer. A computer program capable of recognising the effects of airborne chemicals as sensory irritation, airway obstruction and pulmonary irritation, and a combination of these effects, described earlier was used (Vijayaraghavan *et al.*, 1994). All the respiratory variables were also measured using the computer program.

The concentration that depresses 50 % of the respiratory frequency (RD_{50}) while exposure to SM vapor was calculated (Alarie, 1973; Vijayaraghavan et al., 1994). The lowest respiratory frequency, as percent change of pre-exposure value for the group of four mice, during exposure to various concentrations of SM was taken from the SigmaPlot (version 2.0) worksheet.

RD_{50} was calculated by linear regression analysis. The mice were exposed for 1 hr duration. They were weighed daily and observed for mortality for 14 days. For the RD_{50} and LC_{50} determinations, 3 to 4 groups were used with each group consisting of four mice.

PROTECTION EFFICACY STUDIES

Efficacy of the Prophylactic Agents against Percutaneously Administered SM

Amifostine and DRDE-07 were evaluated as prophylactic agents by administering them 30 min before SM application in mice and rats. The antidotes were administered through oral route at a dose of 0.1 and 0.2 LD_{50} (mice: amifostine = 105 and 210 $mg.kg^{-1}$; DRDE-07 = 125 and 249 $mg.kg^{-1}$; rat: amifostine = 276 and 452 $mg.kg^{-1}$; DRDE-07 = 160 and 320 $mg.kg^{-1}$). SM diluted in PEG 300 was applied on the back of the animals and spread on an area of about 2 cm diameter. The animals were held for a minute before leaving them in the respective cages. In each cage four mice or two rats were kept. Three to four doses were used and for each dose four animals were used. The animals were weighed daily and observed for 14 days for mortality. The PI was calculated as per the following formula:

$$PI = \frac{LD_{50} \text{ of SM with the prophylactic agent}}{LD_{50} \text{ of SM alone}}$$

Efficacy of the Prophylactic Agents against SM Administered through Inhalation

Amifostine and DRDE-07 were evaluated as prophylactic agents by administering them 30 min before SM inhalation in mice. The antidotes were administered through oral route at a dose of 0.2 LD_{50} (amifostine = 210 $mg.kg^{-1}$ and DRDE-07 = 249 $mg.kg^{-1}$). SM diluted in acetone was vaporized and the mice were exposed for the vapors for one hr. During the course of the exposure all the respiratory variables were monitored. RD_{50} was calculated as described earlier. The animals were weighed daily and observed for 14 days for mortality.

Efficacy of the prophylactic agents against SM administered through various routes

Amifostine and DRDE-07 were evaluated as prophylactic agents by administering them 30 min before SM administration in mice. The antidotes were administered through oral route at a dose of 0.2 LD_{50} (amifostine = 210 $mg.kg^{-1}$ and DRDE-07 = 249 $mg.kg^{-1}$). SM diluted in PEG 300 was either applied on the back of the mice or administered orally or subcutaneously. To avoid variations, the same solution was used, and one dose and all the three routes were used per day. The animals were held for a minute before leaving them in the respective cages. In each cage four mice were kept. Three to four doses were used and for each dose four animals were used. The animals were weighed daily and observed for 14 days for mortality.

Protective Effect on the Decrease in Glutathione (GSH) and DNA Damage Induced by Percutaneously Administered SM

To evaluate the protective effect of amifostine and DRDE-07 on the decrease in GSH and DNA damage induced by SM, the following groups were kept.

Group I	:	Distilled water, p.o.	+	SM (77.5 mg.kg^{-1}) in PEG 300 dermal (SM group)
Group II	:	Amifostine 0.2 LD_{50}, 30 min before, p.o.	+	SM (77.5 mg.kg^{-1}) in PEG 300 dermal
Group III	:	DRDE-07, 0.2 LD_{50}, 30 min before, p.o.	+	SM (77.5 mg.kg^{-1}) in PEG 300 dermal
Group IV	:	Distilled water, p.o.	+	SM (155 mg.kg^{-1}) in PEG 300 dermal (SM group)
Group V	:	Amifostine 0.2 LD_{50}, 30 min before, p.o.	+	SM (155 mg.kg^{-1}) in PEG 300 dermal
Group VI	:	DRDE-07, 0.2 LD_{50}, 30 min before, p.o.	+	SM (155 mg.kg^{-1}) in PEG 300 dermal

Each group consisted of four mice. A control group was also kept without any treatment. After SM application the body weights were recorded daily and the mice were sacrificed seven days post exposure by cervical dislocation. The liver was quickly removed, washed free of blood and frozen at—20° C before assay. The fluorometric method of Hisin and Hilf (1976) was used for the determination of hepatic GSH and GSSG concentration. DNA fragmentation was assayed as previously described (Lakshmana Rao et al., 1999). Briefly, the frozen liver samples were homogenised in ice cold lysis buffer (10 mM Tris, 20 mM EDTA, 0.5 % Triton X 100, pH 8.0) and then centrifuged at 27,000 x g for 30 min. Both pellet (intact chromatin) and supernatant (DNA fragments) were assayed for DNA content using diphenylamine colorimetric reaction (Burton, 1956). The percentage of fragmented DNA was defined as the ratio of the DNA content of supernatant at 27,000 x g to the total DNA in the lysate (Wyllie, 1980).

STATISTICAL ANALYSIS

LD_{50} was determined by the moving average method (Gad and Weil, 1989). Body weight changes, GSH and DNA damage were analysed by one way ANOVA with Dunnett's multiple comparisons procedure. A probability of less than 0.05 is taken as statistically significant. SigmaStat (Jandel Sci., USA) was used for the statistical calculations.

RESULTS

The prophylactic efficacy of orally administered amifostine and DRDE-07 against percutaneously administered SM is given in Table 1. Amifostine and DRDE-07 offered a dose dependent protection in mice. DRDE-07 was more efficacious than amifostine. A dose of 249 $mg.kg^{-1}$ (0.2 LD_{50}) of DRDE-07 gave a protection of 27.0 fold, compared to 210 $mg.kg^{-1}$ (0.2 LD_{50}) of amifostine, which gave a protection of 9.5 fold. But the protection offered by amifostine and DRDE-07 in rat was not dose dependent. Both the prophylactic agents gave a maximum protection of 2.8 fold only, irrespective of the dose. Percutaneously administered SM decreased the body weight dose dependently in mice and rats (Fig.1 and 2). Both amifostine and DRDE-07 protected the decrease in the body weight. The protection was better with DRDE-07 than with amifostine.

Table 1 Protective efficacy of amifostine and DRDE-07 against SM administered through percutaneous route

Species	*Sex*	*Antidote*	*Dose $mg.kg^{-1}$ (oral)*	*LD_{50} $mg.kg^{-1}$*	*Protection Index*
Mouse	F	Water	-	8.1 (4.5 - 14.8)	-
		Amifostine	105 (0.1 LD_{50})	54.6 (31 - 95)	6.7
		Amifostine	210 (0.2 LD_{50})	77.3 (18 - 323)	9.5
		DRDE-07	125 (0.1 LD_{50})	97.3 (18 - 323)	12.0
		DRDE-07	249 (0.2 LD_{50})	218.8 (98 - 487)	27.0
Rat	M	Water	-	2.4 (1.5 - 3.8)	-
		Amifostine	276 (0.1 LD_{50})	6.8 (4.8 - 9.7)	2.8
		Amifostine	452 (0.2 LD_{50})	6.8 (4.8 - 9.7)	2.8
		DRDE-07	160 (0.1 LD_{50})	6.8 (4.8 - 9.7)	2.8
		DRDE-07	320 (0.2 LD_{50})	6.8 (4.8 - 9.7)	2.8

When SM was exposed through inhalation neither amifostine nor DRDE-07 gave significant protection. The RD_{50} of SM was 28.9 $mg.m^{-3}$. The RD_{50} of SM with amifostine or DRDE-07 pretreatment was not altered significantly. The LC_{50} of SM was 22.7 $mg.m^{-3}$. Pretreatment with amifostine or DRDE-07 did not give any significant protection (Table 2).

Table 3 gives the PI of amifostine and DRDE-07 against SM administered by various routes. When SM was administered through percutaneous route amifostine and DRDE-07 gave protection of 4.0 and 22.0, respectively. When SM was administered through oral or subcutaneous routes neither amifostine nor DRDE-07 gave significant protection. Only DRDE-07 gave a protection index of 1.2 when SM was administered through subcutaneous route. SM administered through percutaneous route decreased the body weight more than the oral and subcutaneous routes. The decrease in body weight was

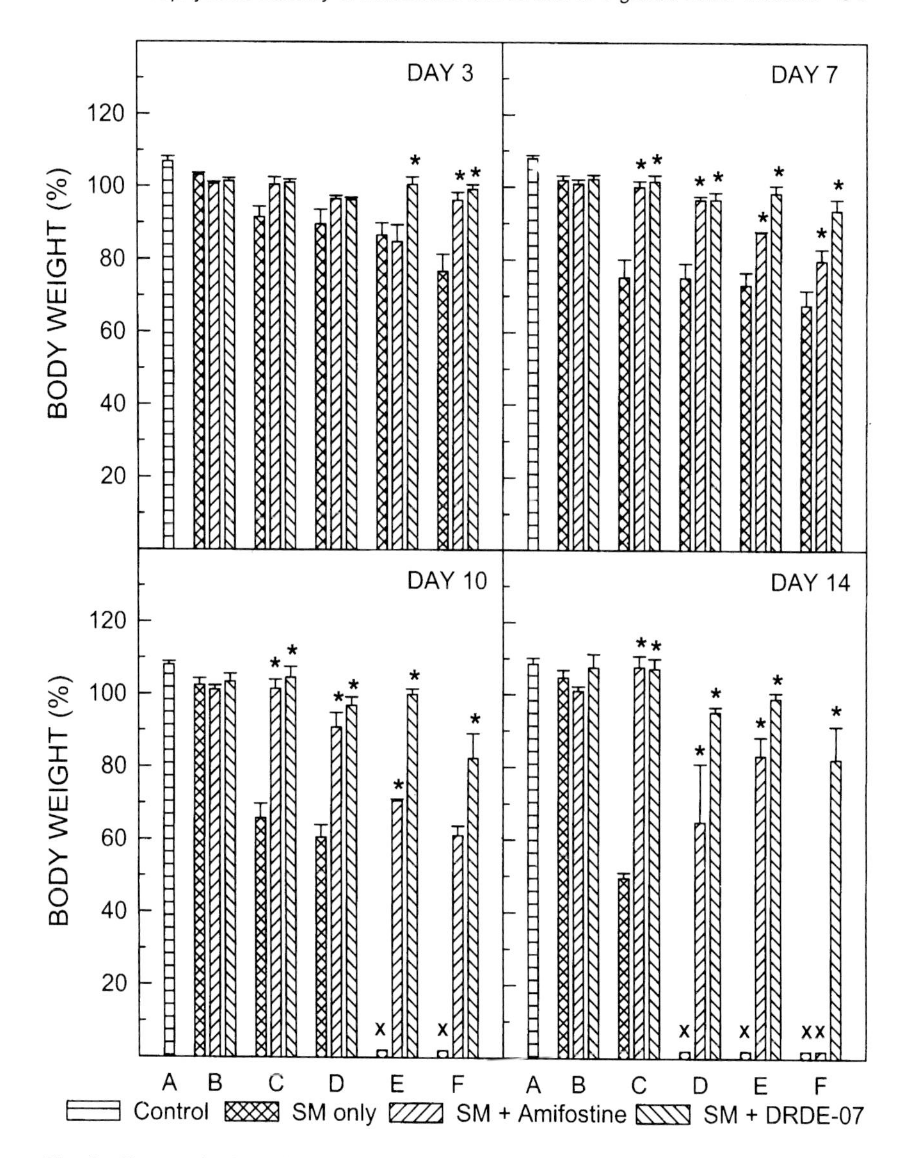

Fig. 1 Percent body weight change in mice after percutaneous administration of SM; SM dose: A=0, B=2.4, C=4.8, D=9.6, E=19.3 and F = 38.7 mg.kg^{-1}. Dose of amifostine = 210 mg.kg^{-1} (0.2 LD_{50}, oral) and DRDE-07 = 249 mg.kg^{-1} (0.2 LD_{50}, oral), 30 min prior to SM administration. Mean ± SE (n=2 to 4). X=animals died. *Statistically significant from corresponding SM group.

progressive in the percutaneous route (Fig. 3). Amifostine and DRDE-07 protected the decrease in the body weight induced by SM.

The most interesting observation was that SM administered through percutaneous route was more toxic than the other routes. The LD_{50} for a 14 day observation period by the percutaneous route in mice was 5.7 mg.kg^{-1},

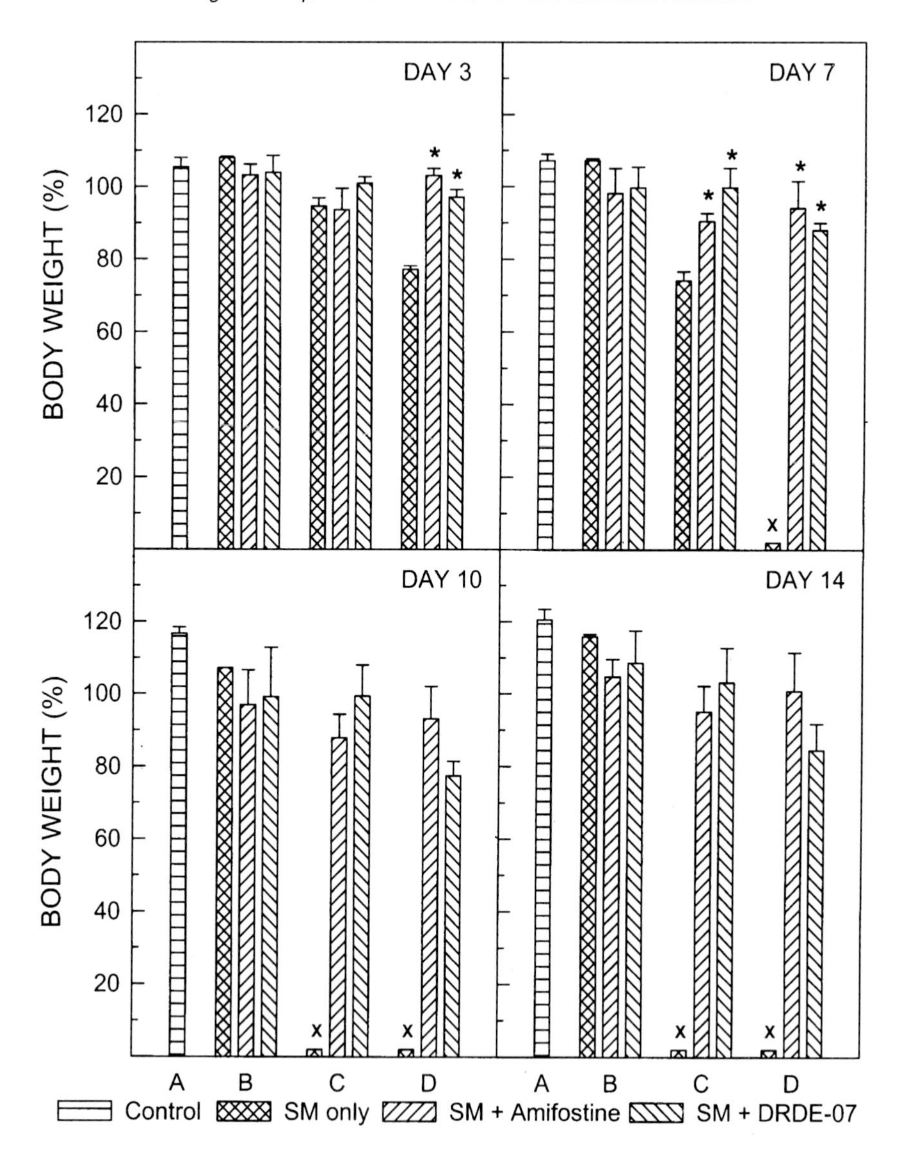

Fig. 2 Percent body weight change in rat after percutaneous administration of SM; SM dose: A=0, B=1.2, C=2.4 and D=4.8 mg.kg^{-1}. Dose of amifostine = 452 mg.kg^{-1} (0.2 LD_{50}, oral) and DRDE-07 = 320 mg.kg^{-1} (0.2 LD_{50}, oral), 30 min prior to SM administration. Mean ± SE (n=2 to 4). X=animals died. *Statistically significant from corresponding SM group.

while in oral and subcutaneous routes the LD_{50} was 8.1 and 23.0 mg.kg^{-1} respectively (Table 4). The results show that the LD_{50} decreased with the period of observation. Following percutaneous and oral administration of SM, mortality was observed up to 14 days and beyond that, the animals survived. In the subcutaneous route the animals died within seven days and hence the

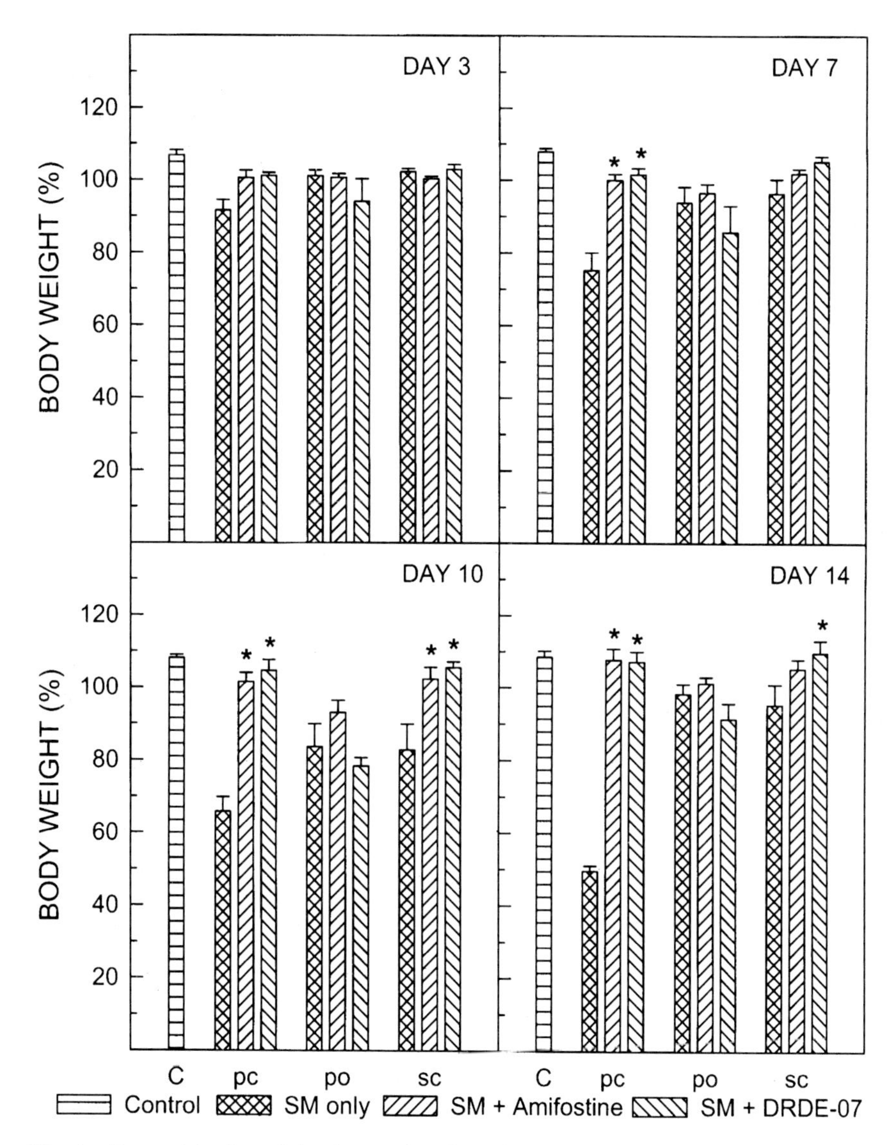

Fig. 3 Percent body weight change in mice after percutaneous (pc), oral (po) and subcutaneous (sc) administration of SM; SM dose = 4.8 mg.kg^{-1}. C = control, no SM. Dose of amifostine = 210 mg.kg^{-1} (0.2 LD_{50}, oral) and DRDE-07 = 249 mg.kg^{-1} (0.2 LD_{50}, oral), 30 min prior to SM administration. Mean ± SE (n=2 to 4). *Statistically significant from corresponding SM group.

seven day and 14 day observation period gave the same LD_{50}. In the percutaneous route the death of the animals were delayed and hence there was a wide variation in the seven day and 14 day observation period LD_{50}. In rat also the percutaneous route was more toxic than the oral and subcutaneous routes.

Table 2 Protective efficacy of amifostine and DRDE-07 against SM administered through inhalation in mice

Antidote	*Dose mg.kg^{-1} (oral)*	*RD_{50} mg.m^{-3}*	*LC_{50} mg.m^{-3}*	*Protection Index*
Water	-	28.9	22.7 (14.1 - 26.6)	-
Amifostine	210 (0.2 LD_{50})	33.8	26.1 (15.5 - 44.0)	1.1
DRDE-07	249 (0.2 LD_{50})	28.7	22.7 (14.1 - 36.6)	1.0

Table 3 Protective efficacy of amifostine and DRDE-07 against SM administered through various routes in female mice

Route of SM	*Dose mg.kg^{-1} (oral)*	*Antidote*	*LD_{50} mg.kg^{-1}*	*Protection Index*
pc	-	Water	5.7 (3.2 - 10.5)	-
	210 (0.2 LD_{50})	Amifostine	23.0 (12.6 - 41.8)	4.0
	249 (0.2 LD_{50})	DRDE-07	130.0 (71.4 - 236.5)	22.8
po	-	Water	8.1 (4.5 - 14.8)	-
	210 (0.2 LD_{50})	Amifostine	6.8 (3.9 - 11.9)	0.8
	249 (0.2 LD_{50})	DRDE-07	5.8 (3.2 - 10.5)	0.7
sc	-	Water	23.0 (15.0 - 35.1)	-
	210 (0.2 LD_{50})	Amifostine	13.6 (9.7 - 19.3)	0.6
	249 (0.2 LD_{50})	DRDE-07	27.3 (2.3 - 60.8)	1.2

Table 4 LD_{50} of SM administered by various routes

Species	*Sex*	*Route of SM*	*LD_{50} 7 days Observation*	*LD_{50} 14 days Observation*
Mouse	F	pc	154.6 (58.1 - 411.7)	5.7 (3.2 - 10.5)
		po	34.4 (16.2 - 73.2)	8.1 (4.5 - 14.8)
		sc	23.0 (15.0 - 35.1)	23.0 (15.0 - 35.1)
Rat	M	pc	4.8 (1.0 - 23.2)	2.0 (1.3 - 3.1)
		po	6.8 (4.8 - 9.7)	2.4 (1.5 - 3.8)
		sc	3.4 (2.4 - 4.8)	3.4 (2.4 - 4.8)

Percutaneous administration of 77.5 and 155.0 mg.kg^{-1} of SM decreased the GSH level. The decrease in GSH was not dose dependent. DRDE-07 significantly protected the decrease in the GSH level, in both the doses of SM. Amifostine did not protect the decrease in GSH level (Table 5). The % DNA fragmentation was increased following percutaneous administration of SM. A 2.2 fold increase in DNA frgamentation was observed at 77.5 mg.kg^{-1} dose and 3.1 fold increase at 155.0 mg.kg^{-1} dose of SM. DRDE-07 significantly protected DNA fragmentation induced by SM. Amifostine also protected the DNA fragmentation but it was significant only at 155.0 mg.kg^{-1} dose of SM.

Table 5. Protective efficacy of amifostine and DRDE-07 on GSH and DNA damage induced by percutaneously administered SM in mice

SM dose $mg.kg^{-1}$	*Group*	*GSH μ moles/g tissue*	*% DNA fragmentation*
-	Control	8.05 ± 0.51	14.3 ± 1.6
77.5	SM	3.71 ± 0.18[a]	31.5 ± 3.4[a]
77.5	SM + amifostine	3.87 ± 0.57[a]	24.2 ± 1.8[a]
77.5	SM + DRDE-07	6.76 ± 1.08[b]	19.2 ± 2.0[b]
-	Control	6.61 ± 0.32	10.4 ± 1.4
155.0	SM	3.35 ± 0.13[a]	32.5 ± 2.6[a]
155.0	SM + amifostine	4.48 ± 0.60[a]	25.9 ± 1.2[ab]
155.0	SM + DRDE-07	5.52 ± 0.63[b]	16.3 ± 2.0[ab]

Dose : amifostine = 210 $mg.kg^{-1}$ (0.2 LD_{50}, oral): DRDE-07 = 249 $mg.kg^{-1}$ (0.2 LD_{50}, oral).
Animals were sacrificed 7 days after SM administration.
Values are mean ± SE (n =4).
[a]Statistically significant from control; [b]Statistically significant from SM group.

DISCUSSION

Antidotes to SM can act by four different mechanisms. (a) Prevention of SM from entering the system (personnel decontamination at the site of contact), (b) prevention of SM from alkylating critical target molecules mainly DNA, (c) retrieval of SM alkylated DNA and (d) prevention and reversal of the cascade of secondary biochemical reactions of alkylation (Papirmeister *et al.*, 1991). SM is known to react with a variety of macromolecules viz., DNA, proteins and several thiol containing molecules including GSH (Somani and Babu, 1989). Decrease in the GSH content was reported following percutaneous or inhalation exposure to SM (Lakshmana Rao et al., 1999). If thiol containing compounds are administered as prophylactic agents they can spare GSH by neutralizing SM. It is expected that amifostine can neutralizc and reduce the concentration of SM inside the cell. Amifostine is dephosphorylated to its free thiol molecule by membrane bound alkaline phosphatase which enters normal tissues and provides protection against alkylating agents and radiation (Capizzi, 1999, Hospers *et al.*, 1999). When amifostine is given as a prophylactic agent for chemotherapeutic agents like cisplatin and cyclophosphamide it selectively protects normal tissues without reducing the effect of the anticancer agents on the cancer cells (Foster-Nora and Siden, 1997; Srivastava *et al.*, 1999; Links and Lewis, 1999; Castiglione *et al.*, 1999; Wasserman, 1999).

Analogues of amifostine were synthesised having better lipophilicity, expecting a better efficacy. Among the several compounds synthesised DRDE-07 gave better protection when it was administered through oral route against percutaneously administered SM (Vijayaraghavan et al., 2001; Kumar et al., 2002). In the present study it is again proved that orally administered DRDE-07 is more efficacious than amifostine against percutaneously administered SM. DRDE-07 is effective only when it is given as a pretreatment or simultaneous treatment, and is not effective as a post treatment. Moreover, DRDE-07 do not bind to SM directly, showing that an active metabolite of DRDE-07 interacts with SM, before SM can cause the damage (Bhattacharya et al., 2001). Percutaneously administered SM induces a progressive decrease in body weight. The body weight reduction is dose dependent and in some cases 50 % reduction in the initial body weight was observed. This was partially due to reduced intake of food and water as a result of the toxic effect of SM. Amifostine and DRDE-07 restored the body weight, and DRDE-07 was more efficacious than amifostine.

Sulfur mustard decreased the GSH content and increased DNA fragmentation. The decrease in GSH may be due to direct interaction of SM or may be due to increased requirement due to SM induced lipid peroxidation (Papirmeister et al., 1991; Vijayaraghavan et al., 1991). Since the available SM inside the body was reduced by amifostine and DRDE-07 when they were given as a prophylactic agent, protection was observed. Amifostine has to be converted to its free thiol metabolite (WR-1065) by the membrane bound alkaline phosphatase for its cytoprotective action (Spencer and Goa, 1995; Capizzi, 1999, Hospers *et al.*, 1999). Since DRDE-07 does not have a phosphate group its further metabolism and mechanism of protection is not understood. DRDE-07 is more efficacious than amifostine as a prophylactic agent against the lethality of SM. Due to the presence of an aryl group in DRDE-07 the lipophilicity is increased with a better bioavailability.

The major routes of entry of SM are the skin and respiratory tract. In the present study SM was exposed as a vapor in mice, pretreated with amifostine and DRDE-07. Amifostine and DRDE-07 did not protect the mice exposed to SM vapor. The primary target of inhalation exposure to SM vapor is the lung parenchyma (Pant et al., 2000). The active metabolite of amifostine and DRDE-07 are expected to be present in the lung parenchyma so as to neutralize the SM vapor. It appears that sufficient concentration of the active metabolite is not reached and hence, there was no protection. Chemicals that can interact with SH, NH_2 and OH groups generally cause a sensory irritation and one of the effects of sensory irritation is decrease in respiratory frequency (Neilsen et al., 1993). RD_{50} is the concentration that decreases 50 % of the respiratory frequency. Pretreatment with amifostine and DRDE-07 did not alter the RD_{50} of SM vapor showing no protective effect.

Amifostine and DRDE-07 did not protect when SM was administered by oral and subcutaneous routes. When SM was administered through percutaneous route amifostine and DRDE-07 were effective, showing that the mechanism of toxic effect of SM varies with different routes. Interestingly, percutaneously administered SM was more toxic in the rodent species viz., mice and rats examined in the present study. There was a wide variation in the LD_{50} values of seven day and 14 day observation periods in the case of percutaneously administered SM. SM's toxic effects are mostly on the rapidly dividing cells. Since maximum cell divisions takes place in skin, it is possible that some more toxic metabolites of SM are formed when it passes through the skin (Emmett, 1991). It appears that the active metabolite produces a progressive toxic effect, inducing depletion of GSH. Amifostine and DRDE-07 were able to neutralize the toxic metabolite of SM. The formation of the toxic metabolite in the skin may be immediate as amifostine and DRDE-07 can protect only when they are given as pretreatment or simultaneous treatment (Bhattacharya et al., 2001). One of the active forms of SM is the sulfonium ion and it is possible that its formation may be more through the percutaneous route. The greater toxicity of SM by the percutaneous route over the subcutaneous route needs further evaluation.

Acknowledgements

The authors are grateful to Mr. K. Sekhar, Director, DRDE for his constant encouragement. Thanks are also due to Dr. R.C. Malhotra and Mr. K. Ganesan for providing sulfur mustard.

References

Alarie Y. Sensory irritation of airborne chemicals. CRC Crit. Rev. Toxicol. **2:** 299-363, 1973.

Bhattacharya R., Rao P.V.L., Pant S.C., Kumar P., Tulsawani R.K., Pathak U., Kulkarni A. and Vijayaraghavan R. Protective effects of amifostine and its analogues on sulfur mustard toxicity in vitro and in vivo. Toxicol. Appl. Pharmacol. **176:** 24-33, 2001.

Burton K. A study of the condition and mechanism of the diphenylamine reaction for the colorimetric estimation of deoxyribonucleic acid. Biochem. J. 62, 315-523, 1956.

Callaway S. and Pearce K.A. Protection against systemic poisoning by mustard gas di(2-chloroethyl) sulphide by sodium thiosulphate and thiocit in albino rat. Br. J. Pharmcol. **13:** 395-399, 1958.

Capizzi R.L. The preclinical basis for broad spectrum selective cytoprotection of normal tissues from cytotoxic therapies by amifostine. Semin. Oncol. **26:** 3-21, 1999.

Castiglione F., Dalla Mola A. and Porcile G. Protection of normal tissues from radiation and cytotoxic therapy: the development of amifostine. Tumori. **85:** 85- 91, 1999.

Dacre J.C. and Goldman M. Toxicology and pharmacology of the chemical warfare agent sulfur mustard. Pharmacol. Rev., **48:** 290-326, 1996.

Emmett E.A. Toxic response of the skin, Chapter 15, In: Casarett and Doull's toxicology (Eds: Amdur, M.O., Doull, J. and Klaassen, C.D.), Fourth Edition, Pergamon Press, New York, 463-483, 1991.

Eisenmenger W., Drasch G., Von Clarmann M., Kretschmer E. and Roider G. Clinical and morphological findings on mustard gas [bis(2-chloroethyl) sulphide] poisoning. J. Forensic. Sci. **36:** 1688-1698, 1991.

Foster-Nora J.A. and Siden R. Amifostine for protection from antineoplastic drug toxicity. Am. J. Health Syst. Pharm. **54:** 787-800, 1999.

Gad S.C. and Weil C.S. Statistics for toxicologists. In: Hayes AW (Ed) Principles and methods of toxicology, 2nd edition. Raven Press, New York, 463-467, 1989.

Hisin P.J. and Hilf R. A fluorometric method for determination of oxidized and reduced glutathione in tissues. Anal. Biochem. **74:** 214-226, 1976.

Hospers G.A., Eisenhauer E.A. and de Vries E.G. The sulfhydryl containing compounds WR-2721 and glutathione as radio and chemoprotective agents. A review, indications for use and prospects. Br. J. Cancer **80:** 629-638, 1999.

Joshi U., Raza S.K., Pravin Kumar, Vijayaraghavan R. and Jaiswal D.K. A process for preparation of S-(ω-aminoalkylamino) alkylaryl sulphide dihydrochloride. Indian patent filed, Patent Office, New Delhi, India, 1999.

Krutzsch W. and Trapp R. A commentary on the chemical weapons convention. Martinus Nijhoff Publishers, London, 543, 1994.

Kumar P., Vijayaraghavan R., Kulkarni A.S., Pathak U., Raza S.K. and Jaiswal D.K. In vivo protection by amifostine and DRDE-07 against sulphur mustard toxicity. Hum. Exp. Toxicol. **21:** 371-376, 2002.

Lakshmana Rao P.V., Vijayaraghavan R. and Bhaskar A.S.B. Sulphur mustard induced DNA damage in mice after dermal and inhalation exposure. Toxicology 139, 39-51, 1999.

Links M. and Lewis C. Chemoprotectants: a review of their clinical pharmacological and therapeutic efficacy. Drugs **57:** 293-308, 1999.

Marrs T.C., Maynard R.L. and Sidell F.R. Chemical warfare agents; toxicology and treatment. John Wiley and Sons, Chichester, 162-163, 1996.

Momeni A.Z., Enshaeih S., Meghdadi M. and Amindjavaheri M. Skin manifestations of mustard gas. A clinical study of 535 patients exposed to mustard gas. Arch Dermatol. **128:** 775-780, 1992.

Nielsen G.D. Mechanisms of activation of the sensory irritant receptor by airborne chemicals. Crit. Rev. Toxicol. **21:** 183-208, 1991.

Om Kumar, Sugendran K. and Vijayaraghavan R. Protective effect of various antioxidants on the toxicity of sulphur mustard administered to mice by inhalation or percutaneous routes. Chemical. Biol. Int. **134:** 1-12, 2001.

Pant S.C., Vijayaraghavan R., Kannan G.M., Ganesan K. Sulphur mustard induced oxidative stress and its prevention by sodium 2,3-dimercapto propane sulphonic acid (DMPS) in mice. Biomed. Environ. Sci. **13:** 225-232, 2000.

Papirmeister B., Feister A.J., Robinson S.I. and Ford R.D. Medical defense against mustard gas: toxic mechanisms and pharmacological implications. CRC Press, Boca Raton, p 359, 1991.

Pechura C.M. and Rall D.P. Veterans at risk: The health effects of mustard gas and lewisite. National Academy Press, Washington DC, p 428, 1993.

Shih M.L., Korte W.D., Smith J.R. and Szafraniec L.L. Reactions of sulfides with S-330, a potential decontaminant of sulfur mustard in formulations. J. Appl. Toxicol. **19:** S83-S88, 1999.

Smith W.J. and Dunn M.A. Medical defense against blistering chemical warfare agents. Arch Dermatol, **127:** 1207-1213, 1991.

Somani S.M. and Babu S.R. Toxicodynamics of sulfur mustard. Int. J. Clin. Pharmacol. Ther. Toxicol. **27:** 419-435, 1989.

Spencer C.M. and Goa K.L. Amifostine: a review of its pharmacodynamic and pharmacokinetic properties, and therapeutic potential as a radioprotector and cytotoxic chemoprotector. Drugs **50:** 1001-1031, 1995.

Srivastava A., Nair S.C., Srivastava V.M., Balamurugan A.N., Jeyseelan L., Chandy M. and Gunasekaran S. Evaluation of uroprotective efficacy of amifostine against cyclophosphamide induced hemorrhagic cystitis. Bone Marrow Transplant **23:** 463-467, 1999.

Vijayaraghavan R., Sugendran K., Pant S.C., Husain K. and Malhotra R.C. Dermal intoxication of mice with bis (2-chloroethyl) sulphide and the protective effect of flavonoids. Toxicology **69:** 35-42, 1991.

Vijayaraghavan R., Thomson R., Schaper M., Lee Ann B., Stock M.F., Luo J. and Alarie Y. Computer assisted recognition and quantification of sensory irritation, airway constriction and pulmonary irritation, Arch Toxicol. **68:** 490-499, 1994.

Vijayaraghavan R. Modifications of breathing pattern induced by inhaled sulphur mustard in mice. Arch. Toxicol. **71:** 157-64, 1997.

Vijayaraghavan R., Kumar P., Joshi U., Raza S.K., Lakshmana Rao P.V., Malhotra R.C. and Jaiswal D.K. Prophylactic efficacy of amifostine and its analogues against sulphur mustard toxicity. Toxicology **163:** 83-91, 2001.

Vijayaraghavan R., Kumar P., Dubey D.K. and Singh R. Evaluation of CC2 as a decontaminant in various hydrophilic and lipophilic formulations against sulphur mustard. Biomed. Environ. Sci. **15:** 25-35, 2002.

Vojvodic V., Milosavljevic Z., Boskovic B. and Bojanic N. The protective effect of different drugs in rats poisoned by sulfur and nitrogen mustards. Fundam. Appl. Toxicol. **5:** S160-168, 1985.

Wasserman T. Radioprotective effects of amifostine. Semin. Oncol. **26:** 89-94, 1999.

Wormser U. Toxicology of mustard gas. Trends Pharmacol. Sci. **12:** 164-167, 1991.

Wyllie A.H. Glucocorticoid induced thymocyte apoptosis is associated with endonuclease activation. Nature **284:** 555-556, 1980.

Pharmacological Perspectives of Toxic Chemicals and Their Antidotes
Editors: S J S Flora, J A Romano, S I Baskin and K Sekhar

Detection, Decontamination, and Detoxification of Chemical Warfare Agents Using Polyurethane Enzyme Sponges

R.K. Gordon[a], A.T. Gunduz[a], L.Y. Askins[a], S.J. Strating[a], B.P. Doctor[a], E.D. Clarkson[b], J.P. Skvorak[c], D.M. Maxwell[b], B. Lukey[b], M. Ross[c]

[a]Walter Reed Army Institute of Research, Division of Biochemistry
Department of Biochemical Pharmacology 503 Robert Grant Ave
Silver Spring, MD 20910-7500, USA
[b]US Army Medical Research Institute of Chemical Defense
Drug Assesment Division
Basic Assessment Branch and Division of Pharmacology
Aberdeen Providing Ground, Aberdeen Providing Ground, MD 2101005400, USA
[c]United States Army Medical Research Medical Command, FT Detrick, MD 21702, USA

During combat, personnel have been exposed to organophosphates (OPs). Other exposures to chemical toxins include pesticides or terrorist acts in subways or sports events. Thus, we are developing enzyme-immobilized polyurethanes configured as (1) biosensors for OPs or (2) as sponges to soak up and inactivate the OPs.

Enzyme sensors have the advantage of selectivity, sensitivity and, most important, specificity, ease and portability, and markedly simplified instrumentation. Our immobilized enzymes will not leach from the polyurethane support so that the product - an OP badge - can now be used to sample anything from soil, water, to air. In the second configuration, polyurethane sponges are synthesized with enzymes and agents for external treatment of OP contaminated skin and other sensitive and exposed surfaces. To detoxify OPs, the cholinesterase is combined with oximes so the catalytic activity of OP-inhibited enzyme is continuously restored.

As a biosensor for OPs, the polyurethane matrix is composed of cholinesterase or other OP hydrolyzing enzymes to both indicate the presence of the OP agents, and to differentially indicate the type of OP present in the field. Resulting sponges for decontamination provided protective ratios of about 15 and 100-fold for soman and VX, respectively, when tested in a guinea pig model, and reduces methylene blue uptake in mustard exposed animals.

In conclusion, these immobilized enzyme biosensors and sponges, by virtue of their high capacity for enzymes, stability, specificity, sensitivity, and resistance to harsh environmental conditions, can be used under diverse conditions encountered by troops and civilian first responders.

INTRODUCTION

During combat, personnel have been exposed to organophosphates (OPs). Other exposures to chemical toxins include pesticides or terrorist acts in subways or sports events. For successful survival of exposed persons with minimal adverse effects, it is important to have rapid and simple detection of the OPs and also uncomplicated decontamination and detoxification procedures. To accomplish this, we are developing enzyme-immobilized polyurethanes configured as (1) sponges to soak up and inactivate the OPs or (2) biosensors for OPs.

In the first configuration, polyurethane sponges are synthesized with enzymes and agents for external treatment of OP contaminated skin and other sensitive and exposed surfaces. Reduction in long-term care can best be accomplished by rapid on-site decontamination and detoxification. To accomplish this, the sponges are lightweight and for individual use. To detoxify OPs, the cholinesterase is combined with oximes so the catalytic activity of OP-inhibited enzyme is continuously restored. Additional post-synthesis components include compounds to improve the extraction of OPs from guinea pig skin. Resulting sponges provided protective ratios of about 15 and 100-fold for soman and VX, respectively, when tested in a modified guinea pig model (Marlow et al.). We also show that the sponge reduces methylene blue uptake in mustard-exposed animals, demonstrating efficacy for chemical warfare agents consisting of both OPs and vesicants.

In the second configuration, we have developed polyurethane-immobilized enzyme biosensors. Enzyme sensors have the advantage of selectivity, sensitivity and, most important, specificity, ease and portability, and markedly simplified instrumentation. Biosensors based on cholinesterases (ChEs) immobilized non-covalently have been prepared by a variety of processes. The currently fielded spot M256A1 chemical agent detector kit and the M272 water test kit use dry eel ChE non-covalently applied onto fiber or ion-exchange paper. It can only be exposed to air/vapor environmental conditions. An immobilized enzyme will not leach from the polyurethane support so that the product - an OP badge - can now be used to sample diverse environment. This represents a significant product improvement (Gordon and Doctor). As a biosensor for OPs, the polyurethane matrix is composed of cholinesterase or other OP hydrolyzing enzymes to both indicate the presence of the OP agents, and to differentially indicate the type of OP present in the field. One of the advantages this immobilization technique affords the enzymes is that they are resistant to denaturing events, and are now suitable for sampling OPs in diverse environments such as soil, large bodies of water, as well as conventional airborne contamination (Gordon and Doctor).

MATERIALS AND METHODS

Sponge Synthesis

Details are provided (see Gordon et al.; Wood et al.). Briefly, the polyurethane sponges were molded in a Tupperware® container to the size of a human hand, then cut for use. The final size depended upon its use; for the guinea pigs, the sponges were approximately 7 × 3.5 × 1 cm (figure 1). The immobilized enzymes are stabilized by the polyurethane matrix (figure 2). A version of the sponge for detection is merely smaller in size (e.g., figure 6).

Fig. 1 Sponge product

Back-titration monitoring of Sponge Decontamination of Guinea pig Skin

After the guinea pig skin was wiped with the sponge(s), each sponge was placed in a separate 50 mL capped polypropylene tube and thoroughly mixed by vortexing. Then, an aliquot was removed and placed in 1 mL tubes containing 0.05% bovine serum albumin and 50 mM potassium phosphate buffer pH 8. The samples were sequentially diluted in the same buffer.

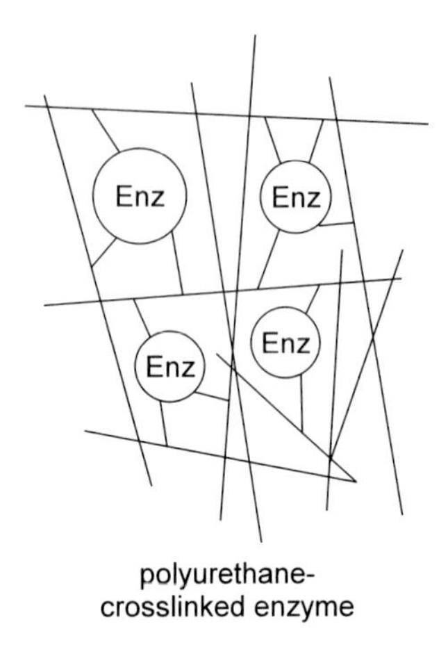

Fig. 2 Schematic of immobilized enzyme

Aliquots of all the dilutions were next transferred to a 96-well microtiter plate containing acetylcholinesterase (typically, 0.055 units). After incubation of the diluted soman or VX with the known quantity of cholinesterase, 10 μl was removed to a second 96-well microtiter plate and inhibition determined using a Molecular Devices Plate Reader and a modified Ellman procedure as previously described (Gordon et al.; Caranto et al. 1994). In this manner, the chemical warfare agent samples were diluted between 105 and 1011 fold, permitting quantification of the resulting inhibition of the cholinesterase (10-90% activity remaining) in at least one of the dilutions.

Animal Use

The protocols for the animal experiments were approved by the U.S. Army Medical Research Institute of Chemical Defense Committee on Animal Care and Use, and research was conducted in compliance with the Animal Welfare Act and other Federal statutes and regulations relating to animals and experiments involving animals, and adheres to principles stated in the Guide for the Care and Use of Laboratory Animals, National Research Council, NRC publication 96-23, 1996 edition.

Sponge Decontamination

The sedated and shaved guinea pigs were cutaneously exposed to neat soman on their sides. One minute after the exposure, a sponge wrapped around a pair of forceps was moved across the guinea pig's side; then the forceps were rotated 180 degrees, so that the clean surface of the sponge was pointed at the animal. Three more passes were taken from the rear towards the front. An identical procedure was used when the protocol required an additional second sponge to decontaminate the animal.

M291 Decontamination Kit

The sedated and shaved guinea pigs were cutaneously exposed to neat soman on their sides. One minute after the exposure, an M291 pad, previously removed from the M291 kit and cut it in half, was held in forceps and the guinea pig was decontaminated using five counter clockwise swipes. The second half of the pad was used to perform an additional five clockwise swipes.

Activity

Several different techniques to determine the activity of immobilized enzymes depending on the biosensor detection scheme. Detection can be performed qualitatively by the human eye for visible chromogens (e.g., Ellman), or dark-adapted eyes for chemiluminescent chromogens, or quantitatively using portable handheld devices, which measure fluorescence, chemiluminescence, and visible chromogens. We found that the TDI sponge (polyurethane matrix, composed of toluene diisocyanate prepolymer, Gordon et al.) has a significantly higher loading capacity for ChEs than the amount of purified BChE or AChE we added. These results showed that sensors with combinations of multiple enzymes, i.e., cholinesterases and OP hydrolases from various sources, can be co-immobilized without reducing the activity of the enzymes when immobilized alone.

RESULTS AND DISCUSSION

Characteristics of polyurethane immobilized enzymes: The longevity of sponges composed of immobilized cholinesterases is more than three years at room temperature (not shown, see Gordon and Doctor). The immobilized enzymes are also very stable in aqueous environments. One significant difference and advantage the immobilized enzymes have compared to the soluble cholinesterases is that immobilized enzymes do not dissociate (leach) from the sponge. Therefore, the immobilized enzymes can be left in the liquid or other environments. For instance, the AChE activity in the immobilized sponge was stable for more than 60 days in continuous immersion in aqueous samples including Allegheny River fresh water or brackish water, making them suitable also for long-term detection (biosensor). The enzymes do not leach from the polyurethane matrix.

OP removal by the sponge and protective ratios: The capacity of sponges to remove GD from guinea pig skin was determined using the back-titration method (see *Methods* section). While we were unable to modify the prepolymer since currently there is no formulation of prepolymer with an increased hydrophobic nature that might be expected to absorb the OP more effectively, we utilized several additives that provided additional ability to remove soman from the skin, protecting guinea pigs about four to five-fold better than the M291 kit. In addition, sponges were synthesized so that activated carbon would be incorporated into the polymer matrix. The addition of carbon did not interfere with the immobilization of ChEs (not shown). Sponges containing the oxime pyridine aldoxime methochloride (2-PAM) or HI-6 also showed increased protection to soman skin toxicity compared to the M291 kit. Next, we evaluated the combination of oximes with additives. In this decontamination treatment of GD, contaminated guinea pigs using sponges containing 2-PAM and tetraglyme, the protective ratio of the sponges over the M291 kit increased to more than 7-fold, and the LD_{50} of soman increased to 135 mg/kg (table 1). This compares to LD_{50} values of 9.9 and 18 mg/kg for untreated animals (not decontaminated) and the M291 kit, respectively. This combination is also effective against VX contaminated guinea pigs: 2-PAM and tetraglyme sponges yielded a protective ratio of approximately 31 for VX (table 1), and tetraglyme and HI-6 sponges yielded a protective ratio of more than 100.

Detoxification of removed OPs by the sponge (Table 1): After the sponge was used to wipe the guinea pig skin and the OP was removed, the sponges were placed into test tubes containing buffer and additional oxime. Then the detoxification of the OP in the test tube containing the sponge was followed by removing aliquots of the supernatant at various times by the back titration assay described in the *Methods section.* For both VX and GD (figure 3), the OP in the test tube in the absence of sponge and oxime (labeled 'buffer' in the figure) showed little change in amount over the time course that was monitored (1.5 h or 2.5 h for VX and GD, respectively). Thus, both GD and VX were stable under these conditions established to evaluate sponge detoxification (Shih and Ellin, 1984). In contrast, all the concentrations of VX

Table 1 Sponge additives protect soman and VX contaminated guinea pigs. PR is the protective ratio observed over untreated and exposed animals. Values are mean for n = 4; SD = 10%.

	GD		*VX*	
Additives to Sponge	*LD_{50} (mg/kg)*	*PR*	*LD_{50} (mg/kg)*	*PR*
HFE	55	5.6		
2-PAM (oxime)	76	7.7		
HI-6 (oxime)	79	8		
Tetraglyme	88	8.9		
2-PAM + Tetraglyme	137	13.8	3.37	24.9
HI-6 + Tetraglyme	156	15.7	15.7	112
Reference Values				
M291 Decon Kit	17.7	1.8	0.14	
OP alone	9.9	-	0.14	-

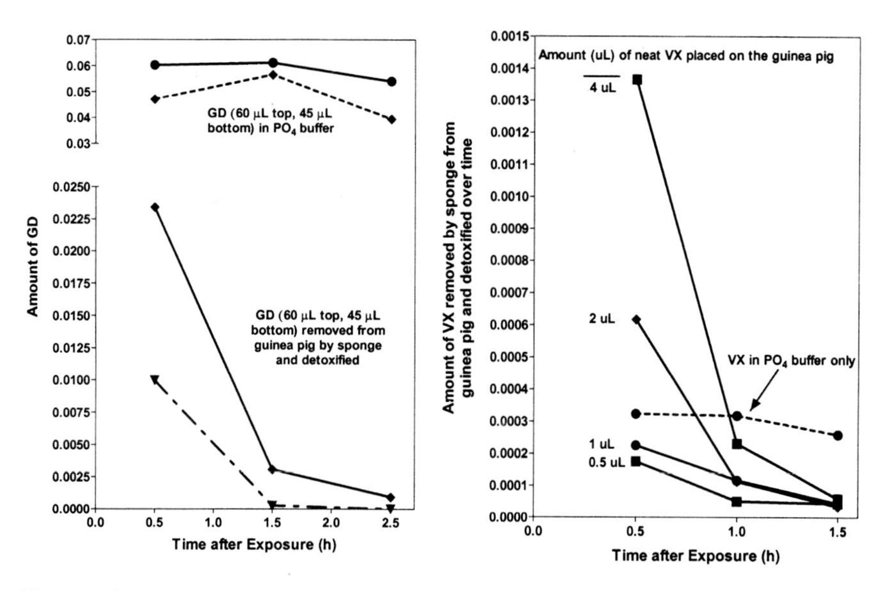

Fig. 3 OP Detoxification in the Sponge after guinea pig Decontamination. Left, detoxification of soman (GD) removed from Guinea Pigs; right, detoxification of VX removed from guinea pigs. Dotted lines are the agent in buffer, and under these conditions shows little hydrolysis. Each test tube is a representative example of at least 2 replicates and contained the sponge, tetraglyme, and the oxime 2-PAM.

and GD were detoxified to very low levels within 1-1.5 h, and were at baseline levels within 1.5-2.5 h, for VX and GD respectively. These results clearly demonstrate that the sponge not only removed OP for the skin of the guinea pigs, but in the presence of oxime effectively and completely detoxified the OPs within hours (Maxwell et al). Thus, these sponges would not pose any additional cross-contamination hazard because they detoxify the chemical warfare agents.

Detoxification of mustard (HD) by the sponge: The sponge was used to wipe guinea pig skin contaminated with neat mustard. The following day, the animals were injected with trypan blue. Those areas representing vesicant injury take up the trypan blue dye because dead and dying cells are metabolically unable to exclude the dye. Live cells with an intact membrane potential do not accumulate the dye and hence are not colored. Thus, these sponges could reduce the damage that HD produced, as shown in figure 4. The area labeled "Positive" shows HD induced dye uptake, while the sponge "Decon" site shows markedly less uptake, and only slightly more than the "Control" or unexposed area of the guinea pig.

Biosensor: The immobilized cholinesterase biosensor, essentially a small sponge the size of a pencil eraser that can have additional enzymes

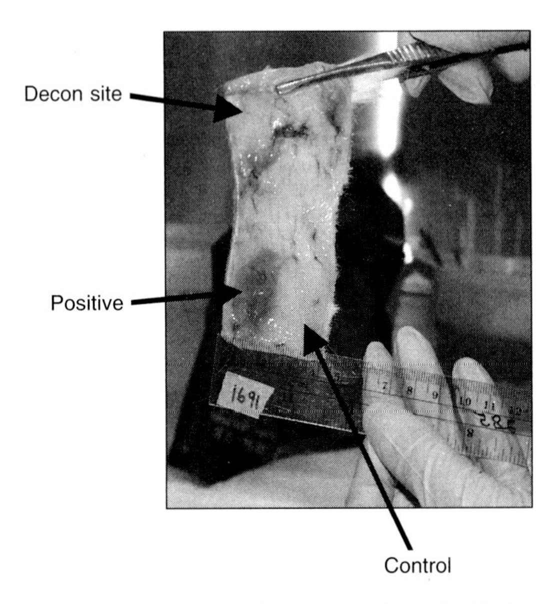

Fig. 4 HD contaminated guinea pig skin decontaminated with the sponge (Decon site), which showed markedly less dye uptake than the HD contaminated site that was not decontaminated (positive). The control site was not exposed to HD.

immobilized (figure 5, table 2), can be constructed for field determination of the type of organophosphate present. This can aid in treatment and tracking the origins of the exposure. We accomplished this as shown in the differential hydrolysis of OP chart (table 2), by first exposing the potentially contaminated solution to one of several discriminating (hydrolyzing) enzymes, then followed by exposure to a ChE biosensor. In the example, if the solution is first exposed to the OP hydrolyzing enzyme OPAA, and then ChE is not inhibited, the agent is soman. Due to the structural integrity of both the immobilized enzyme and matrix, the biosensor can be exposed for an extended period to different environments.

Sensor color reaction: Figure 6 is an example of the sensor (top), the sensor exposed to aqueous solution in the absence of organophosphate (middle), and the sensor poisoned by organophosphate (bottom). The right panel shows a standard M272 ticket where the non-immobilized enzyme is lost when merely exposed to buffer (middle). A positive control (not leached active enzyme) is shown at the top, and inhibited by OP is at the bottom. Note that for the M272 ticket, a false positive indication is observed for the middle ticket. Various color reactions are available, e.g., choline oxidase coupled reaction with amplex red reagent yielding the red chromogen resorufin for a visual indication of enzyme activity. In addition, resorufin is a fluorescent product providing increased sensitivity that could be used in a hand-held unit.

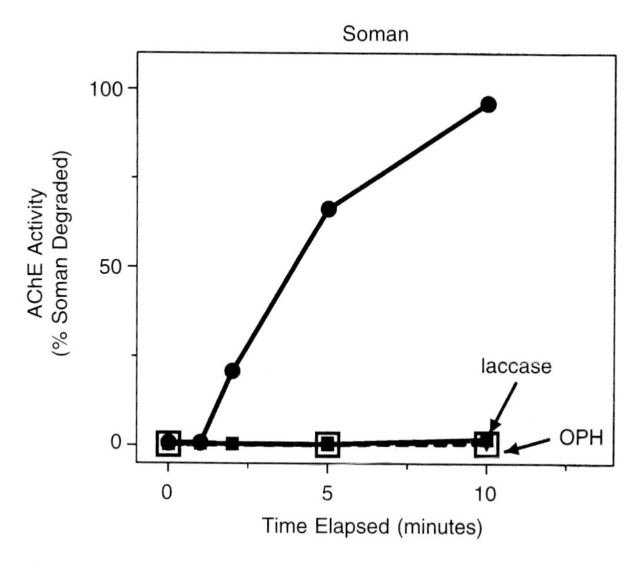

Fig. 5 Immobilized Biosensor Demonstrating Differential Hydrolysis of OPs; CWA was first exposed to either OPAA, laccase (a phenol oxidase from *Pleurotus ostreatus*, Amitai et al.), or OPH, and then the AChE immobilized biosensor.

Table 2 Potential multiple immobilized enzymes in an immobilized biosensor

Enzyme type	*Distinguishing characteristics*
AChE, BChE	Inhibited by OPs
Laccase (from *Pleurotus ostreatus)*	Hydrolyzes VX preferentially with mediator OPH
Human serum	Hydrolyses tabun, VX poorly
Rabbit serum	Hydrolyses sarin preferentially
Pseudomonas	Hydrolyses G agents
Alteromonas undi (OPAA)	Hydrolyses soman preferentially
Squid (*Loligo vulgaris*)	Hydrolyses tabun, VX poorly

SUMMARY

In summary, these immobilized enzyme biosensors and sponges, by virtue of their high capacity for enzymes, stability, specificity, sensitivity, and resistance to harsh environmental conditions, can be used under diverse conditions encountered by troops and civilian first responders. We have demonstrated the rapid *in-situ* copolymerization of ChEs at room temperature and that the ChEs are covalently linked to and do not leach out of the polyurethane matrix, precluding immune reactivity when used on skin. AChE-sponges with HI-6 detoxified the removed OP so that no OP remained. The immobilized ChE also provides a means to monitor the detoxification capacity of the sponge since inhibited enzyme would imply the requirement

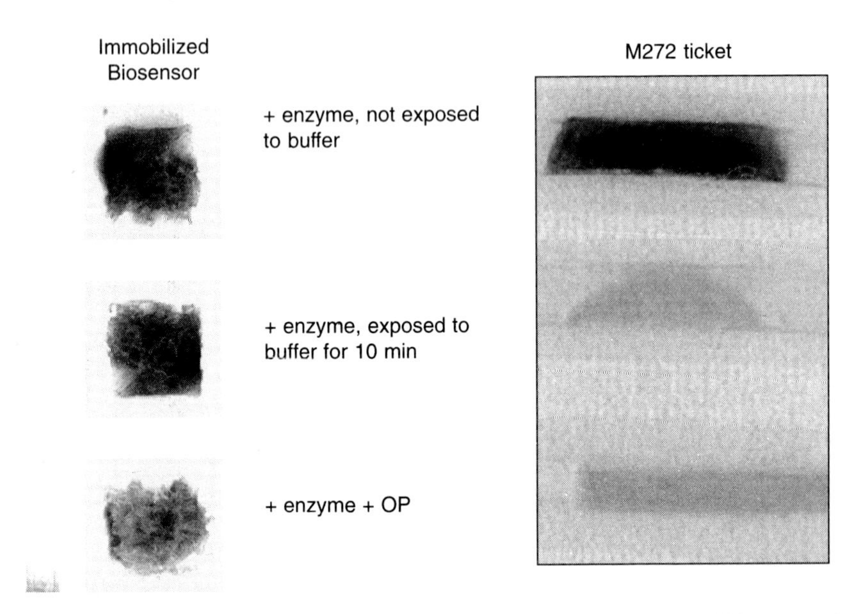

Fig. 6 Comparison of Biosensor and Standard Ticket. The dark color (top of the sensor/ticket panels) indicates a positive response of the enzyme hydrolyzing the substrate (no OP present). The light color demonstrates the loss of enzyme (in the case of the middle panel for the M272 ticket, but the immobilized biosensor is still dark because it retains the enzyme). The bottom sensor/ticket represents the lack of color development because both enzymes were inhibited by OP.

for additional oxime. The sponge should be suitable for a variety of detoxification, decontamination, and detection schemes for chemical weapons of exposed military, civilians, or first responders.

References

Amitai, G., Adani, R., Sod-Moriah, G., Rabinovitz, I., Vincze, A., Leader, H., Chefetz, B., Leibovitz-Persky, L., Friesem, D., and Hadar, Y. (1998) FEBS Lett. 438, 195-200.

Caranto, G. R., Waibel, K. H., Asher, J. M., Larrison, R. W., Brecht, K. M., Schutz, M. B., Raveh, L., Ashani, Y., Wolfe, A. D., Maxwell, D. M., and Doctor, B. P. (1994) Biochem. Pharmacol. **47,** 347-357.

Ellman GL, Courtney KD, Andres V, Featherstone RM. (1961) Biochem. Pharmacol. **7**: 88-95.

Gordon, R. K. Feaster, S. R., Russell, A. J., LeJeune, K. E., Maxwell, D. M., Lenz, D. E., Ross, M., and Doctor, B. P. (1999) Chemico-Biolog. Interac. **119-120**, 463-470.

Gordon, R. K. and Doctor, B. P "Immobilized enzymes as biosensors for chemical toxins". U.S. Patent No: 6,406,876. (2002).

Marlow, D. D., Mershon, M. M., Mitcheltree, L. W., Petrali, J. P., and Jaax, G. P. (1990) J. Toxicol. Cutan. Ocular Toxicol. **13**, 221-229.

Maxwell, D. M., Castro, C. A., De La Hoz, D. M., Gentry, M. K., Gold, M. B. Solana, B. P., Wolfe, A. D., and Doctor, B. P. (1992) Toxicol. Appl. Pharmacol. **115**, 44-49.

Shih, M. L. and Ellin, R. I. (1984) Bull. Environ. Contam. Toxicol. **33**, 1-5.

Wood, L. L., Hardegen, F. J., Hahn, P. A., Enzyme Bound Polyurethane, U.S. Patent 4,342,834 (1982).

Pharmacological Perspectives of Toxic Chemicals and Their Antidotes
Editors: S J S Flora, J A Romano, S I Baskin and K Sekhar

In Search of An Ideal Bioscavenger for Organophosphate Chemical Warfare Agent Toxicity

Bhupendra P. Doctor and Ashima Saxena
Walter Reed Army Institute of Research, Silver Spring, Maryland, 20910-7500, USA

INTRODUCTION

Current antidotal regimens for OP poisoning consist of a combination of pretreatment with spontaneously reactivating AChE (EC 3.1.1.7) inhibitor, i.e., carbamates, (pyridostigmine bromide), to protect it from irreversible inhibition by OP compounds, and post exposure therapy with anti-cholinergic drugs (atropine sulfate) to counteract the effects of excess acetylcholine and oximes (e.g., 2-PAM chloride) to reactivate OP-inhibited AChE (Gray, 1984). Although these antidotal regimens are effective in preventing lethality of animals from OP poisoning, they do not prevent post-exposure incapacitation, convulsions, seizures, performance decrements, or in many cases permanent brain damage (Dirnhuber, et al., 1979; McLeod, 1985; Dunn, et al., 1989). These symptoms are commonly observed in experimental animals and are likely to occur in humans. An anticonvulsant drug diazapam was included as a treatment to minimize convulsions, thereby minimizing the risk of permanent brain damage.

The problems intrinsic to these antidotes stimulated attempts to develop a single protective drug, itself devoid of pharmacological effects, which would provide protection against the lethality of OP compounds and prevent post-exposure incapacitation[4]. One approach is the use of enzymes such as ChEs, b-esterases in general, as single pretreatment drugs to sequester highly toxic OP anti-ChEs before they reach their physiological targets (Wolfe et al., 1987; Raveh et al., 1991; Ashani et al., 1991; Doctor et al., 1991; Broomfield et al.,

1991; Maxwell et al., 1991; Wolfe et al., 1992; Maxwell et al., 1992; Raveh et al., 1997; Allon et al., 1998). This approach turns the irreversible nature of the OP-ChE interaction from disadvantage to an advantage; instead of focusing on the OP as an anti-ChE, one can use the ChE as an anti-OP. Using this approach, it was shown that administration of FBS AChE or equine serum BChE (Eq BChE: EC3.1.1.8.) or human serum BChE (Hu BChE) protected the animals from multiple LD_{50}'s of a variety of highly toxic OPs without any toxic effects or performance decrements (Wolfe et al., 1987; Raveh et al., 1991; Ashani et al., 1991; Doctor et al., 1991; Maxwell et al., 1991; Wolfe et al., 1992; Raveh et al., 1997). Evidence accumulated in last two decades also suggests that exogenous administration of Hu BChE significantly improves the ability of plasma to detoxify certain naturally occurring toxicants (e.g., cocaine, physostigmine etc.), synthetic poisons (nerve agents, pesticides) and several commonly used drugs (neuromuscular blocking agents used in anesthesia and local anesthetics).

Enzymes have been employed as theraputic agents or drugs for many centuries. In the 1930s, Avery and Dubos demonstrated the use of an enzyme that can degrade the capsular polysaccharide of pneumococcus as a protective agent for treating pneumococcal infection in monkeys. In comparison with many drugs, enzymes have many unique advantages; they are specific, catalytically efficient, operate under physiological conditions, and cause essentially no deleterious side effects. Some of the demonstrated uses of enzymes as therapeutic agents include facilitating the digestion of food, wound healing, proteolysis, replacement of defective enzyme in the case of genetic disorders, removal of blood clots, fibrinolysis and depletion of metabolites in cancer. In almost all instances where enzymes have been employed therapeutically, they have been used for their proteolytic/hydrolytic properties, as replacement for defective or deficient enzymes in cases such as congenital diseases, to improve or alter immune properties. Only recently, have enzymes been been employed as scavengers or prophylactic drugs for protection against highly toxic substances.

There are certain requirements for an enzyme to be an effective scavenger for OP toxicity *in vivo*: (1) it should have relatively high turnover number; the higher the turnover number, the faster it will neutralize the toxic agent; (2) it should have a long half life *in vivo* for it to be effective as a scavenger for a prolonged period; (3) the enzyme be readily available in sufficient quantities; (4) it not be immunoreactive, and (5) in cases of enzymes such as esterases, the stoichiometry between scavenger enzyme and toxic agents should approach 1:1. However, it is possible to improve the stiochiomentry in favor of hydrolysis of OPs by including reactivators e.g., oximes, in the treatment regimen.

The bioscavengers that have been explored to date for the detoxification of OPs fall into three categories: (1) The ones that can catalytically hydrolyze

OPs and thus render them non-toxic, such as OP hydrolase and OP anhydrase; (2) the ones that stoichiometrically bind to OPs i.e., one mole of enzyme neutralizes one or two moles of OP inactivating both, OPs and the enzyme e.g., ChEs and related enzymes; (3) the ones generally termed as "pseudo catalytic" e.g. a combination of ChE and an oxime such that the catalytic activity of OP-inhibited ChE can rapidly and continuously be restored in the presence of oxime. This paper describes the progress made in the last two to three decades in exploring the potential use of these enzymes specially ChEs, to counteract the toxicity of OPs. Except for comparison purposes, we will limit our discussion here to BChE only. Since the biochemical mechanism underlying prophylaxis by exogenous esterases such as ChEs is established and tested in several species, including non-human primates, this concept should enable a reliable extrapolation of results from animal experiments to human application.

Enzymes such as OPHs and OPAAs can hydrolyze many of the OP pesticides. The drawback in using these types of enzymes is that although the rate of hydrolysis of pesticides is relatively fast ($k_{cat}/K_m = 10^5 - 10^7$ M^{-1} min^{-1}), it is several thousand folds slower for the hydrolysis of OP nerve agents (Amitai, et al., 1996). Such a rate will be too slow for OPs to compete for binding to ChEs and thus OPs will reach their physiological target, i.e. the AChE in the central nervous system. Many attempts made to modify the catalytic properties of these enzymes toward OP nerve agents by site-specific mutagenesis have met with limited success. The second problem with the use of these types of enzymes is their strict substrate specificity. This will require the use of several enzymes from different sources having different substrate specificities to form a bioscavenger that will be effective against all toxic OPs. For example parathion hydrolase isolated from Pseudomanas Sp. was very effective in protecting mice against ~ 10-12 LD_{50} of tabun for a very short period of time (Broomfield, 1992; Raveh, et al., 1992). However, this enzyme bioscavenger was not effective against soman and sarin. Thirdly, most of these enzymes are derived from bacterial sources and thus are cleared very rapidly form the circulation of mammals. Therefore, none of these catalytic enzyme bioscavengers except paraoxonase 1 (PON 1) present in human plasma are suitable for human use as a single bioscavenger effective against all highly toxic OP chemical warfare agents (Masson et al., 1998).

Attempts were made to confer OP hydrolyzing activity to ChEs by site-directed mutagenesis. In this regard, a Hu BChE mutant, G117H that hydrolyzes the nerve agents soman and VX[19, 20] and a double mutant G117H/E197Q that hydrolyzes soman[21] were produced. The rate of hydrolysis of the OP nerve agents by these mutants of Hu BChE, as it stands to-date, were very slow (Millard et al., 1995; Lockridge et al., 1997; Millard et al., 1998). Thus, they also do not appear to be practical catalytic bioscavengers for animal or human use. Another mutant, E199Q AChE from *Torpedo californica* and the corresponding mammalian E202Q AChE and E197Q Hu BChE are interesting and may be of value as bioscavengers (Saxena et al., 1997; Saxena et al., 1993; Ordentlich et al., 1996). The mutation of glutamic acid located at the N-terminal of the active-site serine to glutamine, significantly slows the process of aging, so that the OP inhibited ChEs can be readily reactivated in presence of an oxime. The sum total effect is that the rate of soman inactivation by the mutant enzyme in *ex-vivo* experiments in presence of an appropriate oxime was three fold higher than that of wild type AChE and it can efficiently catalyze the hydrolysis of all OP nerve agents *in vitro* and *in vivo*.

As of now ChEs are the only enzymes that have been successfully tested and demonstrated to be effective universal bioscavengers against all OP nerve agents in several animal species. Of the ChEs evaluated, Hu BChE is most suitable candidate for human use.

The first and the foremost requirement for developing an effective bioscavenger is the possibility of obtaining sufficient amounts of purified material for testing and evaluation. The large-scale purification of Hu BChE from outdated human plasma has been described (Ralston et al., 1983; Lynch et al., 1997; Grunwald et al., 1997). All these procedures employed affinity binding of plasma BChE to procainamide-Sepharose CL-4B gel followed by salt precipitation and ion exchange chromatography. The yield varied from 20-60 % of the starting BChE activity. The purity of the preparations was also very good (95±5 %). Approximately 150-200 mg of purified enzyme could be obtained from one hundred liters of plasma. Eq BChE can also be purified by essentially any of these procedures (Ralston et al., 1983).

The purification of Hu BChE has been scaled up to almost an industrial level. This is described elsewhere in this book (see Saxena et, al,.). The purity of the final product was ~95 % ± 5 %. The preparation was homogenous as evidenced by single band on SDS-PAGE. The yield was about 30-35 % (~ 5 g). It is hoped that this procedure will be advanced with the starting material of 500 Kg paste which will contain ~ 80-100 g of enzyme and yield 40-50 g of pure Hu BChE. The amount of Cohn Fraction IV-4 generated in USA is estimated to be sufficient to yield Kg quantities of purified enzyme under GMP conditions. Thus this bioscavenger, Hu BChE, has all the potential for early availability in the emergency situations.

CIRCULATORY STABILITY OF BUTYRYLCHOLINESTERASE

In order for bioscavengers, such as ChEs to provide effective protection against toxic OPs for relatively long periods they should remain in circulation for a long time. ChEs such as FBS AChE, Eq BChE and Hu BChE, isolated and purified from mammalian plasma are soluble globular forms and thus have a relatively long mean residence time *in vivo*. The time courses of these three ChEs by three different routes of administration (*i.v. i.p.* or *i.m.*) in, rats, guinea pigs, and rhesus monkeys showed that their mean residence time (MRT) in circulation or biological half life was 20-60 h in mice (Raveh et al., 1991; Raveh et al., 1993; Saxena et al., 1997; Genovese et al., 1995; Allon et al., 1998; Raveh et al., 1997). The routes of administration affected the time at which the maximum concentration of enzyme in circulation was observed, but did not affect MRT and a constant level of enzyme was maintained for a period of approximately 3-10 h for AChE and 3-20 h for BChE. Regardless of the route of administration, 50-90 % of administered enzyme was found in the circulation of the animals. The determination of MRT of all three ChEs in mice suggests that MRT can be correlated with the number of glycosylation sites on these enzymes. There are four glycosylation sites in AChE compared to eight in Eq BChE and nine in Hu BChE. The MRTs of the tetrameric form of native AChEs in mice range from 16-20 h depending on routes of administration and the preparation of enzyme. The MRT of native Eq BChE and native Hu BChE range from 40-60 h. This simple observation led to a further investigation of the role of glycosylation in the circulatory stability of ChEs in mice and other species.

All monomeric forms of ChEs, native or recombinant, tested so far have a relatively short MRT in the circulation of mice (Saxena et al., 1997; Saxena et al., 1998; Kronman et al., 2000; Duysen et al., 2002). With regard to native monomeric and tetrameric forms of FBS AChE, it was shown that both forms of enzyme have essentially similar carbohydrate composition (Saxena et al., 1997; Saxena et al., 1998). However, the MRT of the tetrameric form of FBS AChE injected *i.v.* into mice was six to seven folds higher than the monomeric form. Secondly, desialylation or deglycosylation of tetrameric form of Eq BChE or FBS AChE dramatically decreased their MRT, which was similar to their monomeric form (Saxena, et al., 1997; Saxena, et al., 1998). Similar results were reported using recombinant (r) Hu AChE, where monomeric and tetrameric forms of rHu AChE and rBo AChE showed a very short MRT when injected *i.v.* into mice (Kronman et al., 2000; Duysen et al., 2002; Chitlaru et al., 2001; Chitlaru et al., 1998). Upon completion of sialylation and proper glycosylation, the MRT improved a bit, but tetramerization of these rAChEs was necessary to accomplish the MRT equivalent to that of the native form of FBS AChE. These results lead to the following observations: (a) Sialylation or capping of terminal carbohydrates has, in part, a role in conferring prolonged MRT to these enzymes. Conversely lack or under sialylation influences MRT; (b) similarly proper glycosylation, in part, is essential and under glycosylation reduces the MRT; (c) tetramerization of enzyme is essential for prolonged MRT. In this regard properly sialylated and glycosylated monomeric form of rHu AChE was polymerized using COLQ and tested for its MRT in the circulation of mice (Bon et al., 1997; Saxena et al., 1998). The MRT of this polymerized rHu AChE was approximately 15 h.

In contrast to this preparation of rHu AChE, Duysen et al generated a tetrameric form of rHu BChE (with five N-glycans) without modification of glycosylation of N-glycans (Duysen et al., 2002). The MRT of this COLQ mediated tetrameric rHu BChE was16 h, which is similar to that of native plasma-derived FBS AChE, which has four N-glycans.

Attempt to stabilize rHu AChE in the circulation of mice also involved attachment of polyethylene glycol to properly sialylated and glycosylated rHu AChE polymerized with COLQ (Cohen, et al., 2001). It was reported that such enzyme preparation had a MRT of 26 h in the circulation of mice, which was almost similar to that of native FBS AChE.

The polyethylene glycol is attached to the epsilon amino group of the lysine on the surface of ChEs. AChE has total of eight lysine residues on its surface. Four of these lysine residues are free where as the other four are involved in hydrogen bond or salt bridge formation. BChE has on the other hand thirty-one lysine residues on its surface. Fourteen of these lysine residues are free, where as 16-17 appear to be involved in salt bridge or hydrogen bond formation. At this time, neither the rAChE nor rBChE have been tested as bioscavengers for OP toxicity in any animal species. When the rAChE or rBChE will be available in sufficient quantities for use as a bioscavenger, it will have to be in the tetrameric form, and possibly will have to be PEGylated. Secondly, it will have to be compatible with the human immune system. BChE will be preferred over AChE, for it is a better candidate for PEGylation than AChE.

IMMUNOGENICITY OF BUTYRYLCHOLINESTERASE

In addition to the clearance of ChEs due to receptor binding, antibody-mediated clearance further reduces the level of circulating enzyme below therapeutic levels. This phenomenon usually occurs when a macromolecule isolated from one animal species or from bacterial or viral source is administered to another species. This is because the body immune response elicits the production of antibody as a defense mechanism. Implications of repeated doses of ChE, such as might be needed to counteract effects of multiple or continuous exposures of OPs, were determined. For example, the effect of repeated *i.m.* Injections (twice a week for two weeks) of Eq BChE was examined in rabbits and rats by monitoring blood enzyme and anti-enzyme immunoglobulin G (IgG) levels in both species of animals and IgM levels in rats (Gentry et al., 1993; Genovese et al., 1993). The blood was drawn one day after the injection of Eq BChE. Blood BChE activity on the days following injections (days 2,6,9 and 13) first increased and then decreased. Increases in BChE activity were greatest following the first two injections. On the days following the fourth and fifth injections of BChE, however, activity declined to levels equal to those measured before the first injection.

Increases in both IgG and IgM antibodies were observed following injections of BChE. In the case of the IgM antibody, elevation was observed on the day following the second injection of BChE (day 6). However, little further elevation occurred following subsequent BChE injections. IgM levels remained elevated after BChE injections were discontinued (days 9, 13, 16, 20, and 23). Levels of IgG increased on the day following the third BChE injection (day 9) and further increase were observed on the days following subsequent BChE injections (days 13 and 16), which continued on the days after BChE injections were discontinued (days 20 and 23). A single injection of enzyme was administered to rats on day 26. The enzyme level in blood was elevated somewhat, however, anti body titers were raised more steeply.

Similar investigations were also carried out in rabbits using Eq BChE injected in a fashion similar to rat experiments. The IgG titers elevated after the initial series of enzyme injections were allowed to wash out (~90 days) and a single injection of enzyme was administered. The blood enzyme was elevated somewhat but antibody titers were greatly elevated. The same procedure was repeated again after the second antibody washout (~90 days). Again the enzyme level was barely elevated, however antibody titer was boosted to a high level. The results described here are typical of any proteins eliciting immune reaction when administered in other species. Thus heterologous injection of ChE does generate antibody that clears the enzyme from circulation of animals. It does not, however, cause any severe antibody mediated adverse reaction.

Immunological consequences of administration of purified monkey (macaque) BChE into monkeys of the same species was investigated to confirm the fact that when Hu BChE is administered to humans there will not be any adverse reaction (Rosenberg, et al., 2002). The macaques (four) were given a single injection of ~10 mg of purified macaque serum BChE. These monkeys demonstrated much longer MRT = 225 ± 19 h in plasma compared to those reported for heterologus Hu BChE (33.7 ± 2.9 h). A second (ca. 4.5 mg of the same enzyme preparation) injection given four weeks later attained predicted peak plasma levels of enzyme activity. No antibody response was detected in macaques following either injection of enzyme.

The extended stability of exogenously administered macaque BChE into macaques observed suggests that even a single dose of homologous ChE is sufficient to maintain the enzyme at a long lasting therapeutic level. The results with two injections of BChE have clearly demonstrated the utility of homologous BChE as an effective and safe bioscavenger, exhibiting high stability and low or no immunogenicity in recipient monkeys. In regards to the use of Hu BChE in humans, these results are consistent with the reported half-life *in vivo* of 8-11 days and the absence of reported untoward immunological and physiological side effects following blood transfusions and *i.v.* Injections of partially purified Hu BChE into humans (Jenkins et al., 1967; Stovner et al., 1976; Ostergaard et al., 1988; Cascio et al., 1982).

The Efficacy of Butyrylcholinesterase as A Pretreatment Drug for Orgonophosphate Toxicity.

The efficacy of Eq and Hu BChE as pretreatment drugs was determined in mice, rats, guinea pigs and non-human primates. The OPs used in these investigations were soman, sarin, tabun and VX. In addition to determination of pharmacological and toxicological parameters, the effects on behavior and performance were also evaluated. None of the animals that received either enzyme alone or were challenged with stoichiometric amounts of OPs following enzyme pretreatment, showed any type of toxic signs or performance decrement.

The first successful demonstration of the use of ChEs as a pretreatment drug for OP toxicity was demonstrated in mice (Wolfe et al., 1987; Raveh et al., 1991; Ashani et al., 1991). The pretreatment with BChE sequesters OP compounds rapidly and thus detoxifies them in circulation before they can reach their physiological target. For example, pretreatment of mice with Hu BChE successfully protected the animals against 2-5 x LD_{50} of soman without requiring any additional drug treatment. These studies established a quantitative correlation between the degree of protection against OP compounds and the level of inhibition of administered enzyme, a result consistent with the *in vitro* experiments. The protected mice were not evaluated for potential behavioral incapacitation or for any detrimental immunologic response from administering an exogenous enzyme.

Similarly, it was demonstrated that pretreatment of rats with Hu BChE prevented soman-induced cognitive impairments (Brandeis et al., 1993). Behavioral testing was carried out using the Morris water maze task evaluating learning, memory, and reversal learning processes. Cognitive functioning in rats was significantly impaired following *i.v.* administration of 0.9-1.1 x LD_{50} of soman. Hu BChE significantly prevented the development of soman- induced cognitive decrements. No significant differences were displayed during both retention and reversal learning between soman-challenged animals pretreated with Hu BChE and the control saline-treated rats. However, during part of the training period, the Hu BChE + soman group was somewhat deficient in performance compared to the control (saline) group. These findings are consistent with previous results that have shown that both cognitive functions, retention and reversal learning, when compared to acquisition, are especially sensitive to cholinergic manipulations (Hunter et al., 1988; Smith 1988). Hu BChE treatment alone was devoid of any impairments in either motor or cognitive behavioral performance. These results further support the concept that pretreatment alone with Hu BChE is sufficient to increase not only survival but also to alleviate deficits in cognitive functioning after exposure to a potent nerve agent such as soman.

The use of Hu BChE as a prophylactic measure against inhalation toxicity, which is a more realistic simulation of exposure to volatile OPs, has also been described (Allon, et al., 1998). Hu BChE-treated, awake, guinea pigs were exposed to a controlled concentration of soman vapors ranging from 417 to 430 μg/L, for 45 to 70 sec. The correlation between inhibition of circulating Hu BChE activity and the dose of soman administered by sequential *i.v.* injections or by respiratory exposure indicated that the fraction of the inhaled soman dose that reached the blood was 0.29. A Hu BChE: soman molar ratio of 0.11 was sufficient to prevent toxic signs in guinea pigs following exposure to 2.17 x the inhaled LD_{50} dose of soman (1 LD_{50}, = 101 μg/Kg). Protection was exceptionally high and far superior to the traditional approach. Quantitative analysis of the results suggests that the *in vivo* sequestration of soman by exogenously administered Hu BChE is independent of the species used or the route of entry. Unlike an *i.v.* Bolus injection, inhalation exposure allows soman to enter gradually in circulation. Thus, the initial molar concentration of the challenge in blood should be smaller than the concentration attained by the same dose administered by *i.v.* injection. The slow introduction of the inhaled OP to the circulating scavenger increases the efficacy of soman sequestration to below its toxic levels, relative to a bolus *i.v.* challenge. Since the extent of predicted protection is based on the *i.v.* titration curve, the above-mentioned rationale explains the greater than the calculated values of protection observed in most of the soman-inhaled Hu BChE-pretreated guinea pigs.

Broomfield et al., showed in rhesus monkeys that the toxicity of soman (2 x LD_{50}) can be prevented by pretreatment of appropriate amount of Eq BChE without any performance decrement as measured by Serial Probe Recognition (SPR) Task (Broomfield et al., 1991; Sands et al., 1980; Maxwell et al., 1993; Castro et al., 1991). Also, protection against 3-4 x LD_{50} of soman was obtained with Eq BChE pretreatment followed by atropine post exposure treatment of these animals. These animals were able to perform the SPR task about 9 h post exposure where as animals treated with the conventional atropine/oxime therapy was unable to perform the same task for 14 days. Animals receiving enzyme alone showed only a subtle transient performance decrement on the SPR task.

Another test, the Primate Equilibrium Platform (PEP) task was employed to demonstrate the protection of rhesus monkeys from soman toxicity as high as 5 x LD_{50} by pretreatment with Eq BChE without the occurrence of performance deficits (Farrer et al., 1982; Blick et al., 1989; Blick et al., 1991; Wolfe et al., 1992). This task, in which a seated monkey was trained to manipulate a joy stick to maintain orientation in space, was originally developed as an animal model for evaluating pilot performance for studies in aerospace medicine. When ~ 0.5 mmol of Eq BChE was administered to each of the four animals, blood BChE levels were elevated more than 100-fold. BChE injections alone caused no apparent physiological or neurological effect or deficit, as measured by the PEP task. None of the four monkeys showed any OP toxicity after soman challenges; protection was so complete that there were no fasciculations even at the site of soman injections. Following the first and second soman injections (totaling 25.6 μg/Kg = ~ 4 LD_{50}), the PEP task of all four monkeys was completely normal. Two of the monkeys exhibited significant but minor PEP deficit after the third soman injection; these transient PEP deficits were similar to those observed after a low dose soman challenge (<2.8 μg/Kg) in unprotected monkeys (Blick, et al., 1989; Blick, et al., 1991). Based upon this comparison, the protective ratio afforded by Eq BChE pretreatment can be estimated to be between 10 and 15. In these animals, blood BChE levels measured before and after each soman challenge showed a linear decline with increasing dose of soman. However, ~ 20 % of the initial blood BChE levels remained after administration at the end of the soman challenge. Thus, the extent of protection afforded by exogenous ChE is dependent on the concentration of ChE in the blood, the dose of OP, the rate of circulation, and the clearance times of ChE and OP. The most impressive finding from these investigations is the fact that these monkeys inspite of being exposed to 5 x LD_{50} of soman survived and showed neither any toxic signs nor PEP decrements for upto 6 weeks. During the period following the soman challenge, blood BChE levels initially decreased below the pre-exogenous BChE administration level and then returned to normal levels. Perhaps the exogenous BChE triggered a feedback mechanism responsible for maintaining endogenous BChE levels. During six weeks of post-soman testing, none of the monkeys, including the two that received BChE and had shown PEP deficits on the challenge day, showed any signs of delayed toxicity, convulsions, or other OP symptoms, or any abnormality on PEP task.

Finally, the ability of Hu BChE, purified from human plasma, to prevent the toxicity induced by soman and VX was assessed in rhesus monkeys (Raveh, et al., 1997). The average MRT of Hu BChE in the circulation of monkeys following an *i.v.* loading was 34 h. Also, a high bioavailability of Hu BChE in blood (>80%) was demonstrated after an *i.m.* injection. A molar ratio of Hu BChE:OP of ~1.2 was sufficient to protect monkeys against an *i.v.* bolus injection of 2 x LD_{50} of VX, while a ratio of 0.62 was sufficient to protect monkeys against an *i.v.* dose of 3.3 x LD_{50} of soman, with no additional post exposure therapy. A remarkable protection was also seen against soman-induced behavioral deficits detected in the performance of a spatial discrimination task (Raveh et al., 1997).

These studies firmly established that prophylactially administered Eq or Hu BChE, with no additional treatment, prevent the toxicity induced by highly toxic OP nerve agents in mice, rats, guinea pigs and rhesus monkeys. Not only do these bioscavengers prevent the lethality, but the animals do not show any untoward side effects or performance decrements/deficits determined by SPR task, PEP task or spatial discrimination tasks (Raveh et al., 1997).

The systematic evaluation of the efficacy of Hu BChE in protection of four animal species against nerve agent toxicity offered an extrapolation model from animal to human (Raveh et al., 1991; Ashani et al., 1991; Doctor et al., 1991; Broomfield et al., 1991; Maxwell et al., 1991; Wolfe et al., 1992; Maxwell et al., 1992; Raveh et al., 1997; Allon et al., 1998; Raveh et al., 1997). The stoichiometry of OP sequestration in any given species appears to depend on the concentration of the circulating enzyme at the time of OP exposure. With non-human primates, the 1:1 stoichiometry between BChE and OP dose, and the amount of endogenous scavenger (ChE, CaE, and other 3 unidentified proteins) normally present, the LD_{50} of soman may be extrapolated to be approximately 300 μg/70 Kg (Maxwell et al., 1998). To achieve a protective ratio of 2 in humans, a pretreatment dose of 150-200 mg enzyme should be effective for 3-5 days. Calculations of protective ratios in human required quantitative information on the toxicity of OP in humans. These figures were compiled from the literature describing human volunteer studies with non-lethal doses and accidental exposures to nerve agents that enabled an estimate of sign-free doses as well as toxic doses in humans. Predictions were then made by calculating the amount of Hu BChE required to reduce toxic levels of OP to below the sign-free doses within one blood circulation time (in seconds).

Only Hu BChE at the present time appears to be an appropriate candidate for exploration for human use, for Eq BChE was shown to induce the production of antibody when administered in heterologous species of animals (Gentry et al., 1993; Genovese et al., 1993). The antibody generated by repeated administration of Eq BChE rapidly clears the circulating exogenous BChE from blood indicating that the use of such enzyme in heterologous species may be limited to a single injection. Also, the absence of immunological and physiological side effects following blood and/or plasma transfusions in humans and lack of adverse reaction to partially purified BChE from human plasma administered daily for many weeks suggest that Hu BChE is the most promising prophylactic antidote (Stovner et al., 1976; Ostergaard et al., 1988; Cascio et al., 1982). Secondly, determination of the stability of exogenously administered Hu BChE in individuals (five cases) showed that half life of this enzyme was 8 to 11 days (Stovner et al., 1976; Ostergaard et al., 1988; Cascio et al., 1982).

These observations strongly suggest that the administration of even a single injection of Hu BChE in humans will provide adequate enhancement of the therapeutic properties of the endogenous bioscavenger over a long period of time. It appears that 200 mg/70 Kg Hu BChE would protect against 2 x LD_{50} of VX or soman, without the need of any immediate post-exposure treatment (Ashani, et al., 1998). Combining the enzyme treatment with 2-PAM (the oxime that works best with BChE), or with Huperzine A may reduce the dose of the enzyme to 50-60 mg/70 Kg. These lower levels of enzyme alone are likely to confer protection against long-term exposure to low levels of OPs such as soman or sarin. Further validation in larger animal models needs to be conducted.

Acknowledgements

The author wishes to thank SGT LaTawnya Askins for editing and reformatting this manuscript.

References

Allon N., Raveh L., Gilat E., Cohen E., Grunwald J. and Ashani Y. Prophylaxis against soman inhalation toxicity in guinea pigs by pretreatment alone with human serum butyrylcholinesterase. Toxicol. Sci. **43:** 121-128, 1998.

Amitai G., Moorad D., Adani R., Doctor B.P. Inhibition of Acetylcholinesterase and butyrylcholinesterase by chlorpyrifos-oxon. Biochem. Pharmacol. **56:** 293-299, 1996.

Ashani Y., Grauer D., Grunwald J., Allon N. and Raveh L. Current capabilities in extrapolating from animal to human the capacity of human BChE to detoxify organophosphates in: Structure and function of Cholinesterases and related proteins (B.P.Doctor et al., eds) Plenum Press, NY, 255-260, 1998.

Ashani Y., Shapira S., Levy D., Wolfe A.D., Doctor B.P. and Raveh L. Butyrylcholinesterase and acetylcholinesterase prophylaxis against soman poisoning in mice. Biochem. Pharmacol. **41:** 37-41, 1991.

Blick D.W., Kerenyi S.Z., Miller S.A., Murphy M.R., Brown G.C. and Hartgraves S.I. Behavioral toxicity of anticholinesterases in primates: Chronic pyridostigmine and soman interactions. Pharmacol. Biochem. Behav. **38:** 527-532, 1991.

Blick D.W. Murphy, M.R., Fanton J.W., Kerenyi S.Z., Miller S.A. and Hartgraves S.L. Incapacitation and performance recovery after high-dose soman: Effects of diazepam. Proceedings of the Medical Chemical Defense Bioscience Review, 219-222, 1989.

Bon S., Coussen F., Massoulie J. Quaternary associations of acetylcholinesterase. II. The polyproline attachment domain of the collagen tail. J. Biol. Chem. 272, 3016-3021, 1997.

Brandeis R., Raveh L., Grynwald J. Cohen E. and Ashani Y. Prevention of soman-induced cognitive deficits by pretreatment with human butyrylcholinesterase in rats. Pharmacol. Biochem. Behav. **46:** 889-896, 1993.

Broomfield C.A. A purified recombinant organophosphorous acid anhydrase protrects mice against soman. Pharmacol Toxicol. **70:** 65-66, 1992.

Broomfield C.A., Maxwell D.M., Solana R.P., Castro C.A., Finger A.V. and Lenz D.E., Protection of butyrylcholinesterase against organophosphorus poisoning in nonhuman primates. J. Pharm. Exper. Ther., **259:** 633-638, 1991.

Cascio C., Comite C., Ghiara M., Lanza G. and Popnchione A. The use of serum cholinesterase in severe phosphorus poisoning. Minerva Anestesiol. **54:** 337-338, 1982.

Castro C. and Finger A. The use of serial probe recognition in nonhuman primates as a method for detecting cognitive deficits following CNS challenge. Neurotoxicology, **125:** 125, 1991.

Chitlaru T., Kronman C., Velan B. and Shafferman A. Effect of human acetylcholinesterase subunit assembly on its circulatory residence. Biochem. J. **354:** 613-625, 2001.

Chitlaru T., Kronman C., Zeevi M., Kam M., Harel A., Ordentich A., Velan B. and Shafferman A. Modulation of Circulatory residence of recombinant acetylcholinesterase through biochemical or genetic manipulation of syalation. Biochem. J. **336:** 647-658, 1998.

Cohen O., Kronman C., Chitlaru T., Ordentlich A., Velan B. and Shafferman A. Effect of chemical modification of recombinant human acetylcholinesterase by polyethylene glycol on its circulatory longevity. Biochem. J. **357:** 795-802, 2001.

Cohen O., Kronman C., Chitlaru T., Ordentlich A., Velan B. and Shafferman A., Dirnhuber P., French M.C., Green D.M., Leadbeater I. and Stratton J.A. The protection of primates against soman poisoning by pretreatment with pyridostigmine. J. Pharm. Pharmacol. **31:** 295-299, 1979.

Doctor B.P., Raveh L., Wolfe A.D., Maxwell, D.M. and Ashani Y. Enzymes as pretreatment drugs for organophosphate toxicity. Neurosci. Biobehav. Rev. **15:** 123-128, 1991.

Dunn M.A. and Sidell F.R. Progress in medical defense against nerve agents. J. Am. Med. Assoc. **262:** 649-652, 1989.

Duysen E.G., Bartels C. and Lockridge O. Wild type and A328W mutant butyrylcholinesterase tetramers expressed in Chinese hamster ovary cells have a 16-hour half-life in circulation and protect mice from cocaine toxicity. J. Pharm. Exper. Ther. **302:** 751-758, 2002.

Farrer D.N., Yochmowitz M.G., Mattson J.L., Lof N.E. and Bennett C.T. Effects of benactyzine on an equilibrium and multiple response task in rhesus monkeys Pharmacol. Biochem. Behav. **16:** 605-609, 1982.

Genovese R.F. and Doctor B.P. Behavioral and pharmacological assessment of butyrylcholinesterase in rats. Pharmacol. Biochem. Behavior. **51:** 647-654, 1995.

Genovese R.F., Lu X-C.M., Gentry M.K., Larrison R. and Doctor B.P. Evaluation of purified horse serum butyrylcholinesterase in rats. Bioscience Review, pp. 1993, 1035-1042.

Gentry M., K., Nuwayser E., S. and Doctor B., P., Effect of repeated administration of butyrylcholinesterase on antibody induction in rabbits. Bioscience Review. 1051-1056, 1993.

Gray A.P. Design and structure-activity relationships of antidotes to organophosphorus anticholinesterase agents. Drug Metab. Rev. **15:** 557-589, 1984.

Grunwald J., Marcus D., Papier Y., Raveh L., Pittel Z. and Ashani Y. Large-scale purification and long-term stability of human butyrylcholinesterase: a potential bioscavenger drug. J. Biochem. Biophys. Methods. **34:** 123-135, 1997.

Grunwald J., Marcus D., Papier Y., Raveh L., Pittel Z. and Ashani Y. Large-scale purification and long-term stability of human butyrylcholinesterase: a potential bioscavenger drug. J. Biochem. Biophys. Methods **34:** 123-135, 1997.

Hunter A.J., Roberts F.F. The effect of pirenzepine on spatial learning in the Morris water maze. Pharmacol. Biochem. Behav. **30:** 519-523, 1988.

Jenkins T., Balinski D. and Patient D.W. Cholinesterase in plasma: First reported absence in the Bantu; Half-life determination. Science **156:** 1748-1750, 1967.

Kronman C., Chitlaru T., Elhanay E., Velan B., Shafferman A. Hierchy of post-translational modifications involved in the circulatory longevity of glycoproteins. J. Biol. Chem. **275:** 29488-29502, 2000.

Lockridge O., Blong R.M., Masson P., Froment M.T., Millard C.B. and Broomfield C.A. A single amino acid substitution, Gly117His, confers phosphotriesterase activity on human butyrylcholinesterase. Biochemistry, **36:** 786-795, 1997.

Lynch T.J., Mattes C.E., Singh A., Bradley R.M., Brady R.O. and Dretchen K.L. Cocaine detoxification by human plasma. Toxicol. Appl. Pharmacol. **145:** 363-371, 1997.

Masson O., Josse D., Lockridge O., Viguie N., Taupin C. and Buhler C., Enzymes hydrolyzing organophosphosphates as potential catalytic scavengers against organophosphate poisoning. J. Physol., **92:** 357-363, 1998.

Maxwell D.M., Brecht K.M., Doctor B.P. and Wolfe A.D. Comparison of antidote protection against soman by pyridostigmine, HI-6 and acetylcholinesterase. J. Pharmacol. Exper. Ther. **264:** 1085-1089, 1993.

Maxwell D.M., Brecht K., Saxena A., Feaster S. and Doctor B.P., Comparison of cholinesterases and carboxylesterase as bioscavengers for organophosphorus compounds, in Structure and Function of Cholinesterases and Related Proteins, (Doctor, B.P.,et al., Eds) Plenum Press, New York 387-392, 1998.

Maxwell D.M., Castro C.A., De La Hoz D.M., Gentry M.K., Gold M.B., Solana R.P., Wolfe A.D. and Doctor B.P. Protection of rhesus monkeys against soman and prevention of performance decrement by pretreatment with acetylcholinesterase. Toxicol. Appl. Pharmacol. **115:** 44-49, 1992.

Maxwell D.M., Wolfe A.D., Ashani Y. and Doctor B.P. Cholinesterase and carboxyesterase as scavengers for organophosphorus agents, in Proceedings of the Third International Meeting on Cholinesterase (Massulie et al., Ed), Washington, DC., ACS Books 206-209, 1991.

McLeod C.G. Pathology of nerve agents; Perspectives on medical management. Fundam. Appl. Toxicol. **5:** S10-S16, 1985.

Millard C.A., Lockridge O. and Broomfield C.A., Organophosphorus acid anhydride hydrolase activity in human butyrylcholinesterase: Synergy results in a somanase, Biochemistry **37:** 237-245, 1998.

Millard C.B., Lockridge O., Broomfield C.A., Design and expression of organophosphorous acid anhydride hydrolase activity in human butyrylcholinesterase. Biochemistry **34:** 15925-15933, 1995.

Ordentlich A., Barak D., Kronman C., Ariel N., Segall Y., Velan B. and Shafferman A. The architecture of human acetylcholinesterase active center probed by interactions with selected organophosphate inhibitors. J. Biol. Chem. **271:** 20, 11953-11962, 1996.

Ostergaard D., Viby-Morgensen J., Hanel H.K., Skovgaard L.T. Half-life of plasma cholinesterase. Acta. Anesth. Scand. **32:** 266-269, 1988.

Ralston J.S., Main A.R., Kilpatrick J.L. and Chasson A.L. Use of procainamide gels in the purification of human and horse serum butyrylcholinesterase. Biochem. J. **211:** 243-251, 1983.

Raveh L., Segal Y., Leader H., Rothchild N., Lelevanon D., Henis Y. and Ashani Y. protection against tabun toxicity in mice by prophilaxis with an enzyme hydrolysine organophosphate esters, Biochem Pharmacol **44:** 397-400, 1992.

Raveh L., Ashani Y., Levy D., De La Hoz D. Wolfe A.D. and Doctor B.P. Acetylcholinesterase prophylaxis against organophosphate poisoning. Quantitative correlation between protection and blood-enzyme level in mice. Biochem. Pharmacol. **41:** 37-41, 1991.

Raveh L., Grauer E., Grunwald J., Cohen E. and Ashani Y. The stoichiometry of protection against soman and VX toxicity in monkeys pretreated with human butyrylcholinesterase. Toxicol. Appl. Pharmacol., **145:** 43-53.

Raveh L., Grunwald J., Marcus D., Papier Y., Cohen E. and Ashani Y. Human butyrylcholinesterase as a general prophylactic antidote for nerve agent toxicity. In vivo and in vitro quantitative characterization. Biochem. Pharmacol. **45:** 2465-2474, 1993.

Rosenberg Y., Luo C., Ashani Y., Doctor B.P. et.al. A. Pharmacokinetics and immunologic consequences of exposing macaques to purified homologous butyrylcholinesterase. Life Sci. (Article in Press). 2002.

Sands S.F. and Wright A.A. Primate memory: Retention of serial list items by a rhesus monkey. Science **209:** 938-940, 1980.

Saxena A., Maxwell D.M., Quinn D.M., Radic Z., Taylor, P. and Doctor, B.P. Mutant acetylcholinesterases as potential detoxification agents for organophosphate poisoning. Biochem. Pharmacol. **54:** 269-274, 1997.

Saxena A., Ashani Y., Raveh L., Stevenson D., Patel T. and Doctor B.P. Role of oligosaccharides in the pharmacokinetics of tissue-derived and genetically engineered cholinesterases. Mol. Pharmacol. **53:** 112-122, 1998.

Saxena A., Doctor B.P., Maxwell D.M., Lenz D.E., Radic Z. and Taylor P. The role of glutamate-199 in the aging of cholinesterase. Biochem. Biophys. Res. Commun. **197:** 343-349, 1993.

Saxena A., Raveh L., Ashani Y. and Doctor B.P. Structure of glycan moieties responsible for the extended circulatory life of fetal bovine serum acetylcholinesterase and equine serum butyrycholinesterase. Biochemistry **36:** 7481-7489, 1997.

Smith G. Animal models of Alzheimer's disease: Experimental cholinergic denervation. Brain Res. Rev. **13:** 103-118, 1988.

Stovner J. and Stadsjkeuv K. Suxamethonium apnea terminated with commercial serum cholinesterase. Acta. Anaesth. Scand. **20:** 211-215, 1976.

Wolfe A. D., Blick D.W., Murphy M.R., Miller S.A., Gentry M.K., Hartgraves S.L., and Doctor B.P. Use of cholinesterases as pretreatment drugs for the protection of rhesus monkeys against soman toxicity. Toxicol. Appl. Pharmacol. **117:** 189-193, 1992.

Wolfe A.D., Rush R.S., Doctor B.P., Koplovitz I. and Jones D. Acetylcholinesterase prophylaxis against organosphate toxicity. Fundam. Appl. Toxicol. 266-270, 1987.

Pharmacological Perspectives of Toxic Chemicals and Their Antidotes
Editors: S J S Flora, J A Romano, S I Baskin and K Sekhar

Biomarkers in Detection of Sulfur Mustard

Syed K. Raza and Devendra K. Jaiswal*
Defence Research and Development Establishment
Jhansi Road, Gwalior 474 002, India
*Corresponding author

INTRODUCTION

The detection of exposure to sulfur mustard (SM) can be useful for more than one reason. Firstly, to establish firmly whether casualties have indeed been exposed to the chemical; secondly, these methods will be useful for the verification of alleged non-adherence to the Chemical Weapon Convention (CWC). Especially in the later application it appears that maximal retrospectivity, preferably over a period of several months, is most desirable. Moreover, these methods can be used in a variety of other applications also such as, for health surveillance of workers in destruction facilities of chemical warfare agents (CWAs) and in forensic analyses in cases of suspected terrorist activities.

Sulfur mustard is an alkylating agent that reacts readily with a variety of nucleophiles under physiological conditions. These nucleophilic species include water, the tripeptide glutathione, various amino acid residues present in the protein and DNA as also the purine and pyrimidine residues of DNA. The reaction products of sulfur mustard with these nucleophiles may serve as potential biomarkers which may be quite useful for retrospective detection of SM in alleged use of this agent. The metabolites derived from initial reaction of SM with water and glutathione are excreted in urine, adducts with hemoglobin and albumin are present in blood while the adducts with DNA are present in various tissues, blood and urine. Unchanged sulfur mustard has also been detected in fatty tissues (Drasch et al., 1987) and hair (UN Security Council Report, 1986), where it is protected from reaction with nucleophiles.

Urine is the major route of excretion following percutaneous or intravenous administration of SM (Maisonneuve et al., 1993). The potential for hemoglobin adducts as biomakers of sulfur mustard poisoning is quite significant due to their longer half lives. For an unequivocal identification at trace levels, gas chromatography mass spectrometry (GC/MS) and liquid chromatography mass spectrometry (LC/MS) using Electrospray ionization are the preferred techniques due to the high degree of specificity, selectivity and sensitivity associated with these techniques (Cushnir *et al.*, 1993; Chaudhary et al., 1995; Huand et al., 1998).

DNA ADDUCTS OF SULFUR MUSTARD

Sulfur mustard is a carcinogen by virtue of its ability to alkylate DNA (Brookes and Lawley, 1960 and 1964). SM, at near neutral pH, alkylates free bases, nucleosides and nucleotides, preferentially at ring nitrogen atoms on the N-7 position of guanine and the N-1 position of adenine. Additional reactions have been reported with attached functional groups at the O-6 position of guanine, the N-2 position of guanine and the N-6 position of adenine (Papirmeister et al., 1991). No confirmed alkylation products have been reported for reactions with cytosine, uracil, or thymine under physiological conditions. The reactivity of SM with cytosine under alkaline conditions has been reported and it has been tentatively identified as O-2 adduct (Walker and Watson, 1961).

The alkylation products formed by the reaction of SM with DNA isolated from exposed bacteria, yeast, and bacteriophage are almost similar and same as those formed with mammalian DNA. Three major products, which account for almost all alkylations, are identified in DNA following treatment with bifunctional SM. Out of three, two are monofunctional adducts, N-7[2-[(2-hydroxyethyl) thio] ethyl] guanine, **(I)** (~65%) and N-3-[2-[(2-hydroxyethyl) thio] ethyl] adenine **(II)** (~17%) and one cross linked adduct involving two guanines on the same or on complementary strand, bis-[2-(guanin-7-yl) ethyl] sulphide **(III)** (~17%). The two major products, which accounts for almost all alkylations are recovered from DNA treated with monofunctional SM (half-mustard). They are identical to the monofunctional adducts i.e N7-alkyl-G (~75%) and N3-alkyl-A (~25%) (Scheme 1).

One minor alkylation product, the O-6[2-[(2-hydroxyethyl)thio] ethyl] guanine **(IV)** (0.1%) was identified in chloroethylethylsulfide (CEES) treated DNA. The alkylation at O-6 position of guanine by SM is unusual for two reasons, the general lack of reactivity of SM with oxygen atoms in nucleic acids and the modification of a site involved in Watson-Crick type of hydrogen bonding. The major DNA adducts (N7-G and N3-A) are released spontaneously and/or enzymatically from DNA and therefore, may be excreted in urine. It was not proved experimentally, and this assumption was

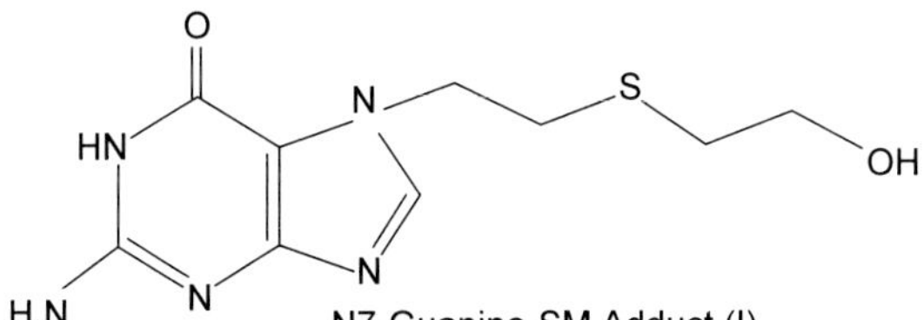

N7-Guanine-SM Adduct (I)

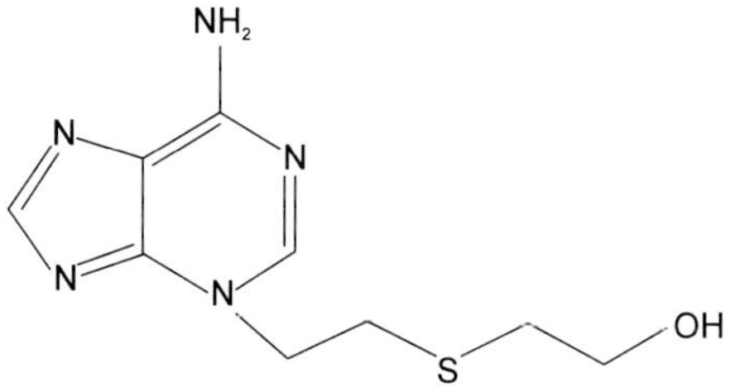

Adenine-SM Adduct (II)

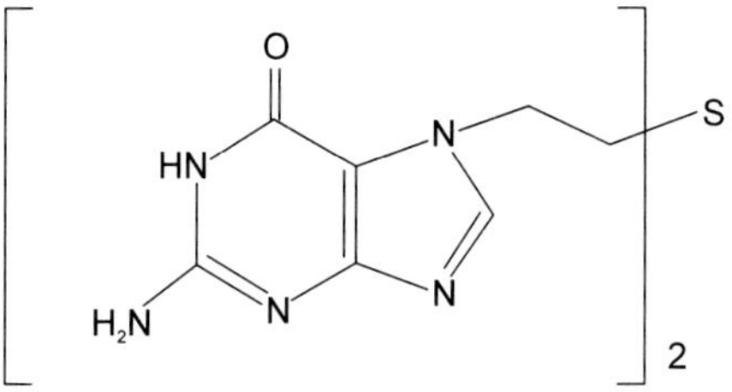

Bis-[2-(Guanine-7-yl)Ethyl] Sulfide (III)

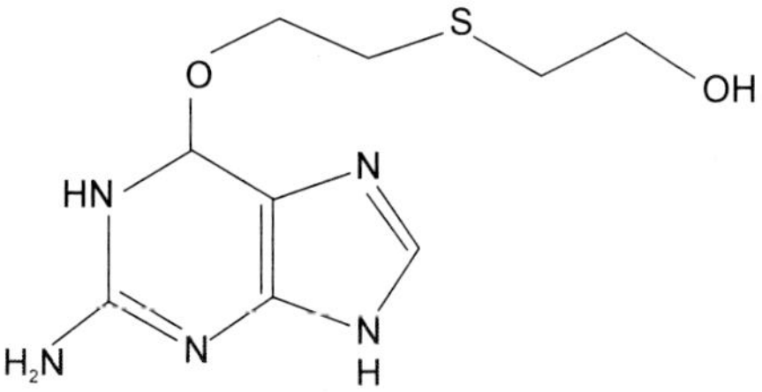

O-6 Guanine-SM Adduct (IV)

Scheme 1 DNA Alkylation Products

made on the increased urinary excretion of N7-methyl guanine after exposure of rats to methyl methanesulfonate (Chu and Lawley, 1973). The N3-alkyl-A is not of much importance due to its spontaneous depurination. Amongst all these, N7-guanine adduct is the major and most important one. This adduct can be formed in proteins found in the skin, blood and urine of animals allegedly exposed to SM and can therefore, serve as a biomarker for the retrospective determination of alleged exposure to SM.

Synthesis of DNA-SM Adducts

The work on alkylation of DNA by sulfur mustard started in 1960's (Brookes and Lawley, 1960 and 1964). Attempts have been made from time to time to develop a synthetic method for the preparation of the synthetic standard of N7-guanine-SM adduct to be used as reference material for its detection in biological samples obtained from the patients allegedly exposed to SM.

The adduct has been synthesized from guanosine-5'-monophosphate (GMP) and SM in water followed by purification on a Q-Sephadex column to remove unreacted SM and thiodiglycol. The depurination of N7-alkylated GMP was achieved by boiling it with 1M HCl, desaltation on a Sephadex G-10 column followed by lyophilization to give the adduct in 15% yield (Fidder et al., 1994). The poor yield of the adduct may be attributed to the non-completion of the reaction and probable hydrolysis of SM or the adduct to thiodiglycol forcing the involvement of purification by Sephadex columns. Recently, a simplified and improved methodology for the synthesis of N7-guanine-SM adduct has been reported (Rao et al., 2002), based on the techniques adopted for the methylation of certain nucleosides (Jones and Robins, 1963) (Scheme 2).

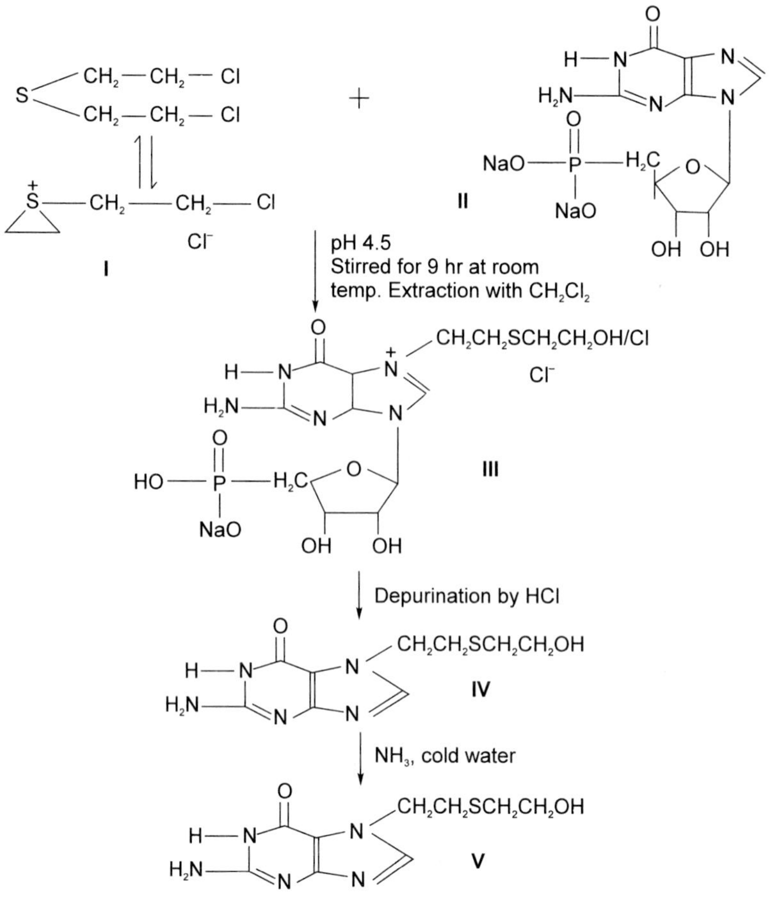

I + II

pH 4.5
Stirred for 9 hr at room temp. Extraction with CH_2Cl_2

III ($CH_2CH_2SCH_2CH_2OH/Cl$, Cl^-)

Depurination by HCl

IV ($CH_2CH_2SCH_2CH_2OH$)

NH_3, cold water

V ($CH_2CH_2SCH_2CH_2OH$)

Scheme 2 Stepwise Formation of N7-Guanine-SM Adduct

It is well documented that alkylation of guanosine using methyl iodide generates N7-methyl guanosine as an iodide salt (Jones and Robins, 1963). This salt on treatment with concentrated aqueous ammonia is converted into a water soluble betaine 7-methyl guanosine intermediate,. It should be noted that this internal betaine (which is not a salt) is a characteristic of N7-alkylated guanosine derivatives. Both N7-methyl guanosine and its iodide salt namely, N7-methyl guanine hydroiodide, can be hydrolyzed or depurinated by acid to yield 7-methyl guanine.

Using the same analogy, sulfur mustard was reacted with guanosine-5'-monophosphate in aqueous medium to generate the chloride salt of N7-[2-[(2-hydroxy/chloro-ethyl)thio]ethyl]guanosine monophosphate. This salt was formed via the interaction between the N7 nitrogen atom of the purine base and SM, presumably existing as sulfonium species, through SN^{i} rearrangement. The pH of the reaction mixture was maintained at 4.5. The hydrochloride salt was then purified by careful extraction with dichloromethane to remove the organic impurities, such as thiodiglycol. The salt thus obtained was subjected to direct depurination by hydrochloric acid to afford hydrochloride salt of the desired N7-guanine-SM adduct. Since GMP was totally consumed, as indicated by HPLC, and the organic impurities removed by extraction, the aqueous extract, containing soluble hydrochloride of N7-guanine-SM adduct, was neutralized with ammonia resulting in the formation of the free insoluble organic base N7-[2-[(2-hydroxyethyl) thio] ethyl] guanine, which was precipitated out and collected.

Experiments showed that hydrolysis was achieved at the salt stage (Scheme 1), ensuring the absence of any unreacted GMP and that this method of hydrolysis did not involve any binding or retention of the product on the column. Furthermore, since the type of hydrolysis does not contain any inorganic salts or thiodiglycol, the purification technique employing Sephadex columns as used in earlier methods was also avoided. This resulted in excellent yields of the final product (70%). The purity of the final product was confirmed by high performance liquid chromatography (HPLC) (99%).

It should be noted that, the solvent, the pH, and the substrate play a major role in the selective synthesis of N7-monoalkylated guanine adduct of SM. The selection of GMP over guanosine is based on its solubility in water and pH 4.5 has been found to be the most effective for exclusively obtaining the N-7 adduct. Alkylation at position 9 through other tautomeric forms of the imidazole ring or the formation of dialkylated product in the ring or a mixture of both is avoided by blocking position 9 with a sugar moiety. Moreover, under this steric model, N7 is the only position favored for alkylation that is not involved in hydrogen bonding in the Watson-Crick model of DNA. At pH 4.5, not only is the nucleophilicity of N-7 nitrogen maintained, but depurination of GMP is also avoided. In addition, an excess of SM is added to avoid the formation of *bis* adduct of purine and SM.

Besides the N7-Guanine-SM adduct, the synthesis of other minor adducts **(II-IV)** have also been reported in the literature (Fidder et al., 1994). All these adducts have been well characterized by modern instrumental techniques such as ultraviolet (UV) spectroscopy, nuclear magnetic resonance (NMR) spectroscopy and mass spectrometry (MS).

Detection of DNA-SM Adducts

The usefulness of the methods in bio-monitoring requires high sensitivity combined with high specificity, because the levels of the adducts formed are very low. The methodology for the determination of DNA adducts is constantly being developed and diversified (Andrews et al., 1999). It mainly depends on the compound or the class of chemical studied. The purification of the adduct helps a lot to boost the sensitivity whereas the amount of DNA available becomes limited. Mass spectrometry coupled to gas chromatography or liquid chromatography is playing a vital role in the analysis of these adducts. These techniques are more powerful and relatively expensive when compared to post labeling and immunoassays. The derivatization of the samples for gas chromatography mass spectrometry (GC/MS) analysis has been one of the main obstacles so far (Allam et al., 1990), but this technique appears most suitable for relatively small and abundant adducts that are analyzed as base derivatives. A number of analytical methodologies have been developed to detect, identify and quantitate DNA adducts (Cushnir et al., 1993; Chaudhary et al., 1995; Huand et al., 1998). The experimental approaches involve isolation and digestion of target DNA after exposure to SM, yielding a complex mixture of nucleic acid fragments. The success of a method is based on the combination of steps, including primary purification, method of detection and use of standards.

The use of urine as a source of adduct for biomonitoring exposure to SM has been reviewed (Shuker and Farmer, 1992; Shuker et al., 1993; Huand et al., 1998). The advantage of using urinary adducts as the measure of exposure to genotoxic compounds include noninvasiveness, the large amount available and simple purification of adducts. The main disadvantage however, is that the origin of the adducts is unknown, whether it is DNA, RNA, free nucleoside or dietary sources. Therefore, blood and tissues are the main biological samples for the detection of DNA adducts as biomarkers.

For the detection of the major adduct, N7-(2-hydroxyethylthioethyl)-2'-deoxy guanosine, a sensitive Enzyme Linked Immuno Sorbent Assay (ELISA) has been developed which allows the detection of adducts in SM-treated calf thymus DNA and white blood cells after exposure of human blood to sulfur mustard (Van der Schans et al., 1994). The optimal conditions for ELISA are listed in Table 1.

Methods based on UV, HPLC, GC/MS and LC/MS have been optimized for easy detection of the DNA adduct in the biological samples. The adducts

Table 1 Optimal Conditions for ELISA

Microtitre plate	:	Polystyrene, high binding (COSTAR)
Precoating	:	Poly(L-lysine) (10mg/mL)
Coating	:	as Calf Thymus DNA treated with 10 mM sulfur mustard for 30 min at 37 °C. (50 μL well of 1 mg of DNA/mL)
Blocking of free binding sites	:	Gelatin instead of FCS that is normally used in this laboratory
Competitive ELISA		
Dilution of rabbit antisera Against SM adducts	:	1 : 40,000
Concentration of monoclonal Antibodies against SM adducts	:	40 ng protein/mL
Dilution of conjugated second Antibodies	:	1 : 1000
Washing steps	:	PBS containing 0.05% Tween 20
Optimization unwinding of DNA	:	Heating at 52oC for 25 min in 0.01 M Tris, 1 mM EDTA, 4% formamide & 0.2% formaldehyde

formed with DNA were identified in double stranded calf thymus DNA and in human blood exposed to SM. An HPLC system was developed for the separation of all the four major nucleosides and the synthesized adducts (Fidder et al., 1994).

Attempts have been made to detect the N7-Guanine-SM adduct by GC/MS but the analysis has been found to be problematic. Derivatization with heptafluorobutyric anhydride and pentafluorobenzyl bromide was abortive (Shuker et al., 1984); while derivatization with N-methyl N-(tert-butyldimethylsilyl) trifluoroacetamide afforded a compound with poor gas chromatographic properties (Cushnir et al., 1993). The adduct was not amenable to GC/MS analysis even after derivatization with N,O-bis(trimethylsilyl) trifuoroacetamide (BSTFA) (Rao et al., 2002). The silylated adduct however could be detected by direct probe MS under both EI and CI modes (Figure 1).

The underivatized compound could be conveniently detected by liquid chromatography mass spectrometry (LC/MS) and liquid chromatography tandem mass spectrometry (LC/MS/MS) (Noort et al., 1996; Rao et al., 2002) (Figure 2). The detection limit achived was 0.2 ng/ml in multiple reaction monitoring (MRM) for m/z 256 → 105. The method has been successfully used for the detection of adduct in the urine samples of SM-treated guinea pigs (Noort et al., 1996). LC/MS with Electrospray ionization has also been

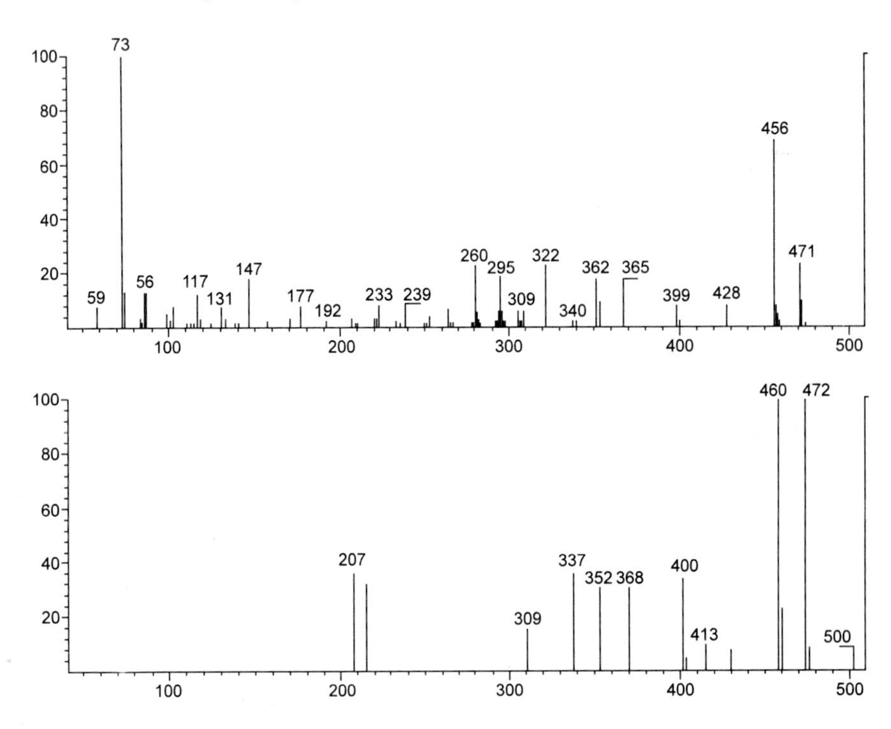

Fig. 1 Direct-probe EI (above) and CI (below) mass spectra of N7-guanine-SM adduct

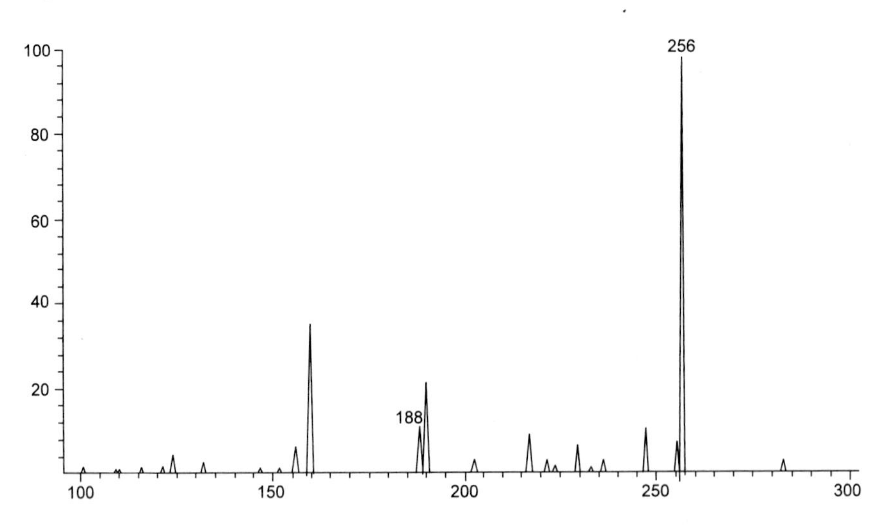

Fig. 2 Electrospray Mass Spectrum of N7-Guanine-SM Adduct showing a $[M+H]^+$ peak at m/z 256

used for the detection of N7-Guanine-SM adduct in liver and spleen of SM-treated rats (Rao et al., 2002).

PROTEIN ADDUCTS OF SULFUR MUSTARD

The reactions of SM or its oxidation products, i.e. SM sulfoxide and SM sulfone, with enzymes and with other proteins have been thought to be responsible for the cytotoxic and tissue-injurant actions of this vesicant. The SM-adducts to the proteins is stable and not likely to be found in urine. The reactive groups, which are of greatest interest from a biochemical point of view, are the sulfhydryl (SH) groups. SM binds competitively to the SH group, under appropriate conditions, one molecule of SM can alkylate and thus cross-link two molecules of SH-containing compound (e.g., bis-S-cysteinyl derivative of SM). The products of the reaction of SM with sulfhydryl compounds are so stable at physiological pH and temperature that it is unlikely to dissociate under these conditions. SM binds to all the proteins, which are available with varying competition factors (Stein, 1946; Stein and Fruton, 1946; Moore, 1946; Goodlad, 1956). Because of the affinity of the highly electrophilic cyclic ethylene sulfonium ion, the alkylation reaction of SM with proteins is greatly dependent on pH. At pH 6.0-8.0, carboxyl, imidazole, thiomethyl and sulfhydryl groups would be expected to react to a significant extent. At pH 6.0, SM reacts with most of the carboxyl groups of proteins. No reaction of SM with amino group occurs at pH 6.0, although some amino groups will react in a more alkaline medium. Other functional groups in proteins, which have been reported to react with SM, are the amino, hydroxyl, and tyrosine groups (Davis and Ross, 1947).

The growing understanding of the mechanism of chemical carcinogenesis and realization of the extent to which chemicals were implicated in the causation of cancer, has generated a new imperative risk-assessment. The identification of covalent interactions with macromolecules such as hemoglobin, as biological markers of exposure, has attracted increasing attention in recent years (Osterman-Golkar et al., 1976; Ehrenberg and Osterman-Golkar, 1980; Farmer et al., 1980; Neumann, 1984; Tannenbaum and Skipper, 1994), because determination of hemoglobin adducts has additional advantages compared to determination of DNA adducts. These advantages are addressed in the following paragraph.

Lars Ehrenberg first recognized the potential of protein adducts to serve as surrogates for DNA adducts (Ehrenberg and Osterman-Golkar, 1980). He and his associates pioneered this area of research and have continued to make important contributions. The group specifically focused on hemoglobin (Hb) and derived the mathematical description of steady state adduct levels that result from continuous exposure to an Hb alkylating agent. Hb has the unique biological property of having a life span equivalent to the life span of the erythrocyte, which in humans is ~120 days. The age distribution is uniform, and thus if a carcinogen forms adduct that is chemically stable and if exposure to the carcinogen is chronic, then the amount of the adduct will be proportional to Hb age. The Hb dosimeter therefore, reflects a four months record of the average exposure. Concomitantly, when exposure ceases, the adduct level drops rapidly, since the initial class of erythrocyte that is lost is the oldest and thus has the highest level of adduct. The occupational and environmental exposure to alkylating agents has been demonstrated by the identification of alkylated amino acid residues isolated from globin hydrolysates obtained from exposed individuals (Wogan, 1989). For example, the adduct of ethylene oxide and acrylonitrile with N-terminal valine was shown to be present in the blood of cigarette smokers (Bailey et al., 1988; Osterman-Golkar et al., 1994). Since hemoglobin is much more abundant than DNA and adducts to human hemoglobin have *in-vivo* life span of several months (Skipper and Tannenbaum, 1990), the hemoglobin-sulfur mustard adducts might be excellent biomarkers for exposure to sulfur mustard. The alkylation of proteins by sulfur mustard was investigated in 1940s and 1950s by various groups (Wheeler, 1962). For instance, mustard gas could esterify carboxyl groups in serum albumin and horse hemoglobin and that alkylation has also occurred with histidine residues (Davis and Ross, 1947). Furthermore, sulfur mustard reacts with the amino group of amino acids and peptides (Douglas and Heard, 1954). Recent studies into the biological fate of SM in blood have shown that when radiolabelled (^{35}S) SM is incubated *in-vitro* with whole blood (both rat and human), the major portion is taken up by the erythrocytes, and that this portion is compared with that remaining in the plasma, increases with increasing concentration of SM (Hambrook et al., 1993; Maisonneuve et al., 1993). It was further observed that most of the radioactivity in the red blood cells (RBC) was associated with the globin fraction rather than with the heme fraction. The administration of ^{35}S-SM percutaneously to the rat *in-vivo*, showed that the distribution of radioactivity in whole blood corresponds to the pattern observed in *in-vitro* experiments. N1-(2-Hydroxyethylthioethyl)-4-methyl imidazole was detected in urine, which is probably a degradation product of histidine alkylated by SM (Sandelowsky et al., 1992). Recently, it has been demonstrated that hemoglobin was efficiently alkylated by SM (Noort et al., 1996). In this study, they have identified various specific residues in haemoglobin which are prone to alkylation by SM using LC tandem MS. Furthermore, they have identified the alkylated N-terminal valine by GC/MS using modified Edman degradation procedure (Fidder et al., 1996).

Synthesis of Protein-SM Adducts

Cysteine-SM Adduct: The fact that mustard reacts readily with sulfur containing nucleophiles suggested that a likely site of alkylation in a readily isolable protein, such as hemoglobin in blood, would be the -SH group on a cysteine residue. The reaction of sulfur mustard with cysteine in aqueous medium with vigorous stirring, gave the bis-cysteine adduct in good yield

(70%). The formation of bis-adduct rather than mono-adduct is primarily due to the presence of two active electron deficient sites in SM which are susceptible to be attacked by the -SH nucleophile through the intermediate three membered cyclic sulfonium species. The extreme insolubility of the product in all common solvents under neutral conditions makes its isolation easy. The adduct has been characterized using both ^{1}H NMR and ^{13}C NMR. The values are well in agreement with the literature values (Black et al., 1992). This reaction was originally examined by Hartwell (1946) and subsequently by Roberts and Warwick (1961), but spectroscopic data were not reported in either case.

Recently, the cysteine-SM mono adduct was prepared according to a method described earlier (Fidder et al., 1996) but with improved yields (Scheme 3). The adduct was characterized using ^{1}H NMR, ^{13}C NMR, IR and mass spectral data (Table 2).

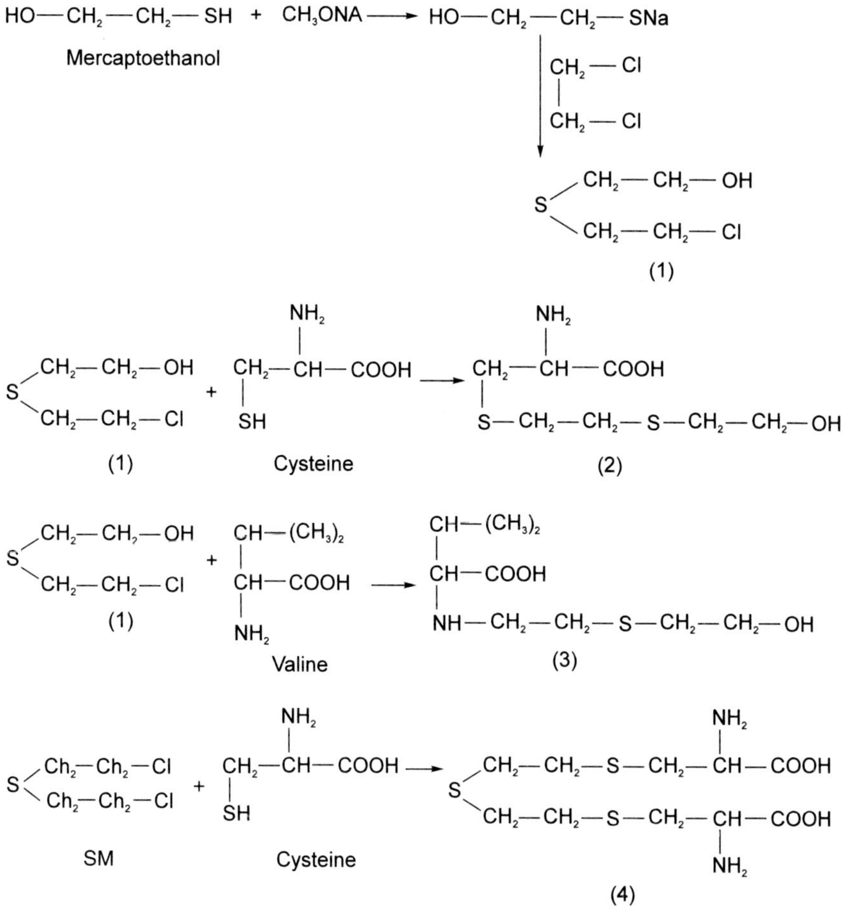

$HO{-}CH_2{-}CH_2{-}SH$ + CH_3ONA → $HO{-}CH_2{-}CH_2{-}SNa$

Mercaptoethanol

↓ $CH_2{-}Cl$ | $CH_2{-}Cl$

$S(CH_2{-}CH_2{-}OH)(CH_2{-}CH_2{-}Cl)$

(1)

$S(CH_2{-}CH_2{-}OH)(CH_2{-}CH_2{-}Cl)$ + $CH_2(SH){-}CH(NH_2){-}COOH$ → $CH_2{-}CH(NH_2){-}COOH$ | $S{-}CH_2{-}CH_2{-}S{-}CH_2{-}CH_2{-}OH$

(1) Cysteine (2)

$S(CH_2{-}CH_2{-}OH)(CH_2{-}CH_2{-}Cl)$ + $CH{-}(CH_3)_2$ | $CH(NH_2){-}COOH$ → $CH{-}(CH_3)_2$ | $CH{-}COOH$ | $NH{-}CH_2{-}CH_2{-}S{-}CH_2{-}CH_2{-}OH$

(1) Valine (3)

$S(Ch_2{-}Ch_2{-}Cl)(Ch_2{-}Ch_2{-}Cl)$ + $CH_2(SH){-}CH(NH_2){-}COOH$ → $S(CH_2{-}CH_2{-}S{-}CH_2{-}CH(NH_2){-}COOH)(CH_2{-}CH_2{-}S{-}CH_2{-}CH(NH_2){-}COOH)$

SM Cysteine (4)

Scheme 3 Scheme for the synthesis of (1) Half-Mustard (2) Cysteine mono-adduct (3) Valine adduct (4) Bis-Cysteine adduct

Table 2 NMR and Mass Spectral Data for Cysteine-SM and Valine-SM Adducts

Compound	*NMR and mass spectral data*
Cysteine-SM adduct	^{1}H NMR (d_6-DMSO) δ (ppm): 2.59 (2H, t, SCH_2CH_2OH), 2.70 (4H, broad s, SCH_2CH_2S), 2.83 and 3.00 (each 1H of SCH_2CH, each dd, AB part of ABX system), 3.55 (2H, t, CH_2OH), 4.2 (1H, NH_2CHCH_2 dd, X part of ABX system).
	^{13}CNMR (d_d-DMSO) δ (ppm): 32.30, 32.39, 33.62, 33.66 ($CH_2SCH_2CH_2SCH_2$), 54.54 (SCH_2CHNH_2), 62.19 (CH_2OH), 169.61 (COOH).
	Mass (Silylated adduct) (EI): m/z 218 (M-$CH_2SCH_2CH_2SCH_2CH_2$OTMS, 100%), 232 (M-$SCH_2CH_2SCH_2CH_2$OTMS, 9%), 264 (M-$CH_2CH_2SCH_2CH_2$OTMS, 6%), 292 (M-SCH_2CH_2OTMS, 20%), 324 (M-CH_2CH_2OTMS, 15%).
	Mass (Silylated adduct) (CI, Isobutane): m/z 442 $(M+H)^-$
Valine-SM adduct	^{1}H NMR (D_2O) δ (ppm): 0.84 (3H, d, CH_3), 0.89 (3H, d, CH_3), 2.15 (1H, m, $CH(CH_3)_2$), 2.71 (t, 2H,SCH_2CH_2OH), 2.84 (2H, m, $NHCH_2CH_2S$), 3.19 (2H, m, $NHCH_2$), 3.55 (d, 1H, NHCH), 3.60 (2H, t, CH_2OH)
	^{13}C NMR (D2O) δ (ppm): 18.11, 18.84 (2 CH_3) 27.94 (SCH_2CH_2NH) 30.03 ($CH(CH_3)_2$), 33.96 ($HOCH_2CH_2S$), 47.36 ($HNCH_2$), 61.01 (CH_2OH), 69.07 (CHNH), 170.12 (COOH).
	PFTH deravatives of valine adduct:
	^{19}F NMR ($CDCL_3$) δ (ppm): Aromatic Fluorines 146.5 [2F, d, Ortho], 157.2 [1F, t, Para], 162.8 [2F, t, Meta].
	Mass (EI): m/z 428 (M^+), 40%), 410 (M-H_2O, 11%), 383 (M-CH_2CH_2OH, 100%), 351 (M-SCH_2CH_2OH, 67%).
	HFBA derivative of valine adduct:
	Mass (NCI, Methane): m/z 624 (M^-), 604 (M-HF), 584 (M-2HF), 564 (100%, M-3HF), 544 (M-4HF), 524 (M-5HF), 504 (M-6HF) 427 (M-COC_3F_7)
	Mass (PCI, Methane): m/z 625 $(M+H)^+$

Valine-SM Adduct: The valine-SM adduct has been synthesized by reacting half-mustard with valine in aqueous solution (Rao et al., 2003). The method involved a multi-step process culminating in a very poor yield (23%) of the adduct. Recently, an improved methodology for the synthesis of valine-SM adduct has been described which gave the adduct in 70% yield (Scheme 3).

The cysteine-SM and valine-SM adducts were characterized on the basis of ^{1}H NMR; ^{13}C NMR and mass spectral data (Table 2).

Glutamic Acid-SM Adduct: The adduct of glutamic acid to SM was synthesized by dicyclohexyl carbodiimide (DCC) mediated esterification of suitably protected derivatives (Noort et al., 1997) with thiodigycol (Scheme 4). Deprotection with trifluoroacetic acid (TFA) gave the desired reference

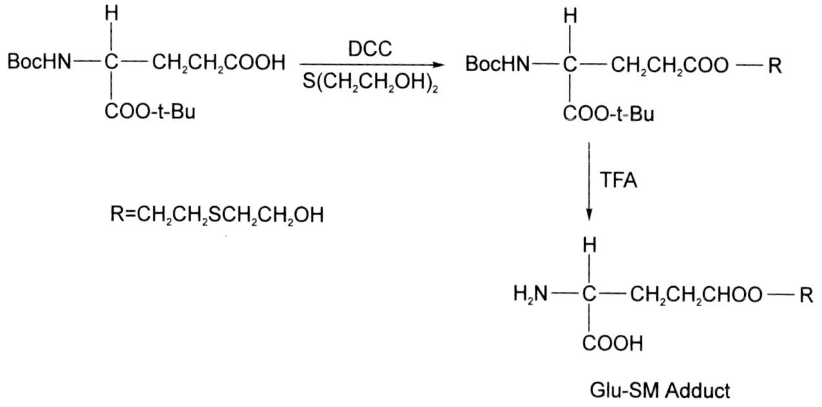

H | H
BocHN—C—CH_2CH_2COOH —DCC / $S(CH_2CH_2OH)_2$→ BocHN—C—CH_2CH_2COO—R
COO-t-Bu | COO-t-Bu

↓ TFA

R=$CH_2CH_2SCH_2CH_2OH$

H
H_2N—C—CH_2CH_2CHOO—R
COOH

Glu-SM Adduct

Scheme 4 Synthesis of Glutamic Acid-SM Adduct

compound. The reaction was monitored by GC/MS and the attachment of hydroxyethylthioethyl (HETE) moiety to the excess carboxyl group of glutamic acid was confirmed by GC/MS analysis of the protected Glu-SM adduct, after derivatization with N-methyl N-(t-butyldimethylsilyl) trifluoroacetamide (MTBSTFA).

Detection of Protein-SM Adducts

The protein adducts may be present in very small amounts in the biological samples and hence there is a need to develop a highly sensitive methods for the detection of these adducts. Methods based on GC/MS with negative chemical ionization (NCI-GC/MS), after converting the adduct into an oxazolidinone using 1,3 dichloro tetrafluoroacetone as the derivatizing agent has been used for the detection of cystein-SM adduct. In order to increase the sensitivity, the oxazolidinone has been derivatized further with heptafluoro butyric anhydride.

For derivatization of the adducts for NCI-GC/MS analysis, mild procedures were applied as described in the literature (Husek et al., 1979). After reaction of the reference compounds with dichlorotetrafluoroacetone (DCTFA) (Scheme 5) followed by derivatization of the obtained oxazolidinone with heptafluorobutyric anhydride (HFBA), NCI-GC/MS analysis showed one major peak having expected mass in each case. S-(HETE)-cysteine gave $M^{-\cdot}$ ion at m/z 601 and an abundant (M^{-}HF-CO_2) ion at m/z 537. It also gave an adduct ion at m/z 636/638 (M+Cl^-). The analysis of 5-(HETE)-L-glutamate gave molecular ion at m/z 627 ($M^{\cdot-}$) besides the other ions at m/z 607 ($M^{\cdot-}$-HF), 563 ($M^{\cdot-}$-HF-CO_2) and an adduct ion at m/z 662/664 (M+Cl^-).

The heptafluorobutyric ester of PFTH-HETE-Val gave a molecular ion M^- at m/z 624 and an abundant ion at m/z 564 (M-3HF), followed by ions at m/z

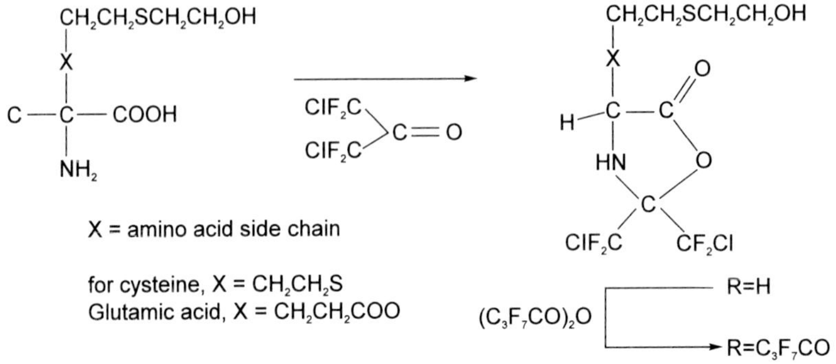

$CH_2CH_2SCH_2CH_2OH$
X
C — C — COOH
NH_2

ClF_2C C=O ClF_2C

X = amino acid side chain

for cysteine, X = CH_2CH_2S
Glutamic acid, X = CH_2CH_2COO

$CH_2CH_2SCH_2CH_2OH$
X
H — C — C=O
HN O
C
ClF_2C CF_2Cl

$(C_3F_7CO)_2O$ → R=H; R=C_3F_7CO

Scheme 5 Scheme for the Derivatization of Cysteine-SM and Glutamic acid- SM Adducts with DCTFA and HFBA.

604 (M-HF), 584 (M-2HF), 544 (M-4HF), 524 (M-5HF) and 504 (M-6HF). It also gave an ion at m/z 427 due to the loss of COC_3F_7 from the HETE moiety. The relative intensities of the ions in the NCI spectrum are particularly dependent on the source temperature and source pressure and may vary from experiment to experiment. The positive chemical ionization mass spectrometry (PCI-MS) of the derivatized adduct gave $(M+H)^+$ at m/z 625 confirming thereby the formation of the adduct.

When blood was incubated with SM, it was found that an approximately proportional amount of SM gets bound to globin (Noort et al., 1996), indicating a linear relationship between exposure and adduct level. It is anticipated that several SM adducts are acid labile, e.g. the adducts formed with cysteine and glutamic acid. Moreover, it is well established that synthetic S-(HETE)-L-cysteine readily decomposes in refluxing 6N HCl (Noort et al., 1997). Therefore, enzymatic hydrolysis by pronase E was applied. The recovery of the amino acids and their adducts after pronase E digestion and derivatization was almost quantitative. The final proof for the formation of the adducts was achieved by NCI-GC/MS analysis of acidic/enzymatic hydrolysates of globin, obtained from blood after exposure to 1-10 mM of SM. It was found that the hydrolysis using enzymes invariably gave reproducible results.

The two ions at m/z 564 and 544 were monitored, but only m/z 564 was sufficiently sensitive for the detection of (HETE)-Val in globin isolated from blood treated with sulfur mustard at micro molar concentrations levels. The reaction of SM with N-terminal valine of globin and subsequent modified Edman degradation followed by reaction with HFBA was shown in Scheme 6. NCI-GC/MS was however, the most sensitive method for the detection and analysis of the valine-SM adduct in derivatized form. This method gave sensitivity of the order of pg levels. This analytical method was successfully applied to the analysis of blood samples after exposure to SM. The control

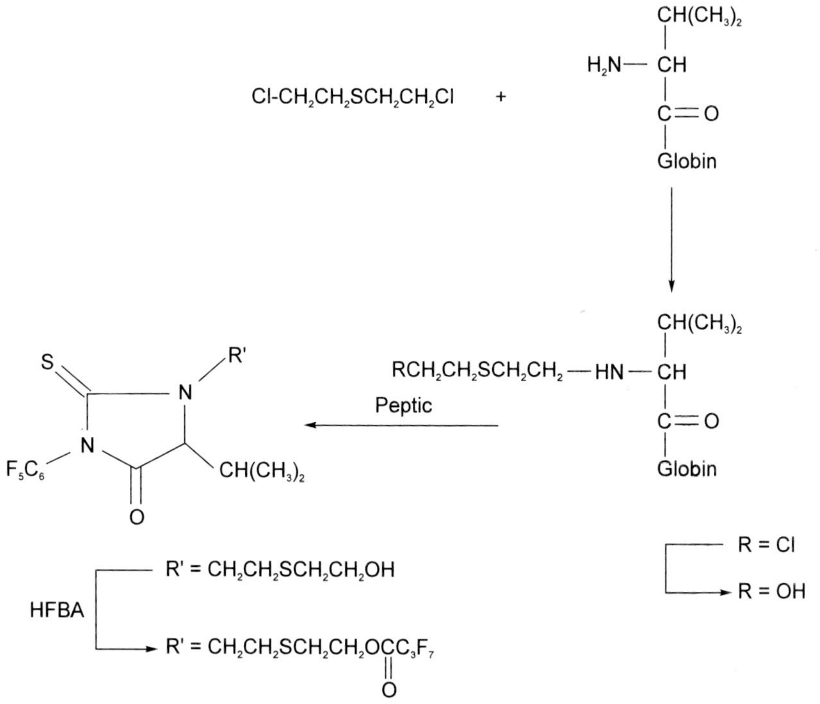

$Cl\text{-}CH_2CH_2SCH_2CH_2Cl$ + $H_2N\text{—}CH(CH(CH_3)_2)\text{—}C{=}O\text{—}Globin$

↓

$RCH_2CH_2SCH_2CH_2\text{—}HN\text{—}CH(CH(CH_3)_2)\text{—}C{=}O\text{—}Globin$

R = Cl → R = OH

Peptic ←

F_5C_6-substituted thiohydantoin (S, N, R', $CH(CH_3)_2$, O)

HFBA: $R' = CH_2CH_2SCH_2CH_2OH$ → $R' = CH_2CH_2SCH_2CH_2OC(=O)C_3F_7$

Scheme 6 Reaction of SM with N-terminal Valine of Globin and Subsequent Modified Edman Degradation followed by Reaction with HFBA.

blood samples gave no response at the appropriate retention times. In order to develop the standard operating procedure for the determination of N-terminal valine adduct in hemoglobin, we have tried several procedures. The first step in the analysis of N-terminal valine adduct in hemoglobin is the isolation of globin. By avoiding the isolation step, the hemolysate of blood that was exposed to SM was treated with the modified Edman reagent pentafluorophenyl isothiocyanate (PFPITC). However, the SM-valine adduct could not be detected by NCI-GC/MS analysis. Next, some modifications were introduced into the modified Edman procedure in order to simplify the procedure. The degradation step was performed by reaction for 2 hrs at 60 °C, instead of incubation overnight at room temperature, followed by reaction for 2 hrs at 45 °C. Furthermore, the reaction mixture was worked up by extraction with toluene only, leaving out the first extraction step with ether. Here the separation of the toluene/formamide layers was achieved by freezing in liquid nitrogen. Both the original and simplified procedures were used for processing of globin that had been isolated from blood exposed to SM.

Conclusion

The use of highly toxic sulfur mustard in wars and terrorist attacks has stressed the need to develop reliable analytical methods for the retrospective determination of SM in biological samples collected from the subjects allegedly exposed to it. There has been therefore, considerable stimulus to develop monitoring methods in order to detect SM, its metabolite(s) or its *in-vivo* reaction product(s) within the body. SM causes structural alterations in DNA and proteins of target cells. Reaction of SM with the nucleophilic sites in macromolecules results in covalent reaction products (adducts) which are thought to be responsible for the toxicity produced by it and are termed as biomarkers. Measurement of such biomarkers is considered to be a suitable monitor of SM exposure. The presence or absence of biomarkers indicates whether exposure to SM has taken place or not and will thereore, be quite useful in verification activities also.

References

Allam K., Saha M. and Giese R.W. Preparation of electrophoric derivatives of N7-(2-hydroxyethyl) guanine. An ethylene oxide DNA adduct, J. Chromatogr. A. **499:** 571-578, 1990.

Andrews C.L., Vouros P. and Harsch A. Analysis of DNA adduct using high performance separation technique coupled to electrospray ionization mass spectrometry, J. Chromatogr.A. **856:** 515-526, 1999.

Bailey E., Brookes A.G.F., Dollery C.T., Farmer P.B., Passinghan B.J., Sleightholm M.A. and Yates D.W. Hydroxyethylvaline adduct formation in hemoglobin as a biologicalmonitor of cigarette smoke intake, Arch. Toxicol. **62:** 247-258, 1988.

Black R.M., Brewster K., Clarke R.J. and Harrison J.M. Chemistry of 1,1'-thiobis(2-chloroethane) (sulfur mustard) Part II. The synthesis of some conjugates with cysteine, N-acetylcysteine and N-acetylcysteine methyl estaer, Phosphorous, Sulfur, and Silico., **71:** 49 -59, 1992.

Brookes P and Lawley P.D. Reaction of mustard gas with nucleic acids in- vitro and in-vivo, Biochem. J. **77:** 478-484, 1960.

Brookes P and Lawley P.D. Reaction of some mutagenic and carcinogenic compounds with mucleic acids, J. Cellular Comp. Physiol. **64:** 111-118, 1964.

Chaudhary A.K., Nokubo M., Oglesby T.D., Marnett L.J. and Blair I.A. Characterization of endogenous DNA adducts by liquid chromatography electrospray ionization mass spectrometry, J. Mass spectrom. **30:** 1157-1160, 1995.

Chu B.C.F. and Lawley P.D. Increased urinary excretion of pyrimidine and nicotinamide derivatives in rats treated with methyl methanesulfonate, Chem. Biol. Interactions. **8:** 65-73, 1974.

Cushnir J.R., Naylor S., Lamb J.H. and Farmer P.B. Tandem mass spectrometric approaches for the analysis of alkylguanines in human urine, Org. Mass spectrom. **28:** 552-558, 1993.

Davis S.B. and Ross W.F. Reaction of mustard gas with proteins, J. Am. Chem. Soc. **69:** 1177-1181, 1947.

Day B.W., Skipper P.L., Wishnock J.S., Coghlin J., Hammond S.K., Gann P. and Tannenbaum S. R. Identification of an in-vitro chrysenediol epoxide adduct in human hemoglobin, Chem. Res. Toxicol. **3:** 340-343, 1990.

Douglas D.E. and Heard R.D.H. Reaction of â- â'-dichloroethyl sulfide with glycine and with glycylglycine, Can. J. Chem. **32:** 221-226, 1954.

Drasch G., Kretschmer E., Kauert G. and Von Meyer L. Concentrations of mustard gas [bis2(chloroethyl)sulfide]in the tissues of a victim of a vesicant exposure, J. Forensic Sci. **32:** 1788-1793, 1987.

Ehrenberg L. and Osterman-Golkar S. Alkylation of macromolecules for detecting mutagenic agents, Teratog.Carcinog. Mutag. **1:** 105-127, 1980.

Farmer P.B., Bailey B., Lamb J.H and Connors T.A. Approach to the quantitation of alkylated amino acids in hemoglobin by gas chromatography mass spectrometry, Biomed. Mass Spectrom. **7:** 41-46, 1980.

Farmer P.B., Bailey E., Naylor S., Anderson D., Brooks A., Cushnir J., Lamb J.H., Sepai O. and Tang Y.C. Identification of endogenous electrophiles by means of mass spectrometric determination of protein and DNA adducts, Environ. Health Perspect. **99:** 19-24, 1993.

Fidder A., Moes G.W.H., Scheffer A.G., van der Schans G.P., Baan R.A., deJong L.P. A. and Benschop H.P. Synthesis, characterization and quantitation of the major adducts formed between sulphur mustard and DNA of calf thymus and human blood, Chem. Res. Toxicol. **7:** 199-204, 1994.

Fidder A., Noort D., DeJong A.L., Trap H.C., deJong L.P.A. and Benschop H.P. Monitoring of in-vitro and in-vivo exposure to sulfur mustard by GC/MS determination of the N-terminal valine adduct in hemoglobin after a modified Edman degradation, Chem. Res. Toxicol. **9:** 788, 1996.

Osterman-Golkar S., Ehrenberg L., Segerback D. and Hallstrom I. Evaluation of genetic risks of alkylating II. Hemoglobin as dose monitor, Mutat. Res. **34:** 1-10, 1976.

Osterman-Golkar S., MacNeela J.P., Turner M.J., Walker V.E., Swenberg J.A., Sumner S.J., Youtsey N. and Fennell T.R. Monitoring exposure to acrylonitrile using adducts with N-terminal valine in hemoglobin, Carcinogenesis. **15:** 2701-2707, 1994.

Goodlad G.A.J. Technical Paper No. 575, Chemical Defence Establishment, Porton, Wilts., DTIC No. AD-123 274, 1956.

Hambrook J.L., Howells D.J. and Schock C. Biological fate of sulfur mustard [1,1'-thiobis(2-chloroethane)] : Uptake, distribution and retention of S-35 in skin and in blood after cutaneous application of ^{35}S-sulfur mustard in rat and comparison with human in-vitro, Xenobiotica. **23:** 537-561, 1993.

Hartwell J.L. Reaction of bis(2-chloroethyl) sulfide and some of its derivatives with proteins and amino acids, J. Nat. Cancer. Inst. 319-324, 1946.

Huand H., Jemal A., David C., Barker S.A., Swenson D.H. and Means J.C. Analysis of DNA adduct, S-[2-(N7-guanyl) ethyl] glutathione by liquid chromatography mass spectrometry and liquid chromatography tandem mass spectrometry, Anal. Biochem. **265:** 139-150, 1998.

Husek P., Felt V. and Matucha M. Cyclic amino acid derivatives in gas chromatography, J. Chromatogr. **180:** 53-68, 1979.

Jones J.W. and Robins R.K. Purine nucleosides III. Methylation studies of certain naturally occurring purine nucleosides, J. Am. Chem. Soc. **81:** 193-201, 1963.

Maisonneuve A., Callebat I., Debordes L. and Coppet L. Biological fate of sulfur mustard in rat : toxicokinetics and disposition, Xenobiotica. **23:** 771-780, 1993.

Moore S., Stein W.H. and Fruton J.S. Chemical reactions of mustard gas and related compounds II. The reaction of mustard gas with carboxyl groups and amino groups of amino acids and peptides, J. Org. Chem. **11:** 675-680, 1946.

Neumann H.G. Analysis of hemoglobin as a dose monitor for alkylating and arylating agents, Arch. Toxicol. **56:** 1-6, 1984.

Noort D., Verheij E.R., Hulst A.G., deJong L.P.A. and Benschop H.P. Characterization of sulfue mustard induced modifications in human haemoglobin by liquid chromatography-tandem mass spectrometry, Chem. Res. Toxicol. **9:** 781-787, 1996.

Noort D., Hulst A.G., Trap H.C., De Jong L.P.A. and Benschop H.P. Synthesis and mass spectrometric identification of the major amino acid adducts formed between sulfur mustard and hemoglobin in human blood., Arch. Toxicol. **71:** 171-178, 1997.

Papirmeister B., Feister A.J., Robinson S.I. and Ford R.D. (Ed.) Chemistry of sulfur mustard, in Medical defence against mustard gas: Toxic mechanisms and pharmacological implications, CRC Press, Boca Raton, FL, USA, 1991.

Rao M.K., Bhadury P.S., Sharma M., Dangi R.S., Bhaskar A.S.B., Raza S.K. and Jaiswal D.K. A facile methodology for the synthesis and detection of N7-guanine adduct of sulfur mustard as biomarker, Can. J. Chem. **80:** 504-509, 2002.

Rao M.K., Sharma M., Raza S.K. and Jaiswal D.K. Synthesis, characterization and mass spectrometric analysis of cysteine and valine adducts of sulfur mustard, Phosphorus, Sulfur and Silicon. **178:** 559-566, 2003.

Roberts J.J. and Warwick G.P. Mode of action of alkylating agents II. Studies on metabolism of Myleram. Reactions of Myleram with some naturally occurring thiols in-vivo, Biochem. Pharmacol. **6:** 205-216, 1961.

Sandelowsky I., Simon G.A., Bel P., Barak R. and Vincze A. N'-(2-hydroxyethylthioethyl)-4-methylimidazole (4-met-1-imid-thiodiglycol) in plasma and urine : A novel metabolite following direct exposure to sulfur mustard, Arch. Toxicol. **66:** 296-297, 1992.

Shuker D.E.G., Bailey E., Gorf S.M., Lamb J. and Fraser P.B. Determination of N-7-[2H3] methylguanine in rat urine by gas chromatography mass spectrometry following administration of trideuteromethylthylating agents or precursors, Anal. Biochem. **140:** 270-275, 1984.

Shuker D.E.G. and Farmer P.B. Relevance of urinary DNA adducts as markers of carcinogen exposure, Chem. Res. Toxicol. **5:** 450-460, 1992.

Shuker D.E.G., Prevost V., Friesen M.D., Lin D., Ohshima H. and Bartsch H. Urinary markers for measuring exposure to endogenous and exogenous alkylating agents and precursors, Environ. Health. Perspect. **99:** 33-37, 1993.

Skipper P.L. and Tannenbaum S.R. Protein adducts in the molecular dosimetry of chemical carcinogens, Carcinogenesis. **11:** 507-518, 1990.

Stein W.H. "Chemical reactions of Sulphur and Nitrogen mustards"- Summary Technical Report: National Defence Research Committee of the Office of Scientific Research and Development, Washington, D.C., DTIC No.AD-234249, 1946.

Stein W.H. and Fruton J.S. Chemical reactions of mustard gas and related compounds IV. Chemical reaction of â -chloroethyl- â'-hydroxyethylsulfide, J. Org. Chem. **11:** 686-691, 1946.

Tannenbaum S.R. and Skipper P.L. Quantitative analysis of hemoglobin-xenobiotica adducts, Methods Enzymol. **231:** 625-632, 1994.

UN Security Council Report, Mission dispatched by the Secretary-General to investigate allegations of the use of chemical weapons in the conflict between the Islamic Republic of Iran and Iraq. Report No. S/17911, 1986.

van der Schans G.P., Scheffer A.G., Mars-Groenedijk R.H., Fidder A., Benschop H.P. and Baan R.A. Immunochemical detection of adduct of sulfur mustard to DNA of calf thymus and human white blood cells, Chem. Res. Toxicol. **7:** 408-413, 1994.

Walker I.G. and Watson W.J. The reaction of mustard gas with purines and pyrimidines, Can. J. Biochem. Physiol. **39:** 377-393, 1961.

Wheeler G.P. Studies related to mechanism of action of cytotoxic agents, Cancer Res., **22:** 651-688, 1962.

Wogan G.N. Markers of exposure to carcinogens, Environ. Health Perspect., **81:** 9-17, 1989.

Pharmacological Perspectives of Toxic Chemicals and Their Antidotes
Editors: S J S Flora, J A Romano, S I Baskin and K Sekhar

Bacterial Enzymes—Potential Applications for Personnel/Casualty Decontamination Against G, V, and HD Chemical Agents

Tu-chen Cheng, Steven P. Harvey, Joseph J. DeFrank, Ilona Petrikovics, and Vipin K. Rastogi*

Biotechnology Team, US Army-ECBC, AMSRD-ECB-RT-BT
APG, MD 21010, USA
*Corresponding author

INTRODUCTION

Since World War II, a diverse number of neurotoxic organophosphate compounds (OPs) have been synthesized for use as agricultural pesticides or chemical warfare agents (CWA). Over the years, a large number of phosphotriesters, phosphothiol esters, and phosphorothioates (such as parathion, diazinon, and malathion) have been used to control insects and animal pests. Some of the phosphofluoridates and phosphothioates (such as soman, sarin, and VX) are extremely toxic cholinergic inhibitors, and have been stockpiled as CWA. The CWA are broadly classified into G-type (sarin, tabun, and soman) and the V-type (VX and Russian VX). The G-agents are sometimes mixed with thickening agents that results in increased environmental persistence. The V-agents are much more persistent in nature and are known to be some of the most toxic synthetic chemicals. The respiratory inhalation or dermal absorption of various OP neurotoxins in dosages ranging from 5-100 mg can kill a human within minutes. The principal toxic effect of OPs is exerted on the central and parasympathetic nervous system via inhibition of B-esterases, acetylcholinesterase (AChE) and butyrylcholinesterase (BChE) activity. This results in the accumulation of acetylcholine (ACh) at vital cholinergic, muscarinic, and nicotinic sites. Traditional clinical treatment following OP poisoning includes artificial respiration and repeated intravenous injection (2-4 mg) of atropine sulfate at 5- 10-min intervals accompanied by the injection of pralidoxime chloride (1 mg). Although these conventional regimens have been demonstrated to show increased survival in model animals, they did not prevent post-exposure symptoms such as convulsions. In recent years, enzymes have been used for providing enhanced protection against CWA.

Alternative hydrolytic enzymes, bacterial organophosphorus hydrolase (OPH) and organophosphorus acid anhydrolase (OPAA) have been extensively developed in past few years. The OPH enzyme was originally isolated from soil bacteria, *Pseudomonas* and *Flavobacterium* (McDaniel et al., 1988; Mulbry and Karns, 1989; Serder, et al., 1989) and the OPAA enzyme was identified from marine halophilic bacteria, *Alteromonas* (DeFrank and Cheng, 1991; Cheng et al., 1993; Cheng et al., 1997). These enzymes are capable of hydrolyzing a variety of CW neurotoxins, including G-type, V-type, pesticides, and diisopropylfluorophosphate (DFP). Cheng et al. (1997-99, and 2000), have cloned the OPAA encoding gene in a high-expression vector and developed a recombinant cell line derived from *Escherichia coli*. Optimization of growth conditions and purification protocol has resulted in the production of soluble recombinant OPAA, accounting for 50% of the total cell protein. The OPH encoding gene has been also been expressed in *E. coli* (McDaniel et al., 1988; Mulbry and Karns, 1989, and Rastogi, unpublished results). However, the OPH-expression levels are considerably lower (up to 5% of total cell protein). Both enzymes are now routinely prepared in large quantity. Until recently, the major application of these enzymes was targeted for the development of an environmentally safe decontamination solution for CW neurotoxins. In past three years, a wider range of application for OPH and OPAA enzymes has been developed, including their use in fire-fighting foam solution, biosensors (Simonian et al., 2002), surface decon, and *in vivo* protection against OP toxicity (see Figure 1). In this presentation, data supporting potential applications of OPH and OPAA enzymes for personnel/casualty decontamination are summarized.

RESULTS AND DISCUSSION

Chemical Nerve Agent Degrading Enzymes, OPH and OPAA

Recently, the presence of the OPH encoding gene (other than *Pseudomonas* and *Flavobacterium*) has been reported in other bacterial strains, *Pseudomonas monteilli* strain C11 and *Agrobacterium radiobacter* P230 (Horne et al., 2002a, b). The OPH (also known as parathion hydrolase or phosphotriesterase) has optimal activity against a variety of organophosphorus pesticides, in addition to its activity against CWA. The

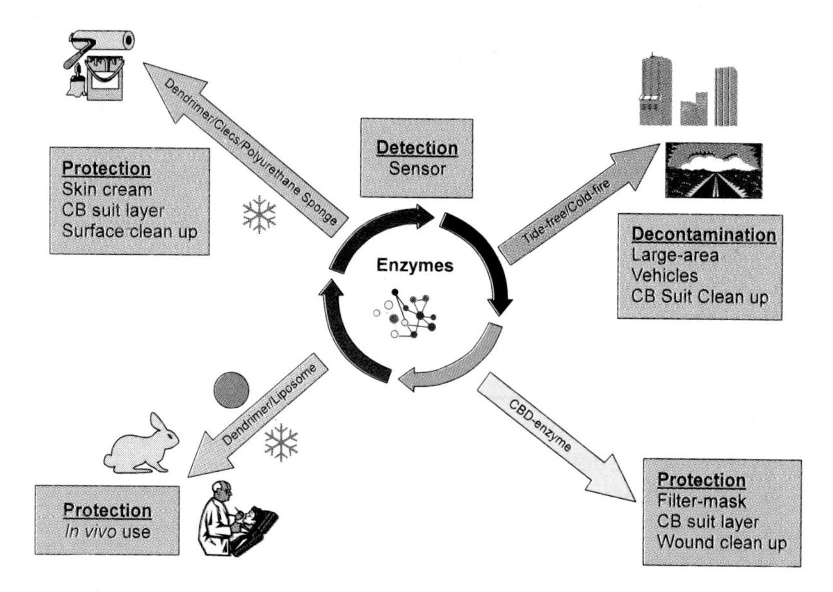

Fig. 1 Wide-range Applications of Bacterial OPH and OPAA Enzymes

three-dimensional crystal structure of OPH has revealed the native enzyme to be a homodimer with two Zn^{2+} ions per subunit. The Co^{2+} substituted enzyme has greater activity on CW agents and substrates with P-F and P-S bonds (Omburo et al., 1992). Although several orders of magnitude slower than its paraoxon activity, the V-agent activity is significant and has been improved by an order of magnitude through site-directed mutagenesis carried out recently by several laboratories (our unpublished results and personal communication from Frank Raushel & Jim Wild, Texas A&M University, TX, USA). Unlike OPH, OPAA has its best activity against G-type nerve agents and negligible activity against pesticides. The enzyme can be freeze-dried and survive for many years at room temperature with no loss of activity. From the amino acid sequence of OPAA and functional studies on a variety of dipeptides, it was identified as an X-Pro dipeptidase (or prolidase, EC 3.4.13.9) with no relationship to phosphorus metabolism. Kinetic studies have shown that VX inhibits the OPAA enzyme in a manner best modeled as competitive, strongly suggesting that VX may be interacting with the active site (Harvey et al., 2000).

Since catalytic, OPH and OPAA are highly efficient and can detoxify many times their own weight of agent in seconds or minutes. The biological properties of OPH and OPAA and their substrate activity are shown in Table 1.

Formulation of Enzymes for Personnel Protection

Both OPH and OPAA retain full catalytic activity in a dried powder form, which can be added to any water-based formulation or solution available to

Table 1 Nerve Agent Degrading Enzymes, OPH and OPAA

Enzyme	*Original Source*	*Mol. Wt. (kDa)*	*Metal Requirement*	*Agent Activity*
Organophosphorus Hydrolase (OPH) or Phosphotriesterase (PTE)	*Pseudomonas diminuta* & *Flavobacterium*)	72	Co^{++}	DFP > GF ˇ GB > GD > VX
Organophosphorus Acid Anhydrolase (OPAA)	*Alteromonas* sp. JD6.5	58	Mn^{++}	GD > GF ˇ DFP > GB > GA

the user. Previously, both OPH and OPAA have been shown to function in a variety of water-based systems such as fire-fighting foams and sprays, aqueous degreasers, odor cleaners, aircraft de-icers, and commercial laundry detergents (Cheng et al., 1998). In addition, both enzymes retain significant enzymatic activity in a skin lotion, Protect-All™ developed by J.G. Worldwide Medical, Rockway, NJ. This lotion contains plant-extracts and a few naturally derived biosurfactants, which are biodegradable and have superior wound-healing capability. The stability of OPAA enzyme was tested in the presence of 80-90% 'Protect-All' solution and lotion. As seen Figure 2 below, the enzyme retained over 90% activity even after a week. Interestingly, after a short-time exposure, i.e. for up to an hour, a stimulation over the control sample was observed. A similar retention of OPH activity was noted in Protect-All solution and lotion (results not shown). This result is strongly suggestive of the fact that a CWA decontaminating safe surface/wound lotion affording protection against these neurotoxic agents can be developed with use of powder formulation of the two enzymes, OPH and OPAA.

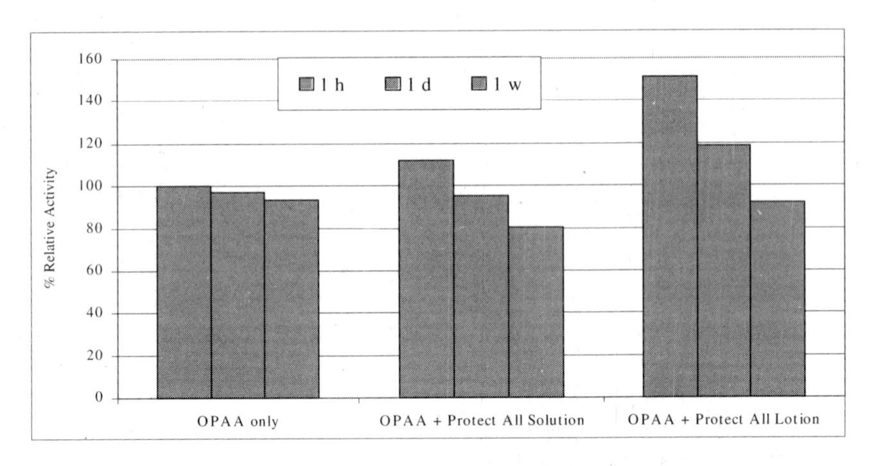

Fig. 2 Stability of OPAA Enzyme in 'Protect-All' Solution and Lotion

Recently, we have tested the compatibility of OPH and OPAA in EcoTru, a surface disinfectant/decontaminant cleaner developed by EnviroSystems Inc. (San Jose, CA). EcoTru is a multi-purpose, broad spectrum, highly effective disinfectant and cleaner with antimicrobial properties. It is effective in disinfecting causative agents of TB, HIV, Hepatitis. In addition, it is also effective against numerous Gram-positive and Gram-negative bacterial strains, viral agents, and fugal organisms. As shown in Figure 3, in the early tests, both OPH and OPAA activities were greatly reduced in the presence of EcoTru. Loss of activity could be attributed to emulsifying and non-polar nature of EcoTru solution. Over one-third of the activity was restored simply by the addition of ammonium carbonate to EcoTru. In our previous studies, we have observed stabilization of OPH and OPAA through nano-encapsulation of the enzyme by use of non-toxic water soluble dendritic polymers (courtesy Ray Yin, Nanopore Inc., MD, USA). The dendritic polymers were tested in this series of experiments. Encapsulation of enzymes in C18D polymer under the same condition (in the presence of EcoTru and ammonium carbonate), restored ~93 and 84 % of the OPAA and OPH activity, respectively. The results demonstrate a potential for development of a catalyst-based dual-use CBW decon formulation with EcoTru.

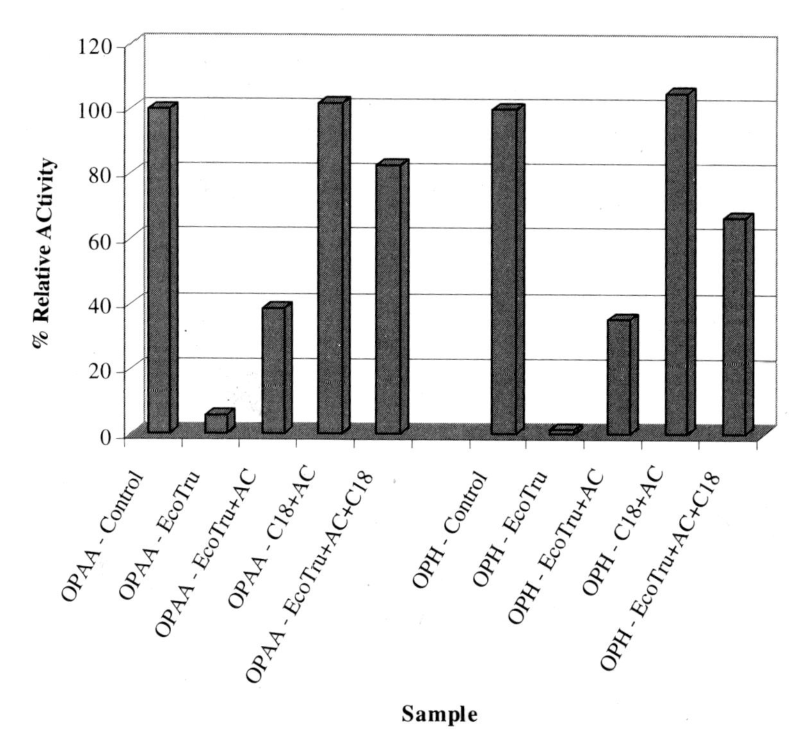

Fig. 3 Effect of Eco Tru and Nanoencapsulation on Enzyme Activity
AC-–ammonium carbonate, C18—dendritic polymer

In vivo Protection Using Liposome-conjugated OPAA.

The cholinergic effect mediated by OP neurotoxins is via irreversible phosphorylation of the esteratic site of AChE. A key step to reducing the toxic effects is through reactivation of cholinesterases with the use of pralidoxime (2-PAM) and atropine. In the last decade or so, experiments designed to address the post-exposure toxicity of OP neurotoxins involved sequestering of OPs in blood by the introduction of purified AChE and BChE enzymes (Ashani et al., 1991; Doctor et al., 1991). The therapeutic pretreatment of mice or rats with human BChE has been shown to increase survival and to prevent soman-induced cognitive impairment (Raveh et al., 1993; Brandeis et al., 1993). Neurotoxic protection was conferred only when the CW concentration was reduced to a level below the median lethal dose. In contrast, similar introduction of purified OPH in the blood was shown to hydrolyze OPs neurotoxins to products that did not inhibit AChE activity (Ashani et al., 1991; Raveh et al., 1992). The potential uses of free bacterial enzyme preparations are limited due largely to unfavorable physiological fate and potential immunological reactions. One approach for partly overcoming these disadvantages is to encapsulate the enzyme within a bioprotective environment, i.e. sterically stabilized long-circulating liposomes. Such liposomes have been extensively used for their potential use as drug carriers (Allen, 1994; Papahadjopoulos et al., 1991).

The OPH (5-10 units) and OPAA (20-30 units) encapsulated in sterically stabilized liposomes were injected into male Balb/C mice (Charles River Breeding Lab, Wilmington, MA) one hour prior to challenging the animals with paraoxon or DFP (diisopropylfluorophosphate). No gross toxic or immunologic effects were noted in mice receiving liposomes containing either 2-PAM and/or enzyme (Petrikovics et al., 1999; 2000). As seen in Table 2, enzymes injected within liposomes, as evident by LD_{50} values, afford increased protection to mice.

Development of Immobilized Enzymes for Surface Clean-up.

Enzymes are inherently sensitive to organic environments and extreme temperature fluctuations. However, a number of approaches generate stable biocatalysts. In elegant experiments reported by Gill and Ballesteros (2000) stabilization of OPH was achieved through use of nano-composite protein-silicone polymers. The resulting complex was found to be a robust and highly active and versatile biocatalyst for both liquid and gas-phase detoxification of OP neurotoxins. Interestingly, this material could also be fabricated as sheets, thick films or macroporous foams.

Immobilization of OPAA has also been achieved through a fusion protein with cellulose-binding domain (CBD). This recombinant enzyme was stable for a 3-month period and was found to bind to cheesecloth better than processed cotton (Table 3). After repeated use, the CBD-OPAA bound to

Table 2 In vivo Protection of Mice Using Steric Liposomes Containing OPH or OPAA (adapted from Petrikovics et al., 1999; 2000)

Experiment	*Treatment*[1]	*LD_{50} (mg/Kg)*[2]
1	Control	1-4
2	Atropine	2-5
3	2-PAM	4-7
4	Atropine + 2-PAM	29-55
4	SL-OPAA	9.6
5	SL-OPAA + Atropine	18.9
6	SL-OPAA + 2-PAM	21.1
7	SL-OPAA + Atropine + 2-PAM	98.6
8	SL-OPH	125
9	SL-OPH + Atropine	540
10	SL-OPH + 2-PAM	550
11	SL-OPH + Atropine + 2-PAM	920

1. For mice given OPAA, DFP (3-120 mg/kg; treatments 4-7), and for mice given OPH, paraoxon (0.6-1200 mg/kg; treatments 8-11) was given sub-cutaneously. Atropine sulfate (10mg/kg) and 2-PAM-Cl (90 mg/kg) were given intraperitoneally. SL, steric liposomes containing OPH or OPAA were given intravenously through tail 1 hour prior to the agent surrogate challenge.
2. Each LD_{50} was obtained from 5 or more graded doses of DFP/paraoxon to 5 or more groups of 6-8 mice.

Table 3 Stable Binding of CBD-OPAA Fusion Protein to Fabric

	% Relative Activity[1]	
Number of Washings	*Cheesecloth*	*Untreated Cotton*
1	100	100
2	98	47
3	93	20
4	90	18
10	80	12

[1]Relative activity was determined using DFP as substrate.

cheesecloth was found to retain 70% of the initial activity compared to only 14% of that to processed cotton. The CBD-OPAA is quite stable even after repeated washing. Analogous experiments with CBD-OPH (Richins et al., 2000; Rastogi et al., unpublished results) also show stabilization of paraoxonase following binding to different cellulose matrices. Self-decontaminating cellulose-based fabric potentially can be used for CB suit/mask and for generation of cotton swabs containing biocatalysts for decontaminating the CW neurotoxins on skin and wound surface.

Recently, in a collaborative effort with Dr. Alok Singh (Naval Research Lab, Washington, DC, USA; Lee et al., 2003, in press), immobilization of OPH and OPAA in polyelectrolyte multilayers deposited on glass beads

(30-50-mm) was achieved. The immobilized enzyme sustained catalytic activity over three months period under ambient conditions. The OPH multilayer consists of two priming layers, branched poly (ethyleneimine) and polystyrene sulfonate, followed by alternating five layers of both components. Such preparations of the immobilized enzyme have applications in designing of improved filter and mask – the first line of defense against CW threat material. Currently available mask and filter afford only mechanical barrier, but the new generation products (in addition to providing mechanical barrier) are expected to provide added safety by detoxifying the CW threat materials.

References

Allen T.M. Long circulating (sterically stabilized) liposomes for targeted drug delivery. Adv. Drug Del. Rev. **13:** 285-309, 1994.

Ashani Y., Shapiro S., Levy D. and Wolfe A.D. Butyrylcholinesterase and acetylcholesterase prophylaxis against soman poisoning in mice. Biochem. Pharmacol. **41:** 37-41, 1991.

Brandeis R., Raveh L., Grunwald J., Cohen E. and Ashani Y. Prevention of soman-induced cognitive deficits by pretreatment with human butyrylcholinesterase in rats. Pharmacol. Biochem. And Behav., **46:** 889-96, 1993.

Cheng Tu-c., Harvey S.P. and Stroup A.N. Purification of and properties of a highly active organophosphorus acid anhydrolase from *Alteromonas undina*. Appl. Environ. Microbiol. **59:** 3138-40, 1993.

Cheng Tu-c., Liu L., Wang B., Wu J.J., DeFrank J.J., Anderson D.M., Rastogi V.K. Nucleotide sequence of a gene encoding an organophosphorus nerve agent degrading enzyme from *Altermonas haloplanktis*. J. Ind. Microbiol. **18:** 49-55, 1997.

Cheng Tu-c., Rastogi V.K., DeFrank J.J. and Sawiris G.P. G-type nerve agent decontamination by *Alteromonas* prolidase. Enzyme Engg. XIV: 253-58, 1998.

Cheng Tu-c., DeFrank J.J. and Rastogi V.K. *Alteromonas* prolidase for organophosphorus G-agent decontamination. Chem.-Bio. Interactions **119/20:** 455-62, 1999.

Cheng Tu-c., DeFrank J.J. and Rastogi V.K. Wide-range application of *Alteromonas* prolidase for decontamination of G-type chemical nerve agents. CBMTS III. (in press), 2000.

DeFrank J.J. and Cheng Tu-c. Purification and properties of an organophosphorus acid anhydrolase from a halophilic bacterial isolate Jour. Bacterial. **173:** 1938-43, 1991.

Doctor B.P., Raveh L., Wolfe A.D., Maxwell D.M. and Ashani Y. Enzymes as pretreatmen for organophosphate toxicity. Neurosci. Biobehav. Rev. 15: 123-28, 1991.

Furlong C., Richter R., Chapline C. and Crabb J. Purification of rabbit and human serum paraoxonase. Biochemistry **30:** 10133-40, 1991.

Gill I. and Ballesterose A. Degradation of organophosphorus nerve agents by enzyme-polymer nanocomposites: efficient biocatalytic materials for personal protection and large-scale detoxification. Biotechnol. Bioeng. **70:** 400-10, 2000.

Horne I., Harcourt R.L., Sutherland T.D., Russell R.J. and Oakeshott J.G. Isolation of a *Pseudomonas monteilli* strain with a novel phosphotriesterase. FEMS Microbiol. Lett. **206:** 51-55, 2002a.

Horne I., Sutherland T.D., Harcourt R.L., Russell R.J. and Oakeshott J.G. Identification of an opd (organophosphate degradation) gene in *Agrobacterium* isolate. Appl. Environ. Microbiol. **68:** 3371-3376, 2002b.

Li W.F., Costa L.G., Richter R.J., Hagen T., Shih D.M., Tward A., Lusis A.J. and Furlong C.E. Catalytic efficiency determines the *in-vivo* efficacy of PON1 for detoxifying organophosphorus compounds. Pharmacogenetics **10:** 767-79, 2000.

McDaniel C.S., Harper L. and Wild J. Cloning and sequencing of a plasmid-borne (opd) encoding a phosphotriesterase. Jour. Bacteriol. **170:** 2306-11, 1988.

Mulbry W. and Karns J. Parathion hydrolase specified by the *Flavobacterium opd* gene. Jour. Bacteriol. **171:** 6740-46, 1989.

Omburo G., Mullins L., Raushel F. Structural characterization of the zinc binding of bacterial phosphotriesterase. J. Biol. Chem. **267:** 13278-13283, 1992.

Papahadjopoulos D., Allen T.M., Gabizon A., Mayhew K., Matthay K., Huang S.K., Lee K.D., Woodle M.C., Lasic D.D., Redemann C. and Martin F.J. Sterically stabilized liposomes: Improvements in pharmacokinetics and antitumor therapeutic efficacy. PNAS (USA) **88:** 11460-64, 1991.

Petrikovics I., Hong K., Omburo G., Hu Q.U., Pei L., McGuinn W.D., Sylvester D., Tamulinas C., Papahadjopoulos D., Jaszberenyi J.C. and Way J.L. Antagonism of paraoxon intoxication by recombinant phosphotriesterase encapsulated within sterically stabilized liposomes. Tox. and App. Pharmacol. **156:** 56-63, 1999.

Petrikovics, I., Cheng, Tu-c., Papahadjopoulos, D., Hong, K., Yin, R., DeFrank, J.J., Jaing, J., Song, Z.H., McGuinn, W.D., Sylvester, Pei, L., Madec, J., Tamulinas, C., Jaszberenyi, J.C. Barcza, T. and Way, J.L. 2000. Long circulating liposomes encapsulating organophosphorus acid anhydrolase in diisopropylfluorophosphate antagonism. Tox. Sciences **57:** 16-21.

Raveh L., Segall Y., Leader H., Rothschild N., Levanon D., Henis Y. and Ashani Y. Protection against Tabun toxicity in mice by prophylaxis with an enzyme hydrolyzing organophosphate esters. Biochem. Pharmacol. **44:** 397-400, 1992.

Raveh L., Grunwald J., Marcus D., Papier Y., Cohen E. and Ashani Y. Human butyrylcholinesterase as a general prophylactic antidote for nerve agent toxicity: In vitro and in vivo. quantitative esters. Biochem. Pharmocol. **45:** 2465-2474, 1993.

Richins R.D., Mulchandani A. and Chen W. Expression, immobilization, and enzymatic characterization of cellulose-binding domain-organophosphorus hydrolase fusion enzyme. Biotech. Bioeng. **69:** 591-596, 2000.

Serder C., Murdock D. and Rohde M. Parathion hydrolase gene from Pseudomonas diminuta MG: Subcloning, complete nucleotide sequence, and expression of the mature portion of the enzyme in *Escherichia coli*. Bio/Technol. **7:** 1151-55, 1989.

Simonian A.L., Grimsley J.K., Flounders A.W., Schoeniger J.S., Cheng T.C., DeFrank J.J. and Wild J.R. An enzyme-based biosensor for the direct detection of G-type chemical warfare agent. Biosensor (in press), 2002.

Pharmacological Perspectives of Toxic Chemicals and Their Antidotes
Editors: S J S Flora, J A Romano, S I Baskin and K Sekhar

Toxicity Evaluation of Thiodiglycol

Gunda Reddy, Michael A. Major, and Glenn J. Leach
Directorate of Toxicology, U.S. Army Center for Health Promotion and Preventive Medicine, Aberdeen Providing Ground
MD 21010 USA

INTRODUCTION

Under multinational agreements, the United States and other countries are demilitarizing their stockpiles of chemical warfare agents, including sulfur mustard (2, 2'-dichlorodiethyl sulfide, HD). Sulfur mustard undergoes hydrolysis to form various products such as thiodiglycol (TG), 1,4-oxathiane, 1,4-dithiane, 2-vinylthionethanol and mustard chlorohydrin in biological or environmental systems (Rosenblatt et al., 1996). Thiodiglycol has been detected as a contaminant of soil and water at certain Army installations. It has been proposed to neutralize HD through a hydrolysis process to convert it to a biodegradable compound, TG (Lee et al., 1996). Thiodiglycol is a precursor in the production of HD and it is considered as a "Schedule 2" compound (chemicals with low to moderate commercial use) within the terms of the Chemical Weapons Convention treaty (Ember, 1993, 1996). It is prepared from ethylene oxide and hydrogen sulfide. It is a water-soluble liquid of low vapor pressure and has industrial use as a solvent for vat, basic and acid dyestuffs (Von Bramer and Davis, 1981). The toxicity data on TG are limited. Therefore the U.S. Army Center for Health Promotion and Preventive Medicine (USACHPPM) conducted several studies to develop the toxicity data base for the evaluation of environmental and health effects. In this paper we summarize the toxicity information on TG that we developed as well as a proposed oral Reference Dose (RfD) for the U.S. Environmental Protection Agency (USEPA) for the inclusion in Integrated Risk Information System (IRIS).

Chemistry

Chemical Name.	Thiodiglycol
Molecular Formula.	$C_4H_{10}O_2S$
CAS Number.	111-48-8
Synonyms.	Thiodiethylene glycol, 2,2'-thiodiethanol, Beta-thiodiglycol, Bis(2-hydroxyethyl) sulfide, Beta-hydroxyethyl sulfide, Glyecine A, Kromfax solvent
Molecular Weight.	122.20
Solubility.	Water-soluble
Specific Gravity.	1.221.

$$S\begin{matrix} CH_2-CH_2-OH \\ CH_2-CH_2-OH \end{matrix}$$

Animal Toxicity

Acute toxicity data on thiodiglycol are limited. Smyth et al. (1941) reported oral LD_{50} values for male Wistar rats, 6610 mg/kg, and guinea pigs of mixed sexes, 3960 mg/kg. The toxic effects reported were similar to those of other glycols, i.e., at higher doses animals displayed sluggish depressed function, digestive tract irritation, and damage to kidneys and the liver. Angerhofer et al. (1998), while studying the approximate lethal dose, administered neat thiodiglycol at a dose of 9900 mg/kg to male and female Sprague-Dawley rats by oral gavage and found no toxic effects or deaths except slight lethargic symptoms in males after 1 hr of dosing and animals recovered within 4 hr. A subcutaneous LD_{50} of 4 gm/ kg for rats and mice and an intravenous LD_{50} of 3 gm/kg for rabbits were reported (Anslow, 1948). The dermal LD_{50} value of TG for rabbits was 20 ml/kg (2440 mg/kg) (Union Carbide, 1971). Thiodiglycol produced mild irritation to skin (500 mg) and moderate irritation to eyes (500 mg) of rabbits (Carpenter and Smyth, 1946; Union Carbide, 1971).

SUBACUTE AND SUBCHRONIC TOXICITY

Recently, Angerhofer et al. (1998, 1999) conducted subacute (14-day) and subchronic (90-day) toxicity studies with thiodiglycol (99.9% pure) in male and female Sprague-Dawley rats. In the 14-day oral toxicity study the dose levels were selected on the basis of the approximate lethal dose. The doses used were 0 (control), 157, 313, 625, 1250, 2500, 5000 and 9999 mg/kg/day. Rats were dosed orally with neat thiodiglycol 5 days per week for two weeks using a stainless steel gavage needle. During the study, food consumption and body weights and any clinical signs were recorded. At the end of the 14-day period, rats were euthanized using CO_2 and blood samples were collected for hematology and clinical chemistry. Gross necropsies were performed and various organs were removed at necropsy for weighing. Thiodiglycol at 9999 mg/kg produced 66% (4/6) and 83% (5/6) mortality in males and females respectively within 1 to 3 days of dosing. Clinical signs observed were lethargy followed by death. Male and female body weights were decreased at 14 days in the high dose 9999 mg kg/day group. Absolute kidney weights were higher than controls in the 5000 and 9999 mg/kg/day groups. In males kidney/body weight ratio in the 5000 and 9999 mg/kg/day dose group and kidney/brain weight ratio in the 9999 mg/kg/day group were significantly higher than in controls. In females these kidney/body and kidney/brain weight ratios were higher but not significant in 2500, 5000 and 9999 mg/kg/day dose groups. There were no meaningful changes in hematological and clinical parameters between treated and control groups. No histopathology was performed on any tissues. Based on these studies, the Low Observed Adverse Effect Level (LOAEL) was 5000 mg/kg/day in the 14-day study, and this dose was selected as the highest dose for the 90-day study.

A 90-day oral toxicity study of thiodiglycol was conducted in male and female Sprague-Dawley rats. Rats were randomly distributed into 4 groups consisting of 10 males and 10 females for each group. The dose levels were 50, 500 and 5000 mg/kg/day and a control with no chemical treatment. Rats were gavaged with neat test compound using a stainless steel feeding needle 5 days per week (excluding weekends) for 91 to 92 days. Control rats were sham treated with an empty needle. Body weights and food consumption were recorded prior to dosing and during dosing days 1, 3, and 7 and thereafter weekly for the remainder of the study. Doses were adjusted weekly to reflect the changes in individual body weights. Animals were observed daily for any clinical signs. Ophthalmic examination was performed prior to dosing and a few days before termination in control and high (5000 mg/kg/day) dose groups. Urine samples also were collected from all rats at the end of the in-life study for routine and microscopic analysis. At the end of the study, blood samples were collected and various tissues were removed for organ weights and histopathological evaluation. There were no consistent signs of toxicity noted during daily observation. Deaths occurred during the study in control and treatment groups, but were considered as gavage treatment related. There were no significant differences in food consumption in either sex. The body weights of both sexes were significantly lower ($p < 0.05$) than control in the 5000-mg/kg/day dose group.

There were no treatment related changes in hematological and clinical chemistry parameters in rats of either sex when compared to controls. There were no gross pathological or histopathological changes in test groups or control. Significant changes were noted in males and females in absolute weights of kidneys, kidney to body weight ratio and kidney to brain weight ratio in the highest dose group, 5000 mg/kg/day. Liver, testes, and brain to body weight ratios in males and adrenal glands to body weight ratios in females of the 5000 mg/kg/day dose group were also significantly ($p<0.01$) higher when compared to control.

The NOAEL determined for oral toxicity in this study was 500 mg/kg/day and the Low Observed Adverse Effect Level (LOAEL) was determined to be 5000 mg/kg/day for the toxic effects observed in body weights and certain organ weights in both sexes.

REPRODUCTIVE TOXICITY

There are no reproductive toxicity studies reported on TG.

DEVELOPMENTAL TOXICITY

Houpt et al., 2001, studied developmental toxicity of thiodiglycol in Sprague-Dawley rats. An initial range-finding study was conducted to select suitable doses for the main developmental toxicity study. In the range finding study, 36 positively mated female rats were randomly placed into six groups of six animals each, and one group served as a sham negative control. The remaining five groups received neat thiodiglycol orally at dose levels of 250, 500, 1000, 2000, 5000 mg/kg/day with gavage needle during the 5th through19th day of gestation. Cesarian-sections were performed on day 20, and litters were examined. There were no consistent dose-related lesions at the necropsies. Rats receiving the high dose (5000 mg/kg/day) had lower food consumption during gestation days 5 to 9, but the body weight gains were not affected. However, the body weights of fetuses derived from females receiving 5000 mg/kg were significantly ($p<0.05$) lower than any other group's mean fetal weight.

On the basis of the range finding study, doses were selected for a subsequent definitive study. The doses were 0 (control), 430, 1290 and 3870 mg/kg/day. The results of this study showed that TG produced maternal toxicity in dams receiving 3870 mg/kg. At this dose, body weights and food consumption were negatively affected during a certain period of gestation. Fetuses derived from those dams exhibited an increase in incidence of variations when compared to controls. However the increase in variation was not significant. Fetal body weights at 3870 mg/kg were significantly ($p<0.05$) lower than control. There was no increase in anomalies in all fetuses from TG treated when compared to controls. It was concluded that TG is not teratogenic at the dose levels tested. The NOAEL for developmental oral toxicity in rats is 1290 mg/kg/day when administered during the major period of organogenesis. The LOAEL for developmental toxicity is 3870 mg/kg. The NOAEL for developmental toxicity is about two and half times higher than the 90-day oral NOAEL (500mg/kg/day) derived from the Angerhofer et al. (1998) study.

GENOTOXICITY

There is no information on genotoxicity of TG. Therefore we evaluated its genotoxic potentials in limited *in vivo* and *in vitro* assays. Stankowski (2001) conducted mutagenicity assays with *Salmonella typhimurium* tester strains,

TA98, TA100, TA1535 and TA1537 and an *Escherichia coli* tester strain (WP2uvrA). The doses used were 33.3, 100, 333, 1000, 3330 and 5000 μg per plate with and without S9 activation and with concurrent vehicle and positive controls. The results showed that TG did not produce mutagenic effects at doses up to 5000 mg per plate in any of the tester strains with and without metabolic activation. Erexson (2001) performed an in *vivo* mouse micronucleus assay with TG. In these studies, five mice were gavaged with TG in sterile water at doses of 500, 1000, and 2000 mg/kg. Clastogenic activity and/or disruption of mitotic activity in the micronuclei in mouse bone marrow were evaluated. The test article TG showed no signs of clinical toxicity and no cytotoxic effects to bone marrow. This test showed that TG is not mutagenic in the mouse bone marrow micronucleus assay under the conditions studied. Trice et al., (1997) evaluated the effects of thiodiglycol on *in vitro* chromosomal aberrations in Chinese Hamster Ovary (CHO) cells at various concentrations with and without a metabolic activation system. The test substance was dissolved in sterile water. Clastogenic activity was evaluated at 3, 4, and 5 mg/ml concentrations and showed a significant increase in the percentage of metaphase cells containing chromosomal aberrations. These induced chromosomal aberrations consisted of chromatid and chromosomal breaks and chromatid type rearrangements. These chromosomal aberrations also were observed without a metabolic activation system. Based on these results the lowest effective dose was 5mg /ml in the absence of metabolic activation and 4 mg /ml in the presence of S9- metabolic activation system. Clark and Donner (1998) conducted mammalian mutagenesis assays with thiodiglycol in cultured mouse lymphoma L5178Ytk+/- cells in the presence and absence of metabolic activation. The test substance was prepared freshly in distilled water on the day of treatment. Five concentrations: 0, 50, 158, 500, 1580 and 5000 μg/ml were tested. Thiodiglycol did not induce a significant increase in mutant frequency in the presence or absence of metabolic activation.

These studies revealed that TG is negative in three assays (Ames test, mouse lymphoma and mouse micronucleus assay) and positive *in vitro* in chromosomal effects in CHO cells. This positive response *in vitro* may not correspond to any change in the *in vivo* system. It has been reported that ethylene glycol (EG) and propylene glycol (PG) are not genotoxic (ATSDR 1997).

DERIVATION OF ORAL REFERENCE DOSE

The review of toxicity data available on TG indicates that this compound is not likely to pose a risk to human health and the environment. This compound is likely to be confined only to the area where HD will be neutralized and at disposal sites. It is a water-soluble compound and has low vapor pressure. The acute toxicity data show that LD_{50} values are high. The oral LD_{50} value ranges from was 6610 mg/kg for male rats to 3960 mg/kg for guinea pigs of mixed sexes (Smyth et al. 1941). The toxicokinetic studies showed that TG is rapidly metabolized and eliminated in 8 days. The major metabolite identified was thiodiglycol sulfoxide (Black et al.. 1993). There are no reports pertaining to reproductive toxicity of TG in animals. However considerable toxicity data are available on ethylene glycol (EG) and propylene glycol (PG). TG may not be a reproductive toxicant, as reported with other glycols (LaKind et al. 1999). Ethylene glycol is used in coolants and as de-icier on airplanes. Propylene glycol has been certified as GRAS (generally recognized as safe) by U.S. Food and Drug Administration and is used in a variety of cosmetics and foods. The oral reproductive NOAEL of EG at 0.5% (approximately 840 mg/kg/day) was observed in CD mice when administered in drinking water, with minimal maternal toxicity. The LOAEL of EG for reproductive effects was approximately 1640 mg/kg/day for mice (Lamb et al. 1985). In a three generation reproductive toxicity study in rats the NOAEL of EG was found to be 1000 mg/kg/day (DePass et al. 1986). EG does not appear to be a reproductive toxicant in rats. The NOAEL of PG for CD-1 mice in drinking water (at 5%) were 10100 mg/kg/day (Gulati et al. 1985). PG was also found to be not a reproductive toxicant. As with other glycols TG does not appear to pose reproductive hazards. The subchronic toxicity and developmental toxicity studies were conducted in compliance with Good Laboratory Practice (GLP). It has been reported that the developmental oral toxicity NOAEL for rats is 1290 mg/kg/day (Houpt et al., 2001). In a subchronic study (Angerhofer et al., 1998), the only effects observed were reduced body weight and changes in certain organ weights at the high dose, 5000 mg/kg, (the LOAEL). The NOAEL of 500 mg/kg is proposed for the derivation of the Reference Dose (RfD).

The RfD for TG is based on the NOAEL of 500 mg/kg/day from subchronic oral toxicity studies (Angerhofer et al., 1998).

The NOAEL is based on a 5-days exposure regimen per week, extrapolated to a 7-day exposure.

$$500 \text{ mg} \times 5/7 = 357 \text{ mg/kg/day}$$

$$\text{Oral chronic RfD} = \text{NOAEL/uncertainty factors}$$

$$357/1000 = 0.36 \text{ mg/kg/day} = 0.4 \text{ mg/kg/day}$$

An uncertainty factor of 1000 was used, (10 for subchronic to chronic, 10 for animal to human extrapolation and 10 for human sensitive populations).

Modifying factors: No modifying factors are used as these studies are conducted according to Health Effects testing guidelines in compliance with GLP.

Oral chronic RfD = 0.4 mg/kg/day or 400 μg/kg/day.

References

Angerhofer R.A., Michie M.W. and Leach G.J. Subchronic oral toxicity of thiodiglycol in rats. U.S. Army Center for Health Promotion and Preventive Medicine, Aberdeen Proving Ground, MD. Report Number: 6415-38-97-05-01, 1998.

Angerhofer R.A., Michie M.W. and Leach G.J. Subchronic oral toxicity of thiodiglycol in rats. Toxicologist, 48, number 1-S March, Page 318 Abstract No. 1501, 1999.

Anslow L.P., Karnofsky B.V., Jager B.V. and Smith H.W. The intravenous, subcutaneous and cutaneous toxicity of bis (beta-chloroethyl) sulfide (mustard gas) and of various derivatives. J. Pharmacol. Exp. Ther. **93:** 1-9, 1948.

ATSDR. Toxicological profile for ethylene glycol and propylene glycol. Agency for Toxic Substances and Disease Registry, U.S. Department of Health and Human Services, Public Health Service, September 1997.

Black R.M., Brewster K., Clarke R.J., Hambrook J.L., Harrison J.M. and Howells D.J. Metabolism of thiodiglycol (2,2-thiobis-ethanol): Isolation and identification of urinary metabolites following intraperitoneal administration to rats. *Xenobiotica* **23:** 473-481, 1993.

Carpenter C.P and Smyth H.F., Chemical burns of the rabbit cornea. Amer. J. Ophthalmology. **29:** 1363-1372, 1946.

Clark S.L. and Donner M.E. Mouse lymphoma mammalian mutagenesis assay. Integrated Laboratory System, Durham, N.C. Contract Number: DADO5-91-C-00018. U.S. Army Center for Health Promotion and Preventive Medicine, Aberdeen Proving Ground, MD. 1998.

DePass L.R., Woodside M.D., Maronpot R.R. and Weil C.S. Three –generation reproduction and dominant lethal mutagenesis studies of ethylene glycol in rat. Fund. Appl. Toxicol. 7: 566-582, 1986.

Ember L.R., Chemical arms treaty makes unprecedented demands of industry. *C&E News (7 June).* 7-18, 1993.

Ember L.R., Failure to ratify chemical arms pact would dampen US chemicals trade. *C&E News (29 Jan).* 19-22.1996.

Erexson L.G. In vivo mouse micronucleus assay with 2,2' –thiodiethanol. Report, Covance Study #22283-0-455OECD, Covance Laboratories Inc, Vienna, VA. 2001. Contract number: DAAD05-99-P-1522., U.S. Army Center for Health Promotion Medicine, Aberdeen Proving Ground, MD 21010.

Gulati D.K., Barnes L.H. and Welch W. Propylene glycol: Reproduction and fertility assessment in CD-1 mice when administered in drinking water (revised September 1985). Report No. NTP-85-321, National Toxicology Program. Research Triangle Park, N.C., 1985.

Houpt J.T., Crouse L.C.B. and Angerhofer R.A. Developmental toxicity of thiodiglycol in rats U.S. Army Center for Health Promotion and Preventive Medicine, Aberdeen Proving Ground, MD. Report No. 7796-52-99-04-05, 2001.

Lamb J.C., Maronpot R.R., Gulati D.K., Russell V.S., Hommel-Barnes L. and Sabharwal, P.S. Reproductive and developmental toxicity of ethylene glycol in the mouse, Toxicol. Appl. Pharmacol. **81:** 100-112, 1985.

LaKind J.S., McKenna E.A., Hubner R.P. and Tardiff R.G. A review of the comparative mammalian toxicity of ethylene glycol and propylene glycol. Cri. Reve. Toxicol. **29:** 331-365, 1999.

Lee T., Pham M.Q., Weigand W.A., Harvey S.P. and Bentley W.E. Bioreactor strategies for the treatment of growth-inhibitory waste: an analysis of thiodiglycol degradation, the main hydrolysis product of sulfur mustard. *Biotechnol. Prog.* **12:** 533-539, 1996.

Rice R.R., Donner M., Udumudi A. and Vasquez M. In vitro chromosomal aberration study in Chinese hamster ovary (CHO)cells. Integrated Laboratory System, Durham, N.C. Contract Number DADO5-91-c-00018. U.S. Army Center for Health Promotion and Preventive Medicine, Aberdeen Proving Ground, MD. 1997.

Rosenblatt D.H., Small M.J., Kimmell T.A. and Anderson A.W. Background chemistry for chemical warfare agents and decontamination processes in support of delisting waste streams at the US Army Dugway Proving Ground, Utah (ANL/EAD/TM-56). Environmental Assessment Division, Argonne National Laboratory, Argonne, IL. 1996.

Smyth H.F., Seaton J. and Fischer L. The single dose toxicity of some glycols and derivatives. J. Indust. Hyg. Toxicol. **23:** 259-268, 1941.

Stankowski L.F. Jr. Salmonella-Escherichia coli/mammalian-microsome reverse mutation assay with confirmatory assay with 2,2'-thidiethanol. Report, Covance study no. 22283-409OECD, Covance Laboratories Inc, Vienna, VA. 2001. Contract number: DAAD05-99-P-1522., U.S. Army Center for Health Promotion and Preventive Medicine, Aberdeen Proving Ground, MD 21010.

Linion Carbide. Linion Carbide Data Sheet 197; (Cited in RTECS 1992).

Von Bramer P. and Davis J.H. Glycols (Ethylene and propylene) pp. 948-949 in; Krik-Othmer Encyclopedia of chemical Technology, 3rd Ed. Vol.11, H.F. Mark, D.F. Othmer, C.G. Overberger and G.T. Seaborg (eds). John Wiley and Sons: New York, 1981.

Pharmacological Perspectives of Toxic Chemicals and Their Antidotes
Editors: S J S Flora, J A Romano, S I Baskin and K Sekhar

Insights on Cyanide Toxicity and Methods of Treatment[1]

S.I. Baskin[a*], I. Petrikovics[a], J.S. Kurche[a***], J.D. Nicholson[b], B.A. Logue[a], B.I. Maliner[c] and G.A. Rockwood[d]**
[a]Pharmacology and [c]Critical Care and [d]Drug Assessment Divisions
United States Army Medical Research Institute of Chemical Defense
Aberdeen Providing Ground, MD USA
[b]KRC, Baltimore, MD
[*]Corresponding author
[**]Senior National Research Council Fellow

INTRODUCTION

Cyanide-mediated histotoxic anoxia is thought to occur by inhibiting the enzyme cytochrome c oxidase (Warburg, 1911; Keilin, 1929). Cytochrome c oxidase is the terminal electron acceptor in the mitochondrial electron transport chain, shuttling reducing equivalents from cytochrome c to diatomic oxygen, producing water. Cyanide inhibits cytochrome c oxidase by covalently binding to the active site iron and copper of the enzyme (Thomson 1985). Inhibition of cytochrome c oxidase is thought to abolish tissue utilization of oxygen, and many symptoms of cyanide intoxication have been reported to stem from ATP depletion as a consequence of that critical step.

Until relatively recently methods for studying cyanide toxicity have focused almost exclusively on cytochrome c oxidase. Cyanide is covalently bound to the enzyme but can be dissociated (Ballantyne, 1987), forming the basis for part of the specific therapeutic regimen: cyanide scavenging.

[1]The opinion or assertions contained in this paper are the private views of the authors and are not to be construed as official or as a reflection of the view of the U.S. Army or Department of Defense

[***]This work was supported in part by an appointment to the Research Participation Program at the U.S. Army Medical Research Institute of Chemical Defense administered by the Oak Ridge Institute for Science and Education through an interagency agreement between the U.S. Department of Energy and USAMRICD.

Induction of methemoglobin formation or use of cobalamins to scavenge cyanide are well-characterized methods of detoxification (for a review, see Ballantyne B, Marrs TC, eds. Clinical and Experimental Toxicology of Cyanides. Bristol, England: Wright;1987; or Vennesland, B., Conn, E.E., Knowles, C.J., Westley, J., Wissing, F., eds. Cyanide in Biology. New York, NY: Academic Press; 1981). Briefly, methemoglobin is formed i.e. sodium nitrite or another member of the methemoglobin-forming class of compounds is administered to change the oxidation state of endogenous heme irons. Cyanide has a high affinity for oxidized hemoproteins, some of which - like methemoglobin - have a higher binding constant for cyanide than cytochrome oxidase (Solomonson, 1981). It is theorized that methemoglobin sequesters cyanide away from cytochrome oxidase, leading to cyanide detoxification. Similarly "cobalamin" and cobaltous ions are thought to act through cyanide sequestration (see Hall and Rumack, 1987). For example, hydroxycobalamin appears to protect against approximately 4 LD_{50}'s of cyanide with a $t_{1/2}$ of ~ 25 hrs.

The other part of cyanide detoxification that of thiocyanate formation deals with the detoxification and excretion of cyanide. Toxicological science owes much to the work of Chen and coworkers (Chen et al., 1933a; Chen et al., 1933b) for the development of nitrite/thiosulfate combination therapy for acute cyanide toxicity, which potently reverses the effects of cyanide and remains the standard method of cyanide antagonism in the United States today. However, advancements in cyanide research have also shown that the proposed mechanisms for cyanide detoxification by the nitrite/thiosulfate regimen may not fully account for the observed pharmacological effects of these drugs (Baskin, 1991; Way et al., 1987; Way et al., 1984a). While conventional wisdom may support the search for a better methemoglobin former or a better sulfur donor, traditional orthodoxy in this area may now be frustrating development of novel treatment alternatives, and in particular, cyanide pretreatments.

Metabolic inhibition by cyanide drives the cell into anaerobic metabolism and, in animals, lactic acid formation. Most animal cell types can tolerate only nominal levels of lactate (a notable exception are erythrocytes, Lehninger et al., 1993). Development of rational cyanide therapies involves recognizing that methemoglobin reductases, the principal enzymes responsible for the conversion of methemoglobin to reduced hemoglobin, are dependent on NADH and NADPH, and that NAD+, the electron acceptor for aerobic glycolysis, may be regenerated in this manner, thereby enhancing non-oxidative ATP production. The same may also be true for other heme oxidoreductases and other hemes that may be oxidized by nitrite. Similarly, development of rational cyanide therapies also involves recognizing that cobalt chloride, an effective but potentially toxic cyanide antidote, is also largely responsible for the induction of Hypoxia-inducible Factor, which has independently been shown to protect against cyanide (Wright et al., 2003) and may be in part responsible for the protective effects of cobalt chloride. Accordingly, it should be the goal of our chapter to supply conventional reasoning regarding cyanide therapies, but also to produce alternative thinking about mechanisms of cyanide therapy in the context of biological information available only recently.

MECHANISMS OF CYANIDE TOXICITY

Cyanide research presents a problem that is increasingly common in biochemistry and toxicology, wherein a small, nucleophilic molecule is associated with a wide range of pathogenic effects that are differentially associated with its perceived mechanism of action. Degrees of toxic effects of cyanide differ from tissue to tissue and from organism to organism. While cyanide tolerance in a given tissue is partially due to the concentration of detoxifying enzymes like rhodanese and 3-mercaptopyruvate sulfurtransferase, acute cyanide toxicity may also follow different pathways within cells, which give rise to differential cyanide resistance. In neurons, for example, NMDA receptor potentiation has been implicated, despite a long time delay, in cyanide toxicity by increasing Ca^{2+} currents, and a direct effect of cyanide on the NMDA receptor has been suggested (Sun et al., 1997; Arden et al., 1998). However, in cardiomyocytes, cyanide inhibits Ca^{2+} currents and stimulates outward K^+ currents (Baskin, 1991). Minute quantities of cyanide are endogenous (Vinnesland et al., 1982; Lundquist et al., 1988), and this has prompted some to hypothesize that cyanide may act as a neuromodulator (Borowitz et al., 1999). Cyanide has also been shown to be produced by myeloperoxidase in activated neutrophils (Lundquist et al., 1988). Furthermore, cyanide toxicity may be, at least partially, genetically regulated. In cancer cells such as rat pheochromocytoma (PC12), which can be terminally differentiated to biochemically and phenotypically resemble sympathetic neurons with administration of nerve growth factor (NGF), undifferentiated cells are highly resistant to cyanide, whereas differentiated cells are not (Mills et al., 1996).

Discerning the disparities in protein expression that underlie cyanide toxicity induced by NGF in PC12 cells would be useful in determining which gene products play a role in mediating cyanide toxicity and designing therapies around them. Indeed, development of a rational cyanide treatment depends on implementing a combined research approach that targets screening studies to known mechanistic details and at the same time uncovers new understanding of pathways of toxicity.

There is no question that the inhibition of cytochrome oxidase is toxic, nor is there any question that hydrogen cyanide (HCN) inhibits cytochrome oxidase. But the debate remains open as to whether this simple mechanism is entirely or even principally responsible for the lethal effects of cyanide at all doses in all tissues. For example, Mills et al. (1996) demonstrated that the neurons of the substantia nigra are more sensitive to the effects of cyanide than cortical neurons, even though the metabolic demands of both neurons are similar. The same group also showed that reactive oxygen species (ROS) and lipid peroxidation products are generated in the presence of cyanide. (Ardelt, 1994) Since the substantia nigra is rich in iron containing proteins, and since Fenton chemistry proceeds in the presence of iron and H_2O_2, it seems likely that cyanide-induced iron disregulation in the presence of ROS (Brazzolotto et al., 1999) may account for the relative sensitivity of the substantia nigra. Having thus qualified the discussion, we proceed to inhibition of cytochrome oxidase as a molecular target.

The oxidative phosphorylation chain is a series of protein complexes and electron carriers, which convert reducing equivalents (such as NADH or succinate) into a proton gradient across the inner membrane of mitochondria. This is a multistep process which, involves the transfer of reducing equivalents from NADH to coenzyme Q to cytochrome c to water, which is formed from molecular oxygen in the final step. Each of these electron transfer steps gives rise to what is called vectorial proton transfer. This refers to the fact that while some protons are transferred non-productively between the aqueous environment and the protein complexes or their substrates, only vectorial protons are moved from the matrix of the mitochondria into the intermembrane space.

There are technically four protein complexes in the oxidative phosphorylation chain. However, Complex II is only included for functional reasons. The other three complexes actually translocate protons from the matrix side to the cytosol as a result of electron transfer and are therefore the oxidative phosphorylation chain proper. These proton pumps are listed as numbered complexes I, III and IV, and are also known as NADH-ubiquinone reductase, dihydroubiquinone-cytochrome c oxidoreductase, and cytochrome oxidase respectively. The terminal protein complex in this chain, cytochrome oxidase, transfers four reducing equivalents (electrons) from cytochrome c to molecular oxygen, which results in two molecules of water. This reaction consumes four electrons and eight protons from the matrix side of the inner mitochondrial membrane. Four of the eight protons are consumed by the O_2 to $2H_2O$ reaction and four are translocated to the cytosol. That implies that one proton is pumped by complex IV per reducing equivalent provided. The experimentally determined number is ~.9 protons per reducing equivalent consumed. This may be due to the fact that the proton pumping mechanism of complex IV depends on the timing of proton entry into the complex or the availability of reducing equivalents. The point mutation N139D of cytochrome oxidase subunit I from Rhodobacter sphaeroides eliminates proton pumping while multiplying oxidation of cytochrome c 1.5 to 3 fold

(Pawate et al., 2002). This demonstrates that the proton translocation is not a function of the oxidation reaction, but depends upon a complex protein-based mechanism.

The proton gradient produced by these complexes is consumed by complex V (F0F1-ATP Synthase) in the creation of ATP from ADP. See Figure 1 for Mini-Map schematic representation of oxidative phosphorylation.

Cyanide inhibits cytochrome oxidase, the terminal protein in this oxidative phosphorylation chain. It binds to the active site of this enzyme and prevents oxygen from accepting electrons. At physiological pH, cyanide exists as HCN. However when bound to cytochrome oxidase, the nitrile CN crosslinks the Fe_{a3}/Cu_B binuclear center in the same way that oxygen does (Li and Palmer, 1993). Unlike oxygen, it cannot be reduced by the normal mechanism of proton addition. As a result, cyanide remains bound across the active site and may only be displaced, probably by spontaneous conversion back to HCN (Figure 2).

Cytochrome oxidase from cow heart is composed of 13 protein subunits (Figure 2), while there are sometimes fewer in other species or tissues. The catalytic activity of this complex requires at minimum the products of the mitochondrial COX1 and COX2 genes, however, other protein subunits seem to be required for proper regulation, transport and assembly of the complex. The mechanism of electron transfer begins with the binding and oxidizing of cytochrome c. The electrons received by the protein complex are then transferred to the CuA (binuclear copper) site in the COX1. Heme a in COX1 then accepts electrons from the CuA site via electron transfer within the protein. Heme a3, also in COX1, accepts electrons from heme a via the protein backbone. Heme and Cu_B taken together are called the binuclear site because they are approximately 4.8 angstroms from one another. The method by which electrons and protons are transferred in the entire complex is still under investigation and what is known is largely beyond the scope of this book (Figure 1).

Understanding cyanide toxicity does not, so far as we know, depend upon knowing all the details regarding oxygen hydrolysis or proton and electron transport in cytochrome oxidase. It does, however, depend upon the oxidation state of the iron atoms in hemes a and a_3 in cytochrome oxidase. A heme group is a modified protoporphyrin bound to an iron atom. Saying that a heme is oxidized implies that the iron is at a +3 oxidation state. Saying that it is reduced implies iron at +2. Thus, saying that cytochrome oxidase is oxidized or reduced implies the oxidation state of the heme groups. To say that its oxidation state is mixed implies that one heme is oxidized while the other is reduced.

Cyanide binds rapidly and reversibly to cytochrome oxidase undergoing turnover of protons and electrons. Stated another way, cytochrome oxidase,

ATP SYNTHESIS from ACETYL-CoA

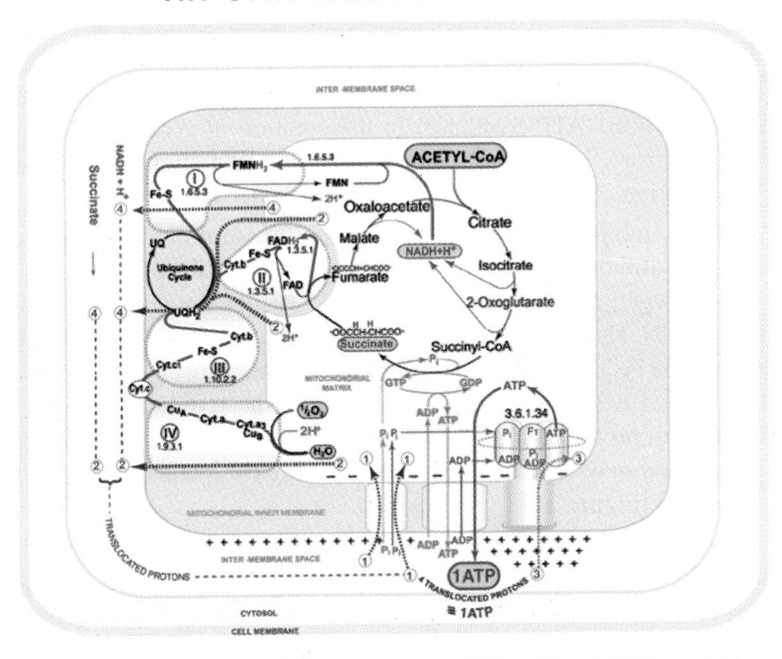

***Anaerobic* Oxidation of Acetyl-CoA in the TCA Cycle** → CO_2, NADH, and Ubiquinol (from succinate)

$CH_3COSCoA + 3NAD^+ + 1UQ$ (FAD) + GDP + P_i + $2H_2O$ → $2CO_2$ + 1HSCoA + 3NADH + $3H^+$ + $1UQH_2$ ($FADH_2$) + 1GTP

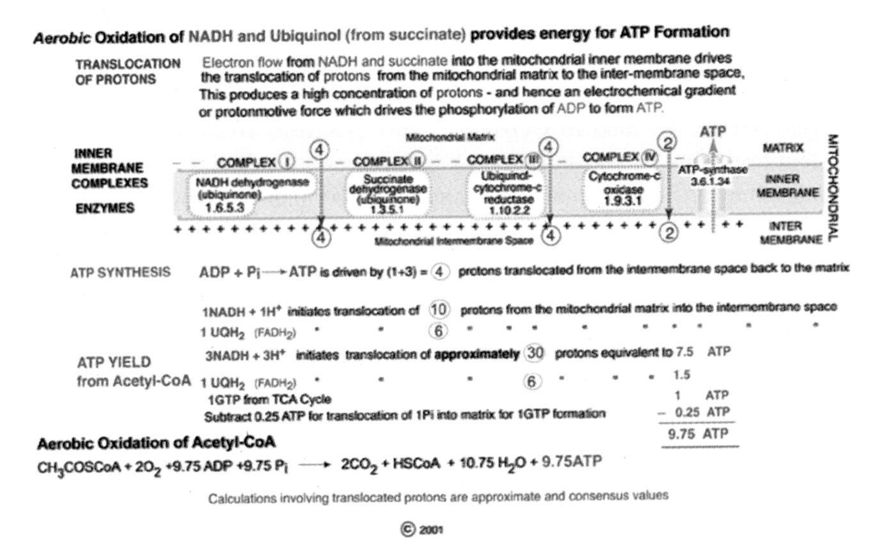

Aerobic Oxidation of Acetyl-CoA

$CH_3COSCoA + 2O_2 + 9.75 ADP + 9.75 P_i$ → $2CO_2$ + HSCoA + 10.75 H_2O + 9.75ATP

Calculations involving translocated protons are approximate and consensus values

© 2001

Fig. 1 Nicholson Mini-Map depicting proton pumping of Oxidative Phosphorylation chain and integration into the Tricarboxylic Acid Cycle. "ATP Synthesis from Acetyl-CoA" (author D. Nicholson). Copyright International Union of Biochemistry and Molecular Biology, published with permission.

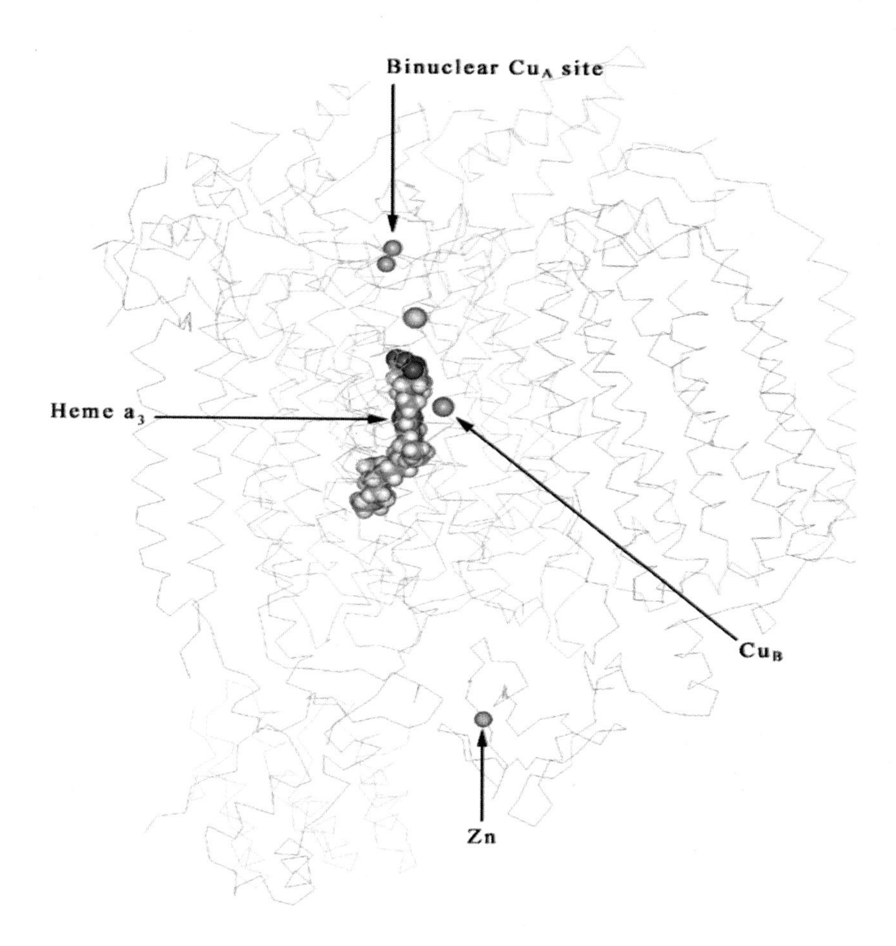

Fig. 2 Human Cytochrome C (by homology to bovine heart). Each of the 13 protein subunits is colored randomly. Heme a3, CuB, CuA, Mg, Zn depicted as CPK spheres.

which is cycled between the oxidized and reduced states, is most susceptible to cyanide inhibition. The statically oxidized form binds cyanide much more slowly, but has a much lower dissociation constant. The rate constants of onset of inhibition for reduced vs. oxidized cytochrome oxidase are 5×10^3 $M^{-1}s^{-2}$ vs. $2\ M^{-1}s^{-1}$ (Moody, 1996)

When cytochrome oxidase is extracted from the mitochondria to perform rate constant determinations, there exist two oxidized forms: one that reacts quickly with cyanide and one that is slow. This fact was not always appreciated, thus, the literature contains many inconsistencies, which may be retrospectively resolved (Moody, 1996). The slow form can be converted into the fast form by cycled reduction and reoxidization. The slow form shows a different Soret spectrum (Absorbance ~410 nm, characteristic of heme iron) than the fast form. It has been suggested that the slow form occurs when the complex undergoes a conformational change, which inserts an aspartate residue into the active site (Moody, 1996). This hypothesis is based on the fact that formate inhibition yields spectra, which are similar to the slow form of the enzyme.

The reduced or mixed valence form is faster than both oxidized form. Thus, from a treatment point of view, the fastest form is actually desirable because cyanide can be washed out of the enzyme. So long as there is a ready supply of electron donors in the form of reduced cytochrome c, the cytochrome oxidase ought to be reduced or of mixed oxidation state and therefore in its fastest form. Under pro-apoptotic conditions which release cytochrome c into the cytosol, it may be that cytochrome oxidase converts a form with a slower off rate for cyanide and becomes refractory to cyanide scavengers. Of course, the release of cytochrome c is a terminal event for many reasons.

Another item of great importance is the pH of cytosol. Cyanide rapidly lowers the pH of the cytosol to 6.9 in PC12 cell culture (Maduh et al, 1990). It has also been reported that the dissociation constant for oxidized CN-cytochrome oxidase in isolated mitochondria rises by an order of magnitude between pH 7.0 and 8.0 (Wilson et al, 1972). Thus, the clinical advice that acidosis should be treated in cyanide intoxication may be more than symptomatic treatment, it may actually reduce inhibition of cytochrome oxidase. An interesting experiment by Maduh et al, 1990 examined PC12 cells in culture media of pH 7.9. When cells in pH 7.9 media were treated with KCN, their pH rose instead of fell. Paradoxically, 1mM KCN exposure caused a pH rise of nearly .35 log units in 12 minutes while .1mM KCN induced only a ~.1 log unit rise. But 10mM KCN caused the typical fall in pH of ~.25 log units in 8 minutes. This might be explained by a pH-dependent reduction of CN inhibition which was overcome at 10mM KCN. On the other hand, alkalosis is reported to be cytotoxic via generation of superoxide (Majima et al., 1998). But alkaline pH is also reported to prevent permeability transition in mitochondria via a direct effect on the permeability transition pore complex (Djerad et al., 2001). Thus, mild, induced alkalosis with antioxidant supplementation could prove to be a treatment option .

The central nervous system is the primary target for cyanide toxicity in the body. Acute inhalation of high concentrations of cyanide provokes a brief central nervous system stimulation followed by depression, convulsion, coma, and death in humans (Bonsall, 1984) and in dogs (Haymaker et al., 1952). The effects are probably due to rapid biochemical changes in the brain, such as changes in ion flux, neurotransmitter release, and possible peroxide formation (Johnson and Isom, 1985). Cyanide is a strong nucleophile with multiple effects including release of secondary transmitters, release of catecholamines from adrenal glands and adrenergic nerves, and it inhibits antioxidant enzymes in the brain (Smith, 1996). Bitner et al. (1997) reported studies regarding the excitotoxic properties of cyanide causing hippocampal

Cyanide exposure also results in adverse effects on the heart, such as decrease in myocardial cardiac contractility and heart failure (Elliot et al 1989). The cyanide-induced heart failure can be attributed to the increase in circulating catecholamines, which stimulates the heart, and the simultaneously impaired energy metabolism. Some of the cyanide-induced decreased myocardial contractility can be attributed to the accumulation of hydrogen ion that contributes to the lack of effectiveness of intracellular calcium in activating the contractile process. The cyanide-induced activation of K_{ATP} channels and K^+ lactate co-transporter, and inhibition of Na^+/K^+ATPase result in K^+ efflux in the heart. Unlike the heart and brain, the skeletal muscle is less active and is also less sensitive to metabolic inhibition by cyanide. Kidney and liver are not considered target organs for cyanide intoxication due to the abundant amounts of rhodanese present to bio-transform cyanide to the less toxic SCN^-.

INSIGHTS ON DETOXIFICATION

The efficacy of treating cyanide intoxication with sulfur donors is well established (Baskin et al., 1999; Baskin et al., 1992; Isom and Johnson, 1987; Vinnesland et al. 1982). Much of the relevant work surrounding the formation and metabolism of sulfane sulfur with respect to cyanide was described and summarized by Westley (1981). Since that time a great deal of biochemical and molecular biological work, combined with a reevaluation of previous discoveries, has combined to shed new light on the biochemistry of sulfane sulfur metabolism. It is the goal of this section to revisit what is known about biological sulfane sulfur and to speculate on the role some recently discovered proteins may play within the context of its formation, purpose in vivo, and source.

Sulfane Sulfur

Sulfane sulfur at its simplest is sulfur-bonded sulfur. The sulfur may be in the form of what is commonly referred to as inorganic or organic persulfides, polysulfides, or elemental sulfur (Figure 3). In mammals sulfane sulfur is commonly found in the form of thiosulfates, persulfides or polysulfides, although elemental sulfur can also be found bound by serum albumin or in cofactors of some enzymes. Sulfane sulfur is labile and shares a common reactivity with various nucleophiles (Westley, 1981), among them the nucleophile of principal interest to this chapter, cyanide. Sulfane sulfur compounds are readily cyanolysable; in the presence of cyanide, persulfides, for example, will react to yield a thiol and thiocyanate, where the sulfane sulfur is transferred as an atom with essentially zero valence (Westley, 1981). This reaction forms the major mechanism for removing cyanide from the body. In humans formation of thiocyanate from persulfide or thiosulfate and

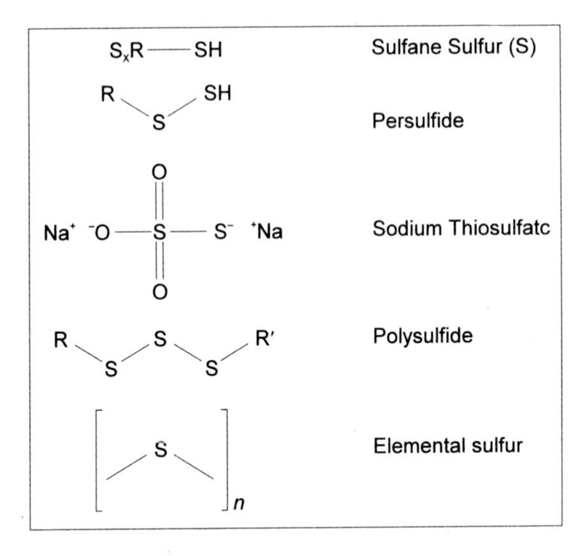

Fig. 3 Examples of molecules that contain sulfane sulfur: Persulfide (R = any side chain) and an example of a persulfide—Sodium Thiosulfate; polysulfide (R, R' = any side chains); and elemental sulfur.

cyanide substrates is largely attributed to the enzyme rhodanese, or thiosulfate sulfur transferase.

In circles of cyanide research, there has been a great deal of effort to describe other possible mechanisms of cyanide detoxification and thiocyanate formation. This effort has focused on contributors to or detractors from the sulfane sulfur pool. Westley noted activity of an enzyme he called sulfur reductase or thiosulfate reductase in prokaryotes, yeast, and higher eukaryotes, which transferred the sulfane sulfur from persulfides to thiol residues and, in a subsequent reaction with a vicinal thiol released organic sulfide forming a dithiol product. Westley predicted that thiosulfate reductase would be useful to mobilize sulfur in the formation of iron-sulfur clusters and would therefore be ubiquitously expressed across a broad array of organisms. He also predicted it may have some function in thiocyanate formation. He was able to purify thiosulfate reductase from yeast but was unable to derive it in purified form from mammals. Other authors conceded the existence of thiosulfate reductase in yeast but attributed the activity of the putative enzyme in mammals to mobilize sulfur to the combined effects of α-cystathionase and other enzymes (Cooper, 1983).

To understand the activity of the enzymes that play a role in sulfur metabolism and may play a role in cyanide detoxification, one must understand the appropriate substrates and catalytic context of their reactions. Since cyanide is produced endogenously (Vinnesland et al., 1982; Lundquist et al., 1988; Borowitz et al., 1997), some have proposed that the primary role of Rhodanese was to mediate cyanide toxicity (Volini and Alexander, 1981). However, it now appears that reactions of sulfane sulfur mediated by rhodanese and other sulfur mobilizing enzymes actually fulfill a diverse range of biological functions, including formation and regulation of iron-sulfur clusters in proteins (Vinnesland et al., 1982; Toohey, 1989).

Sulfane Sulfur and Cysteine

Figure 4 illustrates a summary of known reactions of cysteine to generate sulfane sulfur, elemental sulfur, and other reactions of relevance to cyanide metabolism. Reaction 1 leads to the pathway of methionine metabolism first catalyzed by the addition of homoserine to cysteine to form cystathionine via cystathionine a-synthase. For simplicity, reactions of methionine are largely omitted here. Other pathways involving cysteine will be discussed and in detail in following sections.

Cysteine Metabolism through Mercaptopyruvate Sulfur Transferase (MPST)

The reaction of most significance to cyanide metabolism catalyzed by MPST is the transfer of sulfur from 3-mercaptopyruvate to awaiting nucleophiles, such as cyanide. Sulfate may also be a substrate for sulfur transfer, and this is believed to be one of the primary generators of thiosulfate in vivo. MPST is located in the cytosol and mitochondria and has the capacity to improve the effectiveness of sodium nitrite therapy as much as administration of sodium thiosulfate. Sodium thiosulfate, meanwhile, potentiates the effectiveness of 3-mercaptopyruvate slightly (Way, 1983). This suggests that both pathways may be used to treat cyanide toxicity, however with varying degrees of effectiveness. The reactions of MPST are illustrated in Figure 4, number 3. The reaction using thiosulfate as a substrate is mediated by rhodanese and is depicted in Figure 4, number 4.

3-Mercaptopyruvate is formed through the transamination of cysteine in a pyridoxal phosphate-dependent manner through a classic mechanism that consumes α-ketoglutarate and yields glutamate (Figure 4, number 2). Since free ammonia is highly toxic to most terrestrial animals, transamination is a common method of shuttling ammonia from amino acid metabolism to the urea cycle where it can be safely eliminated. Glutamate may also be transaminated to glutamine, and pyruvate may be transaminated to alanine. These reactions form a major component of amino acid metabolism.

For cyanide, these reactions should be given more than just passing notice because recent evidence (Bhattacharya and Vijayaraghavan, 2002; Bhattacharya et al., 2002; Bhattacharya, 2001) has shown that oral α-ketoglutarate as well as other α-keto-acids are effective treatments for cyanide toxicity. It has been proposed that the mechanism of amelioration of cyanide

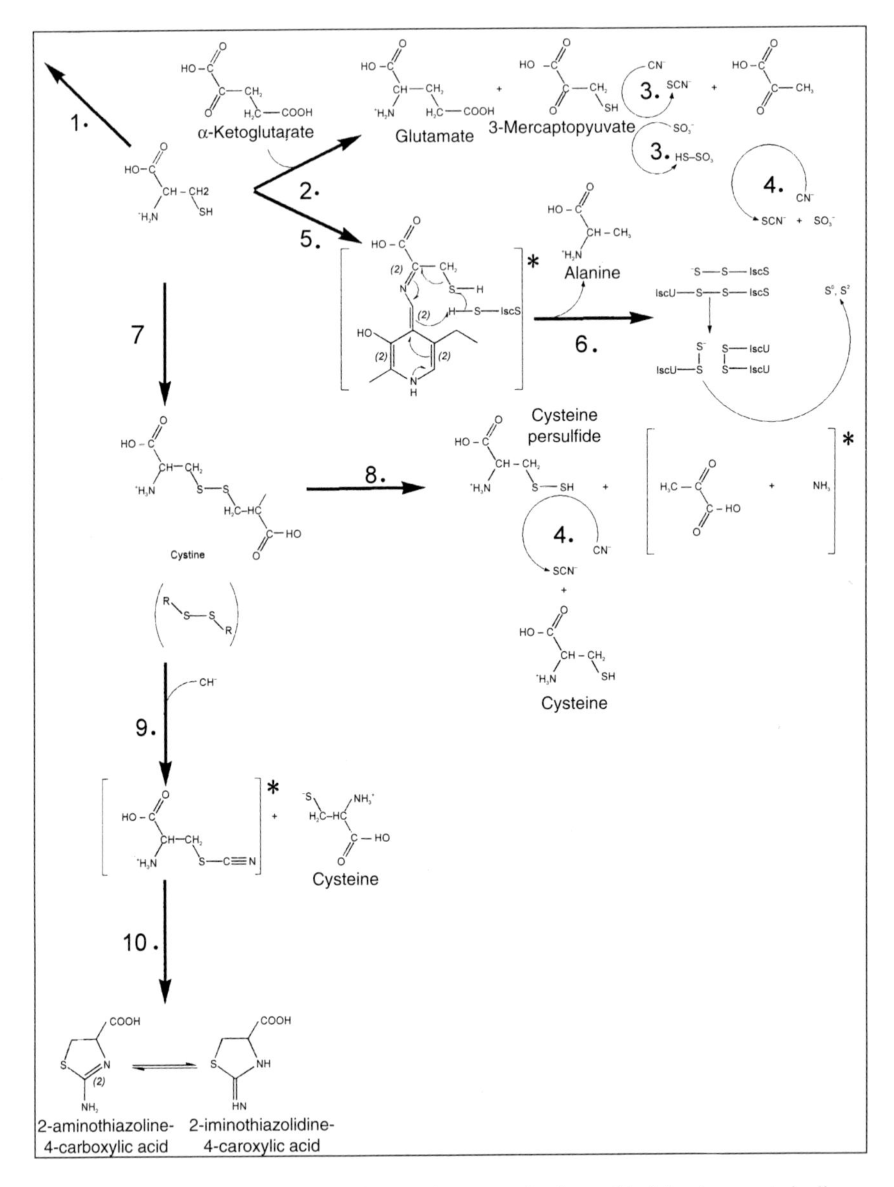

Fig. 4 Alternative pathways of cysteine metabolism. Methionine metabolism, cystathionine α-synthase (1); pyridoxal phosphate-dependent transamination of cysteine (reversable) in the presence of α-ketoglutarate, cysteine deaminase (2); transsulfuration of nucleophiles by mercaptopyruvate sulfur transferase (MPST) (3); Rhodanese (4); pyridoxal phosphate-dependent desulfuration of cysteine by IscS (5) to form a pyridoxal-bound intermediate (*) and to yield alanine; IscU catalyzed transfer and reduction of sulfane sulfur bound to Cys on IscS (6); oxidation of free or bound dithiols by glutathione oxidase, protein disulfide isomerase, or other thiol-disulfide oxidoreductases (7); heterolytic cleavage of cystine (8) to thiocysteine, pyruvate, and ammonia (*); cyanolysis of disulfides by unknown mechanism (9) with unknown intermediates (*); formation of 2-aminothiazoline-4-carboxylic acid (ATCA) by an unknown mechanism (10).

toxicity by α-keto-acids occurs through nucleophylic attack of cyanide on the carbonyl group (Figure 5). This reaction in vivo would bind cyanide into cyanohydrins and ameliorate toxicity by preventing inhibition of cytochrome oxidase.

α-ketoglutarate

α-ketoglutarate cyanohydrin

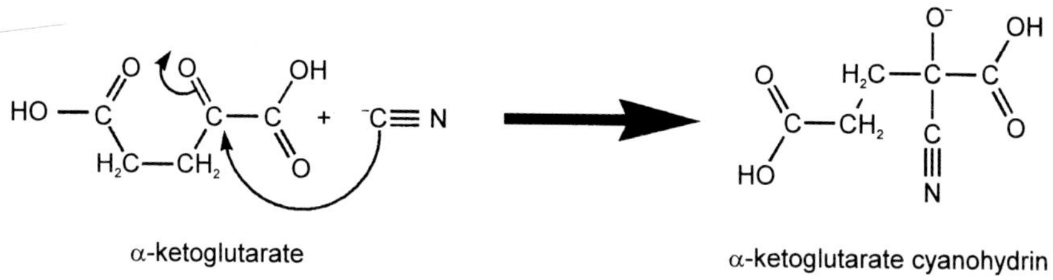

Fig. 5 Nucleophilic attack by cyanide on the carbonyl group of á-keto acids is the proposed mechanism of cyanide detoxification by these compounds in vivo

From available data there is little doubt that α-ketoglutarate affords significant protection against cyanide in vivo (Bhattacharya and Vijayaraghavan, 2002). However, several pieces of evidence support alternative routes of detoxification by α-keto-acids and these warrant further exploration. Norris et al. (1990) is most frequently cited to show the direct mechanism of action of α-ketoglutarate in antagonizing cyanide poisoning. Data obtained for the formation of cyanide in pH 7.4 buffered solution suggests a mild equilibrium reaction of α-ketoglutarate depletion when directly reacted with cyanide (Norris et al., 1990). If this is true α-ketoglutarate would not be expected to outcompete cyanide binding to methemoglobin. Reports differ on whether α-ketoglutarate augments the effects of sodium nitrite. Way (1983) reported that pyruvate, another α-keto acid, did not show any therapeutic effect in conjunction with sodium nitrite above the effects of sodium nitrite. Bhattacharya and Vijayaraghavan (2002), on the other hand, reported that coadministration of nitrite, α-ketoglutarate, and cyanide were only nominally protective, while 10 min. pretreatment with nitrite and a-ketoglutarate, followed by administration of cyanide, were significantly more protective than the effects of nitrite or α-ketoglutarate alone, and significantly more protective than simultaneous administration. That α-ketoglutarate synergistically augments the antagonistic effect of methemoglobin binding of cyanide seems to suggest that it plays a therapeutic role beyond cyanide scavenging.

Other data in Norris et al. (1990), namely the depletion of evaporated cyanide in the headspace of α-ketoglutarate-treated blood, and the amelioration of cytochrome oxidase inhibition in samples treated with cyanide and varying levels of α-ketoglutarate, depict a biphasic behavior, indicative of multiple pathways to produce the given response. The authors also failed to reisolate α-ketoglutarate cyanohydrin after treatment with cyanide and α-ketoglutarate in solution and in vivo. No other authors have shown increased α-ketoglutarate cyanohydrin formation following administration of cyanide and α-ketoglutarate. Other authors have, however, suggested that the effects of α-ketoglutarate may go beyond cyanohydrin formation. Indeed, protective effects are seen at 60 min. pretreatments of α-ketoglutarate, while the half life in vivo has been suggested to be only about 15 min.

We have looked at the effects of various α-keto-acids on the activity of MPST and found them to be mildly inhibitory (Porter and Baskin, 1996), however, to date little work has been done to look at the effect of α-keto-acids on cysteine deamination and subsequent reactions through MPST with cyanide. One must conclude therefore that more work is needed to understand the mechanism of α-keto-acid therapy in cyanide toxicity. Taken with the remarkably high doses used to achieve reported efficacy (as high as 2g/kg, Bhattacharya and Vijayaraghavan, 2002), the benefits of α-keto-acid therapy become occluded.

Cysteine Metabolism through IscS and IscU

The collection of enzymes that play a role in iron sulfur cluster biosynthesis function to mobilize the elements necessary for their construction. Sulfane sulfur appears to be the principal source of elemental sulfur in these cofactors, and rhodanese with thiosulfate and reducing agents, as well as MPST and 3-mercaptopyruvate, have been shown to donate elemental sulfur to iron-sulfur apoproteins in vitro. For a review, see Toohey (1989). Iron sulfur clusters are ubiquitous in living matter (Beinert et al., 1997). They function as prosthetic groups in a wide variety of redox and electron-transfer capacities including nitrogen fixation, carbon monoxide oxidation, sulfite reduction, the catalysis of citrate into isocitrate by aconitase, and possibly as sensors for iron, oxygen, reactive oxygen species, and nitric oxide. The clusters consist of structures with two to eight iron atoms bound by inorganic, acid labile sulfide, which can undergo exchange reactions with free sulfide. In large excesses of free sulfide these clusters are known to collapse. A recent comprehensive review of iron-sulfur clusters and their transformations is provided by Beinert et al. (1997).

Physiological derivation of sulfide for the formation of iron-sulfur clusters is mediated in part by a homodimeric pyridoxal phosphate (PLP) containing cysteine desulfurase enzyme, IscS, and its iron sulfur cluster-containing counterpart, theorized to be a homodimer as well, IscU (Figure 4, number 5 and 6). When combined, the two form a heterotetramer with a covalently bound catalytic intermediate (Kato et al., 2002; Urbina et al., 2001; Agar et al., 2000, Yuvaniyama et al., 1999). These enzymes are conserved across a wide range of cell types, including prokaryotes such as Escherichia Coli, and Azotobacter Vinelandii, as well as lower eukaryotes including yeast, and higher eukaryotes including humans (Agar et al., 2000; Tong and Rouault,

2000; Land and Rouault, 1998). Indeed, IscU is considered to be among the most widely conserved proteins known (Agar et al., 2000). IscU serves as a molecular scaffold and iron chelator. IscS functions by cleaving sulfur from cysteine to yield alanine and elemental sulfur, which is transferred to IscU for cluster formation (Zheng et al., 1994; Zheng et al., 1993; Kato et al., 2002; Urbina et al., 2001; Agar et al., 2000, Yuvaniyama et al., 1999). Since IscS and IscU were first identified in prokaryotes, they may also be referred to in relation to their nitrogen-fixing genetic roots, as NIFS and NIFU, respectively. In humans the genetic loci are referred to as NFS1 (nitrogen fixation 1) and NIFU.

As the intermediate products of IscS and IscU are sulfane sulfur and elemental sulfur and sulfide, it seems prudent to consider their impact on what Westley referred to as the sulfane sulfur pool. Experiments have shown that in the absence of iron for iron-sulfur biosynthesis, the cystenyl persulfide intermediate present in IscS will decompose to elemental sulfur, or, in the presence of reductant, sulfide (Zheng et al., 1993). It is interesting to note that the reaction of IscS proceeds through a PLP-dependent Schiff base intermediate (Figure 4, number 5), under which many bonds are labile, similar to the process catalyzed by γ-cystathionase, discussed below; however what makes the reaction catalyzed by IscS unique and relatively unprecedented in biology is the nucleophylic attack by the protein cystinyl residue (Zheng et al., 1993). It remains unknown whether, under certain conditions, IscU may be able to bind with thiosulfate or protein-conjugated or glutathione-conjugated persulfide to yield elemental sulfur. It also remains unknown whether IscS could generate substrate for the Rhodanese reaction as an intermediate. This activity, however, does seem plausible and its consideration in cyanide metabolism, while speculative, should not be omitted.

To date, cyanide researchers have not specifically referred to the combined action of IscS and IscU enzymes in the cytosol and mitochondria to produce elemental sulfur and sulfide for iron-sulfur clusters, however Westley (1981) hinted that there may be enzymes present in eukaryotes that performed the functions of IscS and IscU which were relevant to cyanide metabolism. An alternative possibility is that production of excess iron-sulfur clusters provides cyanide with a sulfur substrate under which a number of bonds are readily cyanolysable. It remains to be seen whether their actions are relevant in the production of thiocyanate and whether IscU can use other persulfides, such as glutathione-conjugated persulfide, as a substrate.

Cysteine Metabolism through γ-Cystathionase

The enzyme γ-cystathionase (cystathionine-γ-lyase) is interesting in that it is responsible for catalyzing two major steps in the pathway of cysteine metabolism; both the reaction shown above (Figure 4, number 8), where γ-cystathionase heterolytically cleaves the dithiol bond in cystine to form cysteine persulfide, pyruvate, and ammonia, and the cleavage of cystathionine to yield cysteine, ammonia, and α-ketobutyric acid (Westley 1981; Matsuo and Greenberg, 1958; White et al., 1978).

Cysteine persulfide generated by γ-cystathionase may be used as a sulfane sulfur donor by Rhodanese to detoxify cyanide. Whether or not this may be relevant in cyanide detoxification is a subject of some discussion. Westley makes the point that concentrations of free cystine in vivo are normally very low (Westley, 1981). However, cystine is analogous to other oxidized (di-) thiols, such as oxidized glutathione and protein dithiol bonds. Dithiol bond formation is catalyzed in vivo by enzymes such as glutathione peroxidase (Lehninger et al., 1993), protein disulfide isomerase, and other protein-disulfide oxidoreductases (Sevier and Kaiser, 2002). Whether these enzymes can or would use free cysteine as a substrate, whether cystine forms chemically from free cysteine depending on the oxidative environment, and whether γ-cystathionase can use other dithiols as a substrate are issues that seem somewhat unresolved in terms of cyanide toxicity. For small thiols like glutathione, the presence of a large number of reduced equivalents is important for the generation of reduced thiols in proteins (Sevier and Kaiser, 2002). In other words, the cell has an interest in keeping oxidation of glutathione minimal. The issue of low presence of free cystine also complicates the story of 2-amino-thiazoline-4-carboxylic acid (ATCA), discussed below.

The reactions of γ-cystathionase involve cleavage of a carbon-sulfur bond mediated by PLP directed Schiff base formation. Given the model, whereby cysteine and cysteine persulfide are released, followed by release of the α-keto-acid and ammonia (Lehninger et al., 1993), it is interesting that the distance of the α-carbon on the ketone from the site of cleavage is less catalytically relevant than that the labile bond is on the proximal side of a sulfur. It is also interesting that in the reactions catalyzed by γ-cystathionase, following cleavage of the carbon-sulfur bond, the enzyme presumably deaminates the alanine and glutamate products to transform its PLP cofactor into pyridoxamine phosphate, then spontaneously ejects the ammonia group to reset itself. Here again, as in cysteine transaminase above, it seems likely that in vivo and not under the conditions under which the enzyme was first characterized, the enzyme may be aided by α-keto-acids as substrates for transamination activity.

The Role of ATCA in Cyanide Toxicity

A minor, however pathophysiologically relevant pathway of cyanide metabolism includes combination with free cystine to form ATCA, which has been attributed to account for about 20% of cyanide metabolism (Baskin and Brewer, 1997). Others have shown that ATCA formation accounts for 5-15%

of cyanide metabolism, depending on the species studied (Wood and Cooley, 1956; Borowitz et al., 2001).

ATCA is formed when cyanide reacts with cystine through a proposed intermediate beta-cyano-sulfalanine that is transformed to 2-aminothiazoline-4-carboxylic acid (Figure 4, numbers 9 and 10). ATCA is present as a tautomer between itself and 2-iminothiazolidine-4-carboylic acid (ITCA) (Lundquist et. al., 1995a). It appears that at pH 7.5 the amino- rather than the imino- form predominates (Cummings et al., 2003). ATCA's structure resembles a cyclic form of the amino acid cysteine with cyanide inserted to complete the five membered ring. It also appears to have the same general properties as an amino acid with similar functional groups. ATCA does not appear to be metabolized further, but is excreted stoichiometrically to administered cyanide in urine and saliva. The contribution of ATCA metabolism to total cyanide metabolism may increase as cyanide toxicity progresses. Onset of lactic acidosis in cyanide victims favors formation of ATCA rather than thiocyanate, as rhodanese activity rapidly falls off at low pH.

The major target organs for cyanide toxicity are the brain and the heart. In these organs the rhodanese level is low, therefore the thiocyanate formation is suppressed, and the formation of ATCA becomes important. ATCA has been found to have its own intrinsic neurotoxicity and is a convulsant. Interestingly, the neurotoxicity of ATCA and its activity as a convulsant are dissociated, and it has been found that brain damage may occur at low doses, even in the absence of seizures (Bitner et al., 1997). ATCA might play an important role in developing the neurotoxic effects of cyanide (Bitner et al., 1997). The wild running seizures observed with ATCA can be induced with the classic excitotoxins, glutamate, N-methyl-D-aspartate (NMDA) and kainate; therefore, various excitatory amino acid antagonists are expected to decrease ATCA toxicity dramatically. Application of other anti-ATCA agents (e.g. ATCA analogues) along with cyanide antagonists may serve as effective tools in cyanide antagonism.

Conclusions

From these data, a formal picture develops of our current understanding of cyanide toxicity. Predictably, it is intrinsically linked to sulfur metabolism as approximately 79% of cyanide in vivo is metabolized through one of two sulfur-dependent steps (Figure 6). The outcome of cyanide toxicity is ultimately dependent on the metabolic step that predominates. Understanding and enhancing those thiocyanate-forming pathways is of critical import to development of rational cyanide therapies.

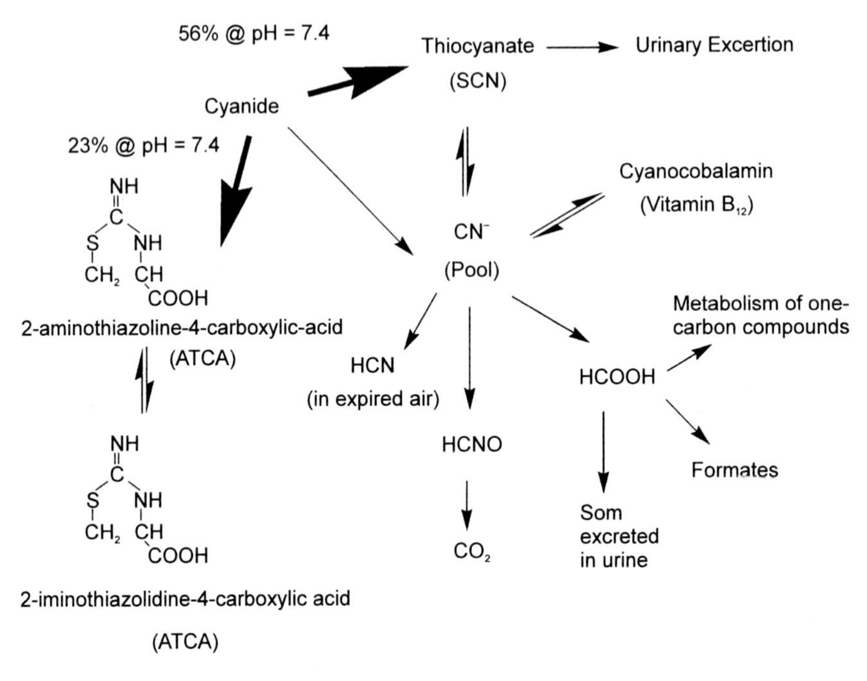

Fig. 6 Basic steps in cyanide metabolism.

ANALYTICAL DETERMINATION OF CYANIDE

Determination of cyanide or its metabolites in biological fluids is necessary for forensic, clinical, military, research, and veterinary purposes. The choice of analytical methods depends on a variety of factors: sensitivity, specificity, sample type, ease of use, facilities, and expertise available. Other factors that influence this choice are sample storage conditions, time of sampling (i.e. how long after exposure), and duration of analysis. There are numerous sensitive analytical techniques to determine cyanide and its metabolite thiocyanate in biological fluids, but very few techniques to analyze for ATCA. Table 1 outlines some recently published methods for determination of cyanide and its metabolites in biological fluids. For excellent reviews of earlier methods for analysis of cyanide and thiocyanate, consult Troup and Ballantyne (1987) and ATSDR (1997).

Factors Affecting Analysis of Cyanide and Its Metabolites in Biological fluids

The initial choice when considering analysis of cyanide in biological fluids is whether to determine cyanide concentrations directly or to analyze for one of its metabolites. Figure 6 illustrates metabolic pathways of cyanide. As seen in Figure 6, exposure to cyanide can be determined directly by analyzing for cyanide that is part of the "pool" or indirectly by analyzing for one of its

Table 1 Some recent analytical methods to determine cyanide and its metabolites in biological fluids[a]

Analyte	*Biological Fluid*	*Analytical Method*	*Detection Limit*		*Reference*
			nM	*ng/ml*	
HCN	Human Blood	Headspace GC[b] with NP[b] detection (also used SPME[b])	511	14	Calafat and Stanfill, 2002
	Human Blood	HPLC[b] with fluorometric detection after derivatization	74	2	Felscher and Wulfmeyer, 1998
	Human Blood	LCMS[b] after derivatization	185	5	Tracqui *et al.*, 2002
	Human Blood	Capillary GC[b] with oven trapping and NP[b] detection	74	2	Ishii *et al.*, 1998
	Human Blood	IC[b] with fluorometric detection after derivatization	3.8	0.10	Chinaka *et al.*, 1998
	Human Blood	GCMS[b] after derivatization with EC[b] detection	10000	270	Kage *et al.*, 1996
	Cow Serum, Rumen, Liver	GCMS[b] after derivatization with selected ion monitoring	700	18	Meiser *et al.*, 2000
	Human Blood	Voltammetry with Ag RDE[b] following trapping of HCN gas	1000	27	Westly and Westly, 1989
	Equine Blood	Spectrophotometry following trapping of HCN gas	74	2	Hughes *et al.*, 2003
SCN^-	Human Urine	Ion interaction LC[b] with UV[b] detection	1720	100	Connolly *et al.*, 2002
	Human Urine	IC[b] with electrochemical detection	500	29	Casella *et al.*, 1998

Table 1 contd.

Table 1 contd.

Analyte	*Biological Fluid*	*Analytical Method*	*Detection Limit*		*Reference*
			nM	*ng/ml*	
	Human Blood	IC[b] with fluorometric detection after derivatization	86	5	Chinaka *et al.*, 1998
	Human Blood	HPLC[b] with fluorometric detection after derivatization	165	9.6	Chen *et al.*, 1996
	Human Blood	GCMS[b] after derivatization with EC[b] detection	3000	174	Kage *et al.*, 1996
	Human Blood and Urine	Spectrophotometry following chlorination	930	54	Lundquist *et al.*, 1995b
	Human Urine and Saliva	Thiocyanate selective poly meric membrane electrode	48	2.8	Ganjali *et al.*, 2002
	Human Blood, Urine, Saliva	Flame atomic absorption spectrometry	69	4	Chattaraj and Das, 1992
ATCA	Human Urine	HPLC[b] with fluorometric detection after derivatization	300	44	Lundquist *et al.*, 1995a

[a] This table is meant to give a general overview of analytical methods in this area and their detection limits. It is not meant to be all-inclusive.

[b] GC—Gas Chromatography, NP—Nitrogen Phosphorous, SPME—Solid Phase Microextraction, IC—Ion Chromatography, HPLC—aHigh Performance Liquid Chromatography, LCMS—Liquid Chromatography Mass Spectrometry, EC—Electron Capture, GCMS—Gas Chromatography Mass Spectrometry, RDE—Rotating Disk Electrode, LC—Liquid Chromatography, UV—Ultraviolet.

metabolites, thiocyanate or ATCA. No matter how sensitive the analytical technique, the physical and chemical properties of the analyte may dictate the relative ease, and sometimes the selectivity, of a particular analytical method.

Accurate determination of hydrogen cyanide concentrations is difficult for a variety of reasons. First, it is extremely volatile, with a vapor pressure of 0.34 atm at 0 °C and 0.98 atm at 25 °C (for comparison, chloroform has a vapor pressure of 0.26 atm at 25 °C). This volatility can cause major problems in the recovery of cyanide from biological samples. With a pK_a of 9.2, maintaining samples under alkaline conditions can minimize the loss of hydrogen cyanide from vaporization. Another factor limiting accurate analysis of cyanide in biological fluids is that cyanide acts as a nucleophile, which may decrease the apparent concentration of cyanide in biological samples. Cyanide binds to tissue (Troup and Ballantyne, 1987) and is sequestered by erythrocytes (McMillan and Svoboda, 1982). Blood is the biological sample of choice when determining cyanide concentrations, yet cyanide is rapidly depleted from blood, generally within the first 20 minutes of exposure (Baskin and Brewer, 1997; Moriya and Hashimoto, 2001; Sylvester et al., 1981). Because of the rapid detoxification of cyanide, urine samples are not normally useful for direct determination of cyanide concentrations (Troup and Ballantyne, 1987), although urine is useful for thiocyanate and ATCA analysis. Moreover, the organ distribution of cyanide varies considerably with the route of administration and the animal species challenged (Ballantyne, 1975). In addition to rapidly decreasing concentrations of cyanide in biological fluids, cyanide sometimes forms as an artifact of storage conditions in a variety of biological samples (Seto, 1996; Ballantyne, 1977a; Curry, 1976; Sunshine and Finkle, 1964). All of these factors necessitate extreme care in the handling and storage of biological samples when analyzing for cyanide directly, and dictate that analysis should be accomplished as soon as possible after sampling. Thiocyanate is the major metabolite of cyanide (Figure 6). One major advantage of thiocyanate over cyanide is that it is found in blood, urine, and saliva in appreciable concentrations after exposure. Thiocyanate also has a relatively long half-life, one week, in humans with normal renal function (Nolan, 1999). Thiocyanate does have some disadvantages as an indicator of cyanide exposure. It is naturally found in human biological fluids, usually in micromolar concentrations (Pettigrew and Fell, 1973; Diem and Letner, 1970). Therefore, it is difficult to gauge the level of exposure to cyanide without prior baseline levels of thiocyanate in an individual. Ballantyne (1977b) found that analytical recovery of thiocyanate from whole blood is not quantitative and thiocyanate concentrations in blood varied inconsistently during storage at various temperatures over a period of two weeks. ATCA accounts for approximately 20% of cyanide metabolism and increases as cyanide toxicity increases (Baskin and Brewer, 1997). ATCA is stable for months in biological samples at freezing and ambient temperatures (Lundquist et al., 1995b). Based on determination of pK_a values (carboxylic acid: 2.03; amine: 8.48) and NMR studies of Gawron et al. (1962), ATCA forms a zwitterion in solution and is therefore not volatile. Although ATCA is a minor metabolite of cyanide exposure, its chemical properties and amenability to sensitive analytical techniques may outweigh its lower initial abundance as compared with cyanide and thiocyanate.

Direct Detection of Cyanide

Analytical methods for direct determination of cyanide in biological fluids are reviewed by Troup and Ballantyne (1987), ATSDR (1997), and a number of recent methods are outlined in Table 1. Historically, colorimetric methods are the most commonly performed analytical technique to determine cyanide concentrations. The most frequently used colorimetric method is the oxidation of cyanide to a cyanogen, which is then reacted with an organic reagent, such as pyridine-benzidine, to form a spectrophotometrically active product (Hughes et al., 2003; Nagashima, 1977; Shanahan, 1973; Epstein, 1947; Aldrich, 1944). These methods are generally not sensitive; they are time consuming and subject to interferences, especially from thiosulfate. Electrodes containing silver can also be used to determine cyanide concentrations in blood (Westly and Westly, 1989; Egekeze and Oehme, 1979). These methods are generally the least sensitive of the analytical methods to determine cyanide in biological fluids, but recent research in microelectrode technology may increase the usefulness of this technique.

Most recent methods of determination of cyanide involve gas chromatography with a variety of detection techniques and liquid chromatography (LC) with fluorometric or mass spectrometric detection (Table 1; Toida et al., 1986). Fluorometry was used as a convenient and sensitive stand-alone method for cyanide measurement in biological fluids prior to combination with HPLC (Groff et al., 1977). Fluorometric methods to determine cyanide are sensitive, but suffer from interferences from natural components of blood. These interfering substances have been mitigated by combining fluorometric detection with chromatographic separation techniques (Table 1; Sano et al., 1992; Sano et al., 1989). Chinaka et al. (1998) reported the lowest detection limit for the analytical determination of cyanide (Table 1) by derivatizing cyanide with 2,3-naphthalene-dialdehyde and taurine and using ion chromatography (IC) in conjunction with fluorometric detection to analyze cyanide in blood. These derivatizing agents were used by others in conjunction with high performance liquid chromatography (HPLC) and fluorometric detection with less success (Felscher and Wulfmeyer, 1998; Sano et al., 1992; Sano et al., 1989). A liquid chromatography mass spectrometric (LCMS) method was developed by Tracqui et al. (2002) also using 2,3-naphthalenedialdehyde and taurine derivatization prior to analysis. This method was even less sensitive. Because cyanide is such a small molecule, the difficulty inherent in analysis of derivatizated cyanide is that cyanide itself makes up very little of the resulting molecule. Therefore, it is difficult to definitively separate the derivatized compound with HPLC (or GC) from other components in the matrix.

Recently, a large number of GC methods to determine cyanide concentrations in biological fluid were developed with limited success (Table 1; Watanabe-Suzuki et al., 2002; Cardeal et al., 1995; Tsukamoto et al., 1994; Maseda et al., 1989). Although GC methods seem to hold great promise for the direct determination of cyanide in biological fluids, in general they lack sensitivity with most detectors (Table 1; Meiser et al., 2000; Odoul et al., 1994; Shiono et al., 1991; Maseda et al., 1990). Only the method of Ishi et al. (1998) approaches the detection limit of LC methods with fluorometric detection. GC methods seem to be plagued by cyanide's extremc volatility. Also, the small molecular weight of cyanide affects the sensitivity and selectivity of mass selective detectors and methods utilizing nitrogen-phosphorous detectors are limited by the natural abundance of nitrogen in biological samples. Overall, GC methods to directly determine cyanide concentrations in biological fluids have not lived up to their promise.

Although many sensitive methods exist for direct cyanide analysis, they all suffer from the practical limitations of the analyte: volatility, nucleophilic properties, rapid detoxification, artifactual formation of cyanide, and differential concentrations of cyanide in biological fluids depending on the route of exposure. Therefore, determination of cyanide metabolites may be more advantageous than direct cyanide determination.

Detection of the Cyanide Metabolite, Thiocyanate

Many methods have been developed for the measurement of thiocyanate in biological fluids, some of which have detection limits close to 1 nM (Grgurinovich, 1982). An abundance of colorimetric methods are reported in the literature (Lundquist et al., 1995b; Falkensson et al., 1988; Grgurinovich, 1982; Lundquist et al., 1979; Ballantyne, 1977b). The colorimetric methods normally involve a chemical complex with ferric ions or cupric ions, pyridine, or are based on the König reaction. Although these methods can be sensitive, they are subject to interferences are time intensive, and require large amounts of sample. Ion selective electrodes have also been used to determine thiocyanate in biological samples (Casella et al., 1998; Below et al., 1990; Westley and Westley, 1989). Although ion selective electrodes generally suffer from lack of sensitivity, Ganjali et al. (2002) reported an extremely sensitive method for thiocyanate determination using graphite electrodes modified with a polymeric membrane containing a nickel(II)-azamacrocycle complex (Table 1).

Recently, methods utilizing HPLC or IC have been developed with IC garnering most of the focus. HPLC with fluorometric detection has been successfully applied to determine thiocyanate in blood and saliva (Table 1; Brown et al., 1995; Tanabe et al., 1988). HPLC with an ion-paring reagent was used to determine thiocyanate concentrations in urine (Connolly et al., 2002). IC has been applied with multiple detection systems: electrochemical detection (Casella et al., 1998), UV absorption (Michigami et al., 1992; Imanari et al., 1982), and fluorometric detection after derivatization (Chinaka et al., 1998). IC can be extremely sensitive, but does suffer from interference from other ions, especially in urine. Compared with LC, few GC methods have been proposed for the determination of thiocyanate. Jacob et al. (1984) used GC with a nitrogen-phosphorous detector for thiocyanate determination following on-column derivatization. Others have used electron capture detection following derivatization with limited success (Kage et al., 1996; de Brabander and Verbeke, 1977). As with cyanide, GC methods to determine thiocyanate generally lack sensitivity compared with IC methods (Table 1). They also involve derivatization of thiocyanate, which carries inherent disadvantages, as discussed above.

As with direct determination of cyanide, methods for determination of thiocyanate in biological fluids suffer from the practical limitations of the analyte: inconsistent concentrations depending on duration and conditions of storage, incomplete recovery from whole blood, and appreciable inherent concentrations in biological fluids.

Detection of Cyanide Metabolite, ATCA

The analytical technique of Lundquist is the only available method for determination of ATCA in biological fluids (Table 1). Lundquist developed a method for ATCA determination in urine that employed derivatization of ATCA by N-(7-dimethylamino-4-methyl-3-coumarinyl)maleimide following thiazolidine ring opening. HPLC with fluorometric detection was subsequently used to analyze the derivative. This method produced acceptable results, but did not reach the sensitivity of some of the methods to determine thiocyanate. The method also requires many time consuming steps to reduce interferences in the matrix. Others have determined ATCA concentrations in aqueous solution. Baskin et al. (2003) and Bradham et al. (1965) developed spectrophotometric methods based on thiazolidine ring-opening, followed by determination of the R–SH, R-NH_2, or R-NH-R functional groups by a suitable chemical reaction. These methods, while relatively simple, are not sensitive and are subject to interference by cysteine and thiocyanate.

As stated above, ATCA approximates a cyclic amino acid in its structure and properties. Therefore, although a fast, reliable, simple analytical method for determination of ATCA in biological samples does not currently exist, many analytical techniques designed for sensitive analysis of amino acids should be readily portable to analysis of ATCA, such as derivatization by a silylating reagent and subsequent GCMS analysis (Chun-quing et al., 2002) or 9-fluorenylmethyl chloroformate derivatization and fluorescent detection after HPLC (Einarsson et al., 1986; Bank et al., 1996). ATCA represents an alternative metabolite of cyanide exposure that may circumvent the numerous disadvantages of cyanide and thiocyanate analysis.

NOVEL METHODS OF CYANIDE ANTAGONISM

Liposomes and Other Drug/Enzyme Carriers in Cyanide Antagonism

A new antidotal mechanism against cyanide involves employment of highly purified recombinant cyanide-metabolizing enzymes with high catalytic activity, encapsulated with an appropriate substrate (Way, 1984b). This antidotal approach is based on the idea of destroying the toxicant before it reaches target organs. However, these enzymes need to be placed in a suitable delivery system so that the physiological disposition of the protein material can be controlled. Various systems like liposomes or nanocapsules can be employed to prevent the metabolizing enzymes from premature degradation and immunologic reaction. Through this approach, a new class of protein delivery system may be developed with important conceptual relevance to biomedical science. The rationale for this approach is that the toxicant can readily diffuse into the carriers and reach the metabolizing enzymes, and the detoxified products can freely diffuse out from the carriers. Employing this approach has been shown to provide a tremendous increase of antidotal protection against intoxication by various chemical warfare agents.

Earlier studies (Way et al., 1984b; Leung et al., 1986; Petrikovics et al., 1994) demonstrated the effectiveness of using resealed, annealed murine bovine erythrocytes (CRBC) containing bovine rhodanese and sodium thiosulfate as a new conceptual approach to antagonize the toxic effects of cyanide (Figure 7). Although this approach proved efficacious, it was limited by the use of inorganic thiosulfate, which does not diffuse easily through cell membranes. This limitation focused attention on other sulfur donors that have better lipid solubility, can penetrate cell membranes, and can serve as better substrates for rhodanese. The encapsulated rhodanese, in combination with other cyanide antagonists, is capable of immediately lowering blood cyanide concentrations in animals exposed to cyanide. The reaction between cyanide and a sulfur donor substrate is fast, and the protection provided by the encapsulated rhodanese is better and can persist for a longer period than any other cyanide antagonists presently known. This enhanced protection can be attributed to a better substrate reactivity of organic thiosulfonates with cyanide, better penetration, minimized product inhibition, and the enhanced

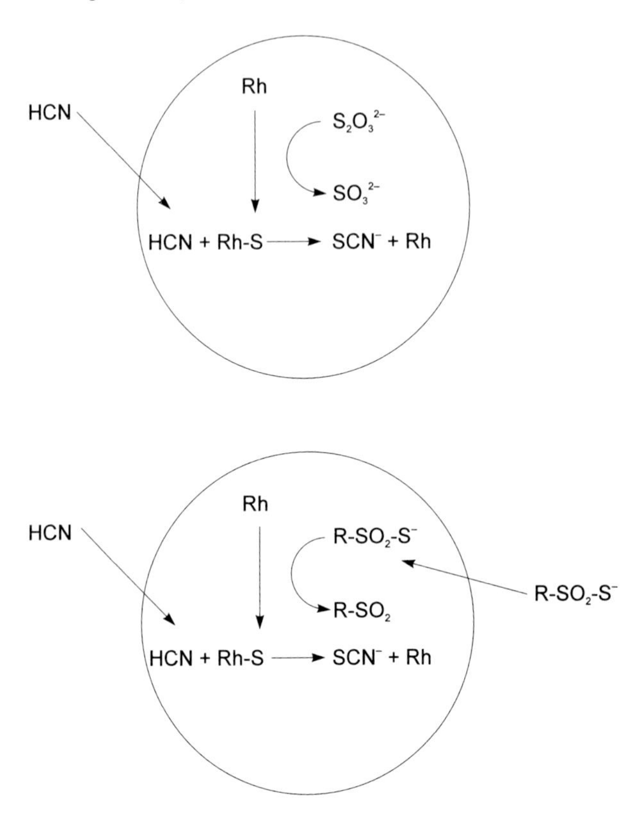

Fig. 7 Rhodanese reaction with sodium thiosulfate and organic tiosulfonates witin carrier cells.

longevity of rhodanese activity in the body by the encapsulation. When butanethiosulfonate (BTS) was employed as a sulfur donor and sodium nitrite was administered, the antidotal system provided a $14 \times LD_{50}$ protection, which is superior to any other known cyanide antidotes (Table 2), (Petrikovics et al., 1995).

CRBCs were the first model carrier system for encapsulating metabolizing enzymes in drug antagonism. While they work successfully, they are not very practical. Long circulating sterically stabilized (stealth) liposomes proved to be better. In long circulating liposomes a protein is encapsulated in a biodegradable cell carrier to circumvent adverse immunologic reactions, thus avoiding the phagocytic activity of the macrophages and the reticuloendothelial system (Papahadjopoulos et al., 1991; Woodle and Lasic, 1992; Allen, 1994). Previous studies have reported the tremendous protective effects with stealth liposomes encapsulating organophosphate (OP) hydrolyzing enzymes in OP antagonism (Petrikovics et al., 1999, 2000).

Liposomes are widely used as a selective drug delivery system for targeted, site-specific, and controlled-release drug delivery. The properties

Table 2 Effects of BTS with/without $NaNO_2$ and/or CRBC Containing Rhodanese on the LD_{50} of KCN in Balb/C Mice

Experiment	Treatments before KCN (sc)[a] BTS (ip) (mmol/kg)	$NaNO_2$ (sc) (mg/kg)	CRBC (iv) (Unit rhodanese/ mice)	LD_{50}[b] (Limits: p=0.05) (mg/kg)
1	0	0	0	9.6(8.6-10.7)
2	0	100	0	28.6(26.8-30.6)
3	6.3	0	0	27.4(26.5-28.2)
4	0	0	1.7-1.9	18.1(16.5-19.9)
5	6.3	0	1.7-1.9	46.0(44.9-47.2)
6	0	100	1.7-1.9	30.9(29.9-32.2)
7	6.3	100	0	95.6(90.1-101.4)
8	6.3	100	1.7-1.9	132.6(130.8-143.4)

[a] CRBC, $NaNO_2$ and BTS were administered approximately 16 hr, 45 min and 30 min, respectively, prior to the administration of KCN.

[b] Each LD50 value was obtained from five or more graded doses of KCN administered to five or more groups of six to eight mice.

and stability of liposomes are determined by the chemical structure (lipid-cholesterol ratio), size and surface modification (polymer coating, incorporated polyethylene glycol). The sterically stabilized liposomes can persist in the body longer, providing an effective biodegradable protective environment for the rhodanese enzyme.

Other drug carriers (e.g. synthetic: polylactides, poly-glycolates, dentrimers, and natural: cellulose, poly-alginates) have a potential use for encapsulating various metabolizing enzymes by forming microspheres. Recent investigation efforts have been focused on hyperbranched dentritic polymer based nanocapsules for protein delivery (Petrikovics et al., work in progress). Biocompatibility of this new class of dentritic polymer represents one of the first successful therapeutic applications of nanotechnology. These nanostructures should play an important future role in drug therapy. This approach is important to prevent chemical nerve agent reintoxication and chronic toxicity.

The Role of Hypoxia-inducible Factor

In 2001 it was shown that the bacteria Pseudomonas aeruginosa PAO1 killed the nematode Caenorhabditis Elegans (C. Elegans) by cyanide poisoning (Gallagher and Manoil, 2001). In the same paper it was also shown that one could obtain a C. Elegans mutant that was resistant to up to four times the lethal dose of cyanide. This mutant was found to be deficient in the egl-9 gene product (Gallagher and Manoil, 2001), a prolyl hydroxylase (PH) (Aravind and Koonin, 2001; Epstein et al., 2001) an enzyme that, in the presence of 2-

oxoglutarate, iron (II), ascorbate, and oxygen (Epstein et al., 2001; Yu et al., 2001) adds a hydroxyl group to a proline residue in a target peptide. Egl-9 represented an entirely new type of prolyl hydroxylase whose substrate was not realized until recently (Bruick and McKnight, 2001).

In humans, egl-9 analogues consist of a family of three prolyl hydroxylases (Epstein et al., 2001; Semenza, 2001; Huang et al., 2002). Their target is HIF-α, or, Hypoxia-Inducible Factor-α, a transcription factor family consisting of three isoforms that, upon translocation to the nucleus and dimerization with HIF-β (also known as the Aryl Hydrocarbon Receptor Nuclear Translocator, or ARNT, for a review see Swanson, 2002), upregulates a number of hypoxia response elements (Jiang et al., 2001; Semenza, 2002; Wang et al., 1995).

The Von Hippel-Lindau factor, VHL, binds to hydroxylated HIF-α and targets it for ubiqutination and degredation by the proteasome, quelching HIF-mediated gene expression (Maxwell et al., 1999; Ivan et al., 2001; Jaakkola et al., 2001). PH will therefore direct HIF to VHL-mediated ubiquitination in the presence of oxygen, limiting expression of HIF-regulated elements. Genes transcribed by HIF include endothelin-1 and tyrosine hydroxylase in the carotid body (Kline et al., 2002), in addition to vascular endothelial growth factor (VEGF), erythropoietin (EPO), adrenomedullin (AM), metallothionein-1 (MT-1), glucose transporter-1 (GLUT-1), and cell adhesion regulator-1 (CAR-1), among many others (Bernaudin et al., 2002).

Upregulation of these genes have been described to confer broad protection and tolerance to ischemia and hypoxia in tissues (Bernaudin et al., 2002; Digicaylioglu and Lipton, 2001). HIF induction would therefore also be expected to confer protection against cyanide-mediated histotoxic anoxia. However since PH is oxygen-dependent, inhibition of cytochrome oxidase by cyanide would not directly activate HIF through the PH/VHL system (Jiang et al., 1996). Indeed, some evidence suggests that cyanide-containing compounds such as sodium nitroprusside and cigarette smoke may actually suppress HIF induction (Kimura, 2002; Michaud, 2003), not enhance it.

The protection against cyanide afforded to C. Elegans egl-9 knockouts (Darby et al., 1999; Gallagher and Manoil, 2001) indicates that a constitutive activation of HIF or ablation of PH's affords protection. If HIF induction alone is sufficient to shield against cyanide it would be expected to work by a mechanism similar to that of hypoxia preconditioning described above. Evidence supports this: At least one lab has shown that hypoxia preconditioning is effective at ameliorating cyanide toxicity in cardiomyocytes (Wright, 2003).

PH inhibition may also work under acute conditions. Cobalt chloride and other cobalt-containing compounds like dicobalt edetate have been used to detoxify cyanide (Way, 1983). The mechanism of action of these compounds has been attributed to their ability to complex cyanide. It is now known, however, that cobalt chloride potently inhibits PH's, and that cobalt provides a better substrate for chelation by PH's than their active site atomic iron (Ivan et al., 2001; Jaakkola et al., 2001; Ivan et al., 2002). Action of cobalt chloride and dicobalt edetate on these PH's and their subsequent inability to hydroxylate HIF should be taken into consideration, particularly given that coadministration of cobalt and nitrite potentiate one another, suggesting different mechanisms of action (Way, 1983).

PH's are deactivated by iron chelators and they require reduced iron (Fe^{2+}), which is provided by the action of ascorbate, an essential but readily dissociable cofactor for the known PH's (Ivan et al. 2002). PH-bound iron may also oxidize more easily with therapeutic administration of methemoglobin-formers; indeed it has been shown that the administration of the reducing compound dithiothreitol can reverse nitric oxide-induced HIF activation (Kimura, 2002). One implication of this result is that administration of oxidants like methemoglobin formers might enhance NO-induced HIF induction, however the data in this respect remains contradictory (Kimura et al. 2002; Salceda and Caro 1997).

While preliminary work has been done to show that hypoxic preconditioning can be effective at reducing toxicity through metabolic inhibition by cyanide in isolated cardiomyocytes (Wright et al., 2003), a proof of concept is still badly needed, a direct relationship to HIF is not yet established, no specific protective HIF-mediated transcript has been yet identified, and no drug development work has yet been done to produce a pretreatment against cyanide toxicity. The relatively recent discovery about the role of PH's and HIF induction in cellular toxicity have, to date, kept them from being incorporated into mainstream cyanide research. The possibility that several heretofore unexplained cyanide interventions might work through the inhibition of PH's lends itself to further study. PH inhibitors based on proline and 2-oxoglutarate analogs are already available commercially (Jaakkola et al., 2001; Ivan et al., 2002; Wright et al., 2003) and may be effective if used as a pretreatment for cyanide toxicity with diminished risk of dangerous side effects exhibited by other therapeutic measures such as methemoglobin formers. They may also be used in concert with other specific therapies to enhance detoxification. Taken together this makes a convincing case for further investigation into mechanisms of prevention and pretreatment against cyanide toxicity associated with hypoxia-inducible-factor.

A FUTURISTIC VIEW OF CLINICAL CYANIDE DIAGNOSTICS AND THERAPEUTICS

Cyanide diagnostic and therapeutic approaches appropriate for the contemporary—or future—battlefields are lacking. As discussed above, our therapeutic strategies have focused on uncoupling cyanide from mitochondrial cytochrome oxidase with nitrites, followed by enhanced excretion with thiosulfate. That strategy is ill-suited to military applications, particularly in light of increasing urbanization of combat combined with continued reduction in concentration of fighting forces and their supporting medical assets. Novel paradigms will also benefit mass casualty management in the event of urban use of cyanide.

Current research strongly suggests that serious cyanide toxicity results from breakdown and imbalance in highly complex interactions involving subcellular signalling and regulation from such compounds as nitric oxide (NO) (Cleeter et al., 1994) and peroxynitrite (ONOO-) (Brown, 1995 and Torres et al., 1998). This contradicts the traditional dictum that cyanide toxicity is a simple problem, but makes the clinical experience more intelligible and likely explains some of the contradictory experiences. Today, many clinical toxicologists doubt the validity of nitrite therapy, in large part because of its perceived capriciousness, expense, side-effect profile, and—from the battlefield medic's perspective—inaccessibility. Continued elucidation of the regulatory systems involved in cellular, particularly excitatory, cell health and death and how they are impacted by cyanide will lead to better interventions that can be delivered by the first contact responder. New research methodologies that can study those systems in non-linear fashion with appreciation for their complexities are needed. Improved understanding of normal clearance of cyanide is also critical. For example, it is not at all clear that the central nervous system relies on transulfuration (Baskin and Rockwood, 2002). Measures of peripheral cyanide levels and clearance (e.g., urinary thiocyanate) likely do not reflect central nervous system experience. Furthermore, transulfuration functions optimally at high pH, not the very low pH conditions experienced by cyanide casualties. Indeed, because the major metabolic pathway for cyanide under conditions of severe metabolic acidosis generates ATCA (Isom and Baskin, 1997) and because ATCA may be a critical neurotoxicant (Bitner et al., 1997), it seems reasonable to pursue enhanced ATCA excretion as a research direction.

Necessary goals, then, for future therapeutics and diagnostics include : (1) cyanide therapies must rapidly and reliably restore critical enzyme systems and cellular homeostasis, resulting in rapid return to normal function. Further, research must target excitable tissues for this benefit, as they seem least well protected by endogenous processes; (2) cyanide therapies should afford neuroprotection so that even quite severely affected individuals recover without neurologic effect; (3) therapies and diagnostics must be user -friendly, have little adverse iatrogenic impact, must have a large margin of safety, and must be useful for all population groups, not just healthy adult males; and (4) therapies and diagnostics must be logistically supportable. Only the first two of these will be discussed here. That clinicians need therapeutic options that are safe and easily used needs no elaboration. Issues of manufacturing, storage, and exact method of delivery are best discussed elsewhere.

Rapid and Reliable Acute Management

The key factor in successful management of severely intoxicated cyanide poisonings—as marked by serious central nervous system effects and/or cardio respiratory effects—is rapid intervention. Both case studies and epidemiologic study support the premise that therapy works fastest and to best result when delivered aggressively and promptly (Yen et al., 1995 and Wurzburg, 1996). Current therapy relies on removal from further exposure, support of critical life systems, administration of antidotes, and symptomatic treatment of seizures. The usual antidotes are nitrites, intended to remove the cyanide from the cytochrome oxidase to restore oxidative energy production, and thiosulfate to enhance excretion. Oxygen is administered as well. Unfortunately, all of this is complicated and challenging on a battlefield. Trauma needs the one-hour solution. Cyanide intoxication needs the three minute solution, something along the lines of the nerve agent antidote kit. It would also make sense to have prophylactics or antidotal enhancers that could be used prior to threatened exposure. One area that might be exploited is the knowledge that cytochrome oxidase exists in both the oxidized (Fe^{3+}) and reduced (Fe^{2+}) states as well as in an intermediate state. Becuase the reduced state releases bound cyanide more easily than does the oxidized, treatment approaches which promote, the reduced form of cytochrome oxidase, even under proapototic conditions, may be useful. Secondly, whilw oxygen is useful therapy, it is award to deliver. In addition, the highly acidotic states typical of cyanide intoxication hinder the transfer of oxygen from hemoglobin to cells. Therefore, it would be helpful to develop medications and/or devices which enhance delivery of oxygen to mitochondria very early after cyanide exposure. Oxygen and cyanide compete for the same binding site on the cytochrome oxidase but the cyanide cannot be reduced and will not exit the site spontaneously. High mitochondrial oxygen levels, rapidly delivered to oxidized cytochrome oxidase, would likely be therapeutically useful. Promising research has begun in this area (Collman and Boulatov, 2002) Finally, it appears that the standard metabolic pathways used to describe and explain the effects of cyanide and its therapies are at best only partial explanations when one considers the organ systems most vulnerable to the toxic effects of cyanide. In particular, we need to know—and then to exploit—cyanide's toxicology, biotransformation, and excretion in order to minimize the impact of excessive exposure on the heart and nervous system.

Neuroprotection

Applying interventions at the time of initial therapy that allow incompletely damaged cells to recover rather than progress to either apoptotic or necrotic death is well known in other clinical areas. This strategy has greatly reduced mortality and morbidity from myocardial infarctions. Much current research seeks to apply the same logic to protect neurons from the effects of cerebrovascular events and from other toxin exposures—such as nerve agents. It makes sense with cyanide intoxication as well. Some severely poisoned persons receiving multiple lethal doses of cyanide survive the incident only to live with serious neural damage. They often sustain serious neurological disability. Specific areas of the brain most likely affected include the putamen, substantia nigra, and the caudate nuclei. These areas have low cell density but high oxygen and glucose utilization, and damage manifests with dystonias and parkinsonism (Borgohain, 1995 and Kasamo et al., 1993). Currently available therapies, however, must rest with rapid restoration of cellular respiration, rapid recovery from severe metabolic acidosis, and rapid interruption of seizures. It is revealing that basal ganglia damage does not manifest until hours or days after recovery from severe cyanide poisonings, leading to the consideration that slowly progressive neuronal damage may be based on release of excitatory amino acids, high calcium influx to neurons, and increased neuronal activity (Kasamo et al., 1993). Such thinking is consistent with research demonstrating relationships between NMDA receptor stimulation by cyanide or its active metabolites resulting in glutamatergic seizures and nitric oxide overproduction, both tied to cortical apoptotosis (Horn et al., 2002).

Without question, the work discussed in this chapter and related works have clear benefit. Research that brings us closer to understanding the full dimensions of mitochondrial oxygen utilization, energy production, and intracellular communication will benefit persons poisoned by toxins such as cyanide and will also benefit persons sustaining other mitochondrial diseases or injuries.

Conclusions

Most of the current therapies for cyanide despite their effectiveness have their shortcomings due to route of administration and/or toxicity. These traditional treatments include methemoglobin formers (e.g., sodium nitrite), sulfane-sulfur compounds (e.g., thiosulfate), oxygen, cyanohydrin formers (e.g., alpha keto glutamic acid), and cobalt salts. This chapter describes possible alternate approaches to develop therapies against cyanide. Several of these methods appear to show promise. Hypoxic-inducible factor–1 (HIF-1), and 2-aminothiazoline 4-carboxylic acid (ATCA) antagonists, may provide other approaches towards the treatment of cyanide poisoning. HIF-1 appears to exert a large dynamic range on the regulation of oxygen in cyanide poisoning models. Administration of ATCA, the cystine metabolite of cyanide, or cyanide itself results in unconsciousness in a dose-related fashion. Further studies need to be performed to establish if these alternative treatments can provide safe, effective therapy for cyanide intoxication.

References

Agar J.N., Krebs K., Frazzon J., Huynh B.H., Dean D.R. and Johnson M.K., IscU as a scaffold for iron-sulfur cluster biosynthesis: Sequential assembly of [2Fe-2S] and [4Fe-4S] clusters in IscU. Biochem. **39:** 7856-7862, 2000.

Aldridge W.N. A new method for the determination of microquantities of cyanide and thiocyanate. Analyst. **69:** 262-265, 1944.

Allen T.M. Long circulating (sterically stabilized) liposomes for targeted drug delivery. Adv. Drug Del. Rev., **13:** 285-309, 1994.

Aravind L. and Koonin E.V. The DNA-repair protein AlkB, EGL-9, and leprecan define new families of 2-oxoglutarate- and iron-dependent dioxygenases. Genome Biol. 2., research0007.1-research0007.8, 2001.

Ardelt B.K., Borowitz J.L., Maduh E.U., Swain S.L. and Isom G.E., Cyanide-induced lipid peroxidation in different organs: subcellular distribution and hydroperoxide generation in neuronal cells. Toxicology. **89:** 127-37, 1994.

Arden S.R., Sinor J.D., Potthoff W.K. and Aizenman, E., Subunit-specific interactions of cyanide with the N-methyl-D-aspartate receptor. J Biol. Chem. **273:** 21505-21511, 1998.

ATSDR. Toxicological profile for cyanide. Agency for Toxic Substances and Disease Registry, Division of Toxicology, Atlanta, GA. 1997.

Ballantyne B. Blood, brain and cerebrospinal fluid cyanide concentrations in experimental acute cyanide poisoning. J. Forensic Sci. Soc., **15:** 51-56, 1975.

Ballantyne B. In vitro production of cyanide in normal human blood and the influence of thiocyanate and storage temperature. Clin. Toxicol. **11:** 173-93, 1977a.

Ballantyne B. Factors in the analysis of whole blood thiocyanate. Clin. Toxicol. **11:** 195-210, 1977b.

Ballantyne B. Toxicology of cyanides. In: Ballantyne B., and Marrs T.C., eds. Clinical and Experimental Toxicology of Cyanides. Bristol, England: Wright; 41-126, 1987.

Bank R.A., Jansen E.J., Beekman B. and Koppele J.M. Amino acid analysis by reverse-phase high-performance liquid chromatography: improved derivatization and detection conditions with 9-fluorenylmethyl chloroformate. Anal. Biochem. **240:** 167-76, 1996.

Baskin S.I. and Rockwood G.A., Neurotoxicological and behavioral effects of cyanide and its potential therapies. Military Psychol. **14:** 159-177, 2002.

Baskin S.I. and Brewer T.G. Cyanide Poisoning. In: Sidell, F.R., Takafuji, E.T., and Franz, D.R., eds. Textbook of Military Medicine, Medical aspects of chemical and biological warfare. Washington: Borden Institute. 271-286, 1997.

Baskin S.I., Horowitz A.M. and Neally E.W. The antidotal action of sodium nitrite and sodium thiosulfate against cyanide poisoning. J Clin, Pharmacol. **32:** 368-375, 1992.

Baskin S.I., Kurche J.S., Petrikovics I. and Way J.L. Analysis of 2-imino-thiazolidine-4-carboxylic acid (ITCA) as a method for cyanide measurement. The Toxicologist. **72:** 164, 2003.

Baskin S.I., Porter D.W., Rockwood G.A., Romano J.A., Patel H.C., Kiser R.C., Cook C.M. and Ternay A.L. In vitro and in vivo comparison of sulfur donors as antidotes to acute cyanide intoxication. J. Appl. Toxicol. **19:** 173-183, 1999.

Baskin S.I. The cardiac effects of cyanide. In: Baskin S.I., ed. Principles of cardiac toxicology. Boca Raton, Florida: CRC Press. 419-429, 1991.

Beinert H., Holm R.H. and Münck E., Iron-sulfur clusters: Nature's modular, multipurpose structures. Science. **277:** 653-659, 1997.

Below H., Muller H., Hundhammer B. and Weuffen W. Determination of thiocyanate in serum with ion-selective liquid membrane electrodes. Pharmazie. **45:** 416-419, 1990.

Bernaudin M., Tang Y., Reilly M., Petit E. and Sharp F.R. Brain genomic response following hypoxia and reoxygenation in the neonatal rat. J. Biol, Chem. **277:** 39728-39738, 2002.

Bhattacharya R., Kumar D., Sugendran K., Pant S.C., Tulsawani R.K. and Vijayaraghavan R. Acute toxicity studies of α-ketoglutarate: A promising antidote for cyanide poisoning. J. Appl. Toxicol. **21:** 495-499, 2001.

Bhattacharya R., Lakshmana Rao P.V. and Vijayaraghavan R. In vitro and in vivo attenuation of experimental cyanide poisioning by α-ketoglutarate. Toxicol. Lett. **128:** 185-195, 2002.

Bhattacharya R., and Vijayaraghavan R. Promising role of α-ketoglutarate in protecting against the lethal effects of cyanide. Hum. Exp. Toxicol. **21:** 297-303, 2002.

Bitner R.S., Yim G.K.W. and Isom G,E. 2-iminothiazolidine-4-carboxylic acid produces hippocampal CA1 lesions independent of seizeure excitation and glutamate receptor activation. NeuroToxl. **18:** 191-200, 1997.

Borgohain R., Singh A.K., Radhakrishna H., Roa V.C. and Mohandas, S. Delayed onset generalized dystonia after cyanide poisoning. Clin. Neurol. Neurosurg. **97:** 213-215, 1995.

Bonsall J.L. Survival without sequelae following exposure to 500 mg/m^3 of hydrogen cyanide. Hum. Toxicol. **3:** 57-60, 1984.

Borowitz J.L., Gunasekar P. and Isom G.E. Hydrogen cyanide generation by mu opiate receptor activation: Possible neuromodulatory role of endogenous cyanide. Brain Res. **768:** 294-300, 1997.

Borowitz J.L., Iso G.E. and Baskin S.I. Acute and Chronic Cyanide Toxicity. In: Somani, S.M., and Romano Jr., J.A., eds. Chemical Warfare Agents: Toxicity at Low Levels. Boca Raton, Florida: CRC Press. 301-319, 1999.

Brandham, L.S., Catsimpoolas, N. and Wood, J.L. Determination of 2-iminothiazolidine-4-carboxylic acid. Anal. Biochem. **11:** 230-237, 1965.

Brazzolotto X., Gaillard J., Pantopoulos K., Hentze M. and Moulis J.M. Human Cytoplasmic aconitase (Iron Regulatory Protein 1) Is converted into Its [3Fe-4S] form by hydrogen peroxide in vitro but Is not activated for iron-responsive element binding. J. Biol. Chem. **31:** 21625-21630, 1999.

Brown D.G., Lanno R.P. van den Heuvel M.R. and Dixon D.G. HPLC determination of plasma thiocyanate concentrations in fish blood: application to laboratory pharmacokinetic and field-monitoring studies. Ecotoxicol. Environ. Saf. **30:** 302-308, 1995.

Brown G.C. Nitric oxide regulates mitochondrial respiration and cell functions by inhibiting cytochrome oxidase. FEBS Letters. **369:** 126-139, 1995.

Bruick R.K. and McKnight S.L. A conserved family of prolyl-4-hydroxylases that modify HIF. Science. **294:** 1337-1340, 2001.

Calafat A.M. and Stanfll S.B. Rapid quantitation of cyanide in whole blood by automated headspace gas chromatography. J. Chromatogr. B. Analyt. Technol. Biomed. Life Sci. **772:** 131-137, 2003.

Cardeal Z.L., Gallet J.P., Astier A. and Pradeau D. Cyanide assay: statistical comparison of a new gas chromatographic calibration method versus the classical spectrophotometric method. J. Anal.Toxicol. **19:** 31-34, 1995.

Casella I.G., Guascito M.R. and De Benedetto G.E. Electrooxidation of thiocyanate on the copper-modified gold electrode and its amperometric determination by ion chromatography. Analyst. **123:** 1359-1363, 1998.

Chattaraj S. and Das A.K. Indirect determination of thiocyanate in biological fluids using atomic absorption spectrometry. Spectrochica. Acta. **47:** 675-680, 1992.

Chen K.K., Rose C.L. and Clowes G.H.A. Amyl nitrite and cyanide poisoning. J. Amer. Med. Assoc. **100:** 1920-1922, 1933a.

Chen S.H., Yang Z.Y., Wu H.L., Kou H.S. and Lin S.J. Determination of thiocyanate anion by high-performance liquid chromatography with fluorimetric detection. J. Anal. Toxicol. **20:** 38-42, 1996.

Chen K.K., Rose C.L. and Clowes G.H.A. Methylene blue, nitrites, and sodium thiosulfate against cyanide poisoning. Proc. Soc. Exp. Biol. Med. **100:** 250-252, 1933b.

Chung-qing S., Chun-hui D. and Yao-ming H. Simultaneous SIM-GC/MS quantititation of phenylalanine and tyrosine as their butyldimethylsilyl derivatives: a new method of newborn screening of phenylketonuria. Chin. Med. J. **115:** 51-55, 2002.

Chinaka S., Takayama N., Michigami Y. and Ueda K. Simultaneous determination of cyanide and thiocyanate in blood by ion chromatography with fluorescence and ultraviolet detection. J. Chromatogr. B. Biomed. Sci. Appl. **713:** 353-359, 1998.

Cleeter M.W.J., Cooper J.M., Darley-Usmar V.M., Moncada S. and Schapira A.H.V., Reversible inhibition of cytochrome c oxidase, the terminal enzyme of the mitochondrial respiratory chain, by nitric oxide, Implications for neurodegenerative diseases. FEBS Letters. **345:** 50-54, 1994.

Collman J.P. and Boulatov R. Electrocatalytic O2 Reduction by synthetic alogues of the heme/Cu site of cytochrome oxidase incorporated in a lipid film. Angew. Chem. Int. Ed., **41:** 18, 2002.

Connolly D., Barron L. and Paull B. Determination of urinary thiocyanate and nitrate using fast ion-interaction chromatography. J. Chromatogr. B. Analyt. Technol. Biomed. Life Sci. **767:** 175-180, 2002.

Cooper A.J. Biochemistry of sulfur-containing amino acids. Annu. Rev. Biochem. **52:** 187-222, 1983.

Cummings S.E., Baskin S. and Nagasawa H.T. The structure of "ITCA", a metabolite of cyanide. J. Minn. Acad. Sci. **67:** 33-34, 2003.

Curry A. Poison Detection in Human Organs. Thomas: Springfield. 97-98, 1976.

Darby C., Cosma C.L., Thomas J.H. and Manoil C. Lethal paralysis of Caenorhabditis elegans by Pseudomonas aeruginosa. Proc. Nat. Acad. Sci. USA. **96:** 15202-15207, 1999.

de Brabander H.C. and Verbeke R. Determination of thiocyanate in tissues and body fluids of animals by gas chromatography with electron-capture detection. J. Chromatogr. **138:** 131-142, 1977.

Diem K. and Lentner C. Scientific Tables, Seventh Edition. Geigy Pharmaceuticals, Ardsely: New York. 1970.

Digicaylioglu M. and Lipton S.A. Erythropoietin-mediated neuroprotection involves cross-talk between Jak2 and NF-kB signaling cascades. Nature. **412:** 641-647, 2001.

Djerad A., Monier C., Houze P., Borron S.W. Lefauconnier J.M. and Baud F.J. Effects of respiratory acidosis and alkalosis on the distribution of cyanide into the rat brain. Toxicol. Sci. **61:** 273-82, 2001.

Egekeze J.O. and Oehme F.W., Direct potentiometric method for the determination of cyanide in biological materials. J. Anal. Toxicol. **3:** 119-124, 1979.

Einarsson S., Folestad S., Josefsson B. and Lagerkvist S. High-resolution reversed-phase liquid chromatography system for the analysis of complex solutions of primary and secondary amino acids. Anal Chem. **58:** 1638-1643, 1986.

Elliott A.C., Smith G.L. and Allen D.G. Simultaneous measurements of action potential duration and intracellular ATP in isolated ferret hearts exposed to cyanide. Circ. Res. **64:** 583-591, 1989.

Epstein J. Estimation of microquantities of cyanide. Anal Chem. **19:** 272-274, 1947.

Epstein A.C.R., Gleadle J.M., McNeill L.A., Hewitson K.S., O'Rourke J., Mole D.R., Mukherji M., Metzen E., Wilson M.I., Dhanda A., Tian Y.M., Masson N., Hamilton D.L., Jaakkola P., Barstead R., Hodgkin J., Maxwell P.H., Pugh C.W., Schofield C.J. and Ratcliffe P.J. C. elegans EGL-9 and mammalian homologs define a family of dioxygenases that regulate HIF by prolyl hydroxylation. Cell. 107: 43-54, 2001.

Falkensson M., Lundquist P., Rosling H. and Sorbo B. A simple method for determination of plasma thiocyanate. Ann. Clin. Biochem. **25:** 422-423, 1988.

Felscher D. and Wulfmeyer M. A new specific method to detect cyanide in body fluids, especially whole blood, by fluorimetry. J. Anal. Toxicol. **22:** 363-366, 1998.

Gallagher L. and Manoil C. *Pseudomonas aeruginosa* PAO1 kills Caenorhabditis elegans by Cyanide poisoning. J. Bacteriol. **183:** 6207-6214, 2001.

Ganjali M.R., Yousefi M., Javanbakht M.J., Poursaberi T., Salavati-Niasari M., Hajiagha-Babaei L., Latifi E. and Shamsipur M. Determination of SCN- in urine and saliva of smokers and non-smokers by SCN-selective polymeric membrane containing a nickel(II)-azamacrocycle complex coated on a graphite electrode. Anal. Sci. **18:** 887-892, 2002.

Gawron O., Fernando J., Keil J. and Weistmann T.J. Zwitterion structure and acylative ring-opening reactions of 2-aminothiazoline-4-carboxylic acid. J. Org. Chem. **27:** 3117-3123, 1962.

Grgurinovich N. A colourimetric procedure for the determination of thiocyanate in plasma. J Anal Toxicol. **6:** 53-55, 1982.

Groff W.A., Cucinell S.A., Vicario P. and Kaminskis A.A., completely automated fluorometric blood cyanide method: a specific assay incorporating dialysis and distillation. Clin. Toxicol. **11:** 159-171, 1977.

Hall A.H. and Rumack B.A., Hydroxocobalamin/sodium thiosulfate as a cyanide antidote. J. Emerg. Med. **5:** 115-121, 1987.

Haymaker W., Ginzler A.M. and Ferguson R.L. Residual neuropathological effects of cyanide poisoning: A study of the central nervous system of 23 dogs exposed to cyanide compounds. Mil. Surg. **3:** 231-246, 1952.

Horn T.F.W., Wolf G., Duffy S., Weiss S., Keilhoff G. and MacVicar B.A. Nitric oxide promotes intracellular calcium release from mitochondria in striatal neurons. FASEB J. **16:** 1611-1622, 2002.

Huang J., Zhao Q., Mooney S.M. and Lee F.S. Sequence determinants in hypoxia-inducible factor-1a for hydroxylation by the prolyl hydroxylases PHD1, PHD2, PHD3. J Biol Chem. **277:** 3972-39800, 2002.

Hughes C. Lehner F., Dirikolu L, Harkins D., Boyles J., McDowell K., Tobin T., Crutchfield J., Sebastian M., Harrison L. and Baskin S.I. A simple and highly sensitive spectrophotometric method for the determination of cyanide in equine blood. Toxicol. Mech. Methods. **13:** 1-10, 2003.

Imanari T., Tanabe S. and Toida T., Simultaneous determination of cyanide and thiocyanate by high performance liquid chromatography. Chem. Pharm. Bull. (Tokyo). **30:** 3800-3802, 1982.

Ishii A., Seno H., Watanabe-Suzuki K., Suzuki O. and Kumazawa T. Determination of cyanide in whole blood by capillary gas chromatography with cryogenic oven trapping. Anal Chem. **70:** 4873-4876, 1998.

Isom G.E. and Johnson J.D. Sulfur donors in cyanide intoxication. In: Ballantyne B, Marrs TC, eds. Clinical and Experimental Toxicology of Cyanides. Bristol, England: Wright. 413-426, 1987.

Isom G.E. and Baskin S.I. Enzymes involved in cyanide metabolism. In: Sipes, I.G., McQueen, C.A., and Gandolfi, A.J., eds. Comprehensive Toxicology. New York, New York: Elsevier Science. 477-488, 1997.

Ivan M., Haberberger T., Gervasi D.C., Michelson K.S., Günzler V., Kondo K., Yang H., Sorokina I., Conaway R.C., Conaway J.W. and Kaelin W.G. Biochemical purification and pharmacological inhibition of a mammalian prolyl hydroxylase acting on hypoxia-inducible factor. Proc. Nat. Acad .Sci. USA. **99:** 13459-13464, 2002.

Ivan M., Kondo K., Yang H., Kim W., Valiando J., Ohh M., Salic A., Asara J.M., Lane W.S. and Kaelin W.G. HIFa targeted for VHL-mediated destruction by proline hydroxylation: Implications for O2 sensing. Science. **292:** 464-468, 2001.

Jaakkola P., Mole D.R., Tian Y.M., Wilson M.I., Gielbert J., Gaskell S.J., vonKriegsheim A., Hebestreit H.F., Mukherji M., Schofield C.J., Maxwell P.H., Pugh C.W. and Ratcliffe P.J. Targeting of HIFa to the von Hippel-Lindau ubiquitylation complex by O2-regulated prolyl hydroxylation. Science. **292:** 468-472, 2001.

Jacob P. III, Savanapridi C., Yu L., Wilson M., Shulgin A.T., Benowitz N.L., Elias-Baker B.A., Hall S.M., Herning R.I. and Jones R.T. Ion-pair extraction of thiocyanate from plasma and its gas chromatographic determination using on-column alkylation. Anal. Chem. **56:** 1692-1695, 1984.

Jiang B.H., Semenza G.L., Bauer C. and Marti H.H. Hypoxia-inducible factor 1 levels vary exponentially over a physiologically relevant range of O2 tension. Am. J. Physiol., **271;** C1172-C1180, 1996.

Jiang H., Guo R. and Powell-Coffman J.A. The Caenorhabditis elegans hif-1 gene encodes a bHLH-PAS protein that is required for adaptation to hypoxia. Proc. Nat .Acad. Sci .USA. **98:** 7916-7921, 2001.

Johnson J.D. and Isom G.E. The oxidative disposition of potassium cyanide in mice. Toxicology. **37:** 215-224, 1985.

Kage S., Nagata T. and Kudo K., Determination of cyanide and thiocyanate in blood by gas chromatography and gas chromatography-mass spectrometry. J. Chromatogr. B. Biomed. Appl. **675:** 27-32, 1996.

Kasamo K., Okuhata Y., Satoh R., Ikeda M., Takahashi S., Kamata R., Nogami Y. and Kojima T. , Chronological changes of MRI findings on striatal damage after acute cyanide intoxication: pathogenesis of the damage and its selectivity, and prevention for neurological sequelae: a case report. Eur. Arch. Psychiatry Clin. Neurosci. **243:** 71-74, 1993.

Kato S., Mihara H., Kurihara T., Takahashi Y., Tokumoto U., Yoshimura T. and Esaki N. Cys-328 of IscS and cys-63 of IscU are the sites of disulfide bridge formation in a covalently bound IscS/IscU complex: Implications for the mechanism of iron-sulfur cluster assembly. Proc. Nat. Acad. Sci. USA. **99:** 5948-5952, 2002.

Kelin D. Cytochrome and respiratory enzymes. Proc. Royal Soc. **104:** 206-252, 1929.

Kimura H., Ogura T., Kurashima Y., Weisz A. and Esumi H. Effects of nitric oxide donors on vascular endothelial growth factor gene induction. Biochem. Biophys. Res. Commun. **296:** 976-982, 2002.

Kline D.D., Peng Y.J., Manalo D.J., Semenza G.L. and Prabhakar N.R. Defective carotid body function and impaired ventilatory responses to chronic hypoxia in mice partially deficient for hypoxia-inducible factor 1a. Proc. Nat. Acad. Sci. USA. **99:** 821-826, 2002.

Land T. and Rouault T.A. Targeting of a human iron-sulfur cluster assembly enzyme, nifs, to different subcellular compartments is regulated by alternative AUG utilization. Molec. Cell. **2:** 807-815, 1998.

Lehninger A.L., Nelson D.L. and Cox M.M., Principles of Biochemistry. New York : Worth; 1993.

Leung P., Ray L.E., Sander C., Leong Way, J., Sylvester D.M. and Way J.L., Encapsulation of thiosulfate: cyanide sulfurtransferase by mouse erythrocytes. Toxicol. Appl. Pharmacol. **83:** 101-107, 1986.

Li W. and Palmer G. Spectroscopic characterization of the interaction of azide and thiocyanate with the binuclear center of cytochrome oxidase: evidence for multiple ligand sites. Biochem. **32:** 1833-43, 1993.

Lundquist P., Kagedal B., Nilsson L. and Rosling H. Analysis of the cyanide metabolite 2-aminothiazoline-4-carboxylic acid in urine by high-performance liquid chromatography. Anal. Biochem. **228:** 27-34, 1995a.

Lundquist P., Kagedal B. and Nilsson L. An improved method for determination of thiocyanate in plasma and urine. Eur. J. Clin. Chem. Clin. Biochem. **33:** 343-349, 1995b.

Lundquist P, Martensson J., Sorbo B. and Ohman S. Method for determining thiocyanate in serum and urine. Clin. Chem. **25:** 678-681, 1979.

Lundquist P., Rosling H. and Sorbo B. The origin of hydrogen CN in breath. Arch, Toxicol. **61:** 270-274, 1988.

Maduh E.U., Borowitz J.L. and Isom G.E. Cyanide-Induced Alteration of Cytosolic pH: Involvement of Cellular Hydrogen Ion Handling Processes, Toxicol. Appl Pharmacol. **106:** 201-208, 1990.

Majima H.J., Oberley T.D., Furukawa K., Mattson M.P., Yen H.C., Szweda L.I. and St Clair D.K. Prevention of mitochondrial injury by manganese superoxide dismutase reveals a primary mechanism for alkaline-induced cell death. J. Biol. Chem. **273:** 8217-24, 1998.

Maseda C., Matsubara K., Hasegawa M., Akane A. and Shiono H., Determination of blood cyanide using head-space gas chromatography with electron capture detection. Nippon Hoigaku Zasshi. **44:** 131-136, 1990.

Maseda C., Matsubara K. and Shiono H. Improved gas chromatography with electron-capture detection using a reaction pre-column for the determination of blood cyanide: a higher content in the left ventricle of fire victims. J. Chromatogr. **490:** 319-327, 1989.

Matsuo Y. and Greenberg D.M. A crystalline enzyme that cleaves homoserine and cystathionine. I. Isolation procedure and some physicochemical properties. J. Biol. Chem. **230:** 545-560, 1958.

Maxwell P.H., Wiesener M.S., Chang G.W., Clifford S.C., Vaux E.C., Cockman M.E., Wykoff C.C., Pugh C.W., Maher E.R. and Ratcliffe P.J. The tumor suppressor protein VHL targets hypoxia-inducible factors for oxygen-dependent proteolysis. Nature. **399:** 271-275, 1999.

McMillan D.E. and Svoboda A.C. IV. The role of erythrocytes in cyanide detoxification. J. Pharmacol. Exp. Ther. **221:** 37-42, 1982.

Meiser H., Hagedorn H.W. and Schulz R. Development of a method for determination of cyanide concentrations in serum and rumen fluid of cattle. Am. J. Vet. Res. **61:** 658-664, 2000.

Michaud S.E., Menard C., Guy L.G., Gennaro G. and Rivard A. Inhibition of hypoxia-induced antiogenesis by cigarette smoke exposure: impairment of the HIF-lalpha/VEGF pathway. FASEB, **17:** 1150-1152, 2003.

Michigami Y., Fujii K., Ueda K. and Yamamoto Y. Determination of thiocyanate in human saliva and urine by ion chromatography. Analyst. **117:** 1855-1858, 1992.

Mills E.M., Gunasekar P.G., Pavlakovic G. and Isom G.E. Cyanide-induced apoptosis and oxidative stress in differentiated PC12 cells. J Neurochem. **67:** 1039-1046, 1996.

Moody A.J. 'As Prepared' forms of fully oxidised haem/Cu terminal oxidases. Biochimica Biophysica Acta. **1276:** 6-20, 1996.

Moriya F. and Hashimoto Y. Potential for error when assessing blood cyanide concentrations in fire victims. J Forensic Sci. **46:** 1421-1425, 2001.

Nagashima S. Spectrophotometric determination of cyanide with g-picoline acid and barbituric acid. Anal. Chim. Acta. **91:** 197-201, 1977.

Nolan C.R. Hypertensive Crisis. In: Wilcox, C.S., ed . Atlas of diseases of the kidney, Volume III. Philadelphia: Current Medicine, Inc. 8.1-8.30, 1999.

Norris J.C., Utley W.A. and Hume A.S. Mechanism of antagonizing cyanide-induced lethality by a-ketoglutaric acid. Toxicol. **62:** 275-283, 1990.

Odoul M., Fouillet B., Nouri B., Chambon R. and Chambon P. Specific determination of cyanide in blood by headspace gas chromatography. J. Anal. Toxicol. **18:** 205-207, 1994.

Papahadjopoulos D., Allen T.M., Gabizon A., Mayhew K., Matthay K., Huang S.K., Lee K.D., Woodle M.C., Lasic D.D., Redemann C. and Martin F.J. Sterically stabilized liposomes: Improvements in pharmacokinetics and antitumor thereapeutic efficacy. Proc. Natl. Acad. Sci. USA. **88:** 11460-11464, 1991.

Pawate A.S., Morgan J., Namslauer A., Mills D., Brzezinski P., Ferguson-Miller S. and Gennis R.B. A Mutation in Subunit I of Cytochrome Oxidase from Rhodobacter sphaeroides Results in an Increase in Steady-State Activity but Completely Eliminates Proton Pumping. Biochemistry. **41:** 13417-13423, 2002.

Petrikovics I., Cannon E.P., McGuinn W.D., Pei L. and Way J.L. Cyanide antagonism with organic thiosulfonates and carrier red blood cells containing rhodanese. Fundamental & Applied Toxicology. **24:** 1-8, 1995.

Petrikovics I., Cheng T.C., Papahadjopoulos D., Hong K., Yin R., DeFrank J.J. and Way J.L. Long circulating liposomes encapsulating organophosphorus acid anhydrolase in diisopropylfluorophosphate antagonism. Toxicological Sciences. **57:** 16-21, 2000.

Petrikovics I., Hong K., Omburo G., Hu Q., Pei L., McGuinn W.D., Sylvester D., Papahadjapoulos D. and Way J.L. Antagonism of paraoxon intoxication by recombinant phosphotriesterase encapsulated within sterically stabilized liposomes. Toxicol. Appl. Pharmacol. **156:** 56-63, 1999.

Petrikovics I., Pei L., McGiunn W.D., Cannon E.P. and Way J.L. Encapsulation of rhodanese and organic thiosulfonates by mouse erythrocytes. Fundamental and Applied Toxicology. **23:** 70-75, 1994.

Pettigrew A.R. and Fell G.S. Microdiffusion method for estimation of cyanide in whole blood and its application to the study of conversion of cyanide to thiocyanate. Clin. Chem. **19:** 466-471, 1973.

Porter D.W. and Baskin S.I. The effect of three alpha-keto acids on 3-mercaptopyruvate sulfurtransferase activity. J. Biochem. Toxicol. **11:** 45-50, 1996.

Salceda S. and Caro J. Hypoxia-inducible factor 1a (HIF-1a) protein is rapidly degraded by the ubiquitin-proteasome system under normoxic conditions. J. Biol. Chem. **272:** 22642-22647, 1997.

Sano A., Takezawa M. and Takitani S. High performance liquid chromatography determination of cyanide in urine by pre-column fluorescence derivatization. Biomed. Chromatogr. **3:** 209-212, 1989.

Sano A., Takimoto N. and Takitani S. High-performance liquid chromatographic determination of cyanide in human red blood cells by pre-column fluorescence derivatization. J. Chromatogr. **582:** 131-135, 1992.

Semenza G.L. HIF-1 and mechanisms of hypoxia sensing. Cell Reg. **13:** 167-171, 2002.

Semenza G.L. HIF-1, O2, and the 3 PHDs: How animal cells signal hypoxia to the nucleus. Cell. **107:** 1-3, 2001.

Seto Y. Stability and spontaneous production of blood cyanide during heating. J. Forensic Sci. **41:** 465-468, 1996.

Sevier C.S. and Kaiser C.A. Formation and transfer of disulphide bonds in living cells. Nat. Rev. Mol. Cell Biol. **3:** 836-847, 2002.

Shanahan R. The determination of sub-microgram quantities of cyanide in biological materials. J. Forensic Sci. **18:** 25-30, 1973.

Shiono H., Maseda C., Akane A. and Matsubara K. Rapid and sensitive quantitation of cyanide in blood and its application to fire victims. Am. J. Forensic Med. Pathol. **12:** 50-53, 1991.

Smith R.P. Toxic responses of the blood. In: Casarett L.J., Amdur M.O., Klaassen C.D. and Doull J., eds., Casarett & Doull's Toxicology: The Basic Science of Poisons, 5th ed. New York: McGraw-Hill. 335-354, 1996.

Solomonson L.P. Cyanide as a metabolic inhibitor. In: Vennesland B., Conn E.E., Knowles C.J., Westley J. and Wissing F. eds. Cyanide in Biology. New York, NY: Academic Press. 11-28, 1981.

Sun P., Rane S.G., Gunasekar P.G., Borowitz J.L. and Isom G.E. Modulation of the NMDA receptor by cyanide: enhancement of receptor-mediated responses. J. Pharmacol. Exp. Ther. **280:** 1341-1348, 1997.

Sunshine I. and Finkle B. The necessity for tissue studies in fatal cyanide poisoning. Int. Arch. Gewerbepath Gewerbehyg. **20:** 558-561, 1964.

Swanson H.I. DNA binding and protein interactions of the AHR/ARNT heterodimer that facilitate gene activation. Chem. Biol. Interact. **141:** 63-76, 2002.

Sylvester D.M., Sandler C., Hayton W.L. and Way J.L. Alteration of the pharmacokinetics of sodium cyanide by sodium thiosulfate. Proc. West. Pharmacol. **24:** 135, 1981.

Tanabe S., Kitahara M., Nawata M. and Kawanabe K. Determination of oxidizable inorganic anions by high-performance liquid chromatography with fluorescence detection and application to the determination of salivary nitrite and thiocyanate and serum thiocyanate. J. Chromatogr. **424:** 29-37, 1988.

Thomson A.J., Greenwood C., Gadsby P.M.. Peterson J. Eglinton D.G., Hill B.C. and Nicolls P. The structure of the cytochrome a3-CuB site of mammalian cytochrome c oxidase as probed by MCD and EPR spectroscopy. J. Inorg. Bichem. **23:** 187-197, 1985.

Toida T., Togawa T., Tanabe S. and Imanari T. Determination of cyanide and thiocyanate in blood plasma and red cells by high-performance liquid chromatography with fluorometric detection. J. Chromatogr. **308:** 133-141, 1986.

Tong W.H. and Rouault T.A. Distinct iron-sulfur cluster assembly complexes exist in the cytosol and mitochondria of human cells. EMBO J. **19:** 5692-5700, 2000.

Toohey J.I. Sulphane sulphur in biological systems: a possible regulatory role. Biochem. J. **264:** 625-632, 1989.

Torres J., Cooper M.S. and Wilson M.T. Reactivity of nitric oxide with cytochrome c oxidase: Interactions with the binuclear centre and mechanism of inhibition. J. Bioenerg. Biomembr. **30:** 63-9, 1998.

Tracqui A., Raul J.S., Geraut A., Berthelon L. and Ludes B. Determination of blood cyanide by HPLC-MS. J Anal Toxicol. **26:** 144-148, 2002.

Troup C.M. and Ballantyne, B., Analysis of cyanide in biological fluids and tissues. In: Ballentync, B., and Marrs, T.C., eds. Clinical and experimental toxicology of cyanides. Bristol, England: Wright. 22-40, 1987.

Tsukamoto S., Kanegae T., Matsumura Y., Uchigasaki S., Kitazawa M., Muto T., Chiba S., Oshida S., Nagoya T. and Shimamura M., Simultaneous measurement of alcohols and hydrogen cyanide in biological specimens using headspace gas chromatography. Nippon Hoigaku Zasshi. **48:** 336-342, 1994.

Urbina H.D., Silberg J.J., Hoff K.G. and Vickery L.E. Transfer of sulfur from IscS to IscU during Fe/S cluster assembly. J Biol Chem. **276:** 44521-44526, 2001.

Vennesland B., Castric P.A., Conn E.E., Solomonson L.P., Volini M. and Westley J. Cyanide metabolism. Fed. Proc. **41:** 2639-2648, 1982.

Volini M. and Alexander K. Multiple forms and multiple functions of the Rhodaneses. In: Vennesland, B., Conn, E.E., Knowles, C.J., Westley, J., Wissing, F., eds. Cyanide in Biology. New York, NY: Academic Press: 61-76, 1981.

Wang G.L., Jiang B.H., Rue E.A. and Semenza G.L. Hypoxia-inducible factor 1 is a basic-helix-loop-helix-PAS heterodimer regulated by cellular O2 tension. Proc. Nat. Acad. Sci. USA. **92:** 5510-5514, 1995.

Warburg O. Inhibition of the action of prussic acid in living cells. Hoppe-Seyler's Z Physiol. Chem. **76:** 331-346, 1911.

Watanabe-Suzuki K., Ishii A. and Suzuki O. Cryogenic oven-trapping gas chromatography for analysis of volatile organic compounds in body fluids. Anal. Bioanal. Chem. **373:** 75-80, 2002.

Way J.L., Leung P., Sylvester D.M., Burrows G., Way J.L. and Tamulinas C. Methaemoglobin formation in the treatment of acute cyanide intoxication. In: Ballantyne B. and Marrs T.C. eds. Clinical and Experimental Toxicology of Cyanides. Bristol, England: Wright. 402-412, 1987.

Way J.L., Sylvester D., Morgan R.L., Isom G.E., Burrows G.E., Tamulinas C.B. and Way J.L. Recent perspectives on the toxicodynamic basis of cyanide antagonism. Fundam Appl. Toxicol. **4:** S231-S239, 1984a.

Way J.L. Mechanism of cyanide intoxication and its antagonism. Annu. Rev. Pharmacol. Toxicol. **24:** 451-481, 1984b.

Way J.L. Cyanide antagonism. Fundam Appl Toxicol. **3:** 383-386, 1983.

Westley A.M. and Westley J. Voltammetric determination of cyanide and thiocyanate in small biological samples. Analytical Biochemistry. **181:** 190-194, 1989.

Westley J. Sulfur-transfer catalysis by enzymes. Bioorganic Chemistry. **1:** 371-390, 1977.

White A., Handler P., Smith E.L., Hill R. and Lehman I.R. Principles of Biochemistry. New York, NY: McGraw-Hill. 733-736, 1978.

Wilson D.F., Erecinska M. and Brocklehurst E.S. The Chemical Properties of Cytochrome c Oxidase in Intact Mitochondria. Arch. Biochem. Biophys. **151:** 180-187, 1972.

Wood J.L. and Cooley S.L., Detoxification of cyanide by cystine. J. Biol. Chem. **218:** 449-457, 1956.

Woodle M.C. and Lassic D. Sterically stabilized liposomes. Biochim. Biophys. Acta. **1113:** 171-199, 1992.

Wright G., Higgin J.J., Raines R.T., Steenbergen C. and Murphy E. Activation of the prolyl hydroxylase oxygen sensor results in induction of GLUT-1, HO-1, NOS-2 proteins, and confers protection from metabolic inhibition to cardiomyocytes. J. Biol. Chem., **278:** 20235-20239, 2003.

Wurzburg, H., Treatment of Cyanide Poisoning in an Industrial Setting. Vet. Human Toxicol. **38:** 44-47, 1996.

Yen D., Tsai J., Wang L.M., Kao W.F., Hu S.C., Lee C.H. and Deng J.F. The clinical experience of acute cyanide poisoning. Am. J. Emerg. Med. **13:** 524-528, 1995.

Yu F., White S.B., Zhao Q. and Lee F.S. HIF-1a binding to VHL is regulated by stimulus-sensitive proline hydroxylation. Proc Nat Acad Sci USA. **98:** 9630-9635, 2001.

Yuvaniyama P., Agar J.N., Cash V.L., Johnson M.K. and Dean D.R. NifS-directed assembly of a transient [2Fe-2S] cluster within the NifU protein. Proc. Nat. Acad. Sci. USA. **97:** 599-604, 1999.

Zheng L., White R.H., Cash V.L. and Dean D.R. Mechanism for the desulfuration of l-cysteine catalyzed by the nifS gene product. Biochemistry. **33:** 4714-4720, 1994.

Zheng L., White R.H., Cash V.L., Jack R.F. and Dean D.R. Cysteine desulfurase activity indicates a role for NIFS in metallocluster biosynthesis. Proc. Nat. Acad. Sci. USA. **90:** 2754-2758, 1993.

Pharmacological Perspectives of Toxic Chemicals and Their Antidotes
Editors: S J S Flora, J A Romano, S I Baskin and K Sekhar

The Virulence Components of Plague Bacilli: A Potent Biological Weapon

H.V. Batra, Rekha Khushiramani, Mahesh Shanmugasundaram, Jyoti Shukla and Urmil Tuteja
Microbiology Division, Defence Research & Development Establishment
Gwalior 474 002, India

INTRODUCTION

The risk of terrorist attacks with biological weapons poses new challenges for governmental agencies that have the responsibility to respond to these threats. For bio-terrorist attacks have multiple dimensions. The release of a biological agent beside affecting troops may also create a public heath emergency. Thus public health, law enforcement and national security agencies all need to re-orient their priorities and prepare and equip themselves for dealing with such a situation. At the same time this being a criminal act perpetrators need to be apprehended.

Implementing these new priorities will require considerable organizational learning and change. Agencies with deeply imbedded professional norms and organizational culture, need to understand the value of co-operation particularly at the time of crisis. Each agency responds with its own routines and its own distinctive view of "the threat," which impedes effective collaboration. Networking of institutions to meet the threats posed by the prepetratoes of bioterrorist/bioweapon attack is the need of the hour.

A biological weapon can be thought of as a 4 component system namely: payload, munition, delivery system, and dispersion system.

Infectious biological payload that could cause a mass casualty by an aerosol route of exposure includes the etiologic agent of plague—*Yersinia pestis* (Hawley and Eitzen, 2001). Plague, the disease caused by the bacteria *Yersinia pestis* (*Y pestis*), has had a profound impact on human history. In AD

541, the first great plague pandemic began in Egypt and swept over the world in the next 4 years.

Population losses attributable to plague during those years were between 50 and 60 percent. In 1346, the second plague pandemic, also known as the Black Death or the Great Pestilence, erupted and within 5 years had ravaged the Middle East and killed more than 13 million in China and 20-30 million in Europe, one third of the European population.

Advances in living conditions, public health and antibiotic therapy make such natural pandemics improbable, but plague outbreaks following an attack with a biological weapon do pose a serious threat.

Plague is one of very few diseases that can create widespread panic following the discovery of even a small number of cases. This was apparent in Surat, India, in 1994, when an estimated 5,00,000 persons fled the city in fear of a plague epidemic.

The intentional use of plague as a bioweapon would have an epidemiology different from the naturally occurring outbreak. It would mostly result in a pneumonic plague if there is an aerosol release. The size of the outbreak would depend on several factors, including the quantity of biological agent used, characteristics of the strains, environmental conditions, munitions used to containerize the payload to maintain its potency during delivery, the delivery system to transport the payload to a susceptible target and methods of aerosolization (Inglesby et al., 2000; Hawley and Eitzen, 2001).

HISTORICAL ASPECT OF POSSIBLE USAGE OF PLAGUE BACILLI AS BIOWEAPON

During medieval times, warriors pressed their catapults into service for spreading pestilence to their enemies. During the 14th century siege of Kaffa, a seaport city in what is now the Ukraine, Tatar forces attacked Kaffa by catapulting deceased plague victims from among their own ranks into the besieged city in order to spread disease and hasten a victory. A plague outbreak was documented in Kaffa, and it has been hypothesized that the fleeing citizens (as well as rats) of Kaffa who escaped via ship to various Mediterranean ports aided in the development of the second plague pandemic in the mid –1300 (Klietmann and Rouff, 2001). In 1710, Russian troops hurled the corpses of plague victims over the city walls of Reval during Russia's war with Sweden (http://www.bioterry.com/history_of_Biological_Terrorism.asp).

During the period between 1932 to 1945 Japan conducted various biological experiments in Manchuria. At the infamous Unit 731, a biological warfare research facility near the town of Ping Fan, prisoners were infected with *Y. pestis*, *B. anthracis* and other disease causing agents. In addition the Japanese attacked a number of Chinese cities. They contaminated the water supplies and food items with *Y. pestis*, *Shigella* and *Salmonella* species,*Vibrio cholerae*. Cultures were tossed into homes and sprayed from aircraft. They dropped small bombs containing plague—infected fleas (Hawlcy and Eitzen, 2001).

In the modern times, the former Soviet Union continued their offensive program clandestinely under the organization called Biopreparat in spite of being a signatory to the 1972 Biological and Toxins Weapons convention. The Soviets started genetic manipulation of the plague strain. The Soviets had finished and accomplished the weaponisation of plague (Hilleman, 2002). The Kirov facility was responsible for storing 20 tons of Plague (http:// www.pbs.org/wgbh/pages/frontline/shows/plague/ interviews/alibekov.html). Domaradskij group had successfully introduced a cloned haemolytic plasmid of an intestinal bacillus—*Escherichia coli* into the plague strain. Researchers were able to produce an EV strain of plague that was resistant to the most widely used antibiotics. The advantage of this vaccine was that it could be employed simultaneously with emergency prevention by means of antibiotics (which kills an ordinary vaccine) (Domaradskij and orent, 2001). The plague bacilli have also been acquired clandestinely for bioterrorism activities. In 1995, on May 5, a lab tech from Ohio (Larry Wayne Harris), ordered the plague bacterium from a Maryland biomedical supply firm, American Type Culture Collection. 3 vials of *Yersinia pestis* were mailed to him using no more than a credit card and a false letterhead. Spurred on by his impatience in receiving the order and in his apparent ignorance in lab techniques, the company contacted federal authorities.

An investigation revealed that he was a member of a white supremacist organization (http://www.bioterry.com/History_of_Biological_Terrorist.asp).

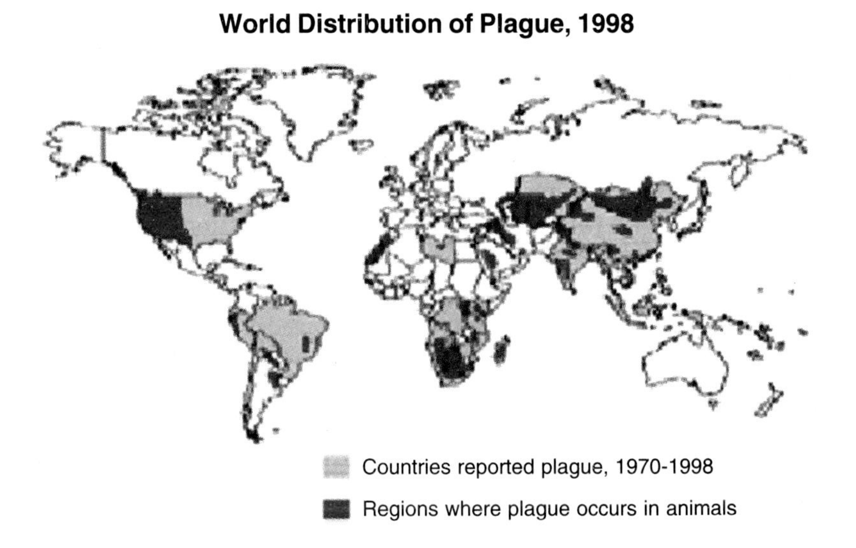

QUANTIFICATION OF THE HAZARD

In 1970, the World Health Organization (WHO) reported that, in a worst case scenario, if 50 kg of *Y. pestis* were released as aerosol over a city of 5 million, pneumonic plague could occur as in as many as 150000 persons, 36000 of whom would be expected to die. The plague bacilli would remain viable as an aerosol for 1 hour for a distance of up to 10 km. significant numbers of city inhabitants might attempt to flee, further spreading the disease (WHO 1970).

During May 20-23, 2000, local, state, and federal officials, and the staff of three hospitals in metropolitan Denver, participated in a bioterrorism exercise called Operation Topoff. As a simulated bioterrorist attack unfolded, participants learned that a *Y. pestis* aerosol had been covertly released 3 days earlier at the city's center for the performing arts, leading to >2,000 cases of pneumonic plague, many deaths, and hundreds of secondary cases. The available medical resources were exhausted virtually within hours (http://www.cdc.gov/ncidod/eid/vol6no6/hoffman.htm).

CHOOSING *Y. PESTIS* AS A BIOWEAPON

The cost of production of a biological weapon is about 0.05% the cost of a conventional weapon to produce a similar amount of mass casualties per square kilometer (Hawley and Eitzen, 2001). The biological characteristics of the plague pathogen make it understandably a reason for use as a biological weapon. Since the pathogen is found on the world's entire continent except Australia, it is easily accessible, simple to reproduce and for weapons purpose it is as economic and efficient. When inhaled, a mere infective aerosol dose of 100 to 500 bacteria are enough to cause pneumonic plague (by comparison it takes between 8000 and 50,000 spores breathed in to cause pulmonary anthrax) (Hilleman, 2002). This has an incubation period of 2-3 days and the person-to-person transmission exists which makes it high on the list of morbidity or mortality if untreated (Klietmann and Rouff, 2001).

For an effective biological attack the aerosol particle size must be in between 1 and 5mm for the particles to settle in the lower respiratory tract and be undetectable by our senses. Smaller particles will be exhaled because of the aerodynamics of particle flow through the respiratory tract. Larger particles will settle on upper respiratory tract, allowing mucocilliary clearance (Hawley and Eitzen, 2002). In short, in a biological attack the main form of pneumonic plague will be on inhalation of aerosolized preparation of *Y. pestis* or person-to-person transmission after infection by respiratory droplets. The chance that pneumonic plague can be spread from person to person by droplet nuclei (aerosol) is very rare (Inglesby et al., 2000).

The disadvantages with these methods are that *Y. pestis* is not stable for long periods in sunlight and high air temperatures. Other factors, which limit the generation of aerosols, include release of the pathogen in extremely small particles with a constant volume, the climatic factors such as wind velocity and direction, humidity, degree of cloud protection from direct sunlight, and rainfall (Beeching et al., 2002).

VIRULENCE FACTORS OF *YERSINIA PESTIS*

To maintain continuous circulation of plague pathogen in natural foci, the pathogen should be capable of invading host organism, resisting the bactericide protective systems of rodent, and reproducing itself to maintain the content of bacteria at a level sufficient for further transmission by fleas to a new host. Each of these stages of the *Y. pestis* circulation is determined by a variety of factors of plague pathogen, which may act either individually or in combination. Each of the factors itself may be involved in the pathological process at different stages of its development or in pathogen transmission. However, it is only the aggregate of the factors (regardless of significant or insignificant individual contribution to the sum effect) that provides persistence of plague pathogen in natural foci. The plague pathogen factors providing its transmission from one host organism to the next as well as correlation of individual factors of pathogenesis and expression of various household genes with plague pathogenesis virulence are considered (Anisimov). How *Y.pestis* adapted to become a pathogen of such ferocity, and that too among the members of *Enterobacteriaceae*, one that makes use of blood sucking insect for its transmission, provides an interesting case study in the evolution of microbial pathogenicity (Hinnebusch, 1997). For a long time, it was a commonly held belief that each invasive pathogenic bacterium has its own lifestyle, and that there is a great diversity of individual bacterial virulence strategies. However, recent data from several laboratories challenges this view and reveal the existence of related major virulence system in various pathogenic bacteria. One of these systems involves the delivery of bacterial proteins inside eukaryotic cell by surface bound bacteria after being close contact with the target cell surface. The Yop virulon of *Y. yersiniae* represents the archetype of this growing family of system (Cornelis and Wolf-watz, 1997). In spite of differences in the infection routes of 3 pathogenic species, a common tropism for lymphoid tissues and a common capacity to resist the non-specific immune response is shared among them (Cornalis and Wolf-watz, 1997). Among the factors associated with virulence, a major virulence determinant of *Y.pestis* is calcium dependence. Virulence and Ca^{2+} dependency turned out to be determined by a 70kb plasmid, which in vitro governs the massive release of a set of approximately 12 proteins called *Yersinia* outer membrane proteins (Yops), 29 Yop secretion proteins (Ysc) and few specific Yop chaperone (Syc) (Cornelis, 2000). Yops were originally designated by a letter, a number or their molecular weights. A uniform nomenclature has been introduced for *yop*B, *yop*D, *yop*E, *yop*H, *yop*M and *lcr*V. YopT was described only recently (Iriarte and Cornelis,

1998). Although initially described as outer membrane proteins, the Yops could also be recovered from the culture supernatant (Heesemann et al., 1984 and Heesemann et al., 1986) and it was later found that they are actually the secreted proteins (Micheils et al., 1990). This secretion occurs by a new pathway now called as type III and requires a specific apparatus called Ysc for Yops secretion, which is also encoded by the same plasmid (Perry and Fetherston, 1997).

To trigger Yops secretion *in-vitro*, *Yersinia* is grown at 28°C in a medium depleted of calcium and then transferred to 37°C. Calcium depletion and temperature, both control transcription of most of the genes involved in Yops synthesis and secretion (Lambert et al., 1992). Genetic analysis indicated that most of the Yop proteins are essential for virulence. In particular, YopE turns out to be responsible for cytotoxicity (Rosqvist et al., 1990). YopH was found to inhibit phagocytosis of bacteria by macrophages (Rosquist et al., 1988) and later was shown to be a protein tyrosine phosphatase (PTPase) related to eukaryotic counterparts. YopB and YopD have hydrophobic domains (Hakansson et al., 1993), this suggests that they could interact with membranes and take part in translocation of other Yops. Yop translocation through the putative pore seems to be controlled by the 21 kDa YopK (Holmstrom et al., 1995). Overproduction of YopK leads to a reduction of the YopB-dependent lytic effect on infected HeLa cell and sheep erythrocytes, probably by influencing the size of the pore (Holmstrom et al., 1997). YopM has thrombin binding activity and competitively inhibits thrombin induced platelet activation *in-vitro*, suggesting that YopM is an extracellular effector (Leung et al., 1990; Reisner and Straley, 1992). YopO shows some similarity to the COT (cancer osaka thyroid) oncogene product, a cytosolic serine or threonine protein kinase expressed in hematopoietic cells and is implicated in signal transduction by growth factors (Higashi et al., 1990). The 32 kDa YopJ is considered a minor Yop in the sense that, in *in-vitro*, it is secreted in smaller amount than most of the other Yops. However, *in-vitro* YopJ induces apoptosis in murine macrophages (Mills et al., 1997; Monack et al., 1997). YopT is a 35.5 kDa Yop effector that has been described and characterized recently (Iriarte and Cornelis, 1998). It induces a cytotoxic effect in HeLa cells and macrophages. The LcrV protein, known since the mid-1950s is a protective antigen against plague. Unlike YopB and YopD, this Yop exhibits a certain degree of polymorphism; in particular, the region between amino acids 225-232 appears to be hypervariable. LcrV has been described as a regulatory protein involved in the calcium response (Cornelis et al., 1998).

The structural genes for plasminogen activator, pesticin and pesticin immunity reside on a 9.5kb plasmid designated pPCP1 (Hoe et al., 1992). The fibrinolytic and coagulase activities are found to reside in a single gene encoding a 34.6kDa outer membrane protein called plasminogen activator (Pla protease). Various roles of Pla have been postulated, including cleaving fibrin deposits that trap the organism, producing excess plasmin that causes poorly organized or ineffective structures between inflammatory cells and fibrin, degrading mechanical constraints to spreading such as extracellular and basement membrane proteins and reducing chemo-attractants at the infection site possibly via inhibition of interleukin-8 production (Perry and Fetherston, 1997). The bacteriocin pesticin is 39.9 kDa protein .The gene coding for pesticin is 1074 bp long and encodes 357 amino acids. It exhibits N-acetylglucosaminidase activity. This bacteriocin has now been clearly shown to lack virulence enhancing effects in organisms after subcutaneous injection. Pesticin synthesis continues during Ca^{2+}-deficient growth restriction, when synthesis of many proteins has ceased (Rakin et al., 1996; Perry and Fetherston, 1997). The 16 kDa pesticin immunity protein was found to be in the periplasm. The location of the immunity protein suggests that imported pesticin is inactivated in the periplasm before it hydrolyses murein, a lipoprotein associated with the cell membrane (Pilsl et al., 1996).

The F1 expressed by 110kb (pMT) mega plasmid, is believed to confer resistance to phagocytosis, possibly by forming aquous pores in the membrane of phagocytic cells. In the recent study Du and coworkers (2002) have addressed the role of F1 antigen in inhibition of phagocytosis by the macrophages like cell line J774. The *Y.pestis* strain EV76 was found to highly resistant to uptake by J774 cells. An in-frame deletion of the cafM gene of the *Y.pestis* strain EV76 was constructed and found to be unable to express F1 polymer on the bacterial surface. This strain had a somewhat lowered ability to prevent uptake by J774 cells. Strain EV76C, which is cured for the virulence plasmid common to the pathogenic *Yersinia* species, was, as expected, much reduced in its ability to resist uptake. A strain lacking both the virulence plasmid and caf1M was even further hampered in the ability to prevent uptake and, in this case, essentially all bacteria (95%) were phagocytosed. Thus, F1 and the virulence plasmid-encoded type III system act in concert to make *Y. pestis* highly resistant to uptake by phagocytes. In contrast to the type III effector proteins YopE and YopH, F1 did not have any influence on the general phagocytic ability of J774 cells. Expression of F1 also reduced the number of bacteria that interacted with the macrophages. This suggests that F1 prevents uptake by interfering at the level of receptor interaction in the phagocytosis process.

The detection of antibodies against F1 is the basis of standard serological tests for the surveillance and diagnosis of plague as infected animals and patients produce a strong antibody response to this antigen. The structural genes for F1 (*caf1*) and the associated genes *caf1M*, *caf1A*, and *caf1R* are encoded on the 110kb plasmid and provides the characteristic signature for *Y.pestis*. Despite this, numerous virulence tests of undefined and defined $F1^-$ mutants indicate that this factor is of little importance in disease in mice, causing at most a slight alteration in time to death. While F1 is a protective antigen, fully virulent $F1^-$ mutants have been isolated from vaccinated mice challenged with $F1^+$ strains. The role of protein capsule F1, was evaluated in virulence and immunity against plague. However, both active immunization with recombinant F1 as well as passive immunization with monoclonal antibody to F1, protected mice from experimental infection with wild type F1 positive strains (Friedlander et al., 1995).

Murine toxin (ymt) refers to the 2 forms of proteins (native polymers of 240 and 120 kDa) toxic for mice and rats but relatively ineffective against guinea pigs, rabbits,dogs, monkeys, and chimpanzees. The structural gene for murine toxin resides on the 110 kb plasmid and probably encodes a 61 kDa protein (Kutyrev et al., 1986). Although some toxins are released during *in-vitro* growth, it is associated primarily with the cell envelope or membrane until cell death and lysis. Murine toxin has been proposed as a b-adrenergic antagonist that may directly block the b-adrenergic receptor after *in-vivo* activation and cause circulatory collapse (Montie, 1981).

At least 3 different proteins: Inv (invasin), Ail (attachment invasion loculs) and YadA can independently promote in vitro invasion of epithelial cell line by Yersinia. Only the first 2 are chromosomally encoded. Although all *Yersinia* harbour an *inv* gene, only enteropathogenic strains synthesize a functional Inv protein.

Iron is an essential growth factor for almost all bacteria. The ability to capture iron in vivo is one of the main factors that differentiate high and low pathogenicity bacteria and is a critical factor in *Yersinia* (Baumler et al., 1993, Perry, 1993). Siderophore—one pathway for iron capture is to secrete small molecules called siderophores. These molecules chelate the iron bound to eukaryotic proteins, bind to specific receptors on the outer membrane, and deliver the metal inside the cytosol of the bacteria. Several siderophores of the hydroxamate and catecholate types are not synthesized but can still be utilized by *Yersinia*. Low pathogenicity *yersiniae* can use this siderophore for capturing iron, thus enhancing their ability to disseminate in the host. High-pathogenicity *Yersinia* do not require this siderophore to cause severe infection. The siderophore identified is yersiniabactin a small particle of 480 daltons that contains a catechol and a noncatechol group but no hydroxamide. This siderophore is synthesized by high-pathogenicity *Yersinia* only and therefore, may be one of the critical factors required for their growth and dissemination in vivo. The second pathway used by *Yersinia* to capture iron is to utilize the iron molecules bound to heme proteins (Carniel, 1995). In *Y.pestis*, a 102kb region of the chromosome termed the pigmentation (*pgm*) locus is bounded by directly repeated copies of IS100. This region is divided in 2 functionally and physically distinct loci (Fetherston and Perry, 1994). One carries the *hms* (hemin storage) locus which confers pigmented phenotype on colonies grown on Congo red-agar plates and enhances transmission of the organism by blockage and death of flea vector. The *hmsHFRST* locus, which contains 5 structural genes encoding mature proteins of HmsH, HmsF, HmsR, HmsS, and HmsT. The second locus carries the genes involved in siderophore mediated iron acquisition, the *ybt* (yersiniabactin biosynthetic) locus which facilitates the bacteria to overcome the otherwise iron deficiency in the mammalian host (Perry and Fetherston, 1997). The genes-*irp1*, *Irp2*, *ybtT*, *ybtA*, *ybtE* and *psn* all lie within the ybt region ®. These genes are responsible for the biosynthesis of yersiniabactin siderophores on an iron deficient growth media. Transport of iron and iron-yersiniabactin complex back into the bacteria cell takes place through a surface receptor termed *psn* in *Y.pestis.*

Yersinia pestis must survive and/or grow in blood to be transmitted between its insect and mammalian hosts. Thus resistance to complement-mediated lysis (serum resistance) is probably required for survival in both hosts. *Yersinia pestis* displays LCR-independent serum resistance after growth at 26°C or 37°C (Perry and Brubaker, 1983). In these studies, complement was activated by the alternate pathway in the absence of specific antibody. *Yersinia pestis* is serum resistant even in the presence of antibodies and that this resistance is encoded chromosomally but outside of the *pgm* locus (Perry and Schuetze-Unpublished observations). The rough or short lipo-polysaccharide of *Y. pestis* appears to mediate this resistance, possibly by causing aberrant attachment of the terminal membrane attack complex as in *Neisseria gonorrhoeae* (Porat et al., 1995).

Catalase activity in *Y. pestis* as measured by decomposition of hydrogen peroxide is high compared to that in a number of other bacteria. Rockenmacher (1949) found a correlation between high catalase activity and virulence. Mehigh and Brubaker (1993) identified a 70 kDa protein that bound hemin and had a modest catalase but no peroxidase or superoxidase dismutase activities. Although this protein cross-reacted with antibody against antigen 5, it represented less than 10 percent of the catalase activity in cell extracts. Intriguingly, this protein was expressed during in vitro Ca^{2+} deficient growth restriction, when protein synthesis is limited primarily to known or putative virulence determinants.

All the 3 pathogenic species of *Yersinia* express fimbriae, which are found, on the surface. These are called pH 6 antigen in *Y. pestis* and *Y. pseudotuberculosis* whereas Myf in *Y. enterocolitica*. This antigen is produced under acidic conditions by *Y. pestis* and *Y. pseudotuberculosis*. Cells grown between pH 5.0-6.7 and at temperatures of 35 to 41°C express this antigen. The antigen is expressed *in-vivo* in the liver and spleen of mice and *in-vitro* during calcium deficient growth restriction (Mehigh and Brubaker, 1993). Intracellular synthesis of pH 6 antigen is dependent upon acidification of the phagolysosome and has been demonstrated in macrophages. Possible extracellular sites where pH 6 antigen could be expressed include abscesses such as buboes and lesions in the liver and spleen (Linder et al., 1990; Linder and Tall, 1993). It has been suggested that the pH 6 fibrillar structure may provide entry into the naive macrophages and participate in the delivery of *Yersinia* outer membrane proteins (Yops) into the phagocytic cells (Perry and Fetherston, 1997).

After 27 years of quiescence, India experienced 2 plague outbreaks in the year 1994, one in the month of August, of bubonic plague from the Beed district of Maharashtra State and another in late September, of pneumonic plague in the city of Surat in neighboring Gujarat State (Campbell and Hughes, 1995). A total of 18 *Y.pestis* isolates that included 11 from pneumonic patients and 7 from rodent could be recovered from outbreak regions (Batra et al., 1996). The initial 3 isolates from the Fort Collins, USA and Stavrapol Anti-plague Research Institute, Russia, brought out an important factor that the infectivity dose to bring out mortality in experimental mice was relatively high even though 3 virulence associated plasmids were demonstrated in these isolates (TAC, Report). Presence of these 3 plasmids was also observed in rest of the *Y.pestis* isolates from outbreak regions (Panda et al., 1996). Further surveillance work yielded a lot of 8 more *Y.pestis* and 4 *Y.pseudotuberculosis* isolates from rodents of Deccan plateau of Southern India (Unpublished work). Recent finding concerning the molecular pathogenesis of *Y.pestis* both in its mammalian and its insect hosts, and genetic comparison of these isolates are beginning to provide some insight into the emergence of modern plague. To ascertain the virulence markers contributed by 3 virulence associated plasmids (110 kb, 70 kb & 9.5 kb) as well as the chromosomal region in the recently recovered Indian strains of *Y.pestis* following work was undertaken in DRDE, Gwalior.

For, 70kb LCR plasmid, 10 important Yops of which 6 effectors (YopE, YopH, YopJ, YopM, YopO, & YopT), 3 translocator (YopB, YopD, & YopK) and 1 regulator (lcrV) were targeted for identification. Identity of these yops was initially attempted by the conventional procedures of partially purifying Omps after cultivation of *Y.pestis* standard and isolates in calcium deficient media. Prominent bands numbering 4-5 between 34-42kDa region corresponding to important Yops were seen in all isolates as well as in other *Yersinia* and non-*Yersinia* species by SDS-PAGE. It may therefore, be difficult to identify the Yops only on basis of molecular mass because certains *Yersinia pestis* proteins (YPP) can overlap Yops. Western blotting with the polyclonal antisera raised against these Omp preparations revealed few immunoreactive bands that appeared to be shared among *Y.pestis*, *Y. pseudotuberculosis*, *Y.enterocolitica*, *Y.fredrekseneri*. *Y.intermedia*, *Y.kristensenerii*, and *E.coli*. It is therefore, evident that under low calcium response, the *Y.pestis* not only activates secretions of Yops but also a large number of other proteins, as per our observations are mostly cross reactive with protein of *Enterobacteriaceae*. Existence of the common epitopes at different antigens in these organisms suggests, the possibility of some homology among the proteins expressed under low calcium response, as all the 4 monoclonal antibodies obtained were cross reactive to the *Enterobacteriaceae* organisms. SDS-PAGE analysis, polyclonal antibodies and the monoclonal antibodies to Omps preparations though confirmed the existence of YPPs, but did not provide a clear indication of presence of specific Yops. In a parallel strategy, PCR was attempted for detection of these genes. For this, primers were designed, PCR conditions were standardized and test performed on all the 26 *Y.pestis* isolates. Sixteen out of 18 *Y.pestis* isolates from the outbreak regions exhibited PCR amplification for all the 10 yop genes tested and product obtained were of expected sizes. Isolate no. 111 was negative for *yopE* gene and isolate no. 115 for *yopH* gene. Isolates recovered from the *Tatera indica* trapped during plague surveillance work from the region of Deccan plateau had lots of variations in the yop genes. One of the 8 isolates, 24H, was found consistently negative for all the 10 yop genes. The *yopD* gene appeared absent in all of these isolates while *yopM* could be detected in one isolate (10R). Another isolate 9R also had negative results in *yopB*. In order to reconfirm the PCR results of different yop genes another experiment was planned, where in, the 3 *yop* genes, *yopM* (effector), *yopB* (translocator), *lcrV* (regulator) were selected. The truncated PCR products of yopM and yopB genes were cloned to express recombinant proteins using pQE32 vector and lcrV was cloned in pQE30 vector. All the recombinant clones were induced and expressed by IPTG and the recombinant proteins were purified by Ni-affinity chromatography. The cloning strategy adapted in the present study employing pQE series of expression vector, produced an N-terminal 6x-histidine fusion proteins with appreciably high yields. Polyclonal antibodies were raised against these recombinant proteins followed by testing on standard *Y.pestis*, other *Yersinia* species and non-*Yersinia* species. Hyper Immune Sera (HIS) to recombinant YopM exhibited reaction at 32kDa, YopB at 28kDa and the LcrV at 31kDa regions in cell lysate antigens of IPTG induced *E. coli* SG13009, also to the purified recombinant proteins. The reaction at expected size of all the 3 antigens was observed with *Y.pestis* standard strains and also in Omp preparations. Host cells as well as the uninduced culture of recombinant *E. coli* SG13009 and other *Yersinia* and non-*Yersinia* species tested did not exhibit any reaction. In order to further establish epitope specific immuno-reactivity of 3 Yop proteins, monoclonal antibodies were generated using standard hybridoma protocol. Ten reactive stable clones were obtained for recombinant YopB, 9 for YopM and 6 for LcrV and specificity of all clones was tested with other *Yersinia* and non-*Yersinia* species in dot-ELISA. All the clones were found reactive only to *Y.pestis* suggesting, therefore, that even at epitope level specificity of these 3 Yops is maintained. All the monoclonal antibodies generated against these 3 recombinant proteins in the present study were tested on Indian *Y.pestis* isolates. Western blot reactions with monoclonal antibodies to recombinant proteins were observed at the expected sizes at 42kDa for YopM, 41kDa for YopB and 37kDa for lcrV region of standard *Y.pestis* A1122 and the isolates. The ELISA results with monoclonal antibodies were found correlating to the PCR results, thus reconfirming that the Indian *Y.pestis* strains are heterogenic in the yop genes of the 70kb plasmid. Also a clear distinction in the isolates of 1994 plague outbreak regions and Deccan plateau surveillance regions could be made on the basis of *yopD* and to an extent *yopM* genes.

Primers were designed and PCRs were standardized for all the important *ybt* genes (*ybtA, ybtE, ybtT, irp1, &irp2*) and *hms* genes (*hmsH, hmsF, hmsR, hmsS & hmsT*) of *pgm* locus. All the PCRs were utilized on the Indian *Y. pestis* strains. The results revealed that the pgm locus is relatively well conserved in isolates recovered from Indian sources. Interestingly, *irp2* region was not located in some of the rodent isolates of outbreak region. There appeared to be an heterogenous population of *Y. pestis* among rodents at the time of outbreak of which the highly virulent strains were responsible for the epidemic. Expression of HMWP also correlated with the PCR results of irp2 gene amplification further reconfirming the above hypothesis. Cloning of *Hms T*-PCR product of *HmsT* gene of *Y.pestis*, was utilized for the cloning and expression purposes. Briefly PCR products were ligated with pUC57/T.Transformed in competent *E. coli* DH5• cells. After blue-white selection, miniprep performed and positive clones were confirmed by PCR.

Subcloning-PCR products of recombinant pUC57 clone of different genes and series of pQE vectors were double digested with K*pnII* and H*indIII*. Ligation was performed at 16°C for overnight. Transformation was performed in *E. coli* SG13009. Antibiotic selection confirmation by PCR and restriction analysis. The recombinant protein was obtained at 44kDa region in SDS-PAGE. The recombinant protein thus obtained had antigenicity similar to that of the native protein. The polyclonal antibodies raised against the recombinant protein exhibited specific immunodominant reaction only with *Y. pestis*. A total of 11 monoclonal antibodies reactive to the protein could be stabilized. Immunoreactivity of these monoclonal antibodies further revealed the epitope specificity of the protein in *Y. pestis*. Result concludes that the hmsT protein appears to be highly conserved among *Y. pestis* strains. One *Y. pestis* isolate recovered from plague endemic region was consistently negative for all the *pgm* locus genes like all the yops studied. A simple sandwich dot ELISA could be standardized for rapid and reliable detection of virulence factor contributed by *hmsT* gene of *pgm* locus. Such an approach would be of immense help in developing immunoassays for other important virulence factors of *Y. pestis*.

Expression of Pla protein in these *Y.pestis* strains of Indian origin was also investigated following similar strategy. For, *pla* gene in pUC 18 vector followed by its subcloning in an expression vector pQE 30 was attempted.

The rPla was induced by addition of IPTG at a concentration of 2 mM to obtain the maximum yields and then the antigen was purified by Ni-NTA affinity chromatography. The yield of purified recombinant protein was found to be 25 mg/litre of culture. McDonough and coworkers (1989) had cloned the 1245 bp *Cla* I and *Hind* III fragment of fibrinolysin (Pla) in pBR322 in *E. coli* DH5a and expressed a 37 kDa protein in the membrane and periplasmic fractions. Both the fibrinolytic and coagulase activity could be seen in these cells. Recently, it was shown that Pla could easily be obtained from *E. coli* LE392/pPCP1 cultivated in enriched medium and that either autolysis or extraction of this isolate with 1.0 M NaCl results in release of soluble alpha and beta forms possessing biological activity (Kutyrev, et al., 1999) .

The molecular weight of histidine tagged rPla was estimated to be around 37 kDa by SDS-PAGE. In Western blot, the polyclonal antisera reacted at 35-37 kDa region only in IPTG induced recombinant *E. coli* and the standard *Y. pestis* indicating that the antigenicity of the recombinant protein. The antisera to rPla was further tested on the isolates of *Y. pestis* from the outbreak as well as from the surveillance regions. Reaction was observed with all the isolates at a similar position of 35-37 kDa region except one isolate, 24H, recovered from the surveillance rodent. This work is also supported by the data earlier obtained wherein PCR using *pla* specific detection primers on all the *Y. pestis* isolates except the isolate 24H, revealed amplification at 478 bp region (unpublished observation). The chances, that the gene being defective or absent in this particular strain of *Y. pestis* are more likely, as it was not only the immuno-reactivity of protein that was found missing, even the PCR failed to amplify the specific sequences.

Like the polyclonal antibodies, all the 9 MAbs failed to react to isolate 24H but reacted to all other isolates. The MAbs against Pla can be used for the detection of *Y. pestis* as an alternative to or in conjunction with the use of MAbs against F1. Being strictly specific, the MAbs against Pla can be useful for differentiation between *Y. pestis* from other *Yersinia* species including *Y. pseudotuberculosis* where high level of genome homology of over 90 percent is shared between the two (Perry and Fetherston 1997). In order to further confirm the presence of the *pla* gene in the *Y. pestis* isolates, the DNA preparations of representative strains were hybridized with the radiolabelled *pla* amplicon from the recombinant *E. coli* SG13009. The radiolabelled 985 bp *pla* probe specifically hybridized to *Y.pestis* strains, recombinant clone *E. coli* SG13009 but not to other *Yersinia* species, *E. coli* and *Y. pestis* 24H. This probe could easily differentiate the *Y. pestis* strain from its closely related species *Y. enterocolitica* and *Y. pseudotuberculosis.* The results are supported by the work of McDonough and coworkers (1988) where they had tested the 900 bp *BamH* I—*Hind* III fragment of the pesticinogenicity island that includes the gene that determines fibrinolysin synthesis.

The specificity and reliability of detection are inherent properties of a DNA probe, whereas its sensitivity is more a function of the system in which the probe is used. In the present study, a sensitivity of 1.95 ng of DNA could be obtained by dot blot type of analysis. As this 985 bp DNA fragment hybridized specifically with *Y. pestis* DNA, there is a possibility of this being useful in detecting DNA fragments of plague bacilli directly in fleas, or in rodent reservoirs. Additional applications for this DNA probe may include plague diagnosis and pathogenesis research.

PATHOGENESIS

Infection by flea bite results in a bubonic or septicemic plague, possibly complicated by secondary pneumonia. The person with pneumonic symptoms may be a source of a droplet-borne inhalatory infection for other people who consequently develop primary pneumonic plague. Despite a clinical form, plague is a severe infection characterized by a short incubation period, rapid onset and quick progress with mortality exceeding 50% if not treated properly. The pneumonic plague is associated with a particularly rapid progress and the mortality rate of almost 100% if not treated properly. As *Y. pestis* can be easily obtained and cultured and is highly pathogenic for humans, it poses a serious threat of being used for bioterrorism purposes.

Following attack by aerosolized *Y. pestis* bacilli primary pneumonic plague develops. The time from exposure to the aerosolized plague bacilli until development of first symptoms has been found to be 1 to 6 days and most often, 2 to 4 days. The first sign of illness would be expected to be fever with cough and dyspnea, sometimes with the production of bloody, watery, or less commonly, purulent sputum.

The ensuing clinical findings of primary pneumonic plague are similar to that of secondary plague pneumonic but with the absence of buboes (rarely, cervical buboes), and on pathological examination, pulmonary disease with areas of profound lobular exudation and bacillary aggregation. Chest radiographic findings are variable but bilateral infiltrates or consolidations are common. The time from respiratory exposure to death in humans is reported to have between 2 to 6 days in epidemics during the pre-antibiotic era, with a mean of 2 to 4 days in most epidemics (Inglesby et al., 2000).

MANAGEMENT OF PLAGUE OUTBREAK

The early diagnosis of plague requires a high index of suspicion in naturally occurring cases and even so following the use of a biological weapon. There are no effective environmental warning systems to detect an aerosol of plague bacilli. Indications that plague had been artificially disseminated would be the occurrence of cases in locations not known to have enzootic infection, among individuals with no known risk factors (e.g., animal contact), and the absence of prior rodent deaths (historically, rats die in large numbers prior to human outbreaks, precipitating the movement of the infesting flea population from the rats to humans). A pneumonic plague outbreak would result in symptoms initially resembling those of other severe respiratory illnesses.

Cultures of sputum, blood, or lymph node aspirate should demonstrate growth approximately 25 to 48 hours after inoculation. Most microbiology laboratories use either automated or semi automated identification systems where the identification may take as many as 6 days and there is some chance that diagnosis may be missed entirely. If a laboratory using automated or non-automated techniques is notified that plague is suspected, it should split the culture: one culture incubated at 28°C for rapid growth and the second culture incubated at 37°C for identification of the diagnostic capsular (F1) antigen. Using these methods, up to 72 hours maybe required following specimen procurement to make the identification (Inglesby et al., 2000). In case where the classical diagnosis using F1 antigen fails, monoclonal antibodies against the Plasminogen activator (Pla) can be used for diagnosis as it can detect the plague bacilli both at 28°C and 37°C by simple dot-ELISA (Feodorava and Devdariani, 2000). Other sophisticated diagnosis using molecular biological tools like PCR has also been used but its availability in peripheral diagnostic laboratories is a limitation in early diagnosis. Antibiotic susceptibility testing should be performed at a reference laboratory because of the lack of standardized susceptibility testing procedures for *Y. pestis* (Inglesby et al., 2000).

In a contained casualty setting, where a modest number of people require treatment a parenteral antibiotic therapy is recommended, preferably parenteral forms of the antimicrobials streptomycin or gentamicin. In a mass casualty setting, intravenous or intramuscular therapy may not be possible, so oral therapy, preferably with doxycycline (or tetracycline) or ciprofloxacin, should be administered. Patients with pneumonic plague will require substantial advanced medical supportive care. Complications of gram-negative sepsis would be expected, including adult respiratory distress syndrome, disseminated intravascular coagulation, shock, and multiorgan failure (Inglesby et al., 2000).

Once plague is confirmed or strongly suspected in a particular area, anyone in that area with fever (of 38.5°C or higher) or cough should immediately be treated with anti-microbials for presumptive pneumonic plague. Delaying therapy until tests confirm plague will greatly decrease the person's chance of survival. Doxycycline is the first-choice antibiotic for post-exposure prophylaxis. Asymptomatic persons who have had household, hospital, or other close contact (2 meters or less) with persons with untreated pneumonic plague should receive post-exposure prophylaxis for 7 days and be monitored for fever and cough. Tetracycline, doxycycline, sulfonamides, and chloramphenicol have been recommended for these individuals. On the basis of mice studies, fluoroquinolones might also be protective. Persons refusing prophylaxis should be closely monitored for the development of fever or cough for the first 7 days after exposure and should be treated immediately if either occurs. Clinical deterioration of patients despite early presumptive therapy could indicate anti-microbial resistance and should be promptly evaluated. Special measures should be taken for treatment or prophylaxis of those unaware of the outbreak or those requiring special assistance, such as persons who are homeless or who have cognitive disorders (Inglesby et al., 2000). (Refer to report by) for the detailed recommendations for treatment with pneumonic plague in the contained and mass casualty settings and for post exposure prophylaxis.

To prevent respiratory droplet infection precautions both the patients and the persons caring for the plague patients should wear surgical masks. Additional precautions such as gloves, gowns, eye protection should be taken care by the patients handling plague victims. Patients with pneumonic plague should be isolated until they have had at least 48 hours of antibiotic therapy and shown clinical improvement. Hospital rooms should receive terminal cleaning consistent with standard precautions; clothing and linens contaminated with the body fluids of pneumonic plague patients should be disinfected per hospital protocol. Laboratories should observe bio-safety level 2 conditions. Activities with a high potential for aerosol or droplet production (centrifuging, grinding, vigorous shaking, animal studies) require bio-safety level 3 conditions. Bodies of patients who have died should be handled with routine strict precautions. Aerosol-generating procedures (bone-sawing associated with surgery or post-mortem examinations) should be avoided. There is no evidence to suggest that environmental decontamination following an aerosol release be warranted. *Y. pestis* is very sensitive to sunlight and heating and does not survive long outside its host . According to the WHO analysis, a plague aerosol would be viable for 1 hour after release, long before the first cases would alert health personnel to a clandestine attack (Inglesby et al., 2000). In case decontamination of the environment is required, it can be discussed from two approaches: surface decontamination and area (space decontamination). Surface decontamination are used primarily for the interior of safety cabinets, room surface wash down, wiping off the exterior of certain items being removed from laboratories etc. There are number of general groups of decontaminants such as alcohols, halogens, quartenary ammonium compounds, phenolics, and glutaraldehyde. The efficacy of decontamination of inanimate surfaces with liquid household bleach is well documented. *Escherichia coli* was used as the gram-negative model in these experiments to simulate *Y. pestis* and other gram-negative bacteria. After contact for 1 minute with the undiluted liquid household bleach, an *E. coli* population was completely inactivated (a 6 $\log_{10}$ reduction in viability) (Weiner, 1996).

PROPHYLAXIS

A killed whole cell plague vaccine has been used in the past, but recent studies in animals have shown that this vaccine offers poor protection against pneumonic disease. The present US licensed plague vaccine is no longer distributed.

The live attenuated vaccine EV76 strain is given as a single dose of $5.8*10^6$ cfu. Immunization of mice with the EV76 vaccine induces an immune response which provides protection against subcutaneous and inhalation challenges with *Y. pestis*. Since it retains some virulence in most countries it is not considered to be suitable for use in humans (Titball and Williamson, 2001).

Recently an improved parenteral vaccination strategy for plague, based on the recombinant subunit approach, has entered clinical trails. The *Y. pestis* subunit antigens (F1 and V) have been successfully incorporated into novel vaccine delivery systems such as biodegradable microspheres composed of poly-L- (lactide) (PLLA). Intranasal and intratracheal administration of PLLA microencapsulated F1 and V serves to protect experimental animals from inhalation and subcutaneous challenge with virulent *Y. pestis* bacilli. Liposomes have also been used to improve the immunogenicity of intranasally administered *Y. pestis* antigens, and the effectiveness of this approach to plague immunization has been evaluated (Alpar et al., 2001). The sub- unit vaccine based on the F1- and V antigens has also tested been tested using live attenuated *Salmonella* as a delivery vector (Titball and Willimson, 2001).

For area or space the most commonly recommended decontaminant is formaldehyde but it is a safety hazard because it is a potential occupational carcinogen. Apart form being a powerful reducing agent, it has limited penetrating ability, and it is potentially explosive. Ozone is also being used as a sterilant. A new sterilization system is based on the vapor phase of hydrogen peroxide. The system provides a rapid, low- temperature technique that, because of its low toxicity, eliminates much of the potential public health hazard associated with decontaminants such as formaldehyde and ethylene oxide. In the cold sterilization process, 30% hydrogen peroxide (300,000 ppm) is vaporized to yield 700-1200 ppm. The hydroxyl radical, a strong oxidant is believed to have microbicidal activity through attack on membrane lipids, DNA and other essential cell components. The hydrogen peroxide vapor is unstable and degrades to the non- toxic residues of water vapor and oxygen. Any sealable enclosures such as small rooms, airlocks, biological safety cabinets, glove boxes, and isolation equipment, (up to 1,200 ft^3) can be sterilized. The process is effective at temperatures ranging from 4°C to 80°C. The vapor phase hydrogen peroxide sterilization appears to be safe and is effective against a variety of microorganisms (Hawley and Eitzen, 2001).

References

Alpar H.O., Eyles J.E, Williamson E.D. and Somavarapu S. Intranasal vaccination against plague, tetanus and diphtheria. Advanced Drug Delivery Reviews. **51:** 173–201, 2001.

Anisimov A.P. Factors of *Yersinia pestis* providing circulation and persistence of plague pathogen in ecosystems of natural foci. Communication 2. [Article in Russian. State Research Center of Applied Microbiology, Obolensk.

Batra H.V., Tuteja U. and Agarwal G.S. Isolation and identification of *Yersinia pestis* responsible for the recent plague outbreaks in India. Curr.Sci. **71:** 787-791, 1996.

Baumler A., Koebnik R., Stojiljkovic I., Heesermann J., Braun V., Hantke K. Survey on newly characterized iron uptake systems of *Yersinia enterocolitica*. Zentralbl Bakteriol. **278:** 416-17, 1993.

Beeching N.J., Dance D.A.B., Miller A.R.O., Spencer R.C. Biological warfare and bioterrorism. BMJ. **324:** 336-339, 2002.

Campbell G.L. and Hughes J.M. Plague in India; a new warning from an old Nemesis. Annals of Internal Medicine. **122:** 151-153, 1995.

Carniel E. Chromosomal virulence factors of *Yersinia*.. Pathogenicity. 13: 218-224, 1995.

Cornalis G.R. and Wolf-watz H. The *Yersinia* Yop Virulon: a bacterial system for subverting eukaryotic cells. Mol Microbio. **23:** 861-7, 1997.

Cornelis G.R., Boland A., Boyd A.P., Geuijen C., Iriarte Neyt C., Sory M.P. and Stainier I. The virulence plasmid of *Yersinia*, an antihost genome. 1998. Microbiol. Mol. Biol. Rev. **62:** 1315-52, 1998.

Cornelis. 2000. Molecular and cell biology aspects of plague. Natl. Acad. Sci. USA **97:** 8778-8783, 2000.

Domaradskij I.V. and Orent W. The Memoirs of an inconvenient man: Revelations about biological weapons research in the Soviet Union. Crit Rev Microbiol. **27(4):** 239-266, 2001.

Du Y., Rosqvist R. and Forsberg A. Role of fraction 1 antigen of *Yersinia pestis* in inhibition of phagocytosis. Infect Immun. **70(3):** 1453-60, 2002.

Feodorava V.A. and Devdariani Z.L. Development, characterisation and diagnostic application of monoclonal antibodies against *Yersinia pestis* fibrinolysin and coagulase. J. Med. Microbiol. **49:** 251-269, 2000.

Fetherston J.D. and Perry R.D. The pigmentation locus of *Yersinia pestis* KIM6+ is flanked by an insertion sequence and includes the structural genes for pesticin sensitivity and HMWP2. Mol. Microbiol. **13:** 697-708, 1994.

Friedlander A.M., Welkos S.L., Worsham P.L., Andrew G.P., Heath D.G., Anderson G.W. Jr, Pitt M.L., Estep J. and Davis K. Relationship between virulence and immunity as revealed in recent studies of the F1 capsule of *Yersinia pestis* . Clin Infect Dis. **21(2):** 178-81, 1995.

Hakansson S., Bergman T., Vanooteghem J.C., Cornelis G., Wolf-Watz H. YopB and YopD constitute a novel class of *Yersinia* Yop proteins. Infect. Immun. **61(1):** 71-80, 1993.

Hawley R.J., Eitzen E.M. Jr. Biological weapons -a primer for microbiologists .Annu Rev Microbiol. **55:** 235-53, 2001.

Health Aspects of chemical and biological weapons: Geneva,Switzerland: World Health Organization. 98-109, 1970.

Heesemann J., Algermissen B. and Laufs R. Genetically manipulated virulence of *Yersinia enterocolitica*. Infect. Immun. **64:** 2308-2314, 1984.

Heesemann J., Gross U., Schmidt N. and Laufs R. Immunochemical analysis of plasmid-encoded proteins released by enteropathogenic *Yersinia* sp. grown in calcium-deficient media. Infect. Immun. **54:** 561-567, 1986.

Higashi T., Sasai H., Suzuki F., Miyoshi J., Ohuchi T., Taikai S., Mori T. and Kakunaga T. Hamster cell line suitable for transfection assay of transforming genes. 1990. Proc. Natl. Acad. Sci. **87:** 2409-2412, 1986.

Hilleman M.R. Overview : cause and prevention in biowarfare and bioterrorism. Vaccine. **20(25-26):** 3055-3067, 2002.

Hinnebusch B.J. Bubonic plague : a molecular genetic case history of the emergence of an infectious disease. J. Mol. Med. **75:** 645-52, 1997.

Hoe N.P., Minion F.C., Goguen J.D. Temperature sensing in *Yersinia pestis*: regulation of YopE transcription by lcrF. J. Bacteriol. **174:** 4275-86, 1992.

Holmstrom A., Rosqvist R., Wolf-Watz H., Forsberg A. Virulence plasmid-encoded YopK is essential for *Yersinia pseudotuberculosis* to cause systemic infection in mice. Infect. Immun. **63:** 2269-2276, 1995.

Holmstrom A., Petersson J., Rosqvist R., Hakannson S., Tafazoll F., Fallman M., Mangnusson K.E., Wolf-Watz H., Forsberg A. YopK of *Yersinia pseudotuberculosis* controls translocation of Yop effectors across the eukaryotic cell membrane. Mol. Microbiol. **24:** 73-91, 1997.

Inglesby T.V., Dennis D.T., Henderson D.A., et al. Plague as a biological weapon: medical and public health. JAMA. **283(17):** 2281-2290, 2000 May.

Inglesby T.V., Henderson D.A., Toole T.O., Dennis D.T. Letters. JAMA. 284(13), 2000.

Iriarte, M., Cornelis G.R. YopT, a new *Yersinia* Yop effector protein, affects the cytoskeleton of host cells. Mol. Microbiol. **29:** 915-929, 1998.

Klietmann W.F. and Ruoff K.L. Bioterrorism: Implications for the clinical microbiologist. Clin Microbiol. Rev. **14(2):** 364-381, 2001.

Kliniki Chorob Zakaznych i Neuroinfekcji Akademii Medycznej w Bialymstoku. Grygorczuk S., Hermanowska-Szpakowicz T. *Yersinia pestis* as a dangerous biological weapon [Article in Polish] Med Pr. **53(4):** 343-8, 2002.

Kukkonen M., Lahteenmaki K., Suomalainen M. et al. Protein regions important for plasminogen activation and inactivation of a_2-antiplasmin in the surface protease Pla of *Yersinia pestis. Mol Microbiol.* **40:** 1097-1111, 2001.

Kutyrev V.V., Papov Yu.A. and Protsenko O.A. Pathogenicity plasmids of the plague microbe (*Yersinia pestis*). Mol. Genet. Mikrobiol. Virusol. **6:** 3-11, 1986.

Kutyrev V.V., Mehigh R.J., Motin V.L., Pokrovskaya M.S., Smirnov G.B., Brubaker R.R. Expression of the plague plasminogen activator in *Yersinia pseudotuberculosis* and *Escherichia coli. Infect Immun.* **67:** 1359-67, 1999.

Lambert de Rouvroit C.L., Sluiters C. and Cornelis G.R. Role of the transcriptional activator, VirF and temperature in the expression of pYV plasmid genes of *Yersinia enterocolitica*. .Mol. Microbiol. **6:** 395-409, 1992.

Leung K.Y., Reisner B.S. and Straley S.C. YopM inhibits platelet aggregation and is necesssary for virulence of *Yersinia pestis* in mice. Infect. Immun. **58:** 3262-3271, 1990.

Lindler L.E., Klempner M.S. and Straley S.C. *Yersinia pestis* pH 6 antigen: genetic, biochemical, and virulence characterization of a protein involved in the pathogenesis of bubonic plague. Infect. Immun. **58:** 2569-2577, 1990.

Lindler L.E., Tall B.D. *Yersinia pestis* pH 6 antigen forms fimbrae and is induced by intracellular association with macrophages. Mol. Microbiol. **8:** 311-324, 1993.

McDonough K.A., Schwan T.G., Thomas R.E. and Falkow S. Identification of a *Yersinia pestis*-specific DNA probe with potential for use in plague surveillance. *J Clin Microbiol.* **26:** 2515-2519, 1988.

McDonough K.A. and Falkow S. A *Yersinia pestis* specific DNA fragment encodes temperature dependent coagulase and fibrinolysin associated phenotypes. *Mol Microbiol* **3:** 767-775, 1989.

Mehigh R.J. and Brubaker R.R. Major stable peptides of *Yersinia pestis* synthesized during the low calcium response. Infect. Immun. **61:** 13-22, 1993.

Michiels T., Wattiau P., Brasseur R., Ruysschaert J.M., Cornelis G. Secretion of Yop proteins by *Yersiniae*. Infect Immun. **58(9):** 2840-9, 1990.

Mills S.D., Boland A., Sory M.P., Vandersmissen P., Kerbourch C., Finlay B.B., Cornelis G.R. *Yersinia enterocolitica* induces apoptosis in macrophages by a process requiring function type lll secretion and translocation mechanisms and involving YopP, presumably acting an effector protein. Proc. Natl. Acad. Sci. U.S.A. **94:** 12638-12643, 1997.

Monack D.M., Mecsas J., Ghori N. and Falkow S. *Yersinia* signals macrophages to undergo apoptosis and YopJ is necessary for this cell death. Proc. Natl. Acad. Sci. **94:** 10385-10390, 1997.

Montie T.C. Properties and pharmacological action of plague Murine toxin. Pharmacol. Ther. **12:** 491-499, 1981.

Panda S.K., Nanda S.K., Ghosh A., Sharma C., Shivaji S., Seshu J.K., Kannan K., Batra H.V., Tuteja U., Ganguly N.K., Chakrabarti A. and Sharatchandra H. The 1994 plague epidemic of India; molecular diagnosis and characterization of *Yersinia pestis* isolates form Surat and Beed. Current Science. **71:** 794-99, 1996.

Perry R.D., Brubaker R.R. Vwa$^+$ phenotype of *Yersinia enterocolitica*. Infect. Immun. **40:** 166-171, 1983.

Perry R.D., 1993. Acquisition and storage of inorganic iron and hemin by the *yersiniae*. Trends Microbiol **1:** 142-147.

Perry RD. and Fetherston J.D. *Yersinia pestis*—etiologic agent of plague. Clin Microbio Rev. **10:** 35-66, 1997..

Perry and Schuetze (Unpublished observation).

Pilsl H., Killmann H., Hantke K. and Braun V. Periplasmic location of the pesticin immunity protein suggests inactivation of pesticin in the periplasm. J. Bacteriol. **178(8):** 2431-2435, 1996.

Porat R., McCabe W.R., Brubaker R.R. Lipopolysaccharide associated resistance to killing of *Yersiniae* by complement. J. Endotoxin Res. **2:** 91-97, 1995.

Rakin A., Boolgakowa E. and Heesemann J. Structural and functional organization of the *Yersinia pestis* bacteriocin pesticin gene cluster. Microbiology. **142:** 3415-3424, 1996.

Reisner B.S. and Straley S.C. *Yersinia pestis* YopM thrombin binding and over expression. Infect. Immun. **60:** 5242-5252, 1992.

Rockenmacher M. Relationship of catalase activity to virulence in *Pasteurella pestis*. Proc. Soc. Exp. Biol. Med. **71:** 99-101, 1949.

Rosquist R., Sturnik M., Wolf-Watz H. Increased virulence of *Yersinia pseudotuberculosis* by two independent mutations. Nature. **334:** 522-524, 1988.

Rosquist R., Forsberg A., Rimpilainen M., Bergman T. and Wolf-Watz H. The cytotoxic protein YopE of *Yersinia* obstructs the primary host defence. Mol. Microbiol. **4:** 657-667, 1990.

Titball R.W. and Williamson E.D. Vaccination against bubonic and pneumonic plague.Vaccine. **19(30):** 4175-4184, 2001.

Wiener S.L. Strategies for the prevention of a successful biological warfare aerosol attack. MilMed. **161(5):** 251-56, 1996.

Pharmacological Perspectives of Toxic Chemicals and Their Antidotes
Editors: S J S Flora, J A Romano, S I Baskin and K Sekhar

Anthrax: Bacteriology, Clinical Presentations and Management

Nancy Khardori
Division of Infectious Diseases, Department of Internal Medicine
Southern Illinois University Springfield, Illinois, USA

HISTORICAL BACKGROUND

The earliest known description of anthrax was made in 1491 BC in writings from Egypt and Mesopotamia and in the Old Testament's description of the Fifth Plague of Egypt(1). There are descriptions of anthrax involving animals and humans in the early literature of Hindus, Greeks and Romans. The first pandemic in Europe known as "Black Bane" was recorded in 1613 and caused more than 60,000 deaths. The first epidemic in the United States occurred in the early 18th century. Outbreaks of occupational cutaneous and respiratory anthrax were reported from Industrial European countries in the mid-1800's. Cutaneous infection was caused by handling wool, hair and hides. Respiratory disease was caused by processes that created aerosol while handling wool, hair and hides.

- Deleford described the microscopic appearance of anthrax bacteria in 1838.
- Devain demonstrated the infectivity of anthrax in 1868.
- Anthrax became the first human disease attributed to a specific etiological agent when Koch showed it to fulfil his "postulates" in 1877.
- Pasteur first tested the attenuated spore vaccines in sheep in 1881.
- Decreased use of imported potentially contaminated animal products and improved industrial and animal husbandry practices led to a steady decrease in annual numbers of cases in the developed countries in the early 1900's.

- Sterne reported the development of an animal vaccine from the spore suspension of an avirulent, noncapsulated live strain of *Bacillus anthracis* in 1939.
- Cell free anthrax vaccine for humans—a sterile filtrate of cultures from an avirulent noncapsulated strain that elaborates protective antigen was licensed in the United States in 1970.
- Both live attenuated and killed vaccines have been developed. In the former Soviet Union, the human live anthrax vaccine has undergone many field trials (2).
- The largest recorded outbreak of anthrax in humans and likely the largest among animals occurred in Zimbabwe in 1978-1980 during the time of its Civil War (3). 10,000 human cases and 151 deaths were documented.

Anthrax-History of Current Threat

- Research on anthrax as a biological weapon started> 80 years ago (4).
- Today 17 nations are believed to have offensive biological weapons programs, that include Anthrax.
- Iraq has acknowledged producing and weaponizing anthrax between 1955 and 1991.
- Aerosols of anthrax bacteria and botulism toxin dispersed in Tokyo on 8 occasions failed to produce illness.
- WHO's expert committee (1970) estimated that an aircraft release of 50 kg of anthrax over a 5 million population would kill 250,000-100,000 of whom many would die without treatment.
- Accidental aerosolized release of anthrax spores in the Soviet Union in 1979 resulted in 79 cases and 68 deaths.
- Outdoor aerosol release could be a threat to people indoors (5).
- US Congressional Office of Technology Assessment (1993) estimated that between 130,000 and 3 million deaths could follow the aerosolized release of 100 kg of anthrax spore upwind of the Washington, D.C. area. The lethality would match or exceed that of a hydrogen bomb.
- The CDC economic model estimated-$26 billion per 100,000 persons exposed.

Microbiology

Bacillus anthracis—anthrakis is the Greek word for coal because anthrax causes black, coal-like skin lesions. All Bacillus species are aerobic, gram-positive, spore forming, non motile, bacteria. The spore size of *B. anthracis* is about 1 μg. Factors that favor sporulation include: alkaline soils (pH greater than 6.0); high nitrogen levels in the soil caused by decaying vegetation; alternating periods of rain and drought and; temperatures in excess of 15.5°C. Spores grow readily at 37°C on ordinary laboratory media with "curled hair" colony morphology and "Jointed Bamboo rod" cellular appearances on staining.

Spores germinate in an environment rich in amino acids, nucleosides and glucose into rapidly multiplying vegetative bacteria. The vegetative cell is nonflagellated and large (1-8 μm in length and 1-1.5 μm in breadth). Full virulence requires the presence of a capsule and a three component toxin-protective antigen, lethal factor and edema factor. The toxin has two enzymatic components. The first or edema factor (EF) is an adenylate cyclase that leads to increase in cyclic AMP resulting in edema at the site of infection. The second or lethal factor (LF) is a protease that appears to alter the production of cytokines by macrophages and to induce macrophage lysis and lethal effects of anthrax in animals.(6, 7). A third non-enzymatic component, the protective antigen (P A) helps in the delivery of the two enzymatic components into the cells (8). Adding a mutant P A can prevent release of EF /LF inside the cells (9). The gene coding for major virulence factors of B. anthracis reside on plasmids (10,11). The sequences of the virulence plasmids, p x 01 and p x 02 in *B. anthracis* have already been completed (12,13). The work on the anthrax genome itself has also been completed. Vegetative bacteria survive poorly outside the animal or human host and form spores after local nutrients are exhausted, e.g., infected body fluids exposed to ambient air. The hardy spores can survive for decades in the environment.

EPIDEMIOLOGY

Anthrax is a disease of herbivores acquired by ingesting spores from the soil However, few if any warm blooded species are entirely immune to it. Prior to animal vaccine and antibiotics, the disease was one of the foremost causes of uncontrolled mortality in cattle, sheep, goats, horses and pigs worldwide (14). Animal vaccination programs have drastically reduced animal mortality. Humans are incidental hosts. Anthrax spores continue to be documented in soil samples from throughout the world.

Sources of Animal Anthrax

1. Grazing in "incubator areas" (soil contaminated with *B. anthracis* spores and organisms).
2. Excreta and saliva from dying or dead animals.
3. Imported bone meals and vegetable protein (e.g. groundnut).
4. Wool, hair wastes.
5. Cleanings used in fertilizers
6. Tannery effluents

7. Commercial animal feed (rare in US - last outbreak among swine in 1952).
8. Blood-sucking flies.
9. Carrion eaters.

Natural disease in humans is acquired by contact with anthrax-infected animals or contaminated animal products. Anthrax remains a problem in developing countries. Human cases occur in industrial or agricultural environments. The incubation period is 2-5 days. Older observations showed that unimmunized workers in wool mills could inhale several hundred spores daily without developing disease. The LD_{50} for aerosolized anthrax spores is around 8000 colony forming units in experiments done by the US Army in cynomolgous monkeys (15). Fatality rate under these experimental conditions is 20-80 %. However, the occurrence of sporadic cases in people with a low dose contact may be explained by differences in the virulence of strains and the role of host factors.

CLINICAL MANIFESTATIONS

Cutaneous Anthrax

- The most common naturally occurring form—95% of anthrax cases in developed countries, 224 cases in the US between 1944 and 1994.
- Exposed areas on the arms and hands followed by face and neck
- Pruritic papule → ulcer surrounded by vesicles → black necrotic central eschar with edema.
- After 1-2 weeks, eschar dries, loosens, separates, leaving a permanent scar.
- Regional lymphangitis and lymphadenitis and systemic symptoms. Mortality rate for untreated disease-20%.
- Antibiotics decrease edema and systemic symptoms.
- Differential diagnosis-Plague and Tularemia.

Respiratory Anthrax

- Inhalational anthrax follows deposition of spore-bearing particles of 1-5 μ into airways (alveolar spaces).
- The size of a B. anthracis spore is 1 μm.
- Macrophages ingest spores resulting in their lysis and destruction.
- Surviving spores are transported to mediastinal lymph nodes.
- Germination may occur up to 60 days.
- Following germination, disease follows rapidly.
- Toxins released by replicating bacteria cause hemorrhage, edema and necrosis.

- Typical bronchopneumonia does not occur. Chest x-ray findings and absence of hemoptysis differentiates inhalational anthrax from pneumonic plague.
- LD_{50} is 2500 to 55,000 inhalcd spores.
- Hemorrhagic thoracic lymphadenitis, hemorrhagic mediastinitis (all patients) and hemorrhagic meningitis (50% of patients) are the pathological hallmarks of disease.
- Clinical presentation shows a biphasic pattern-non specific symptoms followed by fever, dyspnea, diaphoresis and shock.
- Morality rate is 80-90%, when untreated.
- Aggressive, early antimicrobial therapy and improved supportive care improves prognosis.

GASTROINTESTINAL ANTHRAX

Oropharynegeal anthrax

Oral or esophageal ulcer-regional lymphadenopathy, edema and sepsis

Abdominal anthrax

Predominantly terminal ileum or cecum.

Nausea, vomiting, malaise progressing to bloody diarrhea, acute abdomen and sepsis. Mortality rate is high.

DIAGNOSIS

Cutaneous—Vesicular fluid from skin lesions—gram stain and culture

Inhalational—Chest X-ray—widened mediastinum

Blood—Gram stain and culture

Biopsy—histopathology and culture

Gastrointestinal—Biopsy—histopathology and culture

Rapid diagnostic tests—EIA, PCR—for confirmation

New rapid molecular diagnostic tests are being extensively studied. More than 1200 strains of B. anthracis have been identified around the world over the years. Dr. Paul Keim's genetics laboratory at Northern Arizona University in Flagstaff, has used amplified fragment length polymorphism (AFLP) to examine all of them (16). His laboratory has also adapted some precise assays like VNTR (Variable Number Tandem Repeat) and ML V A to study 400 of the 1200 known strains of B. anthracis. It takes 12 hours for this laboratory to analyze an anthrax sample. Detection of *B. anthracis* DNA by light cycler Polymerase Chain Reaction after autoclaving (17) and by Rapid Cycle Real Time Polymerase Chain Reaction - The Mayo Roche Rapid Anthrax Test (18) were reported recently.

Control and Prevention

- Formaldehyde disinfection
- Industrial hygiene-dust collecting equipment and effective environmental clean up procedures.
- Environmental decontamination—paraformaldehyde vapor.
- Spores can persist and remain viable for 36 years.
- Surface contamination—5% hypochlorite or 5% phenol.
- Forbidding the sale of meat from sick animals.
- Cooking all meats thoroughly.
- Control of anthrax in animals—vaccination and reporting of disease.

Immunization

- Human attenuated live anthrax vaccine used in former Soviet Union.
- Human killed anthrax vaccine.
- Sterile filtrate of cultures from an avirulent noncapsulated strain that elaborates protective antigen—human vaccine in use in the US.
- The vaccine was field tested in employees of four textile mills in the US—"Anthrax Vaccine Adsorbed" (A V A).
- Effectiveness—92.5%.
- Given SQ 0.5 ml at 0,2,4 weeks and 6, 12, 19 months followed by annual boosters.
- Used for people exposed to contaminated materials or environments.
- US Armed Forces—1998 Vaccinate every member—1.4 million active duty troops and 1 million reservists.
- The vaccine is produced by Bioport Corp., Lansing, Michigan. Current information available at the Bureau of Disease Control and Laboratory Services, Michigan Department of Public Health, PO Box 30035, 3500 N. Logan Street, Lansing, Michigan.

Post exposure vaccination following a biological attack with anthrax is recommended to protect against residual retained spores after chemoprophylaxis. This approach may also reduce the duration of antibiotic prophylaxis to 30—45 days.

Mycoplasma contamination of A V A had been suggested as a possible cause of Persian Gulf Illness. Recent studies by nonmilitary laboratories did not show any mycoplasma or mycoplasma DNA and did not support its survival in the vaccine (19).

Future vaccines—Recombinant anthrax toxin, P A toxoid vaccines, Pa—producing live vaccines.

Friedlander and others at the USAMRIID in Fort Detrick, Maryland, have shown that recombinant P A, produced by non-spore- forming *B. anthracis* protects rhesus monkeys against inhalational anthrax (20). AVANT Immunotherapeutics, Needham, Massachusetts, is developing an oral one dose anthrax vaccine. This vaccine is made from attenuated V. cholerae that produces P A and acts rapidly.

Chemoprophylaxis/Post -Exposure Prophylaxis*

Drug	*Adults*	*Children*
Oral Fluoroquinolones Ciprofloxacin (Cipro)1	500 mg bid	10-15 mg/kg bid[2]
Oral tetracyclines3 Doxycycline (Vibramycin, others)4	100 mg bid	2.2, g/kg bid2
Oral Penicillins (3.5) Amoxicillin (Amoxil, others)6	500 mg tid	80 mg/kg/day divided into 3 doses

[1] Other tluoroquinolones such as otloxacin (Floxin) 400 mg bid or Levotloxacin (Levaquin) 500 mg once daily may also be effective.

Ciprotloxacin approved by FDA in 2000.

[2] Should be changed to Amoxicillin as soon as susceptibility to penicillin has been confirmed.

[3] Susceptible strains

[4] Tetracycline 500 mg qid should also be effective

[5] Penicillin resistance could emerge during treatment, but should not be a problem in prophylaxis

[6] Penicillin VKI 7.5 mg/kg in adults, or 12.5 mg/kg qid, should also be effective for prophylaxis

* **Medical Letter, October 29, 2001** (21)

Treatment

- No clinical studies of the treatment of inhalational anthrax in humans.
- Most anthrax strains are sensitive to penicillin—preferred antibiotic for treatment in the past (b-lactamase production).
- Penicillin and doxycycline approved by the FDA.
- Engineered vaccine strains resistant to penicillin and tetracycline.
- All fluoroquinolones active in vitro. Ciprofloxacin excellent efficacy in animal models.
- Combination antibiotic therapy may have a role.
- Other antibiotic choices include streptomycin, erythromycin, chloramphenicol, vancomycin, clindamycin, and first generation cephalosporins.
- Treat for 60 days because of risk of delayed germination of spores.
- Treatment of cutaneous anthrax does not alter the evolution of eschar but prevents systemic disease.
- Systemic steroids for cervical edema and meningitis

EMERGING/INVESTIGATIONAL THERAPIES

The CDC and other federal agencies are discussing the use of "antitoxin" therapy as an adjunct to antimicrobial therapy (22). Currently a limited supply of plasma collected from vaccinated military personnel is available. There are plans to collect a second larger batch from vaccinated volunteers for use in treatment and for animal studies. Maynard et al used in vitro DNA manipulation and an E. coli expression system to create an antibody library. Selected antibodies from this library were shown to bind to P A of *B.anthracis* with high affinity, prevented anthrax toxin from binding to macrophages and protected rats from a lethal challenge (23). Iverson and Georgiou at the University of Texas, Austin, have reported the production of a monoclonal antibody against anthrax toxin (unpublished data). This antibody reportedly has 40-fold better affinity for the toxin and protected rats injected with the toxin (22). The other approach would be to design a polyvalent inhibitor of anthrax toxin. Mourez et. al. isolated a peptide from a phage display library that binds weakly to the heptameric cell-binding subunit of anthrax toxin and prevents the interaction between cell-binding and enzymatic moieties. (24) A molecule consisting of multiple copies of this non natural peptide prevented assembly of the toxin and blocked its action in an animal model. A number of investigators have identified the cellular receptor for protective antigen (25) and the crystal structure of the lethal factor (26,27). These advances have the potential of helping design new drugs and or antitoxin therapies.

Hospital Infection Control

- Standard barrier precautions for all forms.
- Contact isolation for draining lesions.
- Notify Microbiology Laboratory—BSL 2 condition.
- Hypochlorite for environmental cleaning.
- Proper burial or cremation of humans and animals.
- Autopsy related instruments and materials autoclaved or incinerated.

FUTURE RESEARCH

- Improved rapid diagnostic techniques.
- Improved prophylactic and therapeutic regimens.
- Improved second generation vaccine.
- Impact of B. cereus genes on vaccine induced immunity.
- Improved capability to distinguish between highly similar types of *B. anthracis*.
- The entire DNA sequence of the "Florida" strain taken from the first 2001 victims has now been read and some rare distinguishing features have been identified (28).

References

Koch R. The etiology of anthrax, based on the life history of Bacillus anthracis. In: Brock T.D. (ed). Milestones in Microbiology. Washington, D.C., American Society for Microbiology, 89-9, 19610.

Turnbull P.C. Anthrax vaccines: past, present and future. Vaccine. **9:** 533-539, 1991.

Kobuch W.E., Davis J., Fleischer K., et al. A clinical and epidemiological study of 621 patients with anthrax in western Zimbabwe. Salisbury Medical Bulletin. Proceedings of the International Workshop on anthrax. **68 (suppl):** 34-38, 1990.

Inglesby T.V., Henderso D.A., Bartlett J.G. et. al. Anthrax as a Biological Weapon. Medical and Public Health Management. JAMA **281:** 1735-1745, 1999.

Cristy G.A. and Chester C.V. Emergency protection against aerosols. Oak Ridge, Tenn:Oak Ridge National Laboratory, 1981.

Hana P.C., Acosta D. and Collier R.J. "On the role of microphage in anthrax." Proc. Natl. Acad. Sci. USA **90:** 10198-10201, 1993.

Erwin J.L. et al. "Macrophage-Derived cell lines do not express proinflammatory cytokines after exposure to Bacillus anthracis lethal toxin" Infect Immun. **69:** 1175-1177, 2001.

Sellman et al. "Dominant-negative mutants of a toxin subunit: an approach to theropy of anthrax" Science. **292:** 647-648, 2001.

Enserink M. This time it was real: Knowledge of anthrax put to the test: Science. **294:** 490-491, 2001.

Mikesell P., Ivins B.E., Ristropn J.D. et al. "Evidence for plasmid-mediated toxin production in Bacillus anthracis." Infect. Immun. **39:** 371-376, 1983.

Green B.D., Battisti C., Koehler T.M., et al. "Demonstration of capsule plasmid in Bacillus anthracis." Infect. Immun. **49:** 291-297, 1985.

Okinaka R.T., et al. "Sequence and organization of pX01, the large Bacillus anthracis plasmid harboring the anthrax toxin gene." J. Bacteriol. **181:** 6509-6515, 1999.

Okinaka R.T., et al. "Sequence, assembly and analysis of pX01 and pX02." J. Appl. Microbiol. **87:** 261-262, 1999.

World Health Organization. Guidelinesfor the Surveillance and Control of Anthrax in Humans and Animals, Geneva Switzerland. 1998 WHO/EMC/ZDI/98.6.

Peters C.J. and Hartley D.M. Anthrax Inhalation and lethal human infection. The Lancet. **359:** 710, 2002.

Dalton R. Genetic sleuths rush to identify anthrax strains in mail attacks. Nature. **413:** 657-658, 2001.

Espy M.J., Uhl J.R., Sloan L.M. et al. Detection of vaccinia virus, herpes simplex virus, varicella-zoster virus, herpes and Bacillus anthracis DNA by light cycler Polymerase Chain Reaction after autoclaving: Implications for biosafety ofbioterrorism agents. Mayo Clin Proc. **77:** 624-628, 2002.

Uhl J.R., Constance B.A., Lynne M.S., et al. Application of rapid-cycle real-time polymerase chain reaction for the detection of microbial pathogens: The Mayo-Roche rapid anthrax test. Mayo Clin Proc. **77:** 673-680, 2002.

Hart M.K., Del Giudice R.A. and Korch G.W. Absence of mycoplasm contamination in the anthrax vaccine. Emerging Infectious Diseases. **8:** 94-96, 2002.

Enserink M. and Marshall E. New anthrax vaccine gets a green light. Science. **296:** 639- 640, 2002.

The Medical Letter. Post-exposure anthrax prophylaxis. **43:** 91-92, 2001.

Enserink M. 'Borrowed Immunity' may save future victims. Science. **295:** 777.

Maynarud J.A. et al. Protection against anthrax toxin by recombinant antibody fragments correlates with antigen affinity. Nature Biotechnol. **20:** 597-601, 2002.

Moure Z.M., Kane R.S. and Mogridge J. Designing a polyvalent inhibitor of anthrax toxin. Nature Biotechnol. **19:** 9580961, 2001.

Bradley K.A., Mogridge J., Maurez M. et al. Identification of the cellular receptor for anthrax toxin. Nature. **414:** 225-229, 2001.

Pannifer A.D., Wong T.Y., Schwarzenbacher R. et a1. Crystal structure of the anthrax lethal factor. Nature. **414:** 229-232, 2001.

Friedlander A. Tackling anthrax. Nature. **414:** 160-161, 2001.

Read T.D. et al. Comparative genome sequencing for discovery of novel polymorphisms in Bacillus anthracis. Sciencexpres, doi, **10:** 1126/Science. 1071737 2002.

Pharmacological Perspectives of Toxic Chemicals and Their Antidotes
Editors: S J S Flora, J A Romano, S I Baskin and K Sekhar

Cyanobacterial Toxins: Effects and Control Measures

P. V. Lakshmana Rao
Division of Pharmacology and Toxicology
Defence Research and Development Establishment
Jhansi Road, Gwalior 474 002, India

INTRODUCTION

Cyanobacteria are prokaryotic, single celled, widespread organisms that grow (a) planktonically in water or (b) benthically on surfaces in water bodies, and (c) on moist terrestrial surfaces. Many species form filaments or colonies, sometimes upto 1 or 2 mm in diameter. Benthic species may form dense mats. Mass aggregations of cyanobacteria have earned the collective term "water blooms". Factors such as nitrogen, phosphorous, temperature, light, micronutrients, pH and alkalinity, buoyancy, hydrologic and meteorological conditions, and the morphology of the impoundment have all been implicated in "bloom" formation (Carmichael, 1997; Chorus et al., 2000). According to the current taxonomy 150 genera with about 2000 species, at least 40 of which are known to be toxicogenic have been identified.

CYANOBACTERIAL TOXINS

Cyanobacteria produce a variety of toxins that are classified functionally into hepato, neuro-, and cytotoxins (Table 1). Defined by their chemical structure, cyanotoxins fall into three groups: cyclic peptides (the hepatotoxins microcystin, nodularin), alkaloids (the neurotoxins anatoxins and saxitoxins) and LPS (Carmichael, 1992). The species most often implicated with toxicity are *Microcystis, Oscillatoria, Nostoc, Nodularia, Aphanizomenon flos-aquae, Anabena flos-aquae, Cylindrospermopsis, Lyngbya,* etc. Animal, bird

Table 1 Some of the toxic cyanobacterial species and their toxins

Toxin	*Organism*	*Chemical Nature*	*LD_{50} (μg/kg i.p. mouse)*
Hepatotoxins			
Microcystin-LR	*Microcystis aeruginosa*	Peptide, MW 994	50
Microcystin-LA	*M. aeruginosa, M. viridis*	Peptide, MW 909	50
Microcystin-YR	*M. aeruginosa, M. viridis*	Peptide, MW 1044	70
Microcystin-RR	*M. aeruginosa, Anabena sp.*	Peptide, MW 1037	600
[Dasp3]micrcys-tin-LR	*M. viridis, O. agardhii*	Peptide, MW 980	300
Nodularin	*Nodularia spumigena*	Pentapeptide, MW 824	50
Cylindrospermo-psin	*C. raciborski*	Alkaloid	2000
Neurotoxins			
Anatoxin-a	*Anabena flosaquae*	Alkaloid, MW 165	200
Homoanatoxin-a	*Oscillatoria formosa*	Alkaloid, MW179	250
Anatoxin-A(s)	*A. flosaquae*	Phosphate ester, MW 252	20
Aphantoxin	*Aphanizomenon flosaquae* (Neo-saxitoxin) Saxitoxin	Purine alkaloid MW 315(neoSTX) MW 299 (STX)	10
Cytotoxins			
Scytophycin	*S. pseudohofmanni*	Methylformamide, MW 821	650
Hapalindole	*H. fontinalis*	Indole alkaloid	NR
Acutiphycin	O. acutissima	Macrolide	NR
Nakienones A-C	Synechocystis	-	NR
Dermatotoxins			
Lyngbyatoxin A	*Lyngbya majuscula*	Indole alkaloid, MW 437	NR
Debromoaply-siatoxin	*L. majuscula, S. calcicola*	Lactone, MW 592	NR
Osillatoxin A	*Oscillatoria nigroviridis*	Phenolic bislactone	NR

NR—Not reported

fish kills, and human illnesses ascribed to cyanobacterial toxins have been reported on a worldwide basis for many years (Bell and Codd, 1994; Rao et al., 1994; Sivonen, 1996; Carmichael, 1997; Hitzfeld et al., 2000). The tragic death of 60 patients in a haemodialysis unit in Caruaru, Brazil due to use of microcystin contaminated water (Pouria et al., 1998; Carmichael et al., 2001) indicates the potential human health hazard due to cyanobacterial toxins. Chemical structures of some of the commonly occurring cyanobacterial toxins are shown in figure 1.

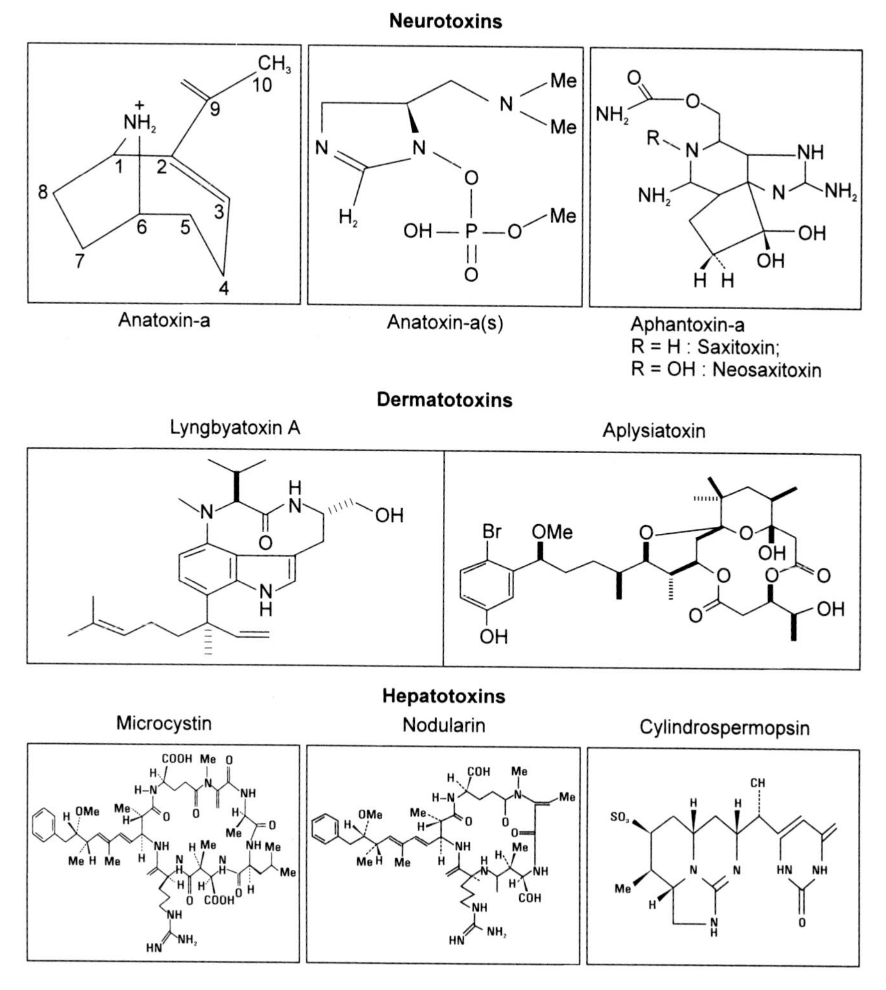

Fig. 1 Chemical structures of some of the cyanobacterial neurotoxins, dermatotoxins and hepatotoxins.

HEPATOTOXINS

Acute hepatotoxicosis involving hepatotoxins (liver toxins) is the most commonly encountered toxicosis involving cyanobacteria. These toxins are produced by strains of species within genera *Microcystis, Anabena, Nodularia, Oscillatoria, Cylindrospermopsis* and *Nostoc*. In addition, chemically undefined hepatotoxins are being studied in Aphanizomenon, *Gleotrichia* and *Coelosphaerium*. The important cyanobacterial hepatotoxins are cyclic peptide microcystins, nodularin and alkaloid toxin cylindrospermopsin.

(a) Microcystins

Cyanobacterial peptide toxins occur worldwide in freshwaters. The general structure of microcystins (MCYST-XY) is cyclo (D-Ala-X-D-MeAsp-Y-Adda-D-Glu-Mdha-) in which X and Y are variable L amino acids, D-MeAsp is, D-erthro-b-methylaspartic acid, Adda is, 3-amino-9-methoxy-2, 6,8-trimethyl-10-phenyldeca-4, 6-dienoic acid and Mdha is N-methyldehydroalanine. The unusual amino acid Adda is essential for expression of biological activity. More than 60 different microcystins are known and their toxicity, as determined by LD_{50} (i.p.) in mice varies between 50 and 800 μg/kg (Dawson, 1998). The most common and toxic among them is microcystin-LR, where the variable L-amino acids are leucine (L) and arginine (R). Microcystins were shown to be chemically very stable (pH, temperature, UV) in water having a half-life of approximately 3 weeks at pH 1 and 40°C (Harada et al., 1996). They are reported to withstand several hours of boiling and may persist for many years when stored at room temperature. It is therefore not readily removed from drinking water by conventional treatment methods. In mice and rats, lethal doses of microcystins lead to death 1-6 hours after intraperitoneal administration. The microcystins are highly liver specific, due to an active uptake into parenchyma liver cells via the multispecific bile acid transport system (Eriksson et al., 1990). The toxicity of microcystins is due to inhibition of catalytic subunit of protein phosphatases 1 and 2A (PP1 and PP2A). PP1 and PP2A inhibition leads to hyperphosphorylation of cytoskeletal proteins resulting in deformation of hepatocytes. The liver swells to nearly double its size, due to massive intrahepatic centrilobular hemorrhaging, which is preceded by hepatocyte rounding and extensive dissociation as well as disruption of sinusoidal epithelium (Beasley et al., 1989; Carmichael, 1992; Rao et al., 1995; Bhattacharya et al., 1996a). Apart from protein phosphatase inhibition, other toxic manifestations mediated by microcystins include inhibition of protein synthesis, oxidative stress (Ding et al., 2000), caspase activation, DNA damage in vivo and in vitro (Rao and Bhattacharya, 1996; Rao et al., 1998) and induction of apoptosis in various cell types (McDermott et al., 1998; Fladmark et al., 2002) Certain chemicals have been used experimentally to prevent microcystin hepatotoxicity in laboratory animals. These include cyclosporin A, rifampin, silymarin (Hermansky et al., 1991). These antagonists have been most successful when given prior to or co-administered with toxin.

(b) Nodularin

The pentapeptide hepatotoxin nodularin has been found in brackish waters in Australia, New Zealand and the Baltic Sea (Carmichael et al., 1988). These blooms have caused numerous cases of animal poisonings. The nodularin is a cyclic pentapeptide and has mouse LD_{50} (i.p.) of 50-70 μg/kg. The toxicity of nodularin results from its ability to inhibit the serine/threonine protein phosphatases PP1 and PP2A (Ohta et al., 1994). Nodularin is a potent tumour promoter and a possible carcinogen. There are no published reports on the chemical antidotes or therapeutic measures against nodularin poisoning.

(c) Cylindrospermopsin

An outbreak of hepatoenteritis at Palm Island in northern Queensland, Australia, led to the finding of a new cyanobacterial toxin, cylindrospermopsin and it is produced by *Cylindrospermopsis raciborskii* (Woloszynska) Seenaya and Subba Raju. It is an alkaloid cytotoxin with molecular weight 415 and it affects the liver, kidneys, thymus and heart. The LD_{50} of cylindrospermopsin in mice by i.p. is 2.1 mg/kg at 24 hr and 0.2 mg/kg at 5-6 days (Ohtani et al., 1992) . Studies using cultured rat hepatocytes have shown that cylindrospermopsin inhibits glutathione synthesis.

NEUROTOXINS

Neurotoxins are produced by species and strains of Anabena, *Aphanizomenon, Oscillatoria* and *Trichodesmium*. Five chemically defined neurotoxins are now known to be produced within these genera. Anatoxin-a, which was the first toxin from a freshwater cyanobacterium to be chemically and functionally defined, has been found in blooms of *Anabena flos-aquae*, *Oscillatoria* mats and *Aphanizomenon flos-aquae* cultures isolated from Finland (Carmichael, 1997). Anatoxin-a(s) with its rare occurrence and inherent chemical instability has limited new investigations on its structure, function properties and its role in water-based diseases. The final group of neurotoxins produced by some cyanobacteria is the saxitoxins and derivatives.

(a) Anatoxin-a

Anatoxin-a (Antx-a) was the first toxin from freshwater cyanobacterium that was chemically and functionally defined. It is the secondary amine, 2-acetyl-9-azabicyclo [4.2.1] non-ene (Devlin et al., 1977) produced by strains of *Anabena* and *Oscillatoria*. The alkaloid neurotoxin Antx-a is a potent post-synaptic depolarizing neuromuscular blocking agent. This toxin causes death within minutes to a few hours depending on the species, the amount of toxin ingested, and the amount of food in the stomach. The LD_{50} intraperitoneal (i.p) mouse for purified toxin is about 200-µg/kg body weight, with a survival time of minutes (Carmichael et al., 1979). Clinical signs of Antx-a poisoning follow a progression of muscle fasciculation, decreased movement, abdominal breathing, cyanosis, convulsions and death. Anatoxin-a induced apoptosis in non-neuronal cells viz., thymocytes and Vero cells was reported recently (Rao et al., 2002) suggesting cytotoxic effects in isolated cells.

Anatoxin-a homologue homoanatoxin-a (MW179) with a mouse i.p. LD_{50} of 250 μg/kg has been characterized from a strain of *Oscillatoria formosa* (Skulberg et al., 1992). No chemical antidotes are available against anatoxin-a.

(b) Anatoxin-a(s)

Anatoxin-a(s) [Antx-a(s)] is a unique phosphate ester of a cyclic N-hydroxyguanine (MW 252) produced by a strain of *Anabena flos-aquae* NRC-525-17. Antx-a(s) induces salivation in mice by which it can be differentiated from other cyanobacterial neurotoxins. It acts as an irreversible peripheral anticholinesterase (Mahmood and Carmichael, 1987). Clinical sign of toxicosis from laboratory experiments involving dosing of antx-a(s) have been observed in ducks and pigs. The signs in pigs include hyper salivation, mucoid nasal discharge, tremors and fasciculation, ataxia, diarrhoea, bruxism, dyspnoea, recumbency and cyanosis. The LD_{50} (i.p.) in mouse for antx-a(s) is about 20 μg/kg body weight. At the LD_{50} level the survival time for mice is 10-30 min. As antx-a(s) has the properties of an organophosphorus insecticide it should be possible to use therapy such as atropine to antagonize its toxicity. In vivo pretreatment with physostigmine and high concentrations of 2-PAM chloride were the only effective antagonists against a lethal dose of anatoxin-a(s).

(c) Aphantoxins

All aphantoxins studied to date have come from water bloom samples of non-fasciculate *Aphanizomenon flos-aquae*. These neurotoxins were shown to be saxitoxins and neosaxitoxins (the LD_{50} i.p. mouse equals 10 μg/kg), the two primary toxins of red tide paralytic shellfish poisoning (PSP) (Mahmood and Carmichael, 1986). These toxins are fast acting neurotoxins that inhibit nerve conduction by blocking sodium channels without affecting permeability to potassium, the transmembrane resting potential, or membrane resistance. These sodium channel blocking agents inhibit transmission by nervous impulses and leading to death by respiratory arrest. For such toxicosis, therapy is best approached by trying to limit the absorption from the gastrointestinal tract by using activated charcoal, and a saline cathartic plus artificial respiration when needed.

DERMATOTOXINS

Swimmers who have come into contact with marine cyanobacterium *Lyngbya majuscula* have contracted acute dermatitis. The inflammatory activity is caused by aplysiatoxins and debromoaplysiatoxin, which were first reported to be causative agents of swimmer's itch (Cardellina et al., 1979). Aplysiatoxin showed biochemical and biological effects of the same potency as TPA and lyngbyatoxin-A, namely irritation on mouse ear, induction of ornithine decarboxylase in mouse skin, induction of HL-60 cell adhesion, specific binding to the phorbol ester receptor, activation of protein kinase C and tumour promotion on mouse skin (Fujiki et al., 1993). In addition to the problem posed by the dermatotoxicity of *Lyngbya majuscula*, the toxins produced by these cyanobacteria could conceivably be involved in the development of human cancer even though there is no direct evidence. An aplysiatoxin-type compound has been found in marine *Schizothrix calcicola*.

LIPOPOLYSACCHARIDES

Cyanobacteria contain lipopolysaccharides (LPS) endotoxins in their cell envelopes. LPS endotoxins were found in several species or strains of cyanobacteria including *Anabena variabilis*, *Phormidium africanum*, *and Schizothrix calcicola*. These LPS molecules have structures similar to the LPS endotoxins isolated from enteric Gram-negative bacteria. LPS of cyanobacteria show lower levels of toxicity in bioassays when compared to *Salmonella* LPS. The contribution of cyanobacterial LPS to cases of fevers and inflammation due to contact with contaminated water is unclear and requires investigation.

BIOACTIVE COMPOUNDS

Cyanobacteria have been known to be a rich source of biologically active peptides, macrolides, alkaloids, sulphur compounds, cytotoxins, fungicides and several enzyme inhibitors (Fujiki et al., 1993; Namikoshi and Rinehart, 1996). Various bioactivities could be assigned to these compounds, and some may prove useful either for development into commercial drugs or as biochemical research tools (Table 2). A wide range of biological activities has been reported for cyanobacterial peptides including cell differentiation promoting activity for microcystilide A, antifungal activity for laxophycin A and B, tyrosine inhibiting activity for microviridin, calcium antagonist activity for scytonemins, insecticidal and anticancer activities for majusculamide C, cardiac activity for puriniphycin, immunosupressive lipopeptides, microcolins A and B and angiotensin converting enzyme inhibitor microginin (Namikoshi and Rinehart, 1996). Cyanobacteria also produce a variety of inhibitors of serine proteases such as trypsin, plasmin, elastase and chymotrypsin. Elastase is suggested to be involved in pulmonary emphysema, rheumatoid arthritis, adult respiratory distress syndrome, and other inflammatory states. Its inhibitors might be useful chemotherapeutic agents for these diseases. Polysaccharides, carrageenans, sulfolipids, and other compounds isolated from photosynthetic cellular microorganisms have antiviral activity including in vitro anti-HIV activity (Ayehunie et al., 1998; Shaeffer et al., 1999). Similarly, Antx-a's greater potency than nicotine in the

Table 2 Some cytotoxic/bioactive compounds of potential pharmacological importance produced by cyanobacteria

Species	*Compound*	*Activity*
Anabena flosaquae	Anabenopeptin A, B,	Protease inhibitor
Anabena circinalis	Circinamide	Papain inhibitor
Anabena laxa	Laxaphysins A-E	Antifungal
Fischerella ambigua	Fisherindole L	Antifungal
Hapalosiphon welwitschii	Hapalosin	Cytotoxic
Hapalosiphon hibermicus	Isonitrile	Antifungal
Lyngbya majuscula	Curacin A	Antineoplastic
	Curacin B-C	Antimitotic
	Microcolin A, B	Immuno suppressive peptide
	Majusculamide	Anti-cancer
Microchaete loktakensis	A90720A	Protease inhibitor
Microcystis viridins	Aeruginosin 102 A,B	Thrombin inhibitor
	Micropeptin 103	Chymotrypsin and thrombin inhibitor
M. aeruginosa	Aeruginosin 98 A,B	Trypsin inhibitor
	Aeruginosin 298 A	Thrombin and trypsin inhibitor
	Micropeptin 90, A , B	Plasmin and trypsin inhibitor
	Micropeptin, 478 A,B	Plasmin inhibitor
	Microginin	Angiotensin converting enzyme inhibitor
	Microviridins B, C	Elastase inhibitor
	Microcystilide A	Cell differentiation promoting peptide
Nostoc minutum	Nostopeptin A, B	Elastase and chymotrypsin inhibitor
	Muscoride A	Antibacterial
Oscillatoria agardhii	Aeruginosin 205 A, B	Serine protease inhibitor
	Anabenopeptin B,E,F	Serine protease inhibitor
	Oscillamide Y	Chymotrypsin inhibitor
	Oscillapeptin	Chymotrypsin inhibitor
Schzotrix sp.	Schizotrin A	Antimicrobial

Central Nervous System (CNS) provides possibility that it could prove a useful CNS agonist for the study of possible treatment of brain disorders such as Alzheimer's disease.

Dynamics of Toxin Production

The occurrence of blooms under natural conditions is determined by biotic and abiotic factors. The environmental or abiotic factors which are known to influence toxic bloom formation are temperature, pH, light intensity, and nutrient concentrations especially nitrogen and phosphorus levels (Sivonen, 1996). These factors are also known to influence toxin production in laboratory cultures of several cyanobacterial (Sivonen, 1996; Rao et al., 1996; Nidhi et al., 2001, 2002). Microcystin and nodularin peptides are produced non-ribosomally via a multifunctional enzyme complex that employs the thio-template mechanisms (Arment and Carmichael, 1996). Most cyanotoxic genes with the exception of microcystin have not been identified. Genes coding peptide synthetases are typically clustered into large operons with repetitive domains in which highly conserved sequences have been identified. Using these sequences it has been possible to locate peptide synthetase genes in *M. aeruginosa* (Dittmann et al., 1997). Insertional mutagenesis of some genes (*mcy A, B* and *D*) produced non-toxic mutants, providing evidence for their involvement in microcystin synthesis (Dittmann et al., 1997). The entire gene cluster (*mcy A-J*) has since been identified and sequenced. The mcyABCDEFGHIJ genes are transcribed as two polycistronic operons, mcyABC and mcyDEFGHIJ, from a central bi-directional promoter between mcyA and mcyD (Kaebernick et al., 2002).

DETECTION METHODS

Methods of environmental detection for the cyanotoxins use bioassay, chemical assay and immunoassays (Codd et al., 1994). The mouse bioassay has been the typical first test for toxicity screening of water blooms and laboratory cultures or cell extracts. The disadvantage of mouse bioassay is its inability to detect low amounts of toxins. The other bioassays include use of bacterial cells, cultured mammalian cells, organ slice (Bhattacharya et al., 1996a,b) immunological and protein phosphatase enzyme inhibition assays (An and Carmichael, 1994). Even plant cells are being used for their susceptibility to the microcystins (Kos et al., 1995). Chemical detection methods are based on HPLC, GC-ECD, and MALDI, LC-MS, and also includes methods that are used for isolation and purification of toxins (Harada, 1994;Lawton and Edwards, 2001). Recently a biosensor with engineered acetylcholinesterases was used to detect anatoxin-a(s) in environmental samples (Devic et al., 2002). The comparability of currtent microcystin analysis methods was evaluated in an International intercomparison exercise focusing mainly on microcystin analysis by HPLC-PDA/UV method (Fastner et al., 2002). However, none of these techniques have been developed into a standard method easily used by monitoring agencies or testing laboratories. Very sensitive immunoassays have been developed for low level detection of microcystins in potable waters. Enzyme-linked immunosorbent assays (ELISAs) using polyclonal and monoclonal antibodies to MCs have been reported for environmental samples (Chu et al., 1989; Nagata et al., 1997). The most promising method for screening for cyanotoxins would be some type of immunoassay coupled with protein phosphatase inhibition assays (Metcalf et al., 2001).

HUMAN HEALTH EFFECTS

The cases of poisoning episodes, laboratory investigation into the toxicity of cyanobacterial toxins, and human exposure assessments, together indicate that cyanobacterial toxins constitute a hazard to human health when present in drinking or recreational waters. Cyanobacterial toxins can cause relatively rapid acute effects such as gastroenteritis, allergic or irritation reactions, pneumonia-like symptoms and hepatoenteritis, and long term chronic effects such as liver damage or liver tumour promotion on prolonged exposure to toxins (Falconer, 1991; Falconer, 1999; Chorus et al., 2000). The potential may exist for freshwater cyanobacterial neurotoxins to be transmitted through food chain, with subsequent intoxication of humans after consuming food contaminated by the toxins (Amorin and Vasconcelos, 1999). Human health hazards occur from three routes of exposure :

1. Direct contact of exposed parts of the body, including the ears, eyes, mouth and throat.
2. Accidental ingestion of water containing cells by swallowing.
3. Uptake of water containing cells by aspiration (inhalation).

(a) Ingestion Related Illness

The majority of reported cases of human illness attributed to cyanobacterial toxins have been episodes of gastroenteritis resulting from ingestion of toxins in drinking water or from accidental ingestion of the toxins in recreational waters. People along the Swedish and Polish coasts were taken ill after eating fish particularly the livers, from waters containing cyanobacterial blooms. The disorder was named Haff disease, and patients exhibited symptoms of vomiting, passing brownish-black urine, muscular pain, and respiratory distress mortalities. The evidence for involvement of cyanobacterial toxins in these cases is circumstantial, but accumulation of toxins in fish cannot be ruled out. Recreational contact with toxic *Microcystis* may, cause blistering of the lips. Human consumption of the dried filamentous cyanobacterium *Spirulina* is widespread, as a result of its marketing as a health food. Other cyanobacteria from natural sources such as freshwater blooms are also marketed as food supplements. These are potentially hazardous products if they contain any of the toxigenic species or strains. The acute toxicity testing alone may not be adequate to safeguard people regularly consuming cyanobacteria in their diet (Schaeffer, 1999).

(b) Allergic and Irritation Reactions

Allergic reactions to cyanobacteria are relatively common and have been described after human contact while swimming in blooms containing several species of both fresh and marine organisms. One of the first reports implicating toxic cyanobacteria as the source of human allergic reactions described symptoms of asthma, hay fever and conjunctivitis due to exposure while swimming in water body containing cyanobacterium *Oscillatoria*. Mittal et al. (1979) showed allergenic properties of cyanobacteria isolated from air samples. Contact dermatitis of varying severity can be caused by a range of freshwater cyanobacteria including *Aphanizomenon, Anabena, Nodularia, Oscillatoria* and *Gleotrichia*. The most common and best documented toxic reaction to marine cyanobacteria is a severe contact dermatitis known as "swimmer's itch". This is a cutaneous inflammation characterized by erythema, followed by blisters and deep desquamation within 12 hours of exposure to algae. The dermatitis is induced by exposure to two compounds, debromoaplysiatoxin and lyngbya toxin-A isolated from *Lyngbya majuscula*.

(c) Liver diseases

Evidence for a link between human liver damage and cyanobacterial toxins has been presented by Falconer (1991). Results of routine liver function tests of hospital patients in Armidale, Australia suggested that there was a significant seasonal and locational increase of activity of the hepatic enzyme γ-glutamyltransferase with the presence of the hepatotoxic cyanobacterium *Microcystis aeruginosa* in the drinking water reservoir. The two most lethal poisonings attributed to cyanobacteria in drinking water occurred in Brazil. A massive *Anabena* and *Microcystis* bloom in Itaparcia Dam was responsible for 2000 gastroenteritis cases resulting in 88 deaths, mostly children (Teixera et al., 1993). In 1996, liver failure and death of 60 patients after exposure to microcystins occurred at a hemodialysis centre in Caruaru, Brazil. The patients displayed choleostatic jaundice with high bilirubin and alkaline phosphatase concentrations and increase in hepatic enzymes (Pouria et al., 1998; Carmichael et al., 2001).

(d) Tumour promotion

One of the most concerning potential threats of cyanobacterial hepatotoxins to human health is tumour promotion after skin contact during bathing, or after prolonged exposure to subacute levels of the toxins in drinking water. Microcystins and nodularin are known to be potent inhibitors of protein phosphatases 1 and 2A as well as tumour promoters in laboratory animals with and without initiators (Nishiwaki-Matsushima et al., 1992; Ito et al., 1997). Cyanobacterial extracts of microcystin-LR in drinking water induce skin tumours in rats and mice after initiation with 7,12-dimethylbenz[a]anthracene (Falconer and Humpage, 1996). Microcystin-LR and nodularin induce the expression of tumour necrosis factor and early response genes (*c-jun*, *jun B*, *jun D*, *c-fos*, *fos B*, *fra-1*) in rat liver and hepatocytes. Mutations in the *K-ras* codon 12 in *Rsa* cell line (Suzuki et al., 1998) and DNA fragmentation have been reported after i.p. injections of cyanobacterial extracts of microcystin-LR in mice (Rao and Bhattacharya, 1996; Rao et al., 1998). The marine cyanobacterium, *Lyngbya majuscula* which causes skin irritation on contact and contains the well-characterized tumour-promoting toxins, lyngbyatoxin A and aplysiatoxin (Fujiki et al., 1993). In summary, whether the tumor-promoting effects or the teratogenic activity of cyanobacteria are of public health significance awaits suitable human epidemiological analysis of cancer deaths and birth defect frequency in populations exposed to this risk.

RISK ASSESSMENT

The most likely route for human exposure is the oral route via drinking water, recreational use of lakes and rivers, consumption of algal health food tablets (Schaeffer et al., 1999), or through food chain (Amorin and Vasconcelos, 1999; Orr et al., 2001). The dermal route may play a role during the recreational use of water bodies for swimming, canoeing, etc. The international discussion on guidelines for cyanotoxins is focusing on microcystins because the neurotoxins are not considered to be as hazardous due to lack of chronic toxicity and widespread distribution. There are insufficient data to derive a guideline value for many of the cyanobacterial toxins other than microcystin-LR. *Cylindrospermopsis,* whose toxin is very hazardous, does not form surface scums, and thus is more of a health risk in drinking water than in bathing waters. Due to growing concern about health effects of microcystins especially via drinking water, WHO has adopted a provisional guideline value of microcystin-LR of 1.0 μg/L in drinking water (WHO 1998). This guideline value is based on a tolerable daily intake (TDI) value derived from two animal studies. This provisional guideline value is applicable only for microcystin-LR, since the database for other microcystin congeners or even other cyanotoxins such as saxitoxin is too small to derive a TDI.

CONTROL MEASURES

Cyanobacterial blooms frequently occur in stratified water bodies with high nutrient content leading to accelerated growth. Control of toxic cyanobacterial blooms is best achieved by prevention rather than eradication. Prevention of bloom formation should include reduction of phosphate inflow into rivers, lakes and water storage bodies, aeration and suppression of phosphate mobilization. Introducing a wetland/swamp area at the inflow region of a reservoir will also help to reduce bloom formation. Algicides as a control measure for cyanobacterial have long been a favorite approach. Copper sulphate in various forms is the most effective algicide but its use on toxic blooms can release the toxins from cells and can enter finished water. Cyanotoxins are water soluble and control measures involve chemical procedures reducing toxicity or completely removing the toxins from drinking water. Studies on microbial degradation of microcystins have been limited. A bacterium belonging to *Sphingomonas* spp. capable of degrading microcystin LR, RR and YR was recently reported by Park et al. (2002). Highest degradation rates of microcystin- RR and –LR were 13 and 5.4 mg/L/day respectively. The degradation rates were strongly dependent on temperature and maximum rate was 30°C.

Coagulation can be an efficient method for eliminating cyanobacterial cells from water, whereas soluble cyanotoxins are not very efficiently removed by this method (Hitzfeld et al., 2000). A study aimed at the removal of cyanobacterial cells with rapid sand filtration and powdered activated carbon found a reduction of only 42%. Detailed studies with activated carbon show that both PAC as well as granular-activated carbon effectively and quickly removed cyanotoxins from water (Lawton and Robertson, 1999). Effectiveness of toxin removal by PAC and GAC will depend on concentration (10 µg/L toxin: > 200 mg/PAC/L), and the choice of the carbon source (coal, wood, peat, coconut) respectively. Rapid and slow sand filtration does not lead to substantial reduction of toxicity. Aqueous chlorine and calcium hypochlorite at ≥ 1 mg/L can remove more than 95% of microcystins or nodularin. Chorine treatment nor change in pH is effective in destroying Anatoxin-a or saxitoxins. Effectiveness of microcfiltration and ultrfiltration in removal of cyanotoxins have not been thoroughly studied though MF and UF can be very efficient in removing (> 98 %) whole cells of *Microcystis aeruginosa* (Hitzfeld et al., 2000). A phtocatalytic process using TiO_2 catalyst and UV radiation quickly decomposed microcystin-LR, YR and YA with half-lives of < 5 min (Shephard et al., 1998).

Studies on effect of ozone on cyanotoxins have shown that up to 800 µg/L microcystin-LR can be oxidized to below the HPLC detection limit by < 0.2 mg/L within seconds to minutes (Rositano et al., 1998). The efficiency of oxidation with ozone with respect to anatoxin-a, anatoxin-a(s), or the saxitoxins (PSPs) has not been well characterized. In order to realistically assess the consequences of ozonation by products have to be identified and their toxicity characterized (Hitzfeld et al., 2000). A chronic exposure to cyanobacterial toxins and/or to the ozonolysis by products should be avoided. The situation for the saxitoxins, anatoxin-a, anatoxin-a(s), and cylindrospermopsin is even less clear. Assessment of some of the physical and chemical water treatment procedures discussed above have shown that most methods would result in a reduction of cyanobacterial toxin concentrations to below acutely toxic levels as well as below the new WHO guideline value of 1 µg/L drinking water. It is important however to structurally characterize the breakdown products generated from chemical treatment methods. Most methods are microcystin-LR centered but it is necessary to investigate the suitability of each process to over 60 variants of microcystin.

CONCLUSION AND FUTURE DIRECTIONS

Cyanobacterial toxin research in the past two decade has advanced from ecological studies on bloom formation, development of analytical techniques for isolation and purification, toxicity evaluation to the realization of importance of these toxins on public health. The mechanisms of tumour promotion by microcystisns and nodularins as well as quantitative relationships have not been properly elucidated. Future efforts should focus on molecular mechanisms of toxin production, development of suitable antidotes, more effective medical counter measures, etc. Another important aspect which needs attention is on the effects of microcystin contaminated water on quality and yield of crop plants and human exposure to toxins by hitherto unexplored and unconventional sources (via food chain). In order to perform an adequate human risk assessment of microcystin exposure via drinking water, the issue of water treatment by products will also have to be addressed in the future.

Acknowledgements

The author is thankful to Mr. K. Sekhar, Director, DRDE, Dr. R. Vijayaraghavan, Head, Division of Pharmacology and Toxicology for providing facilities and Dr. R.V. Swamy, Chief Controller Research and Development, DRDO, for his sustained support.

References

Amorim A., Vasconcelos V. Dynamics of microcystin in the mussel *Mytilus galloprovincialis*. Toxicon **37:** 1041-1052, 1999.

An J., Carmichael W.W. Use of a colorimetric protein phosphatase inhibition assay and enzyme linked immunosorbent assay for the study of microcystins and nodularins. Toxicon **32:** 1495-1507, 1994.

Arment A.R., Carmichael W.W. Evidence that microcystin is a thiotemplate product. J. Phycol. **32:** 591-597, 1996.

Ayehunie S., Belay A., Baba T.W., Ruprecht M. Inhibition of HIV-1 replication by an aqueous extract of *Spirulina platensis*. J. Acquir. Immune. Defic. Syndr. Hum. Retrovirol. **18:** 7-12, 1998.

Beasley V.R., Dahlem A.M., Cook W.O., Valentine W.M., Lovell R.A., Hooser S.B., Harada K.I., Suzuki M. and Carmichael W.W. Diagnostic and clinically important aspects of cyanobacterial (blue-green algal) toxicosis. J. Vet. Diag. Invest. **1:** 359-365, 1989.

Bell S.G., Codd G.A. Cyanobacterial toxins and human health. Rev. Med. Microbiol. **5:** 256-264, 1994.

Bhattacharya R., Rao P.V.L., Bhargava R.K., Pandya G., Tuteja U., Bhaskar A.S.B. Toxicity assessment of freshwater cyanobacterial toxins in mammalian cell cultures. Clin. Chem. Enzymol. Commun. **7:** 221-232, 1996a.

Bhattacharya R, Rao P.V.L., Pant S.C., Bhaskar A.S.B. Liver slice culture for assessing hepatotoxicity of freshwater cyanobacteria. Hum. Exp. Toxicol. **15:** 105-110, 1996b.

Cardellina II J.H., Marner F.J. and Moore F.J. Seaweed dermatitis: structure of lyngbyatoxin A. Science **204:** 193-195, 1979.

Carmichael W.W. Cyanobacteria secondary metabolites-the cyanotoxins. J. Appl. Bacteriol. **72:** 445-459, 1991.

Carmichael W.W. The cyanotoxins. In: Advances in Botanical Research. Callow J.F. (Ed.) London: Academic Press. **27:** 211-256, 1997.

Carmichael W.W., Azevedo M.F.O., An J.S., Molica R.J.R., Jochimsen E.M., Lau S., Rinehart K.L., Shaw G.R. and Eaglesham G.K. Human fatalities from cyanobacteria: Chemical and biological evidence for cyanotoxin, Environ. Health Perspect. **109:** 663-668, 2001.

Carmichael W.W., Biggs D.F., Peterson M.A. Pharmacology of anatoxin-a, produced by the freshwater cyanophyte *Anabaena flos-aquae* NRC-44-1. Toxicon **17:** 229-236, 1979.

Carmichael W.W., Eschedor J.T., Patterson G.M.L., Moore R.E. Toxicity and partial structure of a hepatotoxic peptide produced by the cyanobacterium *Nodularia spumigena* Mertens Emend. L575 from New Zealand. Appl. Environ. Microbiol. **54:** 2257-2263, 1988.

Chorus I., Falconer I.R., Salas H.J., Batram J. Health risks caused by freshwater cyanobacteria in recreational waters. J. Toxicol. Environ. Health **3:** 323-347, 2000.

Chu F.S., Huang X., Wei R.D., Carmichael W.W. Production and characterization of antibodies against microcystins. Appl. Environ. Microbiol. **55:** 1928-1933, 1989.

Codd G.A., Bell S.G., Brooks W.P. Detection methods for cyanobacterial toxins. Cambridge, UK: Royal Society of Chemistry, 1994.

Dawson R.M. The toxicology of microcystin. Toxicon **36:** 953-962, 1998.

Devlin J.P., Edwards O.E., Gorham P.R., Hunter M.R., Pike R.K., Stavric B. Anatoxin-a, a toxic alkaloid from *Anabaena flos-aquae* NRC-44-1. Canadian J. Chem. **55:** 1367-1371, 1977.

Devic E., Li Dunhai Li, Dauta A., Henriksen P., Codd G.A., Marty J.L., Fournier D. Detection of anatoxin-a(s) in environmental samples of cyanobacteria by using biosensor with engineered acetylcholinesterases. Appl. Environ. Microbiol. **68:** 4102-4106, 2002.

Ding W.X., Shen H.M., Ong C.N. Critical role of reactive oxygen species and mitochondrial permeability transition in microcystin-induced rapid apoptosis in rat hepatocytes. Hepatology **32:** 547-555, 2000.

Dittmann E., Neilan B.A., Enhard M.,. Dohren H.V., Borner T. Insertional mutagenesis of a peptide synthetase gene that is responsible for hepatotoxin production in the cyanobacterium *Microcystis aeruginosa* PCC 7806. Mol. Microbiol. **26:** 779-787, 1997.

Eriksson J.E., Toivola D., Meriluoto J.A.O., Karaki H., Han Y.-G., Hartshorne D. Hepatocyte deformation induced by cyanobacterial toxins reflects inhibition of protein phosphatases. Biochem. Biophy. Res. Commun. **173:** 1347-1353, 1990.

Falconer I.R. Tumour promotion and liver injury caused by oral consumption of cyanobacteria. Environ. Toxicol .Water Qual. **6:** 177-184, 1991.

Falconer I. An overview of problems caused by toxic blue-green algae (cyanobacteria) in drinking water. Environ. Toxicol. **14:** 5-12, 1999.

Falocner I.R., Humpage A.R. Tumour promotion by cyanobacterial toxins. Phycologia **35:** 74-79, 1996.

Fastner J., Codd G.A., Metcalf J.S., Woitke P., Wiedner C., Utkilen H. An International intercomparison exercise for the determination of purified microcystin-LR and microcystins in cyanobacterial field material. Anal. Bioanal.Chem. **374:** 437-444, 2002.

Fladmark K.E., Brustugan O.T., Mellgren G., Krakstad C., Boe R., Vintermyr O.K., Schulman H., Doskeland S.O. Ca^{2+}/calmodulin-dependent protein kinase II is reuired for microcystin-induced apoptosis. J. Biol. Chem. **277:** 2804-2811, 2002.

Fujiki H., Suganuma M., Yatsunami J., Komori A., Okabe S., Nishiwaki-R. Matsushima, Ohta T. Significant marine natural products in cancer research. Gaz. Chim. Ital. **123:** 309-316., 1993.

Harada K. Strategy for trace analysis of microcystins in complicated matrix. In: Toxic cyanobacteria-a global perspective. Australian Centre for Water Quality Research, Salisbury, Australia. pp 49-51, 1994.

Harada K.I., Murata H., Qiang Z., Suzuki M., Kondo F. Mass spectrometric screening method for microcystins in cyanobacteria. Toxicon **34:** 701-710, 1996.

Hermansky S.J., Stohs S.J., Eldeen Z.M., Roche V.F., Mereish K.A. Evaluation of potential chemoprotectants against microcystin-LR hepatotoxicity in mice. J. Appl. Toxicol. **11:** 65-74, 1991.

Hitzfeld B.C., Hoger S.J., Dietrich D.R. Cyanobacterial toxins : Removal during drinking water treatment, and human risk assessment. Environ. Health Perspect **108:** 113-122, 2000.

Ito E., Kondo F., Terao K., Harada K.I. Neoplastic nodular formation in mouse liver induced by repeated intraperitoneal injection of microcystin-LR. Toxicon **35:** 453-1457, 1997.

Kaebernik M., Dittman E., Borner T., Neilan B.A. Multiple alternate transcripts direct the biosynthesis of microcystin, a cyanobacterial nonribosomal peptide. Appl. Environ. Microbiol. **68:** 449-455, 2002.

Kos P., Gorzo G., Suranji G., Borbely G. Simple and efficient method for isolation and measurement of cyanobacterial hepatotoxins by plant tests (*Sinapis alba* L.). Anal. Biochem. **225:** 49-53, 1995.

Lawton L.A., Edwards C. Purification of microcystins. J. Chromatogr. A 191-209, 2001.

Lawton L.A., Robertson P.K.J. Physico-chemical treatment methods for the removal of microcystins (cyanobacterial hepatotoxins) from potable waters. Chem. Soc. Rev. **28:** 217-224, 1999.

Mahmood N.A., Carmichael W.W. Paralytic shellfish poisons produced by the cyanobacterium *Aphanizomenon flosaquae* NH-5. Toxicon **24:** 175-186, 1986.

Mahmood N.A., Carmichael W.W. Anatoxin-a(s), an acetylcholinesterase from the cyanobacterium *Anabaena flos-aquae* NRC-525-17. Toxicon **25:** 1221-1227, 1987.

McDermott C.M., Nho C.W., Holtos B. The cyanobacterial toxin microcystin-LR, can induce apoptosis in a variety of cell types. Toxicon **36:** 1981-1996, 1998.

Metcalf J.S., Bell S.G., Codd G.A. Colorimetric immunoprotein phosphatase inhibition assay for specific detection of microcystins and nodularins of cyanobacterial. Appl. Environ. Microbiol. **67:** 904-909, 2001.

Mittal A., Agarwal M.K., Shivpuri D.N. Respiratory allergy to algae : Clinical aspects. Ann. Allergy **42:** 253-256, 1979.

Nagata S., Tsutsumi T., Hasegawa A., Yoshida F., Ueno Y. Enzyme immunoassay for direct determination of microcystins in environmental water. Am. Org. Anal. Chem. Int. **80:** 408-417, 1997.

Namikoshi M., Rinehart K.L . Bioactive compounds produced by cyanobacteria. J. Ind. Microbiol. **17:** 373-384, 1996.

Nidhi Gupta, Bhaskar A.S.B, Dangi R.S., Prasad G.B.K.S., Rao P.V.L. Toxin production in batch cultures of freshwater cyanobacterium *Microcystis aeruginosa*. Bull. Environ. Contam. Toxicol. **67:** 339-346, 2001.

Nidhi Gupta, Bhaskar A.S.B., Rao P.V.L. Growth characteristics and toxin production in batch cultures of *Anabaena flos-aquae*: effects of culture media and duration. World J Microbiol. Biotechnol. **18:** 29-35, 2002.

Nishiwaki-Matsushima R., Ohta T., Nishiwaki S., Suganuma M., Kohyama K., Ishikawa T., Carmichael W.W., Fujiki H. Liver tumour promotion by the cyanobacterial cyclic peptide toxin microcystin-LR. J. Cancer Res. Clin. Oncol. **118:** 420-424, 1992.

Ohta T., Sueoka E., Lida N., Komori A., Suganuma M., Nishiwaki R., Tatematsu M., Kim S.J., Carmichael W.W., Fujiki H. Nodularin, a new potent inhibitor of protein phosphatase 1 and 2A is a new environmental carcinogen in male F344 rat liver. Cancer Res. **54:** 6402-6406, 1994.

Ohtani T., Moore R.E., Runnegar M.T.C. Cylindrospermopsin : a potent hepatotoxin from the blue-green alga *Cylindrospermopsis raciborskii*. J. Am. Chem. Soc. **114:** 7941-7942, 1992.

Orr P.T., Jones G.J., Hunter R.A., Berger K., De Paoli D.A., Orr C.L.A. Ingestion of toxic *Microcystis aeruginosa* by dairy cattle and the implications for microcystin contamination of milk. Toxicon, **39:** 1847-1854, 2001.

Park H.D., Sasaki Y., Maruyama T., Yanagisawa E., Hirashi A., Kato K. Degradation of the cyanobacterial hepatotoxin microcystin by a new bacterium isolated from a hypertrophic lake. Environ. Toxicol. **16:** 337-343, 2001.

Pouria S., de Andrade A., Barbosa J., Cavalcanti R., Barreto V., Ward C., Prieser W., Poon G., Neild G., Codd G. Fatal microcystin intoxication in haemodialysis unit in Caruaru, Brazil. Lancet **352:** 21-26, 1998.

Rao P.V.L., Bhattacharya R. The cyanobacterial toxin micyocystin-LR induced DNA damage in mouse liver in vivo. Toxicology, **114:** 29-36, 1996.

Rao P.V.L., Bhattacharya R., Bhaskar A.S.B. Effects of nutrient media and culture duration on growth, macromolecular composition and toxicity in batch cultures of *Microcystis aeruginosa* (UTEX 2385), Microbios **86:** 95-104, 1996.

Rao P.V.L., Bhattacharya R., Dasgupta S. Isolation, culture and toxicity of cyanobacterium (blue-green alga) *Microcystis aeruginosa* from a freshwater source in India. Bull. Environ. Contam. Toxicol. **52:** 878-885, 1994.

Rao P.V.L., Bhattacharya R., Nidhi Gupta, Parida M.M., Bhaskar A.S.B., Rupa D. Involvement of caspase and reactive oxygen species in cyanobacterial toxin anatoxin-a induced cytotoxicity and apoptosis in rat thymocytes and Vero cells. Arch.Toxicol. **76:** 227-235, 2002.

Rao P.V.L., Bhattacharya R., Pant S.C., Bhaskar A.S.B. Toxicity evaluation of in vitro cultures of freshwater cyanobacterium *Microcystis aeruginosa*: I Hepatic and histo-pathological effects in rats. Biomed. Environ. Sci. **8:** 254-264, 1995.

Rao P.V.L., Bhattacharya R., Parida M.M., Jana A.M., Bhaskar A.S.B. Freshwater Cyanobacterium *Microcystis aeruginosa* (UTEX2385) induced DNA damage in vivo and in vitro. Environ. Toxicol. Pharmacol. **5:** 1-6, 1998.

Rao P.V.L., Nidhi Gupta, Bhaskar A.S.B.,. Jayaraj R. Toxins and bioactive compounds from cyanobacterial and their implications on human health. J. Environ. Biol. **23:** 215-224, 2002.

Rositano J., Nicholson B., Pieronne P. Destruction of cyanobacterial toxins by ozone. Ozone Sci. Eng. **20:** 223-238, 1998.

Shaeffer D.J., Malpas P.B., Barton L. Risk assessment of Microcystin in dietary *Aphanizomenon flos-aquae*. Ecotox. Environ. Safety **44:** 73-80, 1999.

Shephard G., Stockenström S., de Villiers D., Engelbrecht W., Sydenham E., Wessels G. Photocatalytic degradation of cyanobacterial microcystin in water. Toxicon **36:** 1895-1901, 1998.

Sivonen K. Cyanobacterial toxins and toxin production. Phycologia **35:** 12-24, 1996.

Skulberg O.M., Carmichael W.W., Anderson R.A., Matsunaga S., Moore R.E., Skulberg R. Investigations of a neurotoxic oscillatorian strain (Cyanophyceae) and its toxin. Isolation and characterization of homoanatoxin-a. Environ. Toxicol. Chem. **11:** 321-329, 1992.

Suzuki H., Watanabe M., Wu Y., Sugita T., Kita K., Sato T., Wang X.L., Tanzawa H., Sekiya S., Suzuki N. Mutagenicity of microcystin-LR in human Rsa cells. Int. Mol. Med. **2:** 109-112, 1998.

Teixera M., Costa M., Carvalho V., Pereira M., Hage E. Gastroenteritis epidemic in the area of the Itaparica Dam, Bahia, Brazil. Bull. Pan. Am. Health Org. **27:** 244-253, 1993.

WHO (1998) Guidelines for drinking water quality. Recommendations of World Health Organization, WHO, Geneva, Vol-II.

Pharmacological Perspectives of Toxic Chemicals and Their Antidotes
Editors: S J S Flora, J A Romano, S I Baskin and K Sekhar

Toxicology and Pharmacology of Brevetoxins

S. B. Deshpande and J. N. Singh
Department of Physiology, Institute of Medical Sciences
Banaras Hindu University, Varanasi 221 005, India

The rivers, lakes and ponds at times have full of algal growth and contamination. Recently accidental death of tigers and elephants have become major News items in India. No satisfactory explanation for the death of these animals have been given yet, but the consumption of water from the reservoirs containing such algal growth can not be excluded. In the United States and Europe the toxicity by the accidental consumption of infested water by man and livestock have been reported (Baden, 1983, 1989; Poli et al., 2000; Skulberg et al., 1992) and was attributed to the poisoning due to the toxins released from the cyanobacteria, blue-green algae. In China, an epidemiological survey revealed a high incidence of primary liver cancer in isolated pockets of geographical area (Haimen city, and Fusui County) which was attributed to drinking water from the pond and ditches infested with microcystins (Ueno et al., 1996). A hepatotoxic peptide has been isolated in the water by these algal blooms contaminating the drinking water in these areas (Ueno et al., 1996). Thus, there is a need for the understanding of the algal toxicity in perspective to the lack of potable drinking water for human and animal consumption in Indian context.

HABITAT AND REPRODUCTION OF DINOFLAGELLATES

Thousands of living phytoplankton species are found in the marine food web but, only a few dozens of them are toxic (Anderson, 1994). Most are *dinoflagellates*, *prymnesiophytes* or *chloromonads*. A bloom develops when these single-celled algae photosynthesize and multiply by converting dissolved nutrients and sunlight into plant biomass. The mode of reproduction in these organisms is usually asexual. At times of scarcity of nutrients, they switch to the sexual reproduction. Under adverse conditions, they also form thick-walled dormant cells, called cysts, that settle on the seafloor and can survive there for years. When the favourable growth conditions return, the cysts germinate and reinoculate the water that can bloom. Barring a shortage of nutrients or light, or heavy grazing by tiny zooplankton that consume the algae, they grow rapidly. In some cases, a millilitre of sea water can contain tens or hundreds of thousands of algal cells (Anderson, 1994).

RED TIDES

"Red-Tide" a natural phenomenon caused by the blooms of the marine algae. During bloom conditions the tiny pigmented algae grow in such abundance so as to change the colour of the sea water to red, brown, or even green (Skulberg et al., 1984; Anderson, 1994). The name "Red tide" is misleading, however, because many toxic events are called *red tides* even when the water show no discolouration. Likewise, an accumulation of non-toxic, harmless algae can change the colour of ocean water. The picture is even more complicated as some phytoplankton neither discolour the water nor produce the toxins but kill marine animals in other ways. Many diverse phenomena thus fall under "red tide" rubric (Anderson, 1994; Richards et al., 1990).

The red tide blooms, were reported in brackish water fish ponds (Shamsudin, 1996). The dominant dinoflagellate species are comprised of *Peridinium quinquecorne* (> 90% total cell count) with considerable proportion of *Protoperidinium excentricum.* Ciliophora consisting of *Tintinopsis* and *Favella species* were also present during the bloom period.

Harmful effects of these algal blooms or red tides cause most serious impact on human life by the consumption of sea food. Clams, mussels, oysters or scallops ingest the algae as food and retain the toxins in their tissues. Typically the shellfish themselves are only marginally affected, but a single clam can sometimes accumulate enough toxin to kill a human being. These shellfish poisoning syndromes have been described as paralytic (PSP), diarrhetic (DSP) and neurotoxic (NSP) (Chen et al., 1993). In a Canadian outbreak, in which some patients suffered memory losses was appropriately characterized as amnesic shellfish poisoning (ASP; Anderson, 1994). Ciguatera fish poisoning (CFP), causes more human illness than any other kind of toxicity originating by the consumption of sea food. It occurs predominantly in tropical and subtropical islands. Dinoflagellates that are attached to the sea weeds produce the Ciguatera toxins. Herbivorous fishes eat the seaweeds and the attached dinoflagellates as well. Because these toxins are soluble in fat, they are stored in the tissues of the fish who consume them, and through the food chain transferred to carnivores. The most dangerous fish to eat are thus, the largest and the oldest, often considered the most desirable as well (Anderson, 1994). Normally edible fish and shell fish may become poisonous as a result of cumulative feeding on the toxic organisms such as diatoms and dinoflagellates (Shimizu, 1978).

BREVETOXINS

The toxic dinoflagellates which cause red tides are *Gonyaulax Sp* (produced saxitoxins and gonyautoxins), *Ptychodiscus brevis* (produced brevetoxins), etc., (Shimizu, 1978; Alam et al., 1982; Lin et al., 1981). The *Ptychodiscus brevis* toxins (PbTx) are cyclic polyether compounds derived from dinoflagellate algae. These toxins are liberated into water column during bloom conditions to produce red tides (Trieff et al., 1975; Baden, 1983; Shimizu, 1982). Brevetoxins are condensed to form a single hydrocarbon chain to make complex multiring methylated polyether toxin (Mcfarren et al., 1965; Carmichael and Bent, 1981).

The *Ptychodiscus brevis* toxin frequently cause devastating fish kills. As wild fish swim through the *Ptychodiscus brevis* blooms, the fragile algae rupture, releasing neurotoxins on to the gills of the fish. Within a short time the animals asphyxiate (Anderson, 1994). The ways in which algae kill these fish are poorly understood. Some algae produce polyunsaturated fatty acids and galactolipids that destroy the blood cells. Such an effect would explain the ruptured gills, hypoxia and edema in dying fish. Other algal species produce haemolytic compounds and neurotoxins as well. The combination can significantly reduce a fish's heart rate, resulting in reduced blood flow and decrease in oxygen tension. Shell-fish and fish poisoning result from toxic dinoflagellate infestations such as the neurotoxic shellfish poisoning (NSP) by *Ptychodiscus brevis*, the paralytic shellfish poisoning (PSP) by *Protogonyaulax catenella* and *P. tamarensis* and the Ciguatera fish poisoning (CFP) by *Gambierdiscus toxicus* (Sakamoto et al., 1987).

Table 1 The table showing dinoflagellates, the toxins produced and their poisoning [viz., neurotoxic shellfish poisoning (NSP); paralytic shellfish poisoning (PSP); Ciguatera fish poisoning (CFP); diarrheatic shellfish poisoning (DSP); Okadaic acid (OA)]

Species	*Toxins*	*Poisoning*
Ptychodiscus brevis	Brevetoxin	NSP, PSP
Protogonyaulax catenella	-	PSP
Protogonyaulax tamarensis	-	PSP
Gambierdiscus toxicus	Ciguatoxins	CFP
Gonyaulax tamarensis	Saxitoxins	PSP
	Gonyautoxins	PSP
Dinophysis	Dinophysis toxin	DSP
Porocentrum lima	Okadaic acid	DSP

CHEMISTRY OF BREVETOXINS

The taxonomic classification of the *Gymnodinium breve*, organism has been revised to *Ptychodiscus brevis* in the dinoflagellates groups of blue green algae as cyanobacteria (Steidinger, 1983). Structurally, these toxins include alkaloids and peptides. Although the greatest and most dramatic animal losses have been due to the alkaloid toxins. The deaths due to peptide neurotoxins are widespread and of greater concern for animal or human health (Carmichael and Bent, 1981). It has been reported that, the presence of an irritating aerosol which results from contact with *Ptychodiscus brevis* cell particles entrapped in sea spray, may cause deleterious effects in human beings (Baden et al., 1981). Several groups of workers reported isolation of lipophilic cyclic polyether toxins called brevetoxins from *Ptychodiscus brevis* cell cultures. Many toxins with or without phosphorous have been isolated and various effects of these compounds are reported elsewhere (Trieff et al., 1975; Taylor and Seliger, 1979; Shimizu, 1982; Baden, 1983, 1989; Shimizu et al., 1990; Tsai and Chen, 1991). The acyclic phosphorous toxins have an ichthyotoxic property (Steidinger, 1983; Risk et al., 1979; Lin et al., 1981; Shimizu et al., 1986, 1990; Baden et al., 1988; Baden, 1989). In addition to the thiophosphate moiety an oxime group is also present in some toxins (Alam et al., 1982; Kaushik et al., 1988).

Structural correlation of toxins from different laboratories has revealed several nomenclatures reported elsewhere (Shimizu et al., 1986; Baden, 1989). Depending upon the structural backbones (Type-1 or Type-2) and the chemical bond at R_1 and R_2 region of the backbones at N-terminal, more than 10 different types of brevetoxins are synthesized (Baden, 1989). Alam et al., (1982) isolated a phosphorus containing toxin which was further purified and synthesized by Dinovi et al., (1983). Later on, the method for synthesis was modified elsewhere (Kaushik et al., 1988).

STABILITY

The toxins are structurally different and exhibit difference in solubility pattern also. Brevetoxins are either water soluble or soluble in organic solvents (Mcfarren et al., 1965). Structural integrity of brevetoxin can be demonstrated after a month of storage in a dry state or in organic solvent at reduced temperature (Poli et al., 1990). In dry state, brevetoxins are stable at 300 °C (Poli et al., 1990). Brevetoxins are degraded rapidly in 0.1N NaOH and in chlorine solution (Baden, 1989; Poli et al., 1990; Poli, 1988). Further the saponification and ozonolysis also degraded them rapidly (Rodgers et al., 1984). Specific antibodies directed against brevetoxins have been produced in goat, i.e. the haptenic toxin T34 (Baden, et al., 1984; Melinek et al., 1994).

LETHALITY

The brevetoxins (PbTx) are lethal to experimental animals (Alam et al., 1982; Kaushik et al., 1988; Koley et al., 1995; Mazumder et al., 1997c). The lethal doses of PbTx are in the range of nM concentrations correlated with structure but not with hydrophobicity. The signs and symptoms of intoxication from brevetoxins in fish include violent twisting and corkscrew swimming, defecation and regurgitation, pectoral fin paralysis, loss of equilibrium, quiescence, vasodilation, convulsions and culminating in death due to respiratory failure.

Injection of brevetoxins in mice exhibit an immediate irritability followed by lower limb paralysis, dyspnea, autonomic over activity (indicated by salivation, lacrimation, urination and defecation) and ultimately death due to respiratory paralysis. LD_{50} for brevetoxin in mice ranges from 50 μg/kg body weight for oral and i.p. administration (Trieff et al., 1975; Risk et al., 1979; Shimizu, 1982; Baden and Mende, 1982). The lethality does not correlate well with hydrophobicity as, one would expect, the more potent toxin to be more non-polar (Risk et al., 1979; Baden and Mende, 1982; Shimizu et al., 1986). However, the oxidation of aldehyde fraction in PbTx to the carboxylic acid, reduces the *in vivo* potency to non-toxic levels (Poli et al., 1986).

DISTRIBUTION, METABOLISM AND EXCRETION

A major route of human exposure to brevetoxins is reported via the respiratory tract by aerosols or through the consumption of sea food (Benson et al., 1999). Although a majority of the PbTx was cleared rapidly from the gills, lungs, liver, and kidneys, approximately 20% of the initial concentration remain in each organ for 7 days (Washburn et al., 1994). The concentrations of PbTx in brain and adipose tissue were low, but remained relatively constant over time. Approximately twice as much PbTx activity was excreted in faeces than in urine. The majority of excretion occurring within 48 h after instillation. The persons suffering from neurotoxic shellfish poisoning were investigated and the brevetoxins were isolated in the urine samples resembled the toxin present in shellfish extracts (Poli et al., 1990, 2000).

High performance liquid chromatographic (HPLC) analysis of shellfish extracts demonstrated multiple fractions recognized by specific anti-brevetoxin antibodies, suggesting metabolic conversion of parent brevetoxins. Affinity-purification of these extracts yielded four major peaks of activity. One peak was PbTx-3, which was produced metabolically from the dominant parent toxin, PbTx-2 (Poli et al., 2000). Other peaks of activity were determined to consist of compounds of apparent masses and these higher masses are suggestive of conjugated metabolites.

Over 80% of the PbTx-3 is rapidly absorbed from the lungs to the blood and distributed throughout. These results illustrate that, brevetoxin exposure by the respiratory route results in systemic distribution and suggest that the initial respiratory irritation and broncho-constriction may only be a part of the overall toxicological consequences associated with PbTx inhalation (Benson et al., 1999).

Biliary excretion was the major route of detoxification, and *in vivo* extraction ratio was 0.55. Brevetoxins are metabolized to four compounds separable by thin-layer liquid chromatography or HPLC. The metabolism was reproducible and comparable in a perfused liver or hepatocyte models. All of the metabolites were more polar than the parent compounds, all cross-reacted with brevetoxin-specific antiserum, and one was identified as an epoxide. Nearly 90% of the administered toxin was eliminated from the body in a week (Poli et al., 1990).

TOXICITY ON ORGAN SYSTEMS

The systemic toxicology of the brevetoxins reveal deleterious effects on the nervous, neuromuscular, cardiovascular, respiratory and renal systems (Baden et al., 1981, Koley et al., 1995; Mazumder et al., 1997a; Kaushik et al., 1988).

EFFECTS ON NERVOUS SYSTEM

The animals injected with brevetoxins manifested with symptoms of increased parasympathetic activity, fasciculations, convulsions, paralysis and respiratory failure resulting in death (Mahmood and Carmichael, 1987; Skulberg et al., 1992; Baden and Mende, 1982). Atropine pretreatment or decreasing the parasympathetic activities prolonged the survival time (Mahmood and Carmichael, 1987). Further, the algal toxins exhibited acetylcholineesterase inhibitor activity and are implicated for their toxicity (Mahmood and Carmichael, 1987). All these manifestations indicate the involvement of synaptic transmission.

PbTx depressed the synaptic transmission at hippocampus and spinal cord (Apland et al., 1993; Singh and Deshpande, 2002, Singh et al., 2002). In the spinal cord the effect was seen on the segmental monosynaptic reflex (MSR) and polysynaptic reflex (PSR). The depressant effect of PbTx analogs on MSR and PSR varied according to the structures. These analogs have different structures with a common central group i.e. [-P=O-NH-N-] and depressed MSR and PSR differentially. $PbTx_1$ and $PbTx_2$ analogs have aromatic groups on one side and octyl/pentyl group on the other side, respectively. These analogs depressed MSR and PSR to great extent (> 75 %). Whereas, the $PbTx_3$ and $PbTx_4$ have aliphatic groups on both sides and were less potent than $PbTx_1$ for depressing MSR/PSR. Thus, $PbTx_1$ is most potent of all the four analogs for MSR and $PbTx_2$ for PSR (Singh et al., 2002).

Between $PbTx_1$ and $PbTx_2$ analogs, the former exhibited greater depression in terms of MSR and the latter exhibited greater activity on PSR. The $PbTx_1$ had octyl group as against pentyl group with $PbTx_2$. Thus, it appears that the pentyl group of $PbTx_2$ renders it more potent for its action on PSR. PSR is shown to be mediated by NMDA receptors (Singh and Deshpande, 2002; Deshpande, 1993). Thus, it may be possible that the pentyl group of $PbTx_2$ exerts its effect on the NMDA receptors similar to 7-chlorokynurenic acid (Maruoka et al., 1997; Singh and Deshpande, 2002). But, the $PbTx_3$ toxicity on PSR fluctuated between the concentrations. The reasons for these fluctuations can not be ascertained. The toxicity of $PbTx_4$ on MSR and PSR is markedly lower than all the analogs. The $PbTx_4$ differed from $PbTx_3$ by the presence of sulphur in the central group which might have altered its activity. The rank order of potency for the depression of MSR is $PbTx_1 > PbTx_2 >> PbTx_4 > PbTx_3$ and for PSR it is $PbTx_2 > PbTx_1 > PbTx_3 >> PbTx_4$ (Singh et al., 2002).

Extending this work, Singh and Deshpande (2002) performed experiments to discern the mechanisms underlying the depressant actions of the toxin at Ia-α-motoneuron synaptic transmission. They selected potent toxin $PbTx_1$ for their study. The $PbTx_1$ depressed the MSR and the PSR in a concentration-dependent manner as reported earlier (Singh et al., 2002). Since brevetoxins block Na^+ channels, the effect of toxin on nerve action potential and conduction was examined (Caterall, 1992; Trainer et al., 1993). The $PbTx_1$ did not alter the magnitude of the dorsal root or the ventral root potentials. This set of experiments demonstrate the non-involvement of the Na^+ channel at this concentration of $PbTx_1$. Further, there is no alteration in the afferent or efferent axonal activity as seen for diisopropylfluorophospate (Deshpande and Dasgupta, 1997).

The NMDA antagonist (DL-2-amino-5-phosphono valeric acid; APV) or channel blocker (Mg^{2+}), attenuated the $PbTx_1$-induced depression of the MSR. The complete blockade was seen when the cords were simultaneously exposed to 6-cyano-7-nitroquinoxaline-2,3-dione (CNQX) in the presence of APV (Singh and Deshpande, 2002). The involvement of NMDA was confirmed by performing the experiments with submaximal concentrations of NMDA and $PbTx_1$. At concentration where NMDA by itself did not alter the magnitude of MSR or PSR, the submaximal concentration of $PbTx_1$ enhanced the depression of PSR significantly (Singh and Deshpande, 2002). This clearly demonstrated the NMDA receptor involvement for $PbTx_1$ action. Further, 7-chlorokynurenic acid ($glycine_B$ receptor antagonist), failed to block the $PbTx_1$-induced depression. These observations thus indicate that the $PbTx_1$ depressed the spinal reflexes without altering the magnitude of dorsal root or ventral root activity. The depression of the PSR involve NMDA receptors while that of the MSR involve NMDA and non-NMDA receptors. The $PbTx_1$ actions did not involve the $glycine_B$ site of the NMDA receptor (Singh and Deshpande, 2002).

The PbTx produce a differential release of amino acid transmitters from synaptosomes into the medium, which could be effectively antagonised by either tetrodotoxin or verapamil (Risk et al., 1982). Aspartate, glutamate and γ-aminobuturic acid (GABA) release was stimulated 5-, 11- and 4-fold respectively. In view of these observations, the actions of $PbTx_1$ were further explored on the basis of the excitatory and inhibitory transmissions (Singh, 2001, Singh; and Deshpande, 2003). The $GABA_A$ antagonist, bicuculline, blocked the $PbTx_1$-induced depression similar to that observed with APV. The data further demonstrated the dependence of GABA actions over NMDA. The depression was also examined in terms of the excitatory transmitter like 5-HT. The data are consistent for 5-HT involvement at $5\text{-}HT_3$ receptors. Thus, $PbTx_1$ toxicity involve glutamate receptors at NMDA and non-NMDA sites. The NMDA component activate $5\text{-}HT_3$ receptors and GABAergic inhibition (Singh, 2001).

EFFECTS ON NEUROMUSCULAR JUNCTION

Brevetoxin cause the acetylcholine release from guinea pig ileum (Poli et al., 1990) and from rat phrenic nerve hemidiaphragms without altering the clear-core vesicles (Baden et al., 1984). The PbTx stimulate the human bronchial smooth muscle via the cholinergic fibres involving Na^+ channels (Shimoda et al., 1988). In guinea pig ileum, PbTx produced contractions were similar to acetylcholine and the action could be blocked by tubocurarine, atropine and hexamethonium without competing for the binding sites (Chiang et al., 1991). Since the guinea pig ileal contractions were partially blocked by nicotinic antagonists, the involvement of muscarinic and nicotinic receptors is proposed (Chiang et al., 1991).

In rat hemidiaphragm preparation, twitch potentiation and tetanic fading is shown to be due to the excessive cholinergic activity (Mahmood and Carmichael., 1987). The fading of the tetanic contractions may be due to the inhibitory actions of the toxins on the indirect contractions as reported elsewhere (Atchison et al., 1986). However, the concentration-dependent increase in the resting tension of the diaphragm was consistently observed and was effectively blocked by tetrodotoxin (TTX), curare or low amounts of Calcium (Atchison et al., 1986; Baden et al., 1984). The actions thus involve plexus at pre- and post-junctional sites.

The brevetoxins depolarized the isolated nerves and muscle membrane in a concentration-dependent manner (Wu and Narahashi, 1987; Strichartz et al., 1987; Ellis, 1985). However, we did not find any effect of the toxin on compound action potentials either in frog sciatic nerve or in the dorsal/ventral root axons (Singh and Deshpande, 2002).

Miniature end-plate potential (MEPP) frequency increased at higher concentrations. There was depolarization of the muscle fibres at the end-plate and in non-junctional regions (Gallagher and Gallagher, 1980). The effects of brevetoxin on neuromuscular preparations were summarized as, "depolarization of the muscle fibre membrane, increase in MEPP frequency, blockade of the EPP generation, and the biphasic effects on the MEPP amplitude together with depression of the acetylcholine induced depolarization". It was suggested that depolarization of the nerve terminal was sufficient to inhibit transmitters release (Gallagher and Gallagher, 1980, Atchison et al., 1986; Baden et al., 1984). However, these actions at neuromuscular junctions are not due to the involvement of acetylcholinesterase activity (Mazumder et al., 1997c).

EFFECTS ON CARDIOVASCULAR SYSTEM

Intravenous injection of PbTx (1 mg/Kg) killed the animal within 15-20 min. with an instantaneous decrease in heart rate and fall of blood pressure (Deshpande et al., 1996). The fall of pressure progressed until the ventricular asystole. Ellis et al (1979) reported that brevetoxin produced bradycardia and hypotension in anaesthetized dogs in low doses and apnea, cardiac arrythmia, ventricular fibrillation in high doses. Subsequent studies suggested that the bradycardia and hypotension in part due to activation of Bezold-Jarisch reflex (Borison et al., 1980). The toxin by itself produced many characteristics of the Bezold-Jarisch reflex i.e. sinus bradycardia and hypotension and peripheral vasodilation (Borison et al., 1980; Alam et al., 1982; Ellis, 1985; Baden, 1989; Templeton et al., 1989; Koley et al., 1995, 1997). The PbTx did not alter the reflex responsiveness to PDG as seen after scorpion venom or bradykinin (Deshpande et al., 1996, 1999; Bagchi and Deshpande, 2001). Reports indicate that the action of brevetoxin could be partially blocked by cholinergic antagonists (Borison et al., 1985; Johnson et al., 1985). The ventricular asystole indicate the enhanced cholinergic (vagal) activity as mentioned above.

Premature ventricular depolarization, paroxysmal ventricular tachycardia and complete heart block were observed in rat and guinea pig (Poli et al., 1990; Rodgers et al., 1984). The conduction defects in the Purkinje system of the heart were also observed elsewhere (Poli et al., 1990). Heart rate was effected only at higher doses (Borison et al., 1980; Poli et al., 1990). There was no effect on the cardiac, Na^+,K^+-ATPase activity (Rodgers et al., 1984). Early work with neurogenic and myogenic hearts indicated a differential effects. The neurogenic model exhibiting an increase in chronotropy and prolonged tension developments. Myogenic heart models were unaffected by PbTx.

To isolate the systemic effects, the PbTx was examined on the isolated spontaneously beating atrial tissue. PbTx decreased the rate and force of spontaneously beating right atria in a concentration-dependent manner involving different systems. The lower concentration of PbTx involves adrenergic mechanism and higher concentration involves cholinergic mechanisms (Mazumder et al., 1997c). TTX abolished the responses indicating the plexus involvement. Atropine blocked the PbTx-induced changes and physostigmine potentiated the responses (Mazumder et al., 1997c). Similarly, propranolol, ß-adrenergic blocker, blocked the rate and force. These observations indicate that either the ß-adrenergic system is modulated by cholinergic component or *vice versa*. To understand the involvement of the adrenergic system we depleted the adrenergic terminals by reserpinization. Under such circumstances PbTx action remained unaltered. These evidences indicate that the adrenergic system do not mediate the cholinergic component. The nefedipine, L-type Ca^{2+} channel antagonist, blocked the PbTx-induced response either in our study or elsewhere. Thus, the Ca^{2+} channels are involved for the action of toxin (Mazumder et al., 1997a, 1997b).

EFFECTS ON RESPIRATORY SYSTEM

The respiratory abnormalities such as apnea, bradypnea, hyperpnea, etc are consistent features of PbTx toxicity (Poli et al., 1990; Koley et al., 1995; Borison et al., 1985). This feature is compatible with the fish dying due to the paralysis of gills by these toxins while swimming through the blooms (Anderson, 1994). The irritative respiratory symptoms such as coughing, sneezing and asthma are seen in persons inhaling the toxin aerosols accidentally and are attributed to the local actions of the toxin (Sakamoto et al., 1985). There was no instantaneous arrest of respiration after the intravenous injection of toxin in experimental animals but respiratory arrest occurred in these animals after a delay, indicating the time required for the action (Tempelton et al., 1989). These observations indicate that some mechanisms operating for the depression of respiratory centres in the medulla. This point is substantiated by the fact that the threshold concentration to induce respiratory alterations was lower than cardiovascular abnormalities. The toxins also involve different mechanisms for respiratory and cardiovascular actions. Hence, the intoxicated animals were revived only by artificial ventilation not by anticholinergic (Koley et al., 1995; Mazumder et al., 1997c). Thus, the primary cause of lethality is due to the altered neuronal activity of respiratory centres or respiratory motoneurons. This is aggravated further by the fact that PbTx blocks the cholinergic transmission at neuromuscular junction (Gallagher and Gallagher, 1980; Atchison et al., 1986; Baden et al., 1984). Further, there was no alteration in the air way resistance and compliance (Franz and LeClaire, 1989) indicating the noninvolvement of peripheral mechanisms for the toxicity. Thus, the respiratory failure assumes greater importance.

EFFECTS ON RENAL SYSTEM

PbTx has profound toxic effects on kidney functions as it causes reduction in urine flow along with haematuria. The reduction of urine flow is most probably due to the kidney damage as repeated administration of the toxin produced haematuria (Koley et al., 1997). Histological observations reveal significant damage of the glomerular membrane of the cortical Malphigian corpuscles (Koley et al., 1997).

EFFECTS ON ION CHANNELS

Brevetoxins elicited spontaneous train of action potentials in squid giant axons, an effect antagonized by TTX and therefore presumed to originate in the Na^+ channels (Westerfield et al., 1977). This is consistent with the other *in vitro* studies showing the antagonistic effect of the TTX on brevetoxin action (Trieff et al., 1975; Shimizu, 1982; Wu and Narahashi, 1987; Strichartz et al., 1987; Ellis, 1985; Baden et al., 1982; Baden, 1983, Rodgers et al., 1984). Direct evidences of the action of Na^+ fluxes has been subsequently demonstrated in neuroblastoma cells and in rat brain synaptosomes (Risk et al., 1982, Poli et al., 1986; Catterall and Risk, 1981; Catterall and Gainer, 1985). Calcium ion fluxes resulting from a consequences of brevetoxin administration occur indirectly as a result of Na^+ depolarization (Rodgers et al., 1984).

More than five functional domains with α-subunits of the Na^+ channel are being identified using physiological and biochemical cross linking experiments and brevetoxin ligand specific probes (Catterall, 1992). The brevetoxin binding domain is situated in the hydrophobic channel domain and regulates the Na^+ channel inactivation process (Catterall, 1992; Trainer et al., 1993; Adams and Olivera, 1994). Voltage clamp experiments have provided evidences for the specific effect of toxin associated with the opening of Na^+ channels while K^+ channels remain unaffected (Wu and Narahashi, 1987; Huang et al., 1984). Brevetoxin-induced alterations in Na^+ channels do not express fast inactivation thus they remain open for much longer time than the unaltered channels. Further during the activation, the peak Na^+ current is increased while other cation currents selectively remain unchanged.

The different toxins share a common binding site (the receptor-site 5) located on the alpha sub-unit of this neuronal trans-membrane protein. Hence, they also interact with the scorpion toxins (Wada et al., 1992). Electrophysiological studies on the mode of action of PbTx identify that these toxins as a specific Na^+ channel activators. Indeed, during the action of these phycotoxins, Na^+ channels remain permanently open, at the resting membrane potential, which produces a continuous entry of Na^+ ions in most excitable cells. Such a Na^+ entry has various consequences on sodium-dependent physiological mechanisms, resulting in a membrane depolarization which, in turn, causes spontaneous and/or repetitive action potential discharges and thereby increase the membrane excitability. These neuronal discharges may be transient or continuous according to the preparation and the toxin tested. The increase in membrane excitability during the action of brevetoxins is responsible for the different effects exerted by these toxins on various chemical synapses and secretory cells.

Conclusion

In conclusion, the brevetoxins, periodically and fatally contaminate the water supplies of the wild, domestic and marine animals. Thus pose serious health problems to human beings. They can also be a potential biological warfare agents as they produce serious neuronal abnormalities involving NMDA and non-NMDA receptors Na^+ and Ca^{2+} channels. The identification of an agent to oppose the effects of PbTx is a necessity.

References

Adams M.E. and Olivera B.M. Neurotoxins: Overview of an emerging research technology. Trends. Neurosci. **17:** 151-155, 1994.

Alam M.R., Sanduza R., Hossain M.B. and vander Helm D. Gymnodinium breve toxins 1. Isolation and X-ray structure of O, Odipropyl-(E)-2-(1-methyl-2-oxoproylinide phosphorohydro-thioate-(E)-oxime from the red tide dinoflegellate G.breve, J. Am. Chem. Soc. **104:** 5232-5234, 1982.

Anderson D.M. Red Tides. Sci. Am., **271:** 62-68, 1994.

Apland J.P., Alder M. and Sheridan R.E. Brevetoxin depresses synaptic transmission in guinea pig hippocampal slices. Brain Res. Bull. **31:** 201-207, 1993.

Atchison W. D., Luck V.S., Narahasi T. and Vogel S.M. Nerve membrane Sodium channels as the target site of brevetoxins at neuromuscular junctions. Br. J. Pharmacol. **89:** 731-738, 1986.

Baden D.G. Marine food-borne dinoflagellate toxins. Int. Rev. Cytol. **82:** 99-149, 1983.

Baden D.G. Brevetoxin: unique polyether dinoflagellate toxins. FASEB J. **3:** 1807-1817, 1989.

Baden D.G. and Mende T.J. Toxicity of two toxins from Florida's red tide dinoflagellate Ptychodiscus brevis. Toxicon. **20:** 457-462, 1982.

Baden D.G., Mende T.J., Lichter W. and Wellham L. Crystallization and toxicology of T34: a major toxin from Florida's red tide organism (Ptychodiscus brevis). Toxicon, **19:** 455-462, 1981.

Baden D.G., Mende T.J. and Leung I. Bronchoconstriction caused by Florida red tide toxins. Toxicon, **20:** 929-932, 1982.

Baden D.G., Bikhazi G., Decker S.J., Foldes F.F. and Leung I. Neuromuscular blocking action of two brevetoxin from the Florida red tide organism Ptychodiscus brevis. Toxicon. **22:** 75-84, 1984.

Baden D.G., Mende T.J., Szmant A.M., Trainer V.L., Edwards R.A. and Roszell L.E. Brevetoxins binding : Molecular pharmacology versus immunoassay. Toxicon, **26:** 97-103, 1988.

Bagchi S. and Deshpande S.B. Scorpion (Buthus tamulus) venom toxicity on cardiopulmonary reflexes involves kinins via 5-HT_3 receptor subtypes. J. Venom Anim. Toxins. **7:** 25-44, 2001.

Benson J.M., Tischler D.L. and Baden D.G. Uptake, tissue distribution, and excretion of brevetoxin 3 administered to rats by intratracheal instillation. J. Toxicol. Environ. Health. **57:** 345-355, 1999.

Borison H.L., Ellis S. and McCarthy L.E. Central respiratory and circulatory effects of Gymnodinium breve toxin in anaesthetized cats. Br.J. Pharmacol. **70:** 249-256, 1980.

Borison H.L., McCarthy L.E. and Elious S. Neurological analysis of respiratory, cardiovascular and neuromuscular effects of Brevetoxin in cats. Toxicon, **23:** 517-524, 1985.

Carmichael W.W. and Bent P.R. Haemagglutination method for detection of fresh water cyanobacteria (blue-green algae) toxins. Appl. Environ. Microbiol. **41:** 1383-1388, 1981.

Catterall W.A. Cellular and molecular biology of voltage-gated sodium channels. Physiol. Rev. **72:** 15-48, 1992.

Catterall W.A. and Gainer M. Interaction of brevetoxin-A with a new receptor site on the sodium channel. Toxicon. **23:** 497-504, 1985.

Catterall W.A. and Risk M.A. Toxin T46 from Ptychodiscus brevis (formerly Gymnodinium breve) enhances activation of voltage-sensitive sodium channels by veratridine. Mol. Pharmacol **19:** 345-348, 1981.

Chen D.Z.X, Boland M.P., Smillie M.A., Klix H., Ptak C., Anderson R.J. and Holmes C.F.B. Identification of protein phosphatase inhibitors of the microcystin class in the marine environment. Toxicon. **31:** 1407-1414, 1993.

Chiang P.K., Butler D.L. and Brown N.D. Life Sciences **49:** PL13-PL19, 1991.

Deshpande S.B. Significance of monosynaptic and polysynaptic reflexes in rat spinal cord in vitro. Ind. J. Expt. Biol. **31:** 850-854, 1993.

Deshpande S.B. and Dasgupta S. Diisopropylphosphor ofluoridate-induced depression of segmental monosynaptic transmission in neonatal rat spinal cord is also mediated by increased axonal activity. Toxicol. Lett. **90:** 177-182, 1997.

Deshpande S.B., Bagchi S., Tharakan B. and Rai O.P. The phenyldiguanide-induced reflex bradycardia is augmented by scorpion venom but not by the analogue of Ptycodiscus brevis toxin in rats (FIPS congress proceedings, Ed. M. Fahim, pp. 141-148, 1996 published from V.P.Chest Institute, Delhi University Delhi.

Deshpande S.B., Bagchi S., Rai O.P. and Aryya N.C. Pulmonary oedema produced by scorpion venom augments a phenyldiguanide-induced reflex response in anaesthetized rats. J. Physiol. **521:** 537-544, 1999.

Dinovi M., Trainor D.A. and Nakanishi K. The structure of PB-1 an unusual toxin isolated from the red tide dinoflagellate Ptychodiscus brevis. Tetrahedron. Lett. **24:** 855, 1983.

Ellis, S. Introduction to Symposium-brevetoxins: chemistry and pharmacology of red tide toxins from Ptychodiscus brevis (formerly Gymnodinium breve). Toxicon. **23:** 469-472, 1985.

Ellis S., Spikes J.J. and Johnson G.L. In toxic dinoflagellate blooms, Eds. D.I. Taylor and H.H. Seliger (Elsevier North Holand, Amsterdam), 35, 1979.

Franz D.R. and LeClaire R.D. Respiratory effects of brevetoxins and saxitoxins in awake guinea pigs. Toxicon. **27:** 647-654, 1989.

Gallagher J.P. and Gallagher S.P. Effect of Gymnodinium breve toxin in the rat phrenic nerve diaphragm preparation. Br. J. Pharmacol. **69:** 367-372, 1980.

Huang J.M.C., Wu C.H. and Baden D.G. Depolarizing action of a red tide dinoflagellate brevetoxin on axonal membranes. J. Pharmacol. Exp. Ther., **229:** 615-621, 1984.

Johnson G.L., Spikes J.J. and Elious S. Cardiovascular effects of brevetoxin in dogs. Toxicon, **23:** 505-515, 1985.

Kaushik M.P., Parashar B.D. and Swamy R.V. Synthesis of Gymnodinium breve toxin. Ind. J. Chem., **27:** 1150-1151, 1988.

Koley J., Sinha S., Basak A.K., Das M., Dube S.N., Mazumder P.K., Gupta A.K., Das Gupta S. and Koley B. Cardiovascular and respiratory changes following exposure to a synthetic toxin of Ptychodiscus brevis. Eur. J. Pharmacol. **293:** 483-486, 1995.

Koley J., Sinha S., Basak A.K., Das M., Sinha S., Dube S.N., Mazumder P.K., Gupta A.K., Das Gupta S. and Koley B.N. Effect of synthetic toxin of Ptychodiscus brevis on Cardiovascular and renal function. Ind. J. Physiol. Allied Sci. **51:** 119-125, 1997a.

Lin Y.Y., Risk. M.A., Ray S.M. VanEngen D., Clardy J., Golick J., James J.C. and Nakanishi K. Isolation and structure of brevetoxin-B from the red tide dinoflagellates Ptychodiscus brevis (Gymnodinium breve), J. Am. Chem. Soc. **103:** 6773-6775, 1981.

Mahmood N.A. and Carmichael W.W. Anatoxin-a(s), an anticholinesterase from the cyanobacterium Anabaena flos-aquae NRC-525-17. Toxicon. **25:** 1221-1227, 1987.

Mazumder P.K., Gupta A.K., Kumar D. and Dube S.N. Calcium modulatory properties of O,O-Diphenyl O-N-Cyclooctyl phosphoramidate isolated from Ptychodiscus brevis in rat atria and smooth muscle. Ind. J. Physiol. Pharmacol. **41:** 257-262, 1997a.

Mazumder P.K., Gupta A.K., Kumar D., Kaushik M.P. and Dube S.N. Mechanism of cardiotoxicity induced by a marine toxin isolated from Ptychodiscus brevis. Ind. J. Exp. Biol. **35:** 650-654, 1997b.

Mazumder P.K., Gupta A.K., Kaushik M.P., Kumar D. and Dube S.N. Cardiovascular effects of an organophosphate toxin isolated from Ptychodiscus brevis. Bio. Environ. Sci. **10:** 85-92, 1997c.

Maruoka Y., Ohno Y., Tanaka H., Yasuda H., Ohtani K.I., Sakamoto H., Kawabe A., Tamamura C. and Nakamura M. Selective depression of the spinal polysynaptic reflexes by the NMDA receptor antagonists in an isolated spinal cord in vitro. Gen. Pharamcol. **29:** 645-649, 1997.

Melinek R., Rein K.S., Schultz D.R. and Bade D.G. Brevetoxin PbTx-2 immunology: differential epitope recognition by antibodies from two goats. Toxicon, **32:** 883-890, 1994.

McFarren E.F., Tanabe H., Silva F.J., Wilson W.B., Campbell J.E. and Lewis K.H. The occurrence of a ciguatera-like poison in oyesters, clams and Gymnodinium breve cultures. Toxicon **3:** 111-123, 1965.

Poli M.A., Mende T.J. and Baden D.J. Brevetoxins, unique activators of voltage-sensitive sodium channels, bind to specific sites in rat brain synaptosomes. Mol. Pharmacol. **30:** 129-133, 1986.

Poli M.A. Laboratory procedures for detoxification of equipment and waste contaminated with brevetoxins PbTx-2 and PbTx-3. J. Assoc. Off. Anal.Chem. **71:** 1000-1002, 1988.

Poli M.A., Templeton C.B., Pace J.G. and Hines H.B. Detection, metabolism and pathophysiology of brevetoxin. In: marine toxins : origin, structure and molecular pharmacology. Eds. Hall, S. and Strichartz, G., Washington DC:Am.Chem.Soc. pp. 176-191, 1990.

Poli M.A., Musser S.M., Dickey R.W., Eilers P.P. and Hall S. Neurotoxic shellfish poisoning and brevetoxin metabolites: a case study from Florida. Toxicon. **38:** 981-993, 2000.

Richards I.S., Kulkarni A.P. Brooks S.M. and Pierce R. Florida red tide toxins (brevetoxins) produce depolarization of air way smooth muscle. Toxicon, **28:** 1105-1111, 1990.

Risk M.A., Lin Y.Y., Ramanujam V.M.S., Smith L.L., Ray S.M. and Trieff N.M. High pressure liquid chromatographic separation of two major toxic compounds from Gymnodinium breve, Davis. J. Chromatogr. Sci. **17:** 400-405, 1979.

Risk M.A., Orris P.C., Coutinho N.J. and Bradford H.F. Actions of Ptychodiscus brevis red tide toxin on metabolic and transmitter-releasing properties of synaptosomes. J. Neurochem., **39:** 1485-1488, 1982.

Rodgers R.L., Chou H.N., Temma K., Akera T. and Shimizu Y. Positive inotropic and toxic effects of brevetoxin-B on rat and guinea pig heart. Appl. Pharmacol. **76:** 296-305, 1984.

Sakamoto Y., Krzanowski J.J.Jr., Lockey R., Martin D.F., Duncan R., Polson J. and Szentivanyi A. The mechanism of Ptychodiscus brevis toxin-induced rat vas deferens contraction. J. Allergy Clin. Immunol. **76:** 117-122, 1985.

Sakamoto Y., Lockey R.F. and Krzanowski J.J.Jr. Shellfish and fish poisoning related to the toxic dinoflagellates. South. Med. **80:** 866-872, 1987.

Shamsudin L. Lipid and fatty acid contents in red tides from tropical fish ponds of the coastal water of South China Sea. Arch.Physiol.Biochem. **104:** 36-42, 1996.

Shimizu Y. Recent progress in marine toxin research. Pure Appl. Chem., **54:** 1973-1980, 1982.

Shimizu Y. Dinoflagellate Toxin. In Marine natural products, Chemical and Biological Perspectives, Ed. Scheuer, P.J. vol. 1, PP. 1-42, Academic press New York, 1978.

Shimizu Y., Chou H.N., Bando H., VanDuyne G. and Clardy J.C. Structure of brevetoxin-A (Gb-1 toxin), the most potent toxin in the Florida red tide organism Gymnodinium breve (Ptychodiscus brevis). J. Am. Chem. Soc., **108:** 514-515, 1986.

Shimizu Y., Gupta S. and Chou H.N. Biosynthesis of red tide toxin in marine toxin: Origin, structure and molecular pharmacology. Eds. Hall S. and Strichartz G. Published from Am. Chem. Soc., Washington, DC, p21, 1990.

Shimoda T., Krzanowski J. Jr., Nelson R., Martin D.F., Polson J., Duncan R., Locki. In vitro red tide toxin effects on human bronchial smooth muscle. J. Allergy. Clin. Immunol. **81:** 1187-1191, 1988.

Singh J.N. Effect of Ptychodiscus brevis toxin on synaptic transmission in neonatal rat spinal cord in vitro" Ph.D. Thesis of Banaras Hindu University, Varanasi, 2001.

Singh J.N. and Deshpande S.B. Involvement of N-methyl-D-aspartate receptors for the Ptychodiscus brevis toxin-induced depression of monosynaptic and polysynaptic reflexes in neonatal rat spinal cord in vitro. Neuroscience, **115:** 1189-1197, 2002.

Singh J.N. and Deshpande S.B. involvement of the GABAergic system for plychodicus brevis toxin-induced depression of synaptic transmission ericited in isolated spinal cord from neonatal rats, Brain Res. **974:** 243-248, 2003.

Singh J.N., Das Gupta S., Gupta A.K., Dube S.N. and Deshpande S.B. Relative potency of synthetic analogs of Ptychodiscus brevis toxin in depressing synaptic transmission evoked in neonatal rat spinal cord in vitro. Toxicol. Lett. **128:** 177-183, 2002.

Skulberg O.M., Carmichael W.W., Anderson R.A., Matsunaga S., Moore R.E. and Skulberg R. Investigation of a neurotoxic oscillatorialean strain (Cyanophyceae) and its toxin. Isolation and characterization of homoanatoxin-a. Env. Toxicol. and Chem. **11:** 321-329, 1992.

Skulberg O.M., Codd G.A. and Carmichael W.W. Toxic blue-green algal blooms in Europe: A growing problem. Ambio. **13:** 244-247, 1984.

Steidinger K.A. A re-evaluation of toxic dinoflagellate biology and ecology in Progress in Phycological Research. Eds. Round, F.E. and Chapman, D.J., Elsevier Amsterdam. **2:** 147-148, 1983.

Strichartz G., Rando T. and Wang G.K. An integrated view of molecular toxinology of sodium channel gating in excitable cells. Annual. Rev. Neurosci. **10:** 237-267, 1987.

Taylor D.L. and Seliger H.H. Toxic Dinoflagellate Blooms. New York: Elsevier North Holland, 1979.

Templeton C.B., Poli M.A. and Leclair R.D. Cardio-respiratory effects of brevetoxin (PbTx-2) in conscious tethered rats. Toxicon. **31:** 1619-1622, 1989.

Trainer V.L., Moreau E., Guedin D., Baden D.G. and Catterall W.A. Neurotoxin binding and allosteric modulation of receptor sites 2 and 5 on purified and reconstituted rat brain sodium channels. J.Biol. Chem. **268:** 17114-17119, 1993.

Trieff N.M., Ramanujam V.M.S., Alam M. and Ray S.M. Isolation, physicochemical, and toxicological characterization of toxins from Gymnodinium breve Davis. In: Proc.First Intl Conf.toxic Dinoflagellate Blooms, pp. 309-321, 1975. Ed. LoCicero V.M. Wakefield: Mass. Sci. Tech. Found.

Tsai M.C. and Chen M.L. Effects of brevetoxin B on motor nerve terminals of mouse skeletal muscle. Br.J.Pharmacol. **193:** 1126, 1991.

Ueno Y., Nagata S., Tsutsumi T., Hasegawa A., Watanabe M.F, Park H.D., Chen G.C., Chen G. and Yu S.Z. Detection of microcystins, a blue-green algal hepatotoxin, in drinking water sampled in Haimen and Fusui, endemic areas of primary liver cancer in China, by highly sesitive immunoassay. Carcinogenesis. **17:** 1317-1321, 1996.

Washburn B.S., Rein K.S., Baden D.G., Walsh P.J., Hinton D.E., Tullis K. and Denison M.S. Brevetoxin-6 (PbTx-6) a nonaromatic marine neurotoxins, is a ligand of the aryl hydrocarbon receptors. Arch. Biochem. Biophys., **343:** 149-156, 1994.

Wada A., Uezono Y., Arita N., Yuhi T., Kobayashi H., Yanagihara N. and Izumi F. Co-operative modulation of voltage-dependent sodium channels by brev neurotoxins in cultured bovine adrenal medullary cells. J. Pharmacol. Exp. Ther. **263:** 1347-1351, 1992.

Westerfield M., Moore J.W., Kim Y.S. and Padilla G.M. How Gymnodinium breve red-tide toxin(s) produces repetitive firing in squid axons. Am. J. Physiol. **232:** C23-C29, 1977.

Wu C.H. and Narahashi T. Mechanism of action of novel marine neurotoxins on ion channels. Ann. Rev. Pharmacol. Toxicol. **28:** 141-161, 1987.

Section B
Toxic Chemicals and Their Mode of Action

Pharmacological Perspectives of Toxic Chemicals and Their Antidotes
Editors: S J S Flora, J A Romano, S I Baskin and K Sekhar

Non-Cholinesterase Actions of Anticholinesterases

Carey Pope and Jing Liu
Department of Physiological Sciences, College of Vaterinary Medicine
264 McElroy Hall, Oklahoma State University
Stillwater, OK 74078, USA

Anticholinesterases (cholinesterase inhibitors) have been used in the treatment of human and animal diseases, the control of insect pests, and as chemical warfare agents and weapons of terrorism. Diverse clinical conditions including glaucoma, the autoimmune disorder myasthenia gravis, and Alzheimer's dementia have been treated successfully with these agents (Taylor, 1990). Many of the insecticides in use today are organophosphorus (OP) or carbamate cholinesterase inhibitors (Aspelin and Grube, 1999). OP nerve agents were used in both chemical warfare and terrorist attacks during the preceding decade (Macilwain, 1993; Okumura et al., 1996). Paradoxically, the carbamate cholinesterase inhibitor pyridostigmine has been used to enhance the efficacy of antidotes to OP nerve agents by temporarily "protecting" the active site of AChE (Lee, 1997).

The multiple uses of anticholinesterases are based on their interaction with acetylcholinesterase (AChE, EC 3.1.1.7). OP and carbamate anticholinesterases elicit both pharmacological and toxicological effects by covalently binding to the active site serine residue on AChE, thereby antagonizing the catalytic breakdown of the neurotransmitter acetylcholine (Savolainen, 2001). AChE is a very efficient enzyme (Vigny et al., 1978), rapidly hydrolyzing synaptic acetylcholine and thereby terminating the activity of the extracellular signals released by cholinergic neurons. Following extensive AChE inhibition, neurotransmitter molecules accumulate in the synapse leading to excessive stimulation of cholinergic receptors on postsynaptic cells and/or end organs. This action is beneficial in cases where a reduction in cholinergic neurotransmission contributes to symptoms, e.g., muscle weakness in myasthenia gravis due to loss of neuromuscular cholinergic receptors. Under normal conditions however, extensive AChE inhibition leads to signs of "cholinergic" toxicity (autonomic dysfunction, exopthalmus, involuntary movements, muscle fasciculations, changes in atrial heart rate, and in severe cases, respiratory depression) due to excessive stimulation of postsynaptic receptors (Ecobichon, 2001).

Considerable evidence indicates that many OP and carbamate anticholinesterases interact directly with other macromolecules aside from AChE, both within and outside of the cholinergic synapse. These non-cholinesterase actions of OP and carbamate toxicants may contribute both to qualitative and quantitative differences in toxicity exhibited among these compounds. Knowledge of the additional non-cholinesterase actions of anticholinesterases will contribute to a better understanding of mechanisms of toxicity and potentially lead to better strategies for poisoning therapy. The following is a brief review of findings illustrating the wide array of additional macromolecular targets for anticholinesterases.

INTERACTION OF ANTICHOLINESTERASES WITH OTHER ENZYMES BESIDES ACHE

The enzymes most often associated with OP and carbamate compounds are various types of esterases including AChE, neurotoxic esterase, butyrylcholinesterase, carboxylesterases, and A-esterases, e.g., paraoxonase. It has been known for over 70 years that some OP anticholinesterases can induce a delayed neurotoxic syndrome referred to as organophosphorus induced delayed neurotoxicity or OPIDN (Smith et al., 1930). While some OP toxicants that induce OPIDN are potent AChE inhibitors, it was discovered decades ago that AChE inhibition is not involved in the initiation of this disorder (Earl and Thompson, 1952). Rather, inhibition and aging (i.e., a molecular rearrangement leading to loss of an alkyl side chain) of another esterase, referred to as neurotoxic esterase or neuropathy target esterase (NTE) has been shown to be correlated with induction of delayed neurotoxicity (Johnson, 1970). Structure-activity studies (Johnson, 1988; Jokanovic and Johnson, 1993; Singh, 2001) have shown that some OP inhibitors have widely differing degrees of selectivity for AChE and NTE, and that the relative potency toward the two esterases can be used to predict whether a particular toxicant can induce OPIDN at sublethal exposures (Lotti and Johnson, 1978; Makhaeva et al., 1998). NTE represents the most well characterized example of an additional target for anticholinesterases for which direct interaction leads to toxic consequences independent of AChE inhibition.

It has been known for decades that NTE inhibitors that are incapable of aging have no apparent neuropathological capacity on their own, but can prevent the delayed neurotoxicity of OP toxicants when given prior to OP dosing (Johnson, 1980). Some of these same protective agents were later shown to potentiate or promote the expression of OPIDN when given after the OP toxicant (Pope and Padilla, 1990; Lotti et al., 1991; Lotti and Moretto, 1999). The protection afforded by pre-exposure to a non-neuropathic inhibitor is correlated with NTE inhibition, i.e., the pre-exposure masks the active site of NTE and prevents a neuropathic inhibitor from binding. (This is roughly analogous to pyridostigmine prophylaxis of nerve agent toxicity, i.e., pyridostigmine blocks subsequent binding of a nerve agent to AChE.) The potentiation or promotion of OPIDN by post-treatment with a non-neuropathic inhibitor does not, however, appear to involve NTE (Pope et al., 1993). Although the mechanism for exacerbation of delayed neurotoxicity remains unknown, another esterase referred to as M200 may be the target (Moretto et al., 2001; Lotti, 2002).

Many anticholinesterases interact with esterases involved in their own detoxification. Carboxylesterases are distributed widely in mammalian tissues and bind stoichiometrically to some OP anticholinesterases (Clement, 1984; Pond et al., 1995; Maxwell and Brecht, 2001). In contrast, A-esterases enzymatically degrade OP toxicants (Chemnitius et al., 1983; Walker and Mackness, 1987; Li et al., 2000). Both types of esterases can therefore limit the availability of inhibitors for interacting with AChE molecules. While interaction of an anticholinesterase with carboxylesterases and A-esterases can affect its own toxicity, these interactions can also be important with interactive, multiple anticholinesterase exposures. It has been known for decades that pre-exposure to a carboxylesterase inhibitor can markedly potentiate the toxicity of the OP insecticide malathion (Frawley et al., 1957; Murphy et al., 1959). Recent studies suggest that blockade of these esterase-mediated detoxifying pathways can modulate the interactive toxicity of closely timed exposures to multiple anticholinesterases in a complex manner, apparently due to OP-selective differences in biotransformation (Karanth et al., 2001).

Some anticholinesterases can bind to and inhibit other types of enzymes. The prototype OP anticholinesterase O,O'-diisopropylphosphorofluoridate (DFP) has been used for decades to limit degradation of proteins during protein separation techniques (Scopes, 1982). Chymotrypsin is inhibited by a number of OP anticholinesterases (Hamilton et al., 1975; Johnson and Clothier, 1980). The OP insecticide pirimiphos-methyl (10 mg/kg) inhibited several liver proteases (e.g., proline endopeptidase, dipeptidyl aminotransferases I and IV, cathepsin D) (Mantle et al., 1997). Tri-ortho-cresyl phosphate (TOCP), an NTE inhibitor and inducer of OPIDN, was also shown to inhibit liver proteases following *in vivo* exposure in mice (Saleem et al., 1998).

In a comparative study of the *in vitro* inhibitory potencies of 20 OP analogs towards AChE, NTE, trypsin, and an enzyme involved in mitogen-induced activation of T lymphocytes, some were potent inhibitors of all four enzymes whereas others exhibited selectivity towards one or more enzymes (Pruett et al., 1994). Profenofos, tribufos and phenyl saligenin cyclic phosphonate inhibited thrombin, trypsin and elastase, but at higher concentrations than needed to inhibit AChE (Quistad et al., 2000). The prototype OP insecticide parathion (O,O'-diethyl phosphorothioate) can inhibit choline acetyltransferase, the synthetic enzyme for acetylcholine *in vitro*, but only at millimolar concentrations (Murumatsu and Kuriyama, 1976). High dosages of soman (94-120 micrograms/kg) were reported to markedly inhibit (20-50% within 30 minutes and essentially 100% by 3 hours after exposure) brain regional choline acetyltransferase (ChAT) activity in rats (Thompson and Thomas, 1985). As loss of ChAT activity was evident within 30 minutes after exposure, the effect of soman on ChAT appears to be through direct OP-enzyme interaction. Interestingly, Sivam and coworkers (1984) had previously reported that neither DFP, tabun, sarin or soman had any effect on brain regional ChAT activity *in vivo*.

Several reports have shown inhibition of kynurenine formamidase, an enzyme involved in the metabolism of L-tryptophan, by some organophosphorus and carbamate insecticides (Seifert and Casida, 1978; Henderson and Kitos, 1982; Seifert and Pewnim, 1992; Pewnim and Seifert, 1993). Inhibition of this enzyme by diazinon led to reduction in nicotinamide adenine dinucleotide (NAD^+) levels (Henderson and Kitos, 1982). Pirimiphos-ethyl (20 mg/kg) and diazinon (10 mg/kg) almost completely inhibited (>90%) liver kynurenine formamidase in mice, and plasma kynurenine levels and urinary kynurenine metabolites were elevated several fold (Seifert and Pewnim, 1992; Pewnim and Seifert, 1993). Interestingly, the active anticholinesterase diazoxon was much less effective at modifying kynurenine metabolism than diazinon (Pewnim and Seifert, 1993). As metabolites of kynurenine can be both neuroprotective (kynurenic acid, Stone et al., 2000) and neurotoxic (3-hydroxykynurenine and quinolinic acid, Okuda et al., 1998; Vasquez et al., 2000), inhibition of kynurenine formamidase may be a relevant additional target for some anticholinesterases.

Fatty acid amide hydrolase (FAAH) is inhibited by a number of anticholinesterases (Quistad et al., 2001). This enzyme participates in the metabolism of the endocannabinoid, anandamide, and other neuroactive amides, e.g., oleamide (Cravatt et al., 1996; Matsuda et al., 1997). Chlorpyrifos oxon potently inhibited FAAH activity in mouse brain and liver with IC_{50} values of 40-56 nM (Quistad et al., 2001). Furthermore, other OP anticholinesterases evaluated (ethyloctylphosphonofluoridate, oleyl-4H-1,3,2-benzodioxophosphorin 2-oxide, methyl arachidonyl phosphonofluoridate, dodecyl-benzodioxophosphorin oxide, IC_{50} values from 0.08-1.1 nM) were at least 200 times more potent *in vitro* at inhibiting FAAH compared to AChE. While brain FAAH inhibition *in vivo* following chlorpyrifos, diazinon and methamidophos exposure occurred only at high dosages, the OP pesticides profenofos and tribufos inhibited FAAH activity at relatively low dose exposures devoid of cholinergic signs. A number of studies indicate that the major psychoactive agent in marijuana, i.e., Δ-9-tetrahydrocannabinol (Gessa et al., 1998; Carta et al., 1998; Mishima et al., 2002), and other cannabinoids including anandamide (Gifford et al., 1999) can modulate the release of acetylcholine in the mammalian brain. Furthermore, muscarinic receptor activation has been shown to increase secretion of endocannabinoids (Kim et al., 2002). Thus, inhibition of FAAH and the metabolism of endocannabinoids by some anticholinesterases could contribute to selective toxicity.

Acyl peptide hydrolase (APH) catalyzes the removal of the modified residue from N-terminally acetylated peptides, and possibly the degradation of oxidatively damaged proteins (Farries et al., 1991; Fujino et al., 2000). DFP phosphorylated a serine on APH (Scaloni et al., 1992). Richards and colleagues (2000) reported that APH was potently inhibited by methyl chlorpyrifos oxon, dichlorvos and DFP (IC_{50} values from 18-199 nM), with potencies 6-10 times those for inhibiting AChE.

Chlorpyrifos oxon (50 μM) activated extracellular signal-regulated kinases (ERK) 2-3 fold in Chinese hamster ovary cells (Bomser and Casida, 2000). The parent insecticide chlorpyrifos also activated ERK by was less potent than chlorpyrifos oxon. The effects of chlorpyrifos oxon were insensitive to atropine but completely blocked by the phosphatidylinositol-3 kinase inhibitor wortmannin, the protein kinase C inhibitor GF-109203X, and the mitogen activated extracellular signal-regulated protein kinase kinase (MEK) inhibitor PD 098059. A subsequent study (Bomser et al., 2002) reported that chlorpyrifos oxon likely activates ERK by inhibition of diacylglycerol lipase. These findings suggest that signaling pathways utilized by a variety of growth factors and mitogens may be altered by chlorpyrifos oxon and other OP toxicants.

INTERACTION OF ANTICHOLINESTERASES WITH NEUROTRANSMITTER RECEPTORS

Elucidation of the general mechanism of action for the OP insecticides, i.e., inhibition of AChE with the consequent increase in synaptic acetylcholine, excessive stimulation of cholinergic receptors and resulting functional signs of toxicity, was conducted over 50 years ago (DuBois et al., 1949). It has also been known since the 1950's that some some anticholinesterases can directly interact with cholinergic receptors (Frederickson, 1958). Acetylcholine receptors exist in two major subtypes, muscarinic and nicotinic, based on their sensitivity to the naturally occuring toxins muscarine and nicotine. Muscarinic receptors are coupled through G-proteins to second messenger transduction processes whereas nicotinic receptors are ligand-gated ion channels. Five muscarinic cholinergic receptor subtypes (M1-M5) have been identified by cloning (Bonner, 1989). An array of nicotinic receptor subtypes have been reported based on pentameric subunit heterogeneity (McGehee and Role, 1995; Albuquerque et al., 1997).

The OP compounds DFP, phospholine and paraoxon all had rapidly reversible direct antagonistic actions on the nicotinic receptor in *Electrophorus* cells (Bartles and Nachmanson, 1969). These antagonistic actions required considerably higher concentrations, however, than those needed to potentiate the effect of exogenously added acetylcholine by AChE inhibition. At high concentrations (100 μM), azinphos-methyl, dichlorvos, dicrotophos and monocrotophos bound *in vitro* to nicotinic receptors of the electric organ of *Torpedo* (Eldefrawi and Eldefrawi, 1983). Under these same conditions, none of these agents affected total muscarinic receptor ([^{3}H]quinuclidinyl benzilate, QNB) binding in membranes from insect or mammalian brain. Echothiophate and the nerve agent VX also bound directly to neuromuscular nicotinic receptors at relatively high concentrations (Bakry et al., 1988). These studies suggested that a number of OP anticholinesterases could bind directly to nicotinic cholinergic receptors.

Some carbamate anticholinesterases have also been reported to interact directly with nicotinic receptors (Seifert and Eldefrawi, 1974). Using cloned α4β4 receptors expressed in *Xenopus* oocytes, Zwart and colleagues (2000) reported potentiation of acetylcholine-induced ion current at low concentrations (1-10 μM) of physostigmine, but blockade of this current at higher physostigmine concentrations. Binding to the nicotinic ligand [^{3}H]epibatidine was also blocked by physostigmine. Physostigmine was previously reported to bind to a site on nicotinic receptors distinct from the agonist recognition site (Albuquerque et al., 1985). In the frog sciatic nerve-sartorius muscle preparation, physostigmine acted on nicotinic receptors indirectly through AChE inhibition at low concentrations but acted as an open channel blocker at higher concentrations. Using patch clamp, a novel agonist binding site for physostigmine at much lower concentrations was observed (Shaw et al., 1985). Other carbamate inhibitors including pyridostigmine and neostigmine also bound to this allosteric binding site, but at higher concentrations than physostigmine (Akaike et al., 1984; Albuquerque et al., 1984). Nicotinic receptors generally desensitize in a rapid manner following agonist stimulation (Weber et al., 1975; Kirpekar and Prat, 1978). Physostigmine activated nicotinic receptors through this allosteric interaction even following desensitization of the receptors (Kuhlman et al., 1991). Together, these findings indicate that some anticholinesterases may selectively activate nicotinic receptors through allosteric interaction, thus modifying nicotinic receptor activity even when the receptors have been desensitized by endogenous agonist.

[^{3}H]Phencyclidine (PCP) binds to the nicotinic channel in the open position, thus its binding to nicotinic receptors can be used to evaluate both activation of the nicotinic receptor and agonist-induced desensitization (Eldefrawi et al., 1980). Several OP and carbamate anticholinesterases have been reported to modify PCP (or [^{3}H]thienyl-cyclohexylpiperidine [TCP], an analog of phencyclidine) binding to nicotinic receptors. Neostigmine activated [^{3}H]PCP binding to reconstituted nicotinic receptors from *Torpedo nobiliana* (Mansour et al., 1987). Furthermore, both neostigmine and physostigmine were relatively potent inhibitors of agonist-stimulated [^{3}H]PCP binding (K_i = 10-20 μM). Under these same conditions, pyridostigmine and aldicarb were much less potent. DFP also reduced agonist-stimulated [^{3}H[PCP binding to *Torpedo* and antagonized agonist stimulated PCP binding (Eldefrawi et al., 1988). Chlorpyrifos, chlorpyrifos-oxon, parathion and paraoxon all increased TCP binding and decreased agonist-stimulated TCP binding to nicotinic receptors in *Torpedo* membranes in a concentration-dependent and reversible manner (IC_{50} values from 5-300 μM) (Katz et al., 1997). None of these OP toxicants affected equilibrium binding to [a-^{125}I]bungarotoxin in *Torpedo* membranes. Together, these findings indicate that some OP and carbamate anticholinesterases can both activate and desensitize nicotinic receptors.

Effects of anticholinesterases on muscarinic receptor binding have also been evaluated. As noted before, [^{3}H]quinuclidinyl benzilate or QNB is a nonselective muscarinic antagonist that labels all known subtypes of muscarinic receptors with equal binding affinities. Dichlorvos, paraoxon and tetraethylpyrophosphate (TEPP) inhibited QNB binding in membranes from bovine caudate nucleus at low (5-50 nM) concentrations (Volpe et al., 1985). Paraoxon was reported to block QNB binding to M_2 and M_3 subtypes of receptors at extremely low levels (as low as 10^{-15} M)(Katz and Marquis, 1989). [^{3}H]-N-methylscopolamine (NMS) is a nonselective muscarinic antagonist that, due to its polar nature, is often used to label cell surface receptors (Fisher, 1988). Paraoxon and phenyl saligenin phosphate (a potent NTE inhibitor) at high concentrations (about 1 mM) inhibited NMS binding in human neuroblastoma SY5Y cells under saturating conditions, whereas lower concentrations (10 μM) blocked binding when the ligand was at less than saturating concentrations (Ehrich et al., 1994). These cells express predominantly M_3 muscarinic receptors coupled to activation of phosphoinositide turnover. Paraoxon and PSP both reduced basal inositol phosphate levels in a concentration-dependent manner, but these effects were not sensitive to muscarinic or nicotinic antagonists. The inhibition of inositol phosphate production under these conditions was therefore apparently through interaction with another unknown site of action, independent of either target esterases (AChE, NTE) or cholinergic receptors (Ehrich et al., 1994).

A number of studies have evaluated the effects of anticholinesterases on binding of the putative subtype-selective muscarinic agonist, [^{3}H]*cis*-dioxolane (CD) to muscarinic receptors. CD was reported to selectively label the M_2 subtype of receptor in Chinese hamster ovary (CHO) cells transfected with one of either of the muscarinic receptor subtypes (Huff and Abou-Donia, 1994). Echothiophate, paraoxon, VX, soman and tabun all blocked CD binding *in vitro* to membranes from rat heart at sub-micromolar concentrations (Silveira et al., 1990). Paraoxon inhibited CD binding and cAMP formation in an atropine-sensitive manner in rat striatal cells (Jett et al., 1991). Paraoxon and malaoxon (the active metabolite of the insecticide malathion) were relatively potent blockers of CD binding to membranes from rat hippocampus and cortex but the parent compounds (parathion and malathion) were much less potent (Ward et al., 1993). Chlorpyrifos-oxon inhibited CD binding in rat striatum membranes (IC_{50} = 22 nM) and also inhibited forskolin-stimulated cAMP formation (IC_{50} = 155 nM). Interestingly, the inhibition of cAMP formation by chlorpyrifos oxon was reported to be atropine-resistant (Huff et al., 1994). Paraoxon, malaoxon and chlorpyrifos-oxon inhibited CD binding and cAMP formation in a concentration-dependent manner at sub-micromolar concentrations (potencies: chlorpyrifos-oxon > paraoxon > malaoxon) but neither of these toxicants affected basal nor carbachol-stimulated phosphoinositide turnover (Ward and Mundy, 1996). Lack of effects on phosphoinositide turnover under conditions wherein cAMP formation was affected indicated that such changes were direct actions and not a secondary response to AChE inhibition. Chlorpyrifos oxon and paraoxon inhibited forskolin-stimulated cAMP formation in cortical slices from immature and adult rats (Olivier et al., 2001). Atropine was only partially effective at modulating the OP effects on cAMP formation, however, suggesting that these effects were both muscarinic receptor dependent and independent. These findings therefore provide further support that some OP anticholinesterases interact directly with muscarinic M_2 or M_4 receptors, as well as with downstream signaling pathways coupled to those receptors.

Binding of the muscarinic agonist [^{3}H]oxotremorine (OXO) is also sensitive to relatively low concentrations of some OP and carbamate agents (van den Beukel et al., 1997). Paraoxon and physostigmine both displaced OXO binding to rat brain membranes. Interestingly, they were less potent at displacing OXO binding to human muscarinic receptors expressed in CHO cells. The rat cardiac M_2 muscarinic receptor was diethylphosphorylated by chlorpyrifos oxon *in vitro* (Bomser and Casida, 2001). Chlorpyrifos oxon, paraoxon and methyl paraoxon displaced OXO binding to rat cardiac membrane muscarinic receptors (Howard and Pope, 2002). Of these OP toxicants, chlorpyrifos oxon was most potent (IC_{50} = 7 nM) at antagonizing OXO binding. Suggesting covalent modification, the effect of chlorpyrifos oxon on OXO binding was irreversible.

Presynaptic muscarinic autoreceptors inhibit acetylcholine release by a negative feedback mechanism (Weiler, 1989; Feurerstein et al., 1992; Kitaichi et al., 1999). It was proposed that some anticholinesterases may directly bind to muscarinic autoreceptors regulating acetylcholine release in the brain (Katz and Marquis, 1989). The presynaptic regulation of acetylcholine release and its possible direct or indirect modulation by anticholinesterases could be important in the ultimate toxicity of AChE inhibition. If some anticholinesterases directly activate or block presynaptic muscarinic autoreceptors at physiologically relevant concentrations, this could provide a mechanism for selective modulation of cholinergic toxicity.

We have compared the toxicity of the common OP insecticides parathion and chlorpyrifos in a number of studies. Rats treated with high dosages of parathion and chlorpyrifos exhibit similar rates and maximal degrees of brain AChE inhibition (Pope et al., 1991) but different degrees of cholinergic toxicity, with parathion eliciting more extensive signs (Chaudhuri et al., 1993; Liu and Pope, 1998). Qualitative differences in binding to the muscarinic agonist CD were noted in male rats treated with either parathion or chlorpyrifos, i.e., CD binding to brain membranes was reduced following parathion exposure but increased following chlorpyrifos dosing (Chaudhuri et al., 1993). Up-regulation of brain CD binding sites was also reported in female rats following chlorpyrifos exposure (Liu and Pope, 1996). As CD-labeled receptors are located on the presynaptic terminal (Watson et al., 1986) and possibly coupled to muscarinic autoreceptor function, we hypothesized that qualitative differences in modulation of those muscarinic receptors following parathion and chlorpyrifos exposure might be an indicator of selective alteration of presynaptic regulation of acetylcholine release.

Rats treated with parathion (18 mg/kg, sc) or chlorpyrifos (279 mg/kg, sc) exhibited similar degrees of brain regional cholinesterase inhibition. Striatal muscarinic autoreceptor function was reduced in a time-dependent manner by parathion and chlorpyrifos, however (Pope et al., 1995). While autoreceptor function was inhibited within two days after parathion exposure (at a time when cholinergic signs of toxicity were occurring), this regulatory process was still operational at that time point following chlorpyrifos administration. We subsequently evaluated the in vitro effects of paraoxon and chlorpyrifos oxon on striatal muscarinic autoreceptor function (Liu et al., 2002). Chlorpyrifos oxon and paraoxon inhibited acetylcholine release in striatal slices perfused with only physiological buffer, similar to actions of the muscarinic agonists carbachol and cis-dioxolane (CD). When striatal slices were perfused in the presence of physostigmine (to inhibit AChE) and the muscarinic autoreceptor atropine (100 nM to partially block autoreceptors), paraoxon still reduced acetylcholine release but chlorpyrifos oxon increased release. These findings suggested that paraoxon and chlorpyrifos oxon have qualitatively different direct effects on the autoreceptor. We concluded that some OP anticholinesterases can affect presynaptic muscarinic autoreceptor function in a selective manner, independent of AChE inhibition, potentially playing a role in selective toxicity.

Muscarinic autoreceptors in the rat brain are generally either M_2 or M_4 subtypes (Zhang et al., 2002). Agonist-induced regulation of muscarinic M_2 receptors is initiated by phosphorylation *via* the G-protein coupled receptor kinase 2 (GRK2; Nakata et al., 1994; Schlador and Nathanson, 1997). We evaluated the possible effects of paraoxon and chlorpyrifos oxon on GRK2-mediated phosphorylation of purified human recombinant M_2 (hM_2) receptors (Zou and Pope, unpublished observations). Using purified GRK2, chlorpyrifos oxon inhibited the agonist-mediated phosphorylation of hM_2 receptors with an IC_{50} value of 70 μM. Chlorpyrifos also inhibited hM_2 receptor phosphorylation but was less potent. In contrast, paraoxon had no effect on hM_2 receptor phosphorylation under the same conditions. Thus, these results may provide evidence for differential muscarinic receptor regulation by parathion and chlorpyrifos.

Modulation of other presynaptic processes in cholinergic neurons could contribute to selective toxicity of anticholinesterases. High affinity choline uptake (HACU) is the rate-limiting step in acetylcholine synthesis. Selective changes in HACU could therefore alter the consequences of acetylcholinesterase inhibition. Doses of DFP, soman and sarin that elicited similar reductions in brain AChE activity had different effects on striatal and hippocampal choline levels, i.e., soman and sarin increased while DFP decreased choline levels (Flynn and Wecker, 1986). Another study reported that exposure to soman and sarin both affected HACU, with decreases in cortex and hippocampus but increases in striatum (Whalley and Shih, 1989). These effects on choline transport did not appear related to AChE inhibition, however. HACU was differentially reduced by chlorpyrifos and parathion in female rats (Liu and Pope, 1998), i.e., chlorpyrifos reduced HACU early after exposure whereas parathion led to reduction in HACU later. It was proposed that earlier HACU reduction with chlorpyrifos could minimize its acute toxicity by decreasing acetylcholine release. None of these OP anticholinesterases affected HACU *in vitro*, however (soman or sarin, Whalley and Shih, 1989; chlorpyrifos oxon or paraoxon, Liu and Pope, 1998). HACU is regulated by intracellular cAMP through a protein kinase A-mediated cascade (Cancela et al., 1995; Vogelsberg et al., 1997). As cAMP levels can be modulated by action of muscarinic M_2 and M_4 receptors, direct or indirect actions at these receptors could change acetylcholine synthesis and potentially modulate the toxicity of AChE inhibition. Presynaptic muscarinic receptors could therefore be coupled to two distinct neurochemical processes, acetycholine synthesis and acetylcholine release, with the potential for eliciting selective effects by direct binding to those receptors.

Nicotinic autoreceptors are also presynaptically located and control the release of acetylcholine, but in a positive feedback manner (Wilkie et al., 1996; Marchi and Raiteri, 1996; Marchi *et al.* 1999). In synaptosomes preloaded with [^{3}H]choline and stimulated with potassium, exogenous acetylcholine stimulates further release of acetylcholine if atropine is included in the perfusion buffer (to block muscarinic autoreceptors). Using this assay, we recently reported (Wu et al., 2003) that nicotinic autoreceptor function was completely blocked by chlorpyrifos oxon (0.1-10 μM). This represents another possible presynaptic neurochemical process sensitive to modulation by anticholinesterases that may modify the toxic consequences of acetylcholinesterase inhibitors.

Low concentrations of paraoxon (< 1 μM) enhanced the frequency of GABA and glutamate release in cultured hippocampal neurons (Rocha et al., 1996a). At higher concentrations, paraoxon blocked multiple types of postsynaptic receptors, apparently as an open-channel blocker. In these same studies, acetylcholine had no effect, suggesting that paraoxon did not act through AChE inhibition. It was subsequently shown that the nerve agent VX was more potent and efficacious than paraoxon at increasing neurotransmitter release from hippocampal neurons, while soman had no effect under the same conditions (Rocha et al., 1996b). Further studies reported that extremely low levels of VX (0.01 nM) inhibited GABA release in cultured neurons through direct activation of muscarinic receptors (Rocha et al., 1999). Sarin (0.1-3 nM) also blocked the evoked release of GABA in rat hippocampal slices by activation of muscarinic receptors, independent of AChE inhibition (Chebabo et al., 1999).

The above studies suggest that both muscarinic and nicotinic receptor subtypes may be sensitive to direct binding by a number of anticholinesterases. Such direct anticholinesterase-cholinergic receptor interactions may alter ion flux (through nicotinic receptors) and modulate the levels of second messengers (through muscarinic receptors) in either presynaptic or postsynaptic cells and thereby alter cellular functions. Non-cholinergic neurotransmitter receptors may also be targets for anticholinesterases. Cyanofenphos, leptophos, salithion and tri-ortho-cresyl phosphate (TOCP) all blocked [^{3}H]norepinephrine binding *in vitro* in cardiac membranes with potencies similar to that of propranolol, a prototype β-blocker (El-Sebae et al., 1981). DFP, dichlorvos, cyanophos and mipafox inhibited 3-((+)-2-carboxypiperazin-4-yl)-[1,2-^{3}H]propyl-1-phosphonic acid ([^{3}H]CPP) binding to the NMDA receptor complex in rat brain with IC_{50} values around 10 mM. The effect of DFP appeared irreversible (Johnson and Michaelis, 1992). In contrast, binding to other NMDA ligands including kainate- and quisqualate-sensitive [^{3}H]AMPA, strychnine-sensitive

[^{3}H]glycine, and [^{3}H]TCP was not affected. Some OP anticholinesterases are relatively potent inhibitors of $GABA_A$ receptor binding to t-[^{35}S]butyl-bicyclophophorothionate in membranes from *Torpedo californica* (Gant et al., 1987). Soman, sarin and tabun, have all been reported to reversibly block brain and heart adenosine receptor binding *in vitro* (K_i = 37-57 mM) (Lau et al., 1988, 1991).

As noted above, fatty acid amide hydrolase, involved in the catabolism of the endocannabinoids, is inhibited in a potent manner by some OP anticholinesterases. A recent study (Quistad et al., 2002) reported that chlorpyrifos oxon and methyl chlorpyrifos oxon inhibited *in vitro* binding of a cannabinoid CB1 receptor subtype ligand ([^{3}H]CP 55,490) to mouse brain membranes in a potent manner (IC_{50} values 14-64 nM), whereas other OP anticholinesterases (paraoxon, diazoxon and dichlorvos) were less potent (IC_{50} values 1.2-4.2 μM). Tribufos exposure (50 mg/kg) inhibited [^{3}H]CP 55,490 binding to cannabinoid receptors in mouse brain in an apparently covalent manner. Thus, endocannabinoid signaling, both due to disruption of FAAH activity and to modification of cannabinoid receptors, could be affected by some anticholinesterases leading to neurologic and other organ system deficits. Together, the findings noted above indicate that a number of non-cholinergic neurotransmitter receptors may be targeted by anticholinesterases. Some of these additional actions may contribute to the expression of cholinergic toxicity or be important in more subtle neurological changes following low-level, repeated exposures.

OTHER ACTIONS OF ANTICHOLINESTERASES

As parasympathetic neurons regulate heart rate and force of cardiac muscle contraction, changes in cardiac function often occur following anticholinesterase exposures. Some studies suggest, however, that cardiac toxicity may occur independent of acetylcholinesterase inhibition. Marosi and coworkers (1985) reported that lethal exposures to dimethoate in anesthetized guinea pigs led to cardiac failure and extensive EKG disturbances, apparently unrelated to acetylcholinesterase inhibition. Cardiac arrythmias can occur following exposure to the nerve agent VX. Using isolated guinea-pig ventricular muscle, VX elicited delayed after-depolarizations not noted with either the muscarinic agonist carbachol or the carbamate anticholinesterase physostigmine (Corbier and Robineau, 1987). Further studies demonstrated that VX may alter cardiac function through direct inhibition of the alpha-1 Na^+/ K^+-ATPase isoform (Robineau et al., 1991). Takahashi and coworkers (1991) reported that while oral or intravenous exposures to chlorfenvinphos or dichlorvos (direct acting anticholinesterases) or oral dosing with diazinon and fenthion (indirect anticholinesterases) led to typical signs of cholinergic toxicity, respiratory failure and extensive AChE inhibition, intravenous exposure to the indirect anticholinesterases elicited tonic convulsions and opisthotonus with only slight AChE inhibition. Furthermore, cardiorespiratory effects following iv exposures to the indirect toxicants were not blocked by atropine. It therefore appears that a number of OP anticholinesterases may have direct actions on the heart which are independent of AChE inhibition.

A number of studies have concentrated on the effects of the OP insecticide chlorpyrifos on the development and maturation of the rat nervous system. Chlorpyrifos can affect macromolecular (DNA, RNA and protein) synthesis in the postnatally developing brain in a time-dependent and maturational state-dependent manner (Whitney et al., 1995; Campbell et al., 1997; Johnson et al., 1998; Dam et al., 1998). Multiple components of the adenylyl cyclase signaling pathway were disrupted by postnatal chlorpyrifos exposures (Song et al., 1997). Chlorpyrifos (0.5 μg/ml) altered the incidence and orientation of mitotic cells and induced cytoplasmic vacuolation in the cultured rat embryo (Roy et al., 1998). Chlorpyrifos increased production of reactive oxygen species in a concentration-dependent manner in PC12 cells (Crumpton et al., 2000). Effects of chlorpyrifos on DNA synthesis were reported to be more potent in glial than neuronal cells (Garcia et al., 2001; Qiao et al., 2001), and appeared due to chlorpyrifos itself and not to chlorpyrifos oxon. Norepinephrine turnover was extensively reduced in multiple brain regions following early postnatal exposures to chlorpyrifos (Slotkin et al., 2002). Early postnatal exposure to chlorpyrifos altered the expression of c-fos but not p53 in the rat forebrain (Dam et al., 2003). The results from these *in vitro* and *in vivo* studies suggested that chlorpyrifos could potentially alter developmental processes, induce cell death in the developing nervous system and lead to persistent neurochemical and neurobehavioral changes, at least partially independent of acetylcholinesterase inhibition. The selectivity of these actions relative to other anticholinesterases has not been evaluated.

AChE also has different actions besides its role in cholinergic neurotransmission (Small et al., 1996; Robertson and Yu, 1993). AChE expression during development coincides with axonal outgrowth (Kostovic and Goldman-Rakic, 1983; Kristt, 1983; Layer, 1990; Robertson et al., 1991) and the level of AChE correlates with the degree of neurite outgrowth in cultured dorsal root ganglion cells (Dupree and Bigbee, 1994). The putative morphogenic role of AChE is not dependent on its catalytic activity (Layer et al., 1993; Sternfeld et al., 1998; Sharma et al., 2001). Some OP and carbamate anticholinesterases may affect developmental processes associated with this morphogenic function. Physostigmine, neostigmine and edrophonium led to growth cone retraction in primary neurons *in vitro* (Saito, 1998). Neurite extension in C6 glioma cells and N-18 mouse neuroblastoma cells was affected by a number of OP toxicants (Henschler et al., 1992). Tricresyl phosphate, triphenyl phosphite and paraoxon inhibited cell growth in PC12 cells, and tricresyl phosphate reduced neurite density in these cells. Direct-acting NTE inhibitors inhibited neurite outgrowth in PC12 cells at low concentrations but the potent anticholinesterase chlorpyrifos oxon only affected neurite extension at cytotoxic levels (Li and Casida, 1998). Chlorpyrifos and its major metabolite 3,5,6-trichloropyridinol inhibited neurite outgrowth in PC12 cells (Song et al., 1998) in the absence of cholinesterase inhibition. The role of AChE in promoting neurite outgrowth appears due to an adhesive function (Bigbee et al., 1999; Johnson and Moore, 2000; Sharma et al., 2001). While the catalytic action of AChE does not appear important in this function, some anticholinesterases appear to modify adhesion and thus such effects may contribute to an additional mechanism of toxicity.

Lockridge and coworkers have studied the neurodevelopmental and effects of anticholinesterases in AChE knockout mice. Deletion of four exons of the AChE gene totally eliminated AChE activity in nullizygous animals. Mesulam and coworkers (2002) reported normal neurodevelopment in AChE knockout mice, however, leading these investigators to conclude that "acetylcholinesterase is not necessary for the establishment of cholinergic pathways". These AChE knockout mice are actually more sensitive to DFP and to the butyrylcholinesterase-specific inhibitor bambuterol than wild types (Xie et al., 2000). Toxicity of VX was also increased in AChE knockout mice (Duysen et al., 2001). Similar signs of cholinergic toxicity were noted in both wild type and knockout mice exposed to VX. In contrast, atropine protected wild type mice but had no beneficial effect in knockouts. Mesulam and colleagues (2002) showed that butyrylcholinesterase could substitute for acetylcholinesterase in the AChE knockouts and moreover that this enzyme may play a role in the hydrolysis of acetylcholine in the normal brain. Further studies with AChE knockout mice may provide more information regarding multiple roles/actions of both AChE and anticholinesterases.

IMPLICATIONS OF ADDITIONAL ACTIONS OF ANTICHOLINESTERASES

The mechanism of action for this class of chemicals, i.e., inhibition of acetylcholinesterase, accumulation of acetylcholine and consequent changes in cholinergic neurotransmission, has been well described. From the preceeding review, however, it is clear that many of the anticholinesterases have the potential to modify other functions through interaction with additional macromolecules. Table 1 summarizes some of these possible additional sites of action. In some cases, additional effects of anticholinesterases on other processes may only be relevant with high dose exposures. In other cases, interaction with additional target molecules may occur at exposure levels devoid of signs of toxicity or even in the absence of significant AChE inhibition. Some non-cholinesterase actions may contribute to acute toxicity while others may be important in adverse responses to long

Table 1 Macromolecular targets for anticholinesterases and known or possible consequences of their interaction

Known or Putative Target Molecule	*Known or Possible Effects*
Acetylcholinesterase	Cholinergic toxicity, neurodevelopmental deficits
Butyrylcholinesterase	Altered cholinergic neurotransmission, neurodevelopmental deficits
Neurotoxic esterase	Delayed neurotoxicity
M-200	Exacerbation of delayed neurotoxicity
Carboxylesterase	Modulation of toxicity with combined exposures
Proteases	Alteration of protein turnover, modulation of clotting
Choline acetyltransferase	Alteration of choline utilization, acetylcholine synthesis
Kynurenine formamidase	Modulation of L-tryptophan metabolism, production of NAD+
Fatty acid amide hydrolase	Changes in neuroactive amide (e.g., endocannabinoid) metabolism
Acyl peptide hydrolase	Changes in peptide/neuropeptide turnover
Diacylglycerol kinase	Altered growth factor/mitogen signaling
Nicotinic receptors	Variety of central/peripheral neurochemical changes
Muscarinic receptors	Variety of central/peripheral neurochemical changes
Adenylyl cyclase	Altered transduction for a variety of extracellular signals
Cholinergic autoreceptors	Modulation of acetylcholine release
Choline transport	Alteration of acetylcholine synthesis
Cardiac Adrenergic receptors	Modulation of heart function, arrythmia
NMDA receptors	Excitotoxicity, altered neuromodulation
Cannabinoid receptors	Changes in neuromodulation
Na+,K+-ATPase alpha 1	Arrythmias

term, lower-level exposures. Elucidation of the relative roles these additional sites of action have in the expression of selective toxicity could be useful in the design of novel therapeutic strategies.

References

Akaike A., Ikeda S.R., Brookes N., Pascuzzo G.J., Rickett D.L. and Albuquerque E.X. The nature of the interactions of pyridostigmine with the nicotinic acetylcholine receptor-ionic channel complex. II. Patch clamp studies. Mol. Pharmacol. **25:** 102-112, 1984.

Albuquerque E.X., Akaike A., Shaw K.P. and Rickett D.L. The interaction of anticholinesterase agents with the acetylcholine receptor-ionic channel complex. Fundam. Appl. Toxicol. **4:** S27-S33, 1984.

Albuquerque E.X., Deshpande S.S., Kawabuchi M., Aracava Y., Idriss M., Rickett D.L. and Boyne A.F. Multiple actions of anticholinesterase agents on chemosensitive synapses: molecular basis for prophylaxis and treatment of organophosphate poisoning. Fundam. Appl. Toxicol. **5:** S182-S203, 1985.

Albuquerque E.X., Alkondon M., Pereira E.F., Castro N.G., Schrattenholz A., Barbosa C.T., Bonfante-Cabarcas R., Aracava Y., Eisenberg H.M. and Maelicke A., Properties of neuronal nicotinic acetylcholine receptors: pharmacological characterization and modulation of synaptic function. J. Pharmacol. Exp. Ther. **280:** 1117-1136, 1997.

Aspelin A.L. and Grube A.H. Pesticides Industry Sales and Usage: 1996 and 1997 Market Estimates, U.S. EPA, Document #733-R-99-001, 1999.

Bakry N.M., el-Rashidy A.H., Eldefrawi A.T. and Eldefrawi M.E. Direct actions of organophosphate anticholinesterases on nicotinic and muscarinic acetylcholine receptors. J. Biochem. Toxicol. **3:** 235-259, 1988.

Bartles E. and Nachmansohn D. Organophosphate inhibitors of acetylcholine-receptor and –esterase tested on the electroplax. Arch. Biochem. Biophys. **133:** 1-10, 1969.

Bigbee J.W., Sharma K.V., Gupta J.J. and Dupree J.L., Morphogenic role for acetylcholinesterase in axonal outgrowth during neural development. Environ Health Perspect. 107 Suppl. **1:** 81-87, 1999.

Bomser J. and Casida J.E. Activation of extracellular signal-regulated kinases (ERK 44/42) by chlorpyrifos oxon in Chinese hamster ovary cells. J Biochem Mol Toxicol. **14:** 346-353, 2000.

Bomser J.A. and Casida J.E. Diethylphosphorylation of rat cardiac M2 muscarinic receptor by chlorpyrifos oxon in vitro. Toxicol Lett. **119:** 21-26, 2001.

Bomser J.A., Quistad G.B. and Casida J.E. Chlorpyrifos oxon potentiates diacylglycerol-induced extracellular signal-regulated kinase (ERK 44/42) activation, possibly by diacylglycerol lipase inhibition. Toxicol. Appl. Pharmacol. **178:** 29-36, 2002.

Bonner T.I. The molecular basis of muscarinic receptor diversity. Trends Neurosci. **12:** 148-151, 1989.

Campbell C.G., Seidler F.J. and Slotkin T.A. Chlorpyrifos interferes with cell development in rat brain regions. Brain Res. Bull. **43:** 179-189, 1997.

Cancela J.M., Bertrand N. and Beley A. Involvement of cAMP in the regulation of high affinity choline uptake by rat brain synaptosomes. Biochem. Biophys. Res. Commun. **213:** 944-949, 1995.

Carta G., Nava F. and Gessa G.L. Inhibition of hippocampal acetylcholine release after acute and repeated Delta9-tetrahydrocannabinol in rats. Brain Res. **809:** 1-4, 1998.

Chaudhuri J., Chakraborti T.K., Chanda S. and Pope C.N. Differential modulation of organophosphate-sensitive muscarinic receptors in rat brain by parathion and chlorpyrifos. J. Biochem. Toxicol. **8:** 207-216, 1993.

Chebabo S.R., Santos M.D. and Albuquerque E.X. The organophosphate sarin, at low concentrations, inhibits the evoked release of GABA in rat hippocampal slices. Neurotoxicology. **20:** 871-882, 1999.

Chemnitius J.M., Losch H., Losch K. and Zech R. Organophosphate detoxicating hydrolases in different vertebrate species. Comp. Biochem. Physiol. C **76:** 85-93, 1983.

Clement J.G. Role of aliesterase in organophosphate poisoning. Fundam. Appl. Toxicol. 4(2 Pt 2): S96-105, 1984.

Corbier A. and Robineau P. Evidence for a direct noncholinergic effect of an organophosphorous compound on guinea-pig papillary muscles: are ventricular arrhythmias related to a Na+/K+ATPase inhibition? Arch. Int. Pharmacodyn. Ther. **300:** 218-230, 1989.

Cravatt B.F., Giang D.K., Mayfield S.P., Boger D.L., Lerner, R.A. and Gilula, N.B., Molecular characterization of an enzyme that degrades neuromodulatory fatty-acid amides. Nature **384:** 83-87, 1996.

Crumpton T.L., Seidler F.J. and Slotkin T.A. Is oxidative stress involved in the developmental neurotoxicity of chlorpyrifos? Brain Res. Dev. Brain Res. **121:** 189-195, 2000.

Dam K., Seidler F.J. and Slotkin T.A. Developmental neurotoxicity of chlorpyrifos: delayed targeting of DNA synthesis after repeated administration. Dev. Brain Res. **108:** 39-45, 1998.

Dam K., Seidler F.J. and Slotkin T.A. Transcriptional biomarkers distinguish between vulnerable periods for developmental neurotoxicity of chlorpyrifos: Implications for toxicogenomics. Brain Res. Bull. **59:** 261-265, 2003.

DuBois K.P., Doull J., Salerno P.R. and Coon J. Studies on the toxicity and mechanisms of action of *p*-nitrophenyl diethyl thionophosphate (parathion). J. Pharmacol. Exper. Ther. **95:** 79-91, 1949.

Dupree J.I. and Bigbee J.W. Retardation of neuritic outgrowth and cytoskeletal changes accompany acetylcholinesterase inhibitor treatment in cultured rat dorsal root ganglion neurons. J. Neurosci. Res. **39:** 567-575, 1994.

Duysen E.G., Li B., Xie W., Schopfer L.M., Anderson R.S., Broomfield C.A. and Lockridge O. Evidence for nonacetylcholinesterase targets of organophosphorus nerve agent: supersensitivity of acetylcholinesterase knockout mouse to VX lethality. J. Pharmacol. Exp. Ther. **299:** 528-535, 2001.

Earl C.J. and Thompson R.H.S. Cholinesterase levels in the nervous system in tri-ortho-cresyl phosphate poisoning. Br. J. Pharmacol. **7:** 685-692, 1952.

Ecobichon D.J. Toxic effects of pesticides. Pages 763-810 in Casarett and Doull's Toxicology, 6th edition (C.D. Klaassen, ed), McGraw Hill, New York, 2001.

Ehrich M., Intropido L. and Costa L.G. Interaction of organophosphorus compounds with muscarinic receptors in SH-SY5Y human neuroblastoma cells. J. Toxicol. Environ. Health **43:** 51-63, 1994.

Eldefrawi M.E. and Eldefrawi A.T. Neurotransmitter receptors as targets for pesticides. *J.* Environ. Sci. Health [B] **18:** 65-88, 1983.

Eldefrawi M.E., Eldefrawi A.T., Aronstam R.S., Maleque M.A., Warnick J.E. and Albuquerque E.X., [3H] Phencyclidine: a probe for the ionic channel of the nicotinic receptor. Proc. Natl. Acad. Sci. U.S.A. **77:** 7458-7462, 1980.

Eldefrawi M.E., Schweizer G., Bakry N.M. and Valdes J.J. Desensitization of the nicotinic acetylcholine receptor by diisopropylfluorophosphate. J. Biochem. Toxicol. **3:** 21-32, 1988.

El-Sebae A.H., Soliman S.A., Ahmed N.S. and Curley A. Biochemical interaction of six OP delayed neurotoxicants with several neurotargets. J. Environ. Sci. Health [B] **16:** 465-474, 1981.

Farries T.C., Harris A., Auffret A.D. and Aitken A. Removal of N-acetyl groups from blocked peptides with acylpeptide hydrolase. Stabilization of the enzyme and its application to protein sequencing. Eur J Biochem. **196:** 679-685, 1991.

Feuerstein T.J., Lehmann J., Sauermann W., van Velthoven V. and Jackisch R. The autoinhibitory feedback control of acetylcholine release in human neocortex tissue. Brain Res. **572:** 64-71, 1992.

Fisher S.K. Recognition of muscarinic cholinergic receptors in human SK-N-SH neuroblastoma cells by quaternary and tertiary ligands is dependent upon temperature, cell integrity, and the presence of agonists. Mol. Pharmacol. **33:** 414-422, 1988.

Flynn C.J. and Wecker L., Elevated choline levels in brain. A non-cholinergic component of organophosphate toxicity. Biochem Pharmacol. **35:** 3115-3121, 1986.

Frawley J.P., Fuyat H.N., Hagan E.C., Blake J.R. and Fitzhugh O.G. Marked potentiation in mammalian toxicity from simultaneous adininstration of two anticholinesterase compounds. J. Pharmacol. Exp. Ther. **121:** 96-106, 1957.

Frederickson T., Further studies on fluoro-phosphorylcholines. Pharmacological properties of two new analogues. Arch. Int. Pharmacodyn. **115:** 474-482, 1958.

Fujino T., Watanabe K., Beppu M., Kikugawa K. and Yasuda, H. Identification of oxidized protein hydrolase of human erythrocytes as acylpeptide hydrolase. Biochim. Biophys. Acta. **1478:** 102-112, 2000.

Gant D.B., Eldefrawi M.E. and Eldefrawi A.T. Action of organophosphates on GABAA receptor and voltage-dependent chloride channels. *Fundam. Appl. Toxicol.* **9:** 698-704, 1987.

Garcia S.J., Seidler F.J., Crumpton T.L. and Slotkin T.A. Does the developmental neurotoxicity of chlorpyrifos involve glial targets? Macromolecule synthesis, adenylyl cyclase signaling, nuclear transcription factors, and formation of reactive oxygen in C6 glioma cells. Brain Res. **891:** 54-68, 2001.

Gessa G.L., Casu M.A., Carta G. and Mascia M.S. Cannabinoids decrease acetylcholine release in the medial-prefrontal cortex and hippocampus, reversal by SR 141716A. Eur. J. Pharmacol. **355:** 119-124, 1998.

Gifford A.N., Bruneus M., Lin S., Goutopoulos A., Makriyannis A., Volkow N.D. and Gatley S.J. Potentiation of the action of anandamide on hippocampal slices by the fatty acid amide hydrolase inhibitor, palmitylsulphonyl fluoride (AM 374). Eur J Pharmacol. **383:** 9-14, 1999.

Hamilton S.E., Dudman A.P., DeJersey J., Stoops J.K. and Zerner B., Organophosphate inhibitors: the reactions of bis(p-nitrophenyl)methyl phosphate with liver carboxylesterases and alpha-chymotrypsin. Biochim. Biophys. Acta **377:** 282-296, 1975.

Henderson M. and Kitos P.A. Do organophosphate insecticides inhibit the conversion of tryptophan to NAD+ in ovo? Teratology **26:** 173-81, 1982.

Henschler D., Schmuck G., van Aerssen M. and Schiffmann D. The inhibitory effect of neuropathic organophosphate esters on neurite outgrowth in cell cultures: a basis for screening for delayed neurotoxicity. Toxicol. Vitro **6:** 327-325.84, 1992.

Howard M.D. and Pope C.N. *In vitro* effects of chlorpyrifos, parathion, methyl parathion and their oxons on cardiac muscarinic receptor binding in neonatal and adult rats. Toxicology. **170:** 1-10, 2002.

Huff R.A. and Abou-Donia M.B. *cis*-Methyldioxolane specifically recognizes the m2 muscarinic receptor. J. Neurochem. **62:** 388-391, 1994.

Huff R.A., Corcoran J.J., Anderson J.K. and Abou-Donia M.B. Chlorpyrifos oxon binds directly to muscarinic receptors and inhibits cAMP accumulation in rat striatum. J. Pharmacol. Exp. Therap. **269:** 329-335, 1994.

Jett D.A., Abdallah E.A.M., El-Fakahany E.E., Eldefrawi M.E. and Eldefrawi A.T., High-affinity activation by paraoxon of a muscarinic receptor subtype in rat brain striatum. Pest. Biochem. Physiol. **39:** 149-157, 1991.

Johnson D.E., Seidler F.J. and Slotkin T.A. Early biochemical detection of delayed neurotoxicity resulting from developmental exposure to chlorpyrifos. Brain Res. Bull. **45:** 143-147, 1998.

Johnson M.K. Organophosphorus and other inhibitors of brain 'neurotoxic esterase' and the development of delayed neurotoxicity in hens. Biochem J. **120:** 523-531, 1970.

Johnson M.K. The mechanism of delayed neuropathy caused by some organophosphorus esters: using the understanding to improve safety. J. Environ. Sci. Health B. **15:** 823-841, 1980.

Johnson M.K. Sensitivity and selectivity of compounds interacting with neuropathy target esterase. Further structure-activity studies. Biochem Pharmacol. **37:** 4095-4104, 1988.

Johnson M.K. and Clothier B. Biochemical events in delayed neurotoxicity: is aging of chymotrypsin inhibited by saligenin cyclic phosphates a model for aging of neurotoxic esterase? Toxicol. Lett. **5:** 95-98, 1980.

Johnson P.S. and Michaelis E.K. Characterization of organophosphate interactions at N-methyl-D-aspartate receptors in brain synaptic membranes. Mol. Pharmacol. **41:** 750-756, 1992.

Jokanovic M. and Johnson M.K. Interactions in vitro of some organophosphoramidates with neuropathy target esterase and acetylcholinesterase of hen brain. J. Biochem. Toxicol. **8:** 19-31, 1993.

Johnson G. and Moore S.W. Cholinesterases modulate cell adhesion in human neuroblastoma cells in vitro. Int. J. Dev. Neurosci. **18:** 781-790, 2000.

Karanth S., Olivier K., Jr. Liu J. and Pope C. In vivo interaction between chlorpyrifos and parathion in adult rats: sequence of administration can markedly influence toxic outcome. Toxicol. Appl. Pharmacol. **177:** 247-255, 2001.

Katz E.J., Cortes V.I., Eldefrawi M.E. and Eldefrawi A.T. Chlorpyrifos, parathion, and their oxons bind to and desensitize a nicotinic acetylcholine receptor: relevance to their toxicities. Toxicol. Appl. Pharmacol. **146:** 227-236, 1997.

Katz L.S. and Marquis J.K. Modulation of central muscarinic receptor binding in vitro by ultralow levels of the organophosphate paraoxon. Toxicol. Appl. Pharmacol. **101:** 114-123, 1989.

Kim J., Isokawa, M., Ledent C. and Alger B.E. Activation of muscarinic acetylcholine receptors enhances the release of endogenous cannabinoids in the hippocampus. J. Neurosci. **22:** 10182-10191, 2002.

Kirpekar S,M, and Prat J.C. Blockade of desensitization of nicotinic receptors of the cat adrenal medulla by concanavalin A. Br. J. Pharmacol. **62:** 549-552, 1978.

Kitaichi K., Hori T., Srivastava L.K. and Quirion R. Antisense oligodeoxynucleotides against the muscarinic m2, but not m4, receptor supports its role as autoreceptors in the rat hippocampus. Brain Res. Mol. Brain Res. **67:** 98-106, 1999.

Kostovic I. and Goldman-Rakic P.S. Transient cholinesterase staining in the mediodorsal nucleus of the thalamus and its connections in the developing human and monkey brain. J. Comp. Neurol. **219:** 431-447, 1983.

Kristt D.A. Acetylcholinesterase in the ventral thalamus: transience and patterning during ontogenesis. Neuroscience **10:** 923-939, 1983.

Kuhlmann J., Okonjo K.O. and Maelicke A. Desensitization is a property of the cholinergic binding region of the nicotinic acetylcholine receptor, not of the receptor-integral ion channel. FEBS Lett. **279:** 216-218, 1991.

Lau W-M, Freeman S.E. and Szilagyi M. Binding of some organophosphorus compounds at adenosine receptors in guinea pig brain membranes. Neurosci. Lett. **94:** 125-130, 1988.

Lau W-M, Szilagyi M. and Freeman S.E. Effects of some organophosphorus compounds on the binding of a radioligand (8-cyclopentyl 1,3-[3H]dipropylxanthine) to adenosine receptors in ovine cardiac membranes. J. Appl. Toxicol. **11:** 411-414, 1991.

Layer P.G. Cholinesterases preceeding major tracts in vertebrate neurogenesis. BioEssays **12:** 415-420, 1990.

Layer P.G., Weikert T. and Alber R. Cholinesterases regulate neurite growth of chick nerve cells in vitro by means of a non-enzymatic mechanism. Cell. Tissue Res. **273:** 219-226, 1993.

Lee E.J. Pharmacology and toxicology of chemical warfare agents. Ann. Acad. Med. Singapore **26:** 104-107, 1997.

Li W. and Casida J.E. Organophosphorus neuropathy target esterase inhibitors selectively block outgrowth of neurite-like and cell processes in cultured cells. Toxicol. Lett. **98:** 139-146, 1998.

Li W.F., Costa L.G., Richter R.J., Hagen T., Shih D.M., Tward A., Lusis A.J. and Furlong C.E. Catalytic efficiency determines the in-vivo efficacy of PON1 for detoxifying organophosphorus compounds. Pharmacogenetics **10:** 767-779, 2000.

Liu J. and Pope C.N. Effects of chlorpyrifos on high-affinity choline uptake and [^{3}H]hemicholinium-3 binding in rat brain. Fundam. Appl. Toxicol. **34:** 84-90, 1996.

Liu J. and Pope C.N. Comparative presynaptic neurochemical changes in rat striatum following exposure to chlorpyrifos or parathion. J. Toxicol. Environ. Health **53:** 531-544, 1998.

Liu J., Chakraborti T.K. and Pope C.N. *In vitro* effects of organophosphorus anticholinesterases on muscarinic receptor-mediated inhibition of acetylcholine release in rat striatum. Toxicol. Appl. Pharmacol. **178:** 102-108, 2002.

Lotti M. Promotion of organophosphate induced delayed polyneuropathy by certain esterase inhibitors. Toxicology. 181-182: 245-248, 2002.

Lotti M and Johnson MK. Neurotoxicity of organophosphorus pesticides; predictions can be based on in vitro studies with hen and human enzymes. Arch. Toxicol. **41:** 215-221, 1978.

Lotti M., Caroldi S., Capodicasa E. and Moretto A. Promotion of organophosphate-induced delayed polyneuropathy by phenylmethanesulfonyl fluoride. Toxicol. Appl. Pharmacol. **108:** 234-41, 1991.

Lotti M. and Moretto A. Promotion of organophosphate induced delayed polyneuropathy by certain esterase inhibitors. Chem. Biol. Interact. **119-120:** 519-524, 1999.

Macilwain C., Study proves Iraq used nerve gas. Nature **363:** 3, 1993.

Makhaeva G.F., Filonenko I.V., Yankovskaya V.L., Fomicheva S.B. and Malygin V.V. Comparative studies of O,O-dialkyl-O-chloromethylchloroformimino phosphates: interaction with neuropathy target esterase and acetylcholinesterase. Neurotoxicology **19:** 623-628, 1998.

Mansour N.A., Valdes J.J., Shamoo A.E. and Annau Z. Biochemical interactions of carbamates and ecothiophate with the activated conformation of nicotinic acetylcholine receptor. J. Biochem. Toxicol. **2:** 25-42, 1987.

Mantle D., Saleem M.A., Williams F.M., Wilkins R.M. and Shakoori A.R. Effect of pirimiphos-methyl on proteolytic enzyme activities in rat heart, kidney, brain and liver tissues in vivo. Clin. Chem. Acta **262:** 89-97, 1997.

Marchi M. and Raiteri M. Nicotinic autoreceptors mediating enhancement of acetylcholine release become operative in conditions of "impaired" cholinergic presynaptic function. J. Neurochem. **67:** 1974-1981, 1996.

Marchi M., Lupinacci M., Bernero E., Bergaglia F. and Raiteri M. Nicotinic receptors modulating ACh release in rat cortical synaptosomes: role of Ca2+ ions in their function and desensitization. Neurochem. Int. **34:** 319-328, 1999.

Marosi G., Ivan J., Nagymajtenyi L., Csatlos I. and Toszegi A. Dimethoate-induced toxic cardiac failure in the guinea pig. Arch. Toxicol. **57:** 142-143, 1985.

Matsuda S., Kanemitsu N., Nakamura A., Mimura Y., Ueda N., Kurahashi Y. and Yamamoto S. Metabolism of anandamide, an endogenous cannabinoid receptor ligand, in porcine ocular tissues. Exp. Eye Res. **64:** 707-711, 1997.

Maxwell D.M. and Brecht K.M. Carboxylesterase: specificity and spontaneous reactivation of an endogenous scavenger for organophosphorus compounds. J. Appl. Toxicol. 21 Suppl. **1:** S103-107, 2001.

McGehee D.S. and Role L.W. Physiological diversity of nicotinic acetylcholine receptors expressed by vertebrate neurons. Ann. Rev. Physiol. **57:** 521-546, 1995.

Mesulam M.M., Guillozet A., Shaw P., Levey A., Duysen E.G. and Lockridge O. Acetylcholinesterase knockouts establish central cholinergic pathways and can use butyrylcholinesterase to hydrolyze acetylcholine. Neuroscience **110:** 627-639, 2002.

Mishima K., Egashira N., Matsumoto Y., Iwasaki K. and Fujiwara M. Involvement of reduced acetylcholine release in Delta9-tetrahydrocannabinol-induced impairment of spatial memory in the 8-arm radial maze. Life Sci. **72:** 397-407, 2002.

Moretto A., Gardiman G., Panfilo S., Colle M.A., Lock E.A. and Lotti M. Effects of S-ethyl hexahydro-1H-azepine-1-carbothioate (molinate) on di-n-butyl dichlorovinyl phosphate (DBDCVP) neuropathy. Toxicol. Sci. **62:** 274-279, 2001.

Murphy S.D., Anderson, R.L. and DuBois, K.P., Potentiation of toxicity of malathion by triorthotolyl phosphate. Proc. Soc. Exp. Biol. Med. **100:** 482-487, 1959.

Murumatsu M. and Kuriyama K. Effect of organophosphorus compounds on acetylcholine synthesis in brain. Japan. J. Pharmacol. **26:** 249-254, 1976.

Nakata H., Kameyama K., Haga K. and Haga T. Location of agonist-dependent-phosphorylation sites in the third intracellular loop of muscarinic acetylcholine receptors (m2 subtype). Eur. J. Biochem. **220:** 29-36, 1994.

Okuda S., Nishiyama N., Saito H. and Katsuki H. 3-Hydroxykynurenine, an endogenous oxidative stress generator, causes neuronal cell death with apoptotic features and region selectivity. J. Neurochem. **70:** 299-307, 1998.

Okumura T., Takasu N., Ishimatsu S., Miyanoki S., Mitsuhashi A., KumadaK., Tanaka K. and Hinohara S. Report on 640 victims of the Tokyo subway sarin attack. Ann Emerg Med. **28:** 129-135, 1996.

Olivier K., Liu J. and Pope C.N. Inhibition of forskolin-stimulated cAMP formation *in vitro* by paraoxon and chlorpyrifos oxon in cortical slices from neonatal, juvenile and adult rats. J. Biochem. Mol. Toxicol. **15:** 263-269, 2001.

Pewnim T. and Seifert J. Structural requirements for altering the L-tryptophan metabolism in mice by organophosphorous and methylcarbamate insecticides. Eur. J. Pharmacol. **248:** 237-41, 1993.

Pond A.L., Chambers H.W. and Chambers J.E. Organophosphate detoxication potential of various rat tissues via A-esterase and aliesterase activities. Toxicol. Lett. **78:** 245-252, 1995.

Pope C.N. and Padilla S. Potentiation of organophosphorus-induced delayed neurotoxicity by phenylmethylsulfonyl fluoride. J. Toxicol. Environ. Health. **31:** 261-273, 1990.

Pope C.N., Chakraborti T.K., Chapman M.L., Farrar J.D. and Arthun D. Comparison of in vivo cholinesterase inhibition in neonatal and adult rats by three organophosphorothioate insecticides. Toxicology. **68:** 51-61, 1991.

Pope C.N., Tanaka D. Jr. and Padilla S. The role of neurotoxic esterase (NTE) in the prevention and potentiation of organophosphorus-induced delayed neurotoxicity (OPIDN). Chem.-Biol. Interact. **87:** 395-406, 1993.

Pope C.N., Chaudhuri J. and Chakraborti T. K. Organophosphate-sensitive cholinergic receptors: possible role in modulation of anticholinesterase toxicity. Pages 305-312 in Enzymes of the Cholinesterase Family (Quinn D.M., Balasubramanian A.S., Doctor B.P. and Taylor P. eds.), Plenum, New York,

Pruett S.B., Chambers H.W. and Chambers J.E. A comparative study of inhibition of acetylcholinesterase, trypsin, neuropathy target esterase, and spleen cell activation by structurally related organophosphorus compounds. J. Biochem. Toxicol. **9:** 319-327, 1994.

Qiao D., Seidler F.J. and Slotkin T.A. Developmental neurotoxicity of chlorpyrifos modeled in vitro: comparative effects of metabolites and other cholinesterase inhibitors on DNA synthesis in PC12 and C6 cells. Environ Health Perspect. **109:** 909-913, 2001.

Quistad G.B. and Casida J.E. Sensitivity of blood-clotting factors and digestive enzymes to inhibition by organophosphorus pesticides. J. Biochem. Mol. Toxicol. **14:** 51-56, 2000.

Quistad G.B., Sparks S.E. and Casida J.E. Fatty acid amide hydrolase inhibition by neurotoxic organophosphorus pesticides. Toxicol. Appl. Pharmacol. **173:** 48-55, 2001.

Quistad G.B., Nomura D.K., Sparks S.E., Segall Y. and Casida J.E. Cannabinoid CB1 receptor as a target for chlorpyrifos oxon and other organophosphorus pesticides. Toxicol Lett. **135:** 89-93, 2002.

Richards P.G., Johnson M.K. and Ray D.E. Identification of acylpeptide hydrolase as a sensitive site for reaction with organophosphorus compounds and a potential target for cognitive enhancing drugs. Mol. Pharmacol. **58:** 577-583, 2000.

Robertson R.T. and Yu J. Acetylcholinesterase and neural development: new tricks for an old dog? News Physiol. Sci. **8:** 266-272, 1993.

Robertson R.T., Mostamand F., Kageyama G.H., Gallardo K.A. and Yu J. Primary auditory cortex in the rat: transient expression of acetylcholinesterase activity in developing geniculocortical projections. Brain Res. Dev. Brain Res. **58:** 81-95, 1991.

Robineau P., Leclercq Y., Gerbi A., Berrebi-Bertrand I. and Lelievre L.G. An organophosphorus compound, Vx, selectively inhibits the rat cardiac Na+,K(+)-ATPase alpha 1 isoform. Biochemical basis of the cardiotoxicity of Vx. FEBS Lett. **281:** 145-148, 1991.

Rocha E.S., Swanson K.L., Aracava Y., Goolsby J.E., Maelicke A. and Albuquerque E.X. Paraoxon: cholinesterase-independent stimulation of transmitter release and selective block of ligand-gated ion channels in cultured hippocampal neurons. J. Pharmacol. Exp. Therap. **278:** 1175-1187, 1996a.

Rocha E.S., Pereira E.F.R., Swanson K.L. and Albuquerque E.X. Novel molecular targets in the central nervous system for the actions of cholinesterase inhibitors: alterations of modulatory processes. Pages 1635-1643 in Proceedings of the 1996 Medical Defense Bioscience Review III, 1996b.

Rocha E.S., Santos M.D., Chebabo S.R., Aracava Y. and Albuquerque E.X. Low concentrations of the organophosphate VX affect spontaneous and evoked transmitter release from hippocampal neurons: toxicological relevance of cholinesterase-independent actions. Toxicol Appl Pharmacol. **159:** 31-40, 1999.

Roy T.S., Andrews J.E., Seidler F.J. and Slotkin T.A. Chlorpyrifos elicits mitotic abnormalities and apoptosis in neuroepithelium of cultured rat embryos. Teratology. **58:** 62-68, 1998.

Saito S. Cholinesterase inhibitors induce growth cone collapse and inhibit neurite extension in primary cultured chick neurons. Neurotoxicol. Teratol. **20:** 411-419, 1998.

Saleem M.A., Williams F.M., Wilkins R.M., Shakoori A.R. and Mantle D. Effect of tri-o-cresyl phosphate (TOCP) on proteolytic enzyme activities in mouse liver in vivo. J. Environ. Pathol. Toxicol. Oncol. **17:** 69-73, 1998.

Savolainen K. Understanding the toxic actions of organophosphates. Pages 1013-1041 in Handbook of Pesticide Toxicology, 2nd edition (R. Krieger, ed), Academic Press, San Diego, CA, 2001.

Scaloni A., Jones W.M., Barra D., Pospischil M., Sassa S., Popowicz A., Manning L.R., Schneewind O. and Manning J.M. Acylpeptide hydrolase: inhibitors and some active site residues of the human enzyme. J. Biol. Chem. **267:** 3811-3818, 1992.

Schlador M.L. and Nathanson N.M. Synergistic regulation of m2 muscarinic acetylcholine receptor desensitization and sequestration by G protein-coupled receptor kinase-2 and beta-arrestin-1. J. Biol. Chem. **272:** 18882-18890, 1997.

Scopes R.K. Maintenance of active enzymes. Pages 185-200 in Protein Purification, Principles and Practice, Springer Verlag, New York.

Seifert S.A. and Eldefrawi M.E. Affinity of myasthenia drugs to acetylcholinesterase and acetylcholine receptor. Biochem. Med. **10:** 258-265, 1974.

Shaw K.P., Aracava Y., Akaike A., Daly J.W., Rickett D.L. and Albuquerque E.X. The reversible cholinesterase inhibitor physostigmine has channel-blocking and agonist effects on the acetylcholine receptor-ion channel complex. Mol. Pharmacol. **28:** 527-538, 1985.

Silveira C.L., Eldefrawi A.T. and Eldefrawi M.E. Putative M2 muscarinic receptors of rat heart have high affinity for organophosphorus anticholinesterases. Toxicol. Appl. Pharmacol. **103:** 474-481, 1990.

Sharma K.V., Koenigsberger C., Brimijoin S. and Bigbee J.W. Direct evidence for an adhesive function in the noncholinergic role of acetylcholinesterase in neurite outgrowth. J. Neurosci. Res. **63:** 165-175, 2001.

Seifert J. and Casida J.E. Relation of yolk sac membrane kynurenine formamidase inhibition to certain teratogenic effects of organophosphorus insecticides and of carbaryl and eserine in chicken embryos. Biochem. Pharmacol. **27:** 2611-2615, 1978.

Seifert J. and Pewnim T. Alteration of mice L-tryptophan metabolism by the organophosphorous acid triester diazinon. Biochem. Pharmacol. **44:** 2243-2250, 1992.

Singh A.K. QSAR for the organophosphate-induced inhibition and 'aging' of the enzyme neuropathy target esterase (NTE). SAR QSAR Environ Res. **12:** 275-295, 2001.

Sivam S.P., Hoskins B. and Ho I.K. An assessment of comparative acute toxicity of diisopropyl-fluorophosphate, tabun, sarin, and soman in relation to cholinergic and GABAergic enzyme activities in rats. Fundam. Appl. Toxicol. **4:** 531-538, 1984.

Slotkin T.A., Tate C.A., Cousins M.M. and Seidler F.J. Functional alterations in CNS catecholamine systems in adolescence and adulthood after neonatal chlorpyrifos exposure. Brain Res Dev Brain Res. **133:** 163-73, 2002.

Small D.H., Michaelson S. and Sberna G. Non-classical actions of cholinesterases: role in cellular differentiation, tumorigenesis and Alzheimer's disease. Neurochem. Int. **28:** 453-483, 1996.

Smith M.I., Elvove E. and Frazier W.H. The pharmacological action of certain phenol esters, with special reference to the etiology of so-called ginger paralysis. Public Health Rep. **45:** 2509-2524, 1930.

Song X., Seidler F.J., Saleh J.L., Zhang J., Padilla S. and Slotkin T.A., Cellular mechanisms for developmental toxicity of chlorpyrifos: targeting the adenylyl cyclase signaling cascade. *Toxicol. Appl. Pharmacol.* **145:** 158-174, 1997.

Song X., Violin J.D., Seidler F.J. and Slotkin T.A. Modeling the developmental neurotoxicity of chlorpyrifos in vitro: macromolecular synthesis in PC12 cells. *Toxicol. Appl. Pharmacol.* **151:** 182-191, 1998.

Sternfeld M., Ming G-L., Song H-J., Sela H, Timberg R., Poo M-M. and Soreq H. Acetylcholinesterase enhances neurite growth and synapse development through alternative contributions of its hydrolytic capacity, core protein and variable C termini. J. Neurosci. **18:** 1240-1249, 1998.

Stone T.W. Development and therapeutic potential of kynurenic acid and kynurinine derivatives for neuroprotection. Trends Pharmacol. Sci. **21:** 149-154, 2000.

Takahashi H., Kojima T., Ikeda T., Tsuda S. and Shirasu Y. Differences in the mode of lethality produced through intravenous and oral administration of organophosphorus insecticides in rats. Fundam. Appl. Toxicol. **16:** 459-468, 1991.

Taylor P. Anticholinesterase agents. Pages 131-149 in Goodman and Gilman's The Pharmacological Basis of Therapeutics, 8th ed, (A.G. Gilman, T.W. Rall, A.S. Nies, and P. Taylor, eds), Pergammon, New York, 1990.

Thompson T.L. and Thomas W.E. Organophosphate-mediated inhibition of choline acetyltransferase activity in rat brain tissue. Experientia **41:** 1437-1438, 1985.

Van Den Beukel I., Dijcks F.A., Vanderheyden P., Vauquelin G. and Oortgiesen M. Differential muscarinic receptor binding of acetylcholinesterase inhibitors in rat brain, human brain and Chinese hamster ovary cells expressing human receptors. J. Pharmacol. Exp. Therap. **281:** 1113-1119, 1997.

Vasquez S., Garner B., Sheil M. and Truscott R.J. Characterization of the major autooxidation products of 3-hydroxykynurenine under physiological conditions. Free Radic. Res. **32:** 11-23, 2000.

Vigny M., Bon S., Massoulie J. and Leterrier F. Active-site catalytic efficiency of acetylcholinesterase molecular forms in Electrophorus, torpedo, rat and chicken. Biochem. **85:** 317-323,1978.

Vogelsberg V., Neff N.H. and Hadjiconstantinou, M., Cyclic AMP-mediated enhancement of high-affinity choline transport and acetylcholine synthesis in brain. J. Neurochem. **68:** 1062-1070, 1997.

Volpe, L.S., Biagioni, T.M. and Marquis J.K. In vitro modulation of bovine caudate muscarinic receptor number by organophosphates and carbamates. Toxicol. Appl. Pharmacol. **78:** 226-234, 1985.

Walker C.H. and Mackness M.I. "A" esterases and their role in regulating the toxicity of organophosphates. Arch. Toxicol. **60:** 30-33, 1987.

Ward T.R., Ferris D.J., Tilson H.A. and Mundy W.R. Correlation of the anticholinesterase activity of a series of organophosphates with their ability to compete with agonist binding to muscarinic receptors. Toxicol. Appl. Pharmacol. **122:** 300-307, 1993.

Ward T.R. and Mundy W.R. Organophosphorus compounds preferentially affect second messenger systems coupled to M2/M4 receptors in rat frontal cortex. Brain Res. Bull. **39:** 49-55, 1996.

Watson M., Roeske W.R., Vickroy T.W., Smit T.L., Akiyama K., Gulya K., Duckles S.P., Serra M., Adem A., Nordberg A., Gelhert J.K., Wamsley J.K. and Yamamura H.I. Biochemical and functional basis of putative muscarinic receptor subtypes and its implications. Trends Pharmacol. Sci. Suppl. **2:** 44-55, 1986.

Weber M., David-Pfeuty T. and Changeux J.P. Regulation of binding properties of the nicotinic receptor protein by cholinergic ligands in membrane fragments from Torpedo marmorata. Proc Natl Acad Sci U S A. **72:** 3443-3447, 1975.

Weiler M.H. Muscarinic modulation of endogenous acetylcholine release in rat neostriatal slices. J. Pharmacol. Exp. Therap. **250:** 617-623, 1989.

Whalley C.E. and Shih T.M. Effects of soman and sarin on high affinity choline uptake by rat brain synaptosomes. Brain Res. Bull. **22:** 853-858, 1989.

Whitney K.D., Seidler F.J. and Slotkin T.A. Developmental neurotoxicity of chlorpyrifos: cellular mechanisms. Toxicol. Appl. Pharmacol. **134:** 53-62, 1995.

Wilkie G.I., Hutson P., Sullivan J.P. and Wonnacott S. Pharmacological characterization of a nicotinic autoreceptor in rat hippocampal synaptosomes. Neurochem. Res. **21:** 1141-1148, 1996.

Wu Y.-J., Harp,P., Yan X-R and Pope C.N. Nicotinic autoreceptor function in rat brain during maturation and aging: possible differential sensitivity to organophosphorus anticholinesterases. Chem.-Biol. Interactions **142:** 255-268, 2003.

Xie W., Stribley J.A., Chatonnet A., Wilder P.J., Rizzino A., McComb R.D., Taylor P., Hinrichs S.H. and Lockridge O. Postnatal developmental delay and supersensitivity toorganophosphate in gene-targeted mice lacking acetylcholinesterase. J. Pharmacol. Exp. Ther. **293:** 896-902, 2000.

Zhang W., Basile A.S., Gomeza J., Volpicelli L.A., Levey A.I. and Wess J. Characterization of central inhibitory muscarinic autoreceptors by the use of muscarinic acetylcholine receptor knock-out mice. J. Neurosci. **22:** 1709-17, 2002.

Zwart R., van Kleef R.G.D.M., Gotti C., Smulders C.J.G.M. and Vijverberg H.P.M. Competitive potentiation of acetylcholine effects on neuronal nicotinic receptors by acetylcholinesterase-inhibiting drugs. J. Neurochem. **75:** 2492-2500, 2000.

Pharmacological Perspectives of Toxic Chemicals and Their Antidotes
Editors: S J S Flora, J A Romano, S I Baskin and K Sekhar

Toxicological Review of Capsaicin as a Weapon

Anna Barbara Fishcer
Institute of Hygiene and Environmental Medicne
Justus-Liebig-University Giessen, Gerrmany

INTRODUCTION

In various species of *Capsicum,* belonging to the plant family *Solanaceae*, substances are found which are responsible for chilli peppers' pungency. The main active component is capsaicin, of which up to 1% is contained in the wet weight of capsicum fruits (Monsereenusorn et al., 1982). At a closer look there is a mixture of two unsaturated and three saturated homologues (capsaicinoids) constituting the basic structure N-[4-hydroxy-3-methoxybenzyl]-alkylamide (Govindarajan & Sathyanarayana 1991). A spray of the capsicum extract called *Oleoresin capsicum* or of the synthetically produced pelargonic acid vanillamide (PAVA), can be used as a defence weapon. The present literature review has the aim to evaluate the possible health risk by deploying this weapon.

In the USA, Switzerland, Austria and recently also in some German Federal States pepper spray is used by the police. According to police reports this weapon has proved effective against single persons leading to a reduced incidence of physical injuries in police officers as well as in arrested subjects. Compared with the so far employed irritant or tear gases chloroacetophenone (CN) und orthochlorobenzalmalonnitrile (CS) the ingredients of the naturally occurring capsicum appear to possess several advantages. The effects of CN and CS appear only after 20-30 seconds, by contrast those of pepper spray are almost instantaneous. CN and CS are also less effective against animals than pepper-spray. A further disadvantage of CN and CS is their adsorption to different materials necessitating difficult decontamination, while capsaicin is easy to remove and, as a natural substance, is biologically decomposed (Busker & Van Helden, 1996).

Of main interest for the evaluation of the toxicology of pepper spray are the reactions following short-term exposure to skin, eye or respiratory system. In this context the question arises whether there are especially sensitive individuals like asthmatic or heart patients and whether long-term health hazards can be expected. Other points are whether persons under the influence of drugs or alcohol differ in their sensitivity to pepper spray and what effects are to be expected of a combination of pepper-spray, excitement and physical violence. In the last years several reviews have been dealing with the toxicology of capsaicin or PAVA in connection with security use (Ruddick, 1993; Busker & Van Helden, 1996; Hobbs & Rice, 1997).

CHEMICAL AND PHYSICAL PROPERTIES

Besides capsaicin a several other closely related analogues are found in the oleoresin of capsicum, among them norhydrocapsaicin, dihydrocapsaicin, homocapsaicin and homodihydrocapsaicin (Fig. 1).

Capsaicin (CAS number: 406-86-4), with the empirical formula $C_{18}H_{27}NO_3$ and the molecular weight of 305.4 daltons forms white, transparent crystals with a melting point of 64°C. At 210-220°C capsaicin can be distilled without decomposition. Capsaicin is soluble in ether, ethanol, acetone, methanol, hexachlorbenzene, benzene and its analogues, diluted alkalihydroxyd solutions, sparingly soluble in carbon sulfide, hot water, concentrated hydrochloric acid, practically insoluble in cold water; solubility in physiological saline at 40° C: 3×10^{-4} g/L. The synthetic compound pelargonic acid vanillylamide (PAVA, N-vanillylnonanamide, CAS number: 2444-46-4) has the formula $C_{17}H_{27}NO_3$ and a molecular weight of 293.4 daltons.

An US-American preparation (CAP STUN®), based on oleoresin (OC) contains 5.5% OC and 0.92% capsaicin, the OC level reported in the spray cans employed in Austria is 10%, and according to the German regulations (Technische Richtlinie 1999) the aerosol should contain 0.2 ± 0.03 % (w/w) capsacinoids (purity >60%) or 0.3 ± 0.03.% (w/w) PAVA (purity >95%).

TOXICOLOGY

General Character

Capsaicin is a neuropharmacological substance acting on primary afferent neurons which function to receive and conduct information from the external or internal environment of the organism to the brain in order to retain physical homeostasis. The cell bodies of the primary afferent neurons are located in the spinal or cranial sensory ganglia, sending fibres into the central nervous

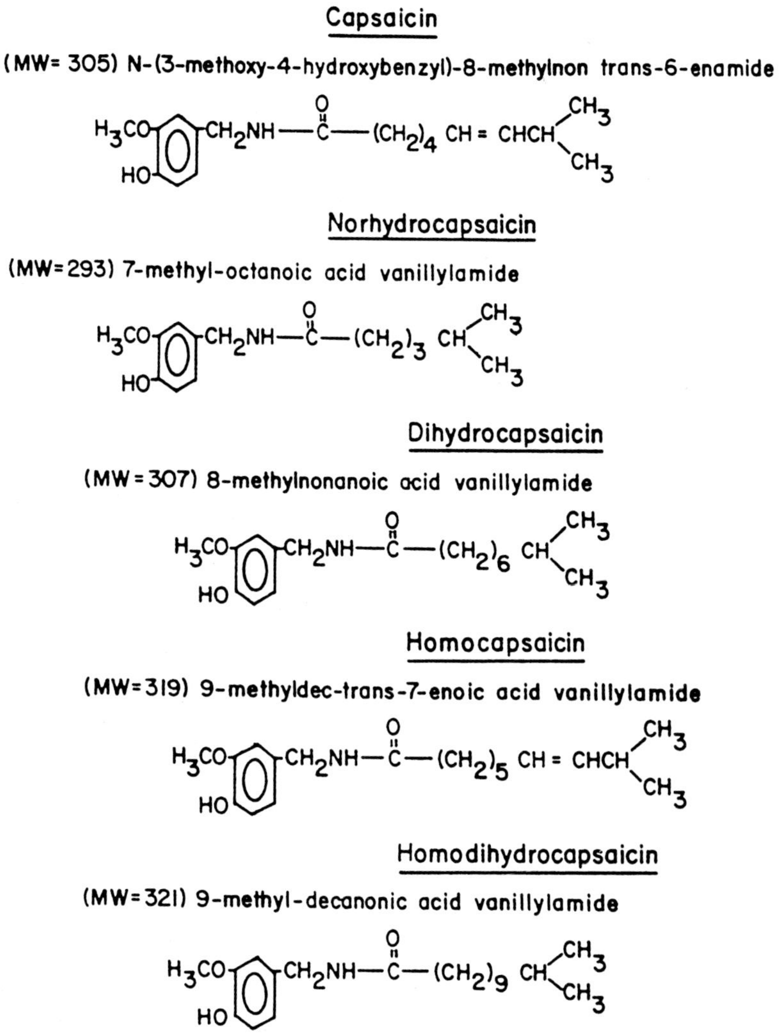

Capsaicin

(MW= 305) N-(3-methoxy-4-hydroxybenzyl)-8-methylnon trans-6-enamide

H_3CO, HO — $CH_2NH—C(=O)—(CH_2)_4\ CH = CHCH(CH_3)_2$

Norhydrocapsaicin

(MW=293) 7-methyl-octanoic acid vanillylamide

H_3CO, HO — $CH_2NH—C(=O)—(CH_2)_3\ CH(CH_3)_2$

Dihydrocapsaicin

(MW = 307) 8-methylnonanoic acid vanillylamide

H_3CO, HO — $CH_2NH—C(=O)—(CH_2)_6\ CH(CH_3)_2$

Homocapsaicin

(MW=319) 9-methyldec-trans-7-enoic acid vanillylamide

H_3CO, HO — $CH_2NH—C(=O)—(CH_2)_5\ CH = CHCH(CH_3)_2$

Homodihydrocapsaicin

(MW=321) 9-methyl-decanonic acid vanillylamide

H_3CO, HO — $CH_2NH—C(=O)—(CH_2)_9\ CH(CH_3)_2$

Fig. 1 Graphic formulas of capsaicin and derivatives (Monseerenusorn et al., 1982)

system and the periphery. Two types of neurons can be stimulated by capsaicin: (a) thin unmyelinated afferent C-type (or group VI) neurons in mammals, (b) in the whole animal kingdom also other, myelinated fibres, mainly A delta fibres (or group III-fibres), although also some other A-fibers might respond to the drug (Fitzgerald 1983, Holzer 1991). Among the fibres and sensory neuronal somata connected to these fibres, however, only nociceptors (which also react to other chemical stimuli and may therefore be designated as chemonociceptors) and some warmth receptors are stimulated by capsaicin (Kaufman et al. 1982, Such & Jancso 1986). The small-diameter neurons transduce the signals into action potentials and transmit this information to the central nervous system, ultimately eliciting a perception of pain or discomfort. The signal transduction occurs by the binding of capsaicin to a specific recognition site referred to as the vanilloid receptor VR1, a sensory neuron-specific ion channel that is structurally related to members of the TRP (transient receptor potential) family of ion channels (Caterina et al., 1997). The binding of capsaicin to the receptor elicits a sodium influx thus evoking an action potential and then an augmentation of intracellular calcium. Depending on capsaicin dosage, as intermediate and long-term effects an impairment of neuronal function (desensitization) and degeneration of the afferent neurons may result (Fitzgerald, 1983, Szallasi, & Blumberg, 1996).

Peptide mediators released by the peripheral nerve axons produce local reactions including vasodilation and altered vascular permeability, changes in the activity of cardiac, bronchial and visceral muscles, in tissue growth and repair, immunological functions and the regulation in postganglionic sympathetic efferent nerves (Holzer 1988, 1991, Herbert and Holzer, 2002b). The results are erythema, flare and, in some species, oedema of the skin, while apnoea, cough, bronchoconstriction, hypotension, and bradycardia are observed in response to irritation of the respiratory tract.

One of the most important mediators released is substance P (SP), a neuropeptide belonging to the group of tachykinins. These proteins have a common carboxyterminal group (Phe-X-Gly-Leu-Met-NH_2). The peripheral release of tachykinins determines a set of responses (known as neurogenic inflammation) that includes vasodilatation, plasma protein extravasation, smooth muscle contraction and stimulation of afferent nerves. SP is released locally by afferent nerves as well as by the spinal ganglion, is responsible for several stimulating effects to central and peripheral neurons, for smooth muscle contraction and bronchoconstriction and is possibly involved in pain transmission as a non-synaptically released diffusible chemical messenger (Gamse et al., 1979, Lundberg et al., 1985, Saria et al., 1986, Helme et al., 1987, Herbert & Holzer, 2002a). However, the primary role of SP in nociception is inconclusive, whereas hyperalgesia may depend on the synthesis and release of this peptide (Lawson et al., 1997, Valtschanoff et al. 2001). SP binds to the NK1 (Neurokinin-1) receptor. The distribution pattern of further neuropeptides like calcitonin gene-related peptide (CGRP), somatostatin, neurokinin A etc. in afferent axons is complex and their role in capsaicin-induced processes is not yet resolved. An interesting field of research is also the involvement of inflammatory cytokines, e.g. IL-6, IL-8 and TNF-alpha in the chain of events leading to the neurogenic inflammation induced by capsaicin (Veronesi et al., 1999, Reilly & Green, 1999, Herbert & Holzer, 2002a).

The mechanisms of desensitization, which can be induced by topical or systemic application of capsaicin are not completely understood. One of the explanations is a depletion of the precursors of substance P in the cells, which occurs already at low capsaicin concentrations and is caused by a decrease of the relevant mRNA levels (Hakanson et al., 1987, Amann, 1990, Holzer, 1991, Szallasi et al., 1999). Elevated systemic doses of capsaicin administered to neonatal animals produce a destruction of C-type neurons leading to permanent desensitization (Lundberg & Saria, 1983). When 2 days old rats were s.c. treated with 50 mg/kg capsaicin, irreversible small B-type sensory ganglion cells were destroyed, while the large A-type neurons were unaffected (Jancso et al., 1977). Within few hours degeneration of the unmyelated fibres in the dorsal root and axon terminals in laminae 1 and 2 of the dorsal horn occurs. Glial elements rapidly appear in the upper dorsal horn and engulf the degenerated axon terminals. A study of the distribution of axon terminal degeneration in the CNS of rats showed that it coincided with the central projections of C fibre primary sensory neurons supplying the skin, mucous membranes and viscera in the Vth, IXth and Xth cranial nerves and the spinal nerves. Degeneration was found in laminae I and II of the spinal cord, superficial layers of the trigeminal *nucleus caudalis*, a small part of the *nucleus oralis*, the nucleus of the solitary tract and the adjoining *nucleus commissuralis* and a small part of the *area postrema*. Degeneration was not seen in any other structures of the CNS. Concomitant to the neuronal destruction neuropeptides, particularly sbstance P, were depleted (Fitzgerald 1983). The pattern of capsicin-induced degeneration of the CNS is dose- and age-dependent. Prenatal rats and rats older than 14 days are less sensitive than neonatal animals. Systemic capsaicin treatment of adult rats results in a decrease of substance P in those regions where a depletion is seen following neonatal treatment (Jancso et al., 1985a, Perfumi & Sparapassi, 1999).

The extent of stimulation by capsacinoids depends on the length and saturation of the alkyl chain of the molecule, but the acylamide linkage can be replaced by an esteric group and the alkyl chain by cycloalkyl rings. In desensitizing action the presence of an alkyl chain is essential and its optimal length corresponds to 10-12 C atoms (Szolcsanyi & Jancso-Gabor, 1975, 1976, Janusz et al., 1993).

The transient desensitization caused by capsaicin is used for the treatment of rheumatic diseases. The possible therapy of other dermatological and neurological complaints with capsaicin and its analogues has also been suggested (Rumsfield & West, 1991, Cordell & Araujo, 1993, Keitel et al., 2001), but the mono-therapy with capsaicin has so far not been sufficient or satisfactory (Rains et al., 1995). However, the treatment of neurogenic inflammation with substances involved in the capsaicin neuro-exitatory pathway offers promises (Herbert & Holzer, 2002b).

Pharmacokinetics

Absorption and metabolism of capsaicin were investigated in the rat. Following intragastric administration radiolabelled ^{3}H-dihydrocapsaicin and unlabelled capsaicin were readily absorbed, but almost completely metabolized in the liver before reaching the general circulation. In the portal vein blood unchanged compounds were identified, but 15 minutes after treatment less than 5% of the substance was detected in trunk blood and brain. Capsaicin is metabolized by hepatic mixed-function oxidase systems to the corresponding catechol metabolite via a ring epoxide capable of reacting covalently with nucleophilic sites of hepatic protein. Cell-free extracts of various tissues of rats contain enzyme activity for hydrolysing capsaicin at the acid-amide bond to produce vanillyl amine and the fatty acid moiety. To a certain degree the substance is already biotransformed by bacteria in the intestinal lumen. It was concluded that the rapid hepatic metabolization limits the systemic effects of enterally absorbed capsaicin and the drug reaches the CNS and other extrahepatic organs almost exclusively as degradation products (Donnerer et al., 1990, Surh & Lee, 1995a).

Acute Toxicity

The concentrations of capsaicin causing 50% lethality (LD_{50}) in mouse, rat, guinea pig and rabbit following intraperitoneal application were ca. 7, 12, 1, and >50 mg/kg body weight (bw). In the mouse the LD_{50} was 0.56 und 1.6 mg/kg bw following intravenous and intratracheal administration, while the intramuscular and subcutaneous treatment yielded similar LD_{50} -values as intraperitoneal treatment. Oral capsaicin is considerably less effective with LD_{50} -values of ca. 100 and 150 mg/kg bw in mice and rats; the intragastrically and intrarectally determined values in the mouse amounted to 190 and >218 mg/kg bw. Following dermal application an LD_{50} of >512 mg/kg bw was found (Nagabhushan & Bhide 1985, Glinsukon et al., 1980, Saito & Yamamoto, 1996). The lethality of pelargonic acid vanillylamide (PAVA) corresponds approximately to that of capsaicin (Hobbs & Rice, 1996).

The symptoms of intoxication in the rat are convulsions within 1-2 minutes, followed by extension of the limbs, apnoea and death by respiratory failure after 2-5 minutes. Survivors of a high single dose had similar symptoms and recovered within half an hour (Glinsukson et al., 1980). Intravenous application to anaesthesized dogs (10-300 µg/kg bw) caused a transient increase in blood pressure followed by long-lasting hypotension, whereas rabbits showed only hypotension. In dogs no effects on heart function were detected (Toda et al., 1972). In rats hypotension, bradycardia and apnoea were observed (Toth et al., 1984).

Thermoregulation is disturbed by different modes of capsaicin treatment, also following the direct injection of capsaicin into the hypothalamus; subsequently the sleep-waking behaviour and caloric food intake do no longer adapt to ambient temperatures (Benedek et al., 1980, Comareche-Leydier, 1981, Monsereenusorn et al., 1982).

Upon oral application of capsaicin a pungent, hot taste is noted up to a dilution of $1{:}3 \cdot 10^6$ (Molnar, 1965). The threshold concentration on the tongue amounts to ca. 0.7 μM/L. The substance causes a burning sensation on the mucous membranes of mouth and respiratory tract which is accompanied by sweating, salivation, and lacrimation (Szolcsany & Jancsó-Gabor, 1975, Rozin et al., 1981, Sizer & Harris, 1985).

Following the topical application of capsaicin to the skin a primary excitation stage is replaced by a phase of resistance. Upon treatment of the human forearm skin (threshold concentration 30 nM/L), a burning, stinging sensation results. Erythema, pain and heat sensation persist for some hours depending on the applied dose. A markedly less intensive reaction occurs following treatment of the same skin area 24 h later (Carpenter & Lynn, 1981, Simone et al., 1989). In human volunteers a 73.3% absorption of capsaicin in ointment (0.2 g containing 0.075% capsicin) was determined; 15-20 min after the application of ointment containing 0.025-0.075% capsaicin, the subjects experienced a burning pain and at this time the skin temperature had increased by 1.30-1.35°C compared to that measured before treatment. The erythema disappeared within 2-4 h. PAVA penetrated the skin faster, but caused less erythema (Fang et al., 1996). An intradermal dose of 30 nM capsaicin is sufficient to activate pain receptors and elicit pain in monkeys and rats (Martin et al., 1987, LaMotte et al., 1988). In humans the pain threshold was noted at an intradermal dose of 0.1 μg injected in 10 μl (Simone et al., 1989).

In an animal experiment ordered by the producer 0.5 ml of a capsaicin-solution (Cap-Stun®, with 56.2 mg/L of the substance) was applied to the shaved back of rabbits which was then covered by an occlusive dressing. One to 48 h after application a well defined erythema was observed, which receded by 72 h and had completely disappeared by day 10. None of the animals developed oedema (United States Testing Company, 1992a). In intradermal tests of guinea pigs for skin sensitization (allergenicity) by the oleoresin of Capsicum (Cap-Stun®) according the method of Magnusson & Kligmann no dermal reactions were observed (United States Testing Company, 1992b). On the other hand allergy to capsaicin has been reported in humans (Monsereenusorn et al., 1982), but in the recent literature has not been further studied.

The eye reacts with strong pain and transient blindness. The pain threshold concentration of the rat eye was 30 nM capsaicin The application of a 1% capsaicin solution to the cornea produced swelling of the mitochondria and reduced the number of microvesicles at nerve endings, but there was no axonal degeneration (Szolcsany & Jancso-Gabor, 1975). In the rat cornea substance P was depleted 4 h after topical treatment; this effect could no longer be detected after three weeks (Gamse et al., 1981).

An eye irritation test was carried out in six rabbits with the Cap-Stun® preparation in the same concentration as used in the USA as a police weapon

(0.1 ml per eye, 56.2 mg capsaicin/L). After one hour all animals displayed conjunctival and chemosis irritation with scattered or diffuse areas of opacity, but details of the iris were still visible. After 24 h the iris was swollen in three animals with the iris still reacting to light. One rabbit developed this iris swelling by 48h and improved by day 10. Complete recovery was seen in the remaining animals by day 10 (United States Testing Company 1992c).

Also pelargonic acid vanillylamid (PAVA) is a fast eye irritant (Legrin, 1996). With 0.15 μM PAVA the concentration leading in 50% of guinea pigs to eye blinking or closing (ED_{50}) was 100 or 350 times as effective as that of the irritant gases CS and CR (Battensby et al., 1981). In concordance with the findings obtained after instillation of capsaicin a similar study with PAVA showed no long-term damage. One and 24 h after treating rabbits' eyes redness and swelling of the conjunctiva was seen which lasted until day 3 and in one rabbit there were signs of iris damage and corneal opacity, but all animals had fully recovered 7 days after application of PAVA (Chevarne, 1995).

In ten human volunteers oleoresin spray produced occasional areas of focal epithelial cell damage that healed within one day. All eyes showed conjunctival hyperaemia and in two subjects, mild chemosis. All except one eye had unchanged best corrected visual acuity (BCVA). Basal epithelial cell morphology suggested temporary corneal epithelial swelling, whereas keratocytes, endothelial cells and subbasal nerves remained unchanged (Vesaluoma et al., 2000).

The inhalation effects of capsaicin on the respiratory tract have been studied in experimental animals as well as in humans. In an acute inhalation test with young rats with the Cap-Stun® preparation used by the US American police, 10 rats were exposed for one hour to an aerosol (whole body exposure to 56.2 mg/L breathing air). All animals appeared incapacitated during the one hour exposure, eye irritation (redness etc.) was noted. The rats appeared normal by the 24 h observation period. After 14 days no anomalies were observed (United States Testing Company 1992d). In further experiments rats inhaled nebulized PAVA in a concentration of 3, 6, 12 und 20 μM/L air for 4 h (mouth-nose exposure). Observation for 7 days showed no appreciable abnormality and autopsy 7 d after exposure gave no indication of inhalation toxicity (Confarma, 1996).

Three studies with guinea pigs addressed the question if chronic exposure to tobacco smoke (TS) and/or atopy increases airway sensitivity to capsaicin. After 90 days of TS exposure, airway reactivity increased in ovalbumin-sensitized (OA) guinea pigs challenged with capsaicin aerosol, but less so in non-sensitized animals (Bergren, 2001a). OA alone also induced capsaicin hyperresponsiveness (Bergren 2001b). Another study showed that peak response in total lung resistance to i.v. capsaicin was significantly greater in guinea pigs chronically exposed to TS than that evoked by the same dose of capsaicin in control (air-exposed) animals (Kwong et al., 2001). Taken together these studies allow the conclusions that OA-sensitized animals show airway hyperirritability to capsaicin and that smoking increased sensitivity to capsaicin, especially in sensitized animals. The increased C-fiber responsiveness may have consequences for atopic individuals.

In human volunteers inhaling aerosols ≥1nM/L for 1 min evoked cough, and a dose-dependent bronchoconstriction was already noted at 0.1 nM/L. In single breath challenges the stock solutions had concentrations of 1-25 mM/L (ca. 0.03-0.75%) of which small volumes of 0.2 or 0.05 ml (ca. 0.2-13.6 nM) were inhaled (Collier & Fuller 1984, Holzer, 1991). At concentrations below the cough threshold a transient increase of the respiratory resistance was measured. This reaction showed no differences between normal subjects and asthmatics without persistant cough (Fuller et al. 1985, Maxwell et al. 1987, Fuller 1991). In two further studies it was reported that an initial bronchoconstriction was followed by bronchodilatation, again without differences between individuals with or without asthma (Lammers et al., 1988, 1989). On the other hand, subjects with asthma and chronic obstructive pulmonary disease (COPD) were both more responsive to capsaicin-induced cough than normal subjects (Doherty et al., 2000a). The cough response could also be elicited by significantly lower capsaicin doses in patients with cryptogenic fibrosing alveolitis (Doherty et al. 2000b). However, capsaicin cough sensitivity was decreased in smokers (Doherty et al. 2001, Millqvist & Bende, 2001). There are studies suggesting that the larynx is the main site of cough stimulation for inhaled capsaicin (Barros et al., 1991; Collier & Fuller, 1984). This might explain why the comparison between asthmatic and non-asthmatic subjects gives different results for the two different endpoints cough and bronchoconstriction.

The potentially severe pulmonary toxicity of high doses of capsaicin was demonstrated by a case of an 11-years-old boy, who purposely inhaled deeply the jet of propelled substance from a capsaicin cartridge several times. After coughing for almost an hour the symptoms subsided for 4 hours, developing towards respiratory arrest, laryngotracheobronchitis and pulmonary oedema necessitating artificial respiration. The patient recovered following adequate therapy (Winsgrad, 1977).

Subchronic Toxicity

In a feeding study where mice received ground red chilli (0.5-10% of the diet), general health, body weight and food intake were not adversely affected at any level of pepper consumption. Histopathological evaluation revealed slight glycogen depletion and anisocytosis of hepatocytes in the 10% group. However, other organs did not reveal any lesions attributable to the chilli exposure (Jang et al., 1992). The application of 50 mg capsaicin/kg bw and day to rats by stomach tube for 60 days caused a significant reduction of growth in spite of elevated food intake (Monsereenusorn, 1983). Chilli and capsaicin in the diet produced cirrhosis of the liver in a dose-dependent fashion (Monsereenusorn et al., 1982), damage of the duodenal mucosa (Naponitaya, 1974) and gastric ulcers (Ootha, 1968). In a 13-week toxicity study, renal toxicity was observed in male mice receiving a 1% mixture of capsaicinoids (64.5% capsaicin and 32.6% dihydrocapsaicin) in their diet (Akagi et al., 1998).

CHRONIC EFFECTS

General Toxicity

No toxicity was detected in experiments with mice receiving 10% capsicum extract in their diet (Jang et al. 1992). No effect on life-span was observed in mice ingesting 0.125-1% capsaicin in their diet (Toth et al., 1984).

The prevalence of cough was increased in hot pepper workers chronically exposed to chilli pepper. Lung function was unchanged in the workers compared to unexposed controls with neither group demonstrating a significant fall in forced vital capacity in response to capsaicin inhalation (Blanc et al., 1991).

Mutagenity and Carcinogenicity

Concerning the mutagenicity of capsaicin there are contradictory results. In some studies the Ames-Test with the strains TA98 und TA100 of *Salmonella typhimurium* no mutagenic effects were noted, in further tests with the same strains slight effects were found (Rockwell & Raw 1979, Buchanan et al. 1981, Toth et al. 1984, Nagabhusan & Bhide 1985; Gannett et al. 1988; Surh & Lee 1995; Azizan and Blevine, 1995). An acetone extract of *Capsicum annuum* reduced the mutagenicity of capsaicin in TA100 with S9 metabolic activation; the paprika extract had also antimutagenic effects against 2-aminoanthracene (Azizan and Blevine, 1995). Capsaicin attenuated the mutagenicity of vinyl carbamate and N-nitroso-dimethylamine in TA100 (Surh et al., 1995). Chromosome aberrations were reported in human lymphocytes *in vitro* (Surh & Lee, 1995). In addition capsaicin induced the formation of micronuclei in a dose-dependent manner and led to a moderate increase of sister chromatid exchanges in human lymphocytes (Marques et al., 2002). In V79 cell cultures of the Chinese hamster ouabain-resistant, but no azaguanidin-resistant mutations were induced (Lawson & Gannett 1989; Nagabhushan & Bhide, 1985).

In vivo micronuclei were detected in the bone marrow of mice fed with capsaicin or chilli (Villasenor & Ocampo, 1994, Nagabhusan & Bhide, 1985). Following i.p. treatment micronuclei as well as sister chromatid exchanges were noted (Diaz Barriga Arceo et al., 1995). Sperm morphology was unchanged in male mice treated with capsaicin and the subsequent mating of the animals yielded negative results in the dominant-lethal assay indicating lack of damage to male germ cells (Muralidhara & Narasimhamurthy, 1988).

Feeding rats with 10% chilli induced hepatomas (Hoch-Ligeti, 1951). Benign adenomas of the duodenum occurred in 10% of mice given 0.125-1% capsaicin in the diet (Toth et al., 1984). The life-long administration of a diet containing 0.03% capsaicin to mice, starting from 6 weeks of age, led to the development of polypoid adenomas of the caecum (Toth & Gannett, 1992). In contrast, a mixture of capsaicinoids (64.5% capsaicin and 32.6% dihydrocapsaicin) was non-carcinogenic. The numbers of tumour-bearing females in the high-dose groups were significantly lower than that in the controls, and the incidence of hepatocellular neoplasms in both sexes was negatively correlated with the dose of capsaicinoids (Akagi et al., 1998). The feeding of chilli powder to rats did not affect number, size and location of benign and malignant duodenal and colonic tumours (Kang et al., 1992).

Several studies were carried out on the modifying effects of capsaicin in combination with carcinogenic substances. Using a protocol of methyl(acetoxymethyl)nitrosamine as initiator and chilli extract as a promoter, induction of stomach tumours was observed in mice. The promoter effect of chilli was also seen in the hexachlorocyclohexane-induced hepato-carcinogenesis system (Agrawal et al., 1986). In contrast, a lack of tumour promoting activity of capsaicin in mouse skin carcinogenesis caused by dimethylbenzanthracene was reported (Park et al., 1998).

Further working groups found predominantly inhibitory effects of capsaicin on tumour formation. Inhibition was reported in the following systems: i.p. benzo(a)pyrene or 9,10-dimethyl-1,2-benzanthracene and capsaicin in the diet and lung cancer in mice (Jang et al., 1989), azoxymethane-induced rat colon adenocarcinoma induction and dietary capsaicin (Yoshitani et al., 2001), vinylcarbamate and topical pretreatment of female mice with capsaicin and skin tumours. The cytochrome P-450 CYP2E1 isoform responsible for activation of these carcinogens was suppressed by capsaicin suggesting a physiological cause for the effects of capsaicin (Surh et al., 1995). Dietary capsaicin following treatment with nitrosamines lowered lung tumour incidence, but increased the incidence of papillary or nodular hyperplasia of the rat urinary bladder, while the tumour incidence of other organs, such as the kidney and thyroid, was not significantly different from that of the corresponding controls (Jang et al., 1991). Capsaicin feeding during either the initiation or promotion phase in combination with 4-nitroquinoline 1-oxide reduced the frequency of tongue carcinoma without statistical significance (Tanaka et al. 2002). In studies of lung tumour formation in the mouse intragastrically administered capsaicin did not inhibit lung tumours induced by the tobacco-specific nitrosamine 4-(methylnitrosamino)-1-(3-pyridyl)-1-butanone. Capsaicin alone did not affect the spontaneous incidence of lung tumours (Teel and Huynch, 1999).

In many tropical countries including India great quantities of *Capsicum* are consumed. Several studies suffering, however, from restricted numbers of patients were devoted to the question whether chilli can cause fibrosis or cancer in the oral cavity, but the results were equivocal, mainly because of the simultaneous exposure to betel (Pindborg et al., 1967; Sirgot & Kanolkar, 1962). A case-control study in South India found an elevated odds ratio for stomach cancer associated with high consumption of chilli (Mathew et al., 2000). Similarly, in a Mexican case-control-study a significant correlation was found between the consumption of pungent chilli paprika and the incidence of stomach tumours (Lopez-Carillo et al., 1989). In four counties with a high concentration of specific ethnic groups (Mexican-American, Cajun, white and black Creole) who use high amounts of pepper or chillies significantly higher rates for stomach and liver cancer were found than in matched control counties strengthening the hypothesis of an association of stomach cancer with capsaicin pepper (Archer & Jones 2002). On the other hand the fact that Indians and Malays, who consume the greatest amount of chilli, evince the lowest rates of stomach ulcers, and colorectal carcinoma among the three main races in Singapore appears to furnish evidence against a carcinogenic influence of chilli, but other genetic or environmental factors may also play a role (Lee et al., 1988; Kang, 1988). A case-control study of diet and gallbladder cancer showed a protective effect of vegetables and fruits on gallbladder carcinogenesis, among them green chilli, but an (insignificant) increase in the odds was observed with consumption of capsicum and red chilli (Pandey & Shukla, 2002).

Embryotoxicity and Teratogenicity

Wistar rats were s.c. injected every other day from day 7-15 of pregnancy, the period of main organogenesis, with 50, 100 und 200 mg/kg capsaicin. Neither in the dams nor in the new-borns were signs of toxicity observed and the growth of the pups was normal (Perfumi & Sparapassi, 1999). This confirms earlier reports (Kirby et al., 1982, Pellicer et al., 1996).

EXPERIENCES WITH POLICE USE OF CAPSAICIN

From 1989 the Firearms Training Unit of the FBI-Academy in Quantico (Virginia) tested an oleoresin capsicum (OC) pepper spray for three years; 828 subjects were sprayed without medical problems. Subsequently the spray was adopted by FBI-agents. Following the introduction of pepper spray in the USA the media reported some in-custody deaths related to OC (Bunting 1993). In the first such case autopsy showed that in the presence of bronchiolitis/bronchitis and cardiomegaly stomach contents had been aspirated and a contribution of pepper spray was concluded. In a second case cardiac disease was accepted as the cause of death (Steffee et al., 1995).

From 1992 the police of Californa has adopted an oleoresin-aerosol as a force option. The American Civil Liberties Union reviewed all reports of police use of OC on a total of 265 humans. In 14.3% a limited effectiveness was observed, mainly in subjects under the influence of drugs or victims of acute mental illness. The seven death cases reported statewide in 1992-1993 were analysed. Autopsy confirmed or suspected drug-overdose in the victims (3 times cocaine, 2 times methamphetamines and 2 times pending suspected methamphetamine overdose). In two cases police restraint techniques contributed or were related to the deaths; pre-existing heart disease and stress were present in three cases (ACLU 1993).

In 1990-1993 thirty incidents happened in 13 states of the USA, in which death of a subject occurred following a spraying with OC (Granfield et al., 1995). An evaluation showed that the victims were 24-53 year old men sharing the following common features: all subjects behaved in a combative and/or bizarre manner and struggled with the police and in at least 23 cases drugs or alcohol were involved. In the majority of these cases OC was of limited effectiveness. A review of the autopsy reports came to the conclusion that OC was in no case responsible for the death. In 18 of the 22 sufficiently documented cases positional asphyxia was the cause of death with drugs and/or disease also being contributing factors. It is known that the oxygen required after great physical exertions, e.g. during fighting or flight, must be met by heavy breathing and that forcible restraint, especially hog-tying, impedes oxygen supply (Bell et al., 1992; O'Halloran & Lewman, 1993). Cocaine-induced delirium with bizarre and violent behaviour necessitating forcible restraint by the police can also lead to sudden death without the use of pepper spray (Wetli, 1987).

In a retrospective study the cases exposed to OC during law-enforcement action and presenting in the emergency department of Kansas-City (Oklahoma) hospital were analyzed (Watson et al., 1996). In the first three years after introduction of the new weapon 908 subjects were exposed to the spray; approximately 10% of these were brought to the emergency department. There were 81 predominantly black patients (Afro-Americans: 73%, male: 91%). In 63% of the patients eye symptoms (burning and redness) were noted, in seven cases corneal abrasions with an unclear aetiology and altered vision in three patients. The second most frequent symptoms were dermal; burning and erythema of the skin occurred in 20 and 12 patients, respectively. Five patients presented with shortness of breath or wheezing, one each with cough and irritation of the throat. Two of the five patients with respiratory complaints were asthmatics, whereas no respiratory symptoms were observed in the remaining ten patients with asthma. Patients stayed in the emergency department for 1.6 ± 0.9 hours. Three were transferred to a psychiatric hospital and two were hospitalized due to complications of drug abuse.

Two reports deal with the experiences with pepper spray in Austria during 1994-1999 (Zehetner, 1998; Anonymous, 1999). The employed aerosol contained 10% oleoresin capsicum. After a pilot phase all police officers were furnished with OC spray in 1997. A total of 438 operations with OC were reported. None of them led to death or long-lasting health impairment of exposed individuals including exposed police officers. Forty-one patients were taken to a hospital, in 50 cases first aid provided by an ambulance was sufficient. It was considered an advantage of this less dangerous weapon that the deployment of other weapons, e.g. of truncheons and fire arms (reduction by 35%), was decreased. But also the number of personal injuries and material damages was reduced by ca. 50% and 35%, respectively. Injuries of the officers decreased in 1997 and 1998 by ca. 75% and 25%, respectively, compared with the preceding year. The effectiveness of OC spray was predominantly rated as good, although in 24 of 136 analysed cases no sufficient effect was achieved. In addition, the spray was in most cases effective against dogs.

Conclusions

1. The deployment of capsaicin in pepper spray usually leads to a single acute exposure, although several spray jets may be directed to individuals. The expected toxic effects relate to skin, eyes and respiratory tract.
2. Skin reactions are transient. Allergic reactions have seldom been reported in humans and were not observed in experimental animals.
3. Effects on eyes (burning, redness, pain, tears, temporary blindness) are transient. Unclear is the causation of corneal abrasions, which appear to heal according to animal experiments.
4. The effects of possibly greatest uncertainty concern the respiratory tract. In studies with volunteers especially cough reflex and a transient bronchoconstriction was described, while other lung function parameters were unchanged. Asthmatic individuals showed no differences in bronchoconstriction, but were more sensitive to capsaicin-induced cough. Smokers were less sensitive in this regard. Fuller, whose group conducted many of the experiments with humans stated in a personal commentary quoted by Ruddick (1993) that more serious bronchoconstriction might possibly result in individuals with severe asthma. It is difficult to relate the doses used for the testing of volunteers to the concentrations occurring during exposure to pepper spray experienced in practice. The highest aerosol concentrations (0.75% capsaicin) employed in studies with humans exceeded the range given in the German regulations for irritant spray cans (0.2 ± 0.03 % capsaicinoids), however very small volumes (0.05-0.2 ml) were delivered. The available reports on the deployment of pepper

spray by the police do not allow conclusions how many of the sprayed individuals inhaled which quantities of the aerosol.

5. A further point in the evaluation of OC safety concerns death cases following law enforcement. In several cases pre-existing heart disease, together with stress, was considered the main cause. Whether individuals with such a basic disease are at special risk has not been elucidated. In several cases positional asphyxia and/or drug over-dose contributed to death.
6. The data concerning the mutagenicity and carcinogenicity of chilli and capsaicin are equivocal. These effects are less relevant for the use of pepper spray, as exposures are normally single and very short.
7. While evaluating the safety of pepper spray the comparative toxicology of the irritants CN and CS employed so far should be considered; these substances are, however, not the subject of this review.
8. The Austrian police emphasizes that pepper spray applicators should be trained to know and take into account practical problems, possible health hazards and be provided with first aid and the necessary medical aftercare. This condition can be met in law enforcement but the licensing as a private defence weapon is more problematical.
9. So far in most countries only the natural substance capsaicin contained in the oleoresin has been employed. According to Hobbs & Rice (1997) synthetically produced PAVA is being tested in Switzerland, but the results have not been available. The main advantage of PAVA is the greater purity of the substance, while the oleoresin contains a host of compounds. Therefore the concentrations of PAVA in spray cans can be analysed and checked better.

References

ACLU (American Civil Liberties Union): Pepper Spray: Magic Bullet under Scrutiny. A Report by the American Civil Liberties Union of Southern California, Fall 1993, 20-33, 1993.

Akagi A., Sano N., Uehara H., Minami T., Otsuka H. and Izumi K.: Non-carcinogenicity of capsaicinoids in B6C3F1 mice. Food Chem. Toxicol. **36:** 1065-1071, 1998.

Agrawal R.C. and Bhide S.V. Histopathological studies on toxicity of chilli in Balb/c mice. Indian J. Med. Res. **86:** 377-382, 1987.

Agrawal R.C., Wiessler M., Hecker E. and Bhide S.V. Tumor promoting effect of chilli extract in Balb/c mice. Int. J. Cancer **38:** 689-695, 1986.

Anonymous: Several letters and reports by the Austrian police 1998.

Azizan A. and Blevins R.D. Mutagenicity and antimutagenicity testing of six chemicals associated with the pungent properties of specific spices as revealed by the Ames Salmonella/Microsomal assay. Arch. Environ. Contam. Toxicol. **28:** 248-258, 1995.

Barros M.J., Zammotto S.L. and Rees P.J. Effect of changes in inspiratory flow rate on cough responses to inhaled capsaicin. Clin. Sci. **81:** 539-542, 1991.

Battensby J., Creasey N.H. and Grace T.J. A comparison and evaluation of the whole body pletysmograph and the guinea pig eye blink procedure for estimating the potency of sensory irritants. CBD Technical Paper 301 (1981), quoted after Hobbs & Rice, 1991.

Bell M.D., Rao V.J., Wetli C.V. and Rodriguez R.N. Positional asphyxiation in adults. A series of 30 cases from the Dade and Broward County Florida Medical Examiner Offices from 1982 to 1990. Am. J. Forensic Med. Pathol. **13:** 101-107, 1992.

Benedek G., Obal F., Jr., Jancso-Gabor A. and Obal F. Effects of elevated ambient temperatures on the sleep-waking activity of rats with impaired warm reception, Waking Sleeping **4:** 87, 1980.

Bergren D.R. Chronic tobacco smoke exposure increases airway sensitivity to capsaicin in awake guinea pigs. J. Appl. Physiol. **90:** 695-704, 2001.

Bergren D.R. Enhanced lung C-fiber responsiveness in sensitized adult guinea pigs exposed to chronic tobacco smoke. J. Appl. Physiol. **91:** 1645-1654, 2001.

Blanc P., Liu D.; Juares C. and Boushey H.A. Cough in hot pepper workers. Chest **99:** 27-32, 1991.

Buchanan R.L., Goldstein S. and Budrow I.D. Examination of chilli pepper and nutmeg oleoresins using the *Salmonella*/mammalian microsome mutagenicity assay. J. Food Sci. **47:** 330-331, 1981.

Bunting S. First death attributed to OC occurs in North Carolina. The ASLET Journal. 13-14, Sept/Oct 1993.

Busker R.W., van Helden H.P.M. Toxicological Evaluation of Pepper Spray as a Possible Weapon for the Dutch Police Forces. TNO Prins Maurits Laboratory **C87:** 1-31, 1996.

Carpenter S.E. and Lynn B. Vascular and sensory responses of human skin to mild injury after topic treatment with capsaicin. Brit. J. Pharmacol. **73:** 755-758, 1981.

Caterina M.J., Schumacher M.A., Tominaga M., Rosen T.A., Levine J.D. and Julius D. The capsaicin receptor: a heat-activated ion channel in the pain pathway. Nature **389:** 816-824, 1997.

Chevarne F.E. Technical/toxicological back-up data to synthetic capsaicin solution (PAVA). Analysis SA Laboratory, San Antonio, Spain, for IDC System AG, Switzerland (1995), quoted after Hobbs & Rice

Collier J.G. and Fuller R.W. Capsaicin inhalation in man and the effects of sodium cromoglycate. Br. J. Pharmacol. **81:** 113-117, 1994.

Comareche-Leydier M. The effect of ambient temperature on rectal temperature, food intake and short term body weight in the capsaicin desensitized rat. Pfluegers Arch. 389, 171, 1981.

Confarma: Prüfung der Inhalationstoxizität an Nonivamid / Nonylsäurevanillylamid. Emil-Frey-Str. 39, Ch-4142 Münchenstein/Basel, Schweiz. 1-24, 1996.

Cordell G.A. and Araujo O.E. Capsaicin: identification, nomenclature and pharmacotherapy. Ann. Pharmacother. **27:** 330-336, 1993.

Diaz Barriga Arceo S., Madrigal-Bujaidar E., Calderon Montallano E., Ramirez Herrera and Diaz Garcia B.D. Genotoxic effects produced by capsaicin in mouse during subchronic treatment. Muation Res. **345:** 105-109, 1995.

Doherty M.J. Capsaicin cough sensitivity is decreased in smokers. Respir. Med. **95:** 768, 2001.

Doherty M.J., Mister R., Pearson M.G. and Calverley P.M. Capsaicin responsiveness and cough in asthma and chronic obstructive pulmonary disease. Thorax **55:** 643-649, 2000a.

Doherty M.J., Mister R., Pearson M.G. and Calverley P.M. Capsaicin induced cough in cryptogenic fibrosing alveolitis. Thorax **55:** 1028-1032, 2000b.

Donnerer J., Amann R., Schuligoi R. and Lembeck F. Absorption and metabolism of capsaicinoids following intragastric administration in rats. Naunyn-Schmiedeberg's Arch. Pharmacol. **342:** 357-361, 1990.

Fang J.Y., Wu P.C., Huang Y.B. and Tsai Y.H. In vivo percutaneous absorption of capsaicin, nonivamide and sodium nonivamide acetate from ointment bases: skin erythema test and non-invasive surface recovery technique in humans. Int. J. Pharmac. **131:** 143-151, 1996.

Fischer A.B. Tabasco als Genussstoff und Waffe. In: M. Götz (Ed.): Der TABASCO-Effekt—Wirkung der Form, Formen der Wirkung . Halle Schwabe & Co Verlag, Basel. 195-199, 1999.

Fitzgerald M. Capsaicin and sensory neurons – a review. Elsevier Biomed. Press **15:** 109-130, 1983.

Fuller R.W. Pharmacology of inhaled capsaicin in humans. Respirat. Med. **85** (Suppl.) 31-34, 1991.

Fuller R.W., Dixon C. and Barnes M.S. Bronchoconstrictor response to inhaled capsaicin in humans. J. Appl. Physiol. **58:** 1080-1084, 1985.

Gamse R., Holzer P. and Lembeck F. Indirect evidence for presynaptic location of opiate receptors on chemosensitive primary sensory neurons. Naunyn-Schmiedebergs Arch. Pharmacol. **308:** 281, 1979.

Gamse R., Lehmann S.E., Holzer P. and Lembeck F. Differential effects of capsaicin on the content of somatostatin, substance P, and neurotensin in the nervous system of the rat. Naunyn-Schmiedebergs Arch. Pharmacol. **317:** 140-148, 1981.

Gannett P.M., Nagel D.L., Redly P.J., Lawson J. Sharp J. and Toth B. The capsaicinoids; their separation synthesis and mutagenicity. J. Org. Chem. **53:** 1064-1071, 1988.

Glinsukon T., Stitmunnaithum V., Toskulkao C., Buranawuti T. and Tangkrisanavinont V. Acute toxicity of capsaicin in several species. Toxicon **18:** 215-220, 1980.

Govindarajan V.S. and Sathyanarayana M.W. Capsaicin: production, technology, chemistry and quality. V. Impact on physiology, pharmacology, nutrition and metabolism—structure, pungency, pain and desensitization sequences. CRC Crit. Rev. Food Sci. Nutr. **29:** 435-474, 1991.

Granfield J., Onne J. and Petty C.S. Pepper spray and in-custody deaths. In: Pepper Spray Evaluation Project. Final Report prepared for National Institute of Justice, US Department of Justice. 1-6, 1995.

Helme R.D., Eglezos A. and Hosking C.S. Substance P induces chemotaxis of neutrophils in normal and capsaicin-treated rats. Immunol. Cell. Biol. **65:** 267-269, 1987.

Herbert M.K. and Holzer P. Neurogenic inflammation. I. Basic mechanisms, physiology and pharmacology. Anaesthesiol. Intensivmed. Notfallmed. Schmerzther. **37:** 314-325, 2002a.

Herbert M.K. and Holzer P. Neurogenic inflammation. II. Pathophysiology and clinical implications. Anaesthesiol. Intensivmed. Notfallmed. Schmerzther. **37:** 386-394, 2002b.

Hobbs M. and Rice P: A review of the toxicology of pelargonic acid vanillylamids (U.). Defense Evaluation and Research Agency, CBD Porton Down, Salisbury, Wiltshire SP4 0JQ: 1-13, 1997.

Hoch-Ligeti G. Production of liver tumours by dietary means: effect of feeding chillies to rats. Acta Unio Intern. contra Cancrum **7:** 606-611, 1951.

Holzer P. Local effector functions of capsaicin-sensitive sensory nerve endings: involvement of tachykinins, calcitonin gene-related peptide and other neuropeptides, Neuroscience **24:** 739-768, 1988.

Holzer P. Capsaicin: cellular targets, mechanisms of action and selectivity for thin sensory nerves. Physiol. Rev. **43:** 143-201, 1991.

Jang J.J., Devor D.E., Logsdon D.L. and Ward J.M. A 4-week feeding study of ground red chili (Capsicum annuum) in male B6C3F1 mice. Food Chem. Toxicol. **30:** 783-787, 1992.

Jang J.J., Kim S.H. and Yun T.K. Inhibitory effect of capsaicin on mouse lung tumor development. In Vivo **3:** 49-53, 1989.

Jang J.J., Cho K.J., Lee Y.S. and Bae J.H. Different modifying responses of capsaicin in a wide-spectrum initiation model of F344 rat. J. Korean Med. Sci. **6:** 31-36, 1991.

Jancso G. and Kiraly E. Distribution of chemosensitive primary sensory afferents in the rat CNS. J. Comp. Neurol. **190:** 781-792, 1980.

Jancso G., Kiraly E. and Jancso-Gabor A. Pharmacologically induced selective degeneration of primary sensory neurons in the adult rat. Neurosci. Lett. **59:** 209-214, 1985a.

Jancso G., Kiraly E., Joo F., Such G. and Nagy A. Selective degeneration by capsaicin of a subpopulation of chemosensitive primary sensory neurons in the adult rat. Neurosci. Lett. **59:** 209-214, 1985b.

Janusz J.M., Buckwater B.L., Young P.A., LaHann T.R. and Farmer R.W. Vanilloids – novel non-narcotic analgesic agents. J. Med. Chem. **36:** 2595-2604, 1993.

Kang J.Y. Surgery for gastric cancer in Singapore with particular reference to racial differences in incidence. Austr. New Zeal. J. Med. **18:** 661-664, 1988.

Kang Y.Y., Alexander B., Barker F., Man W.K. and Williamson R.C.N. The effect of chilli ingestion on gastrintestinal mucosal proliferation and azomethane incuced cancer in the rat. J. Gastroenterol. Hepatol. **7:** 194-198, 1992.

Kaufman M.P., Iwamoto G.A., Longhurst J.C. and Mitchell J.H. Effects of capsaicin and bradykinin on afferent fibers with endings in skeletal muscle. Circ. Res. **50:** 133-139, 1982.

Keitel W., Frerick H., Kuhn, U., Schmidt U., Kuhlmann M. and Bredehorst A. Capsicum pain plaster in chronic non-specific low back pain. Arzneimittelforschung **51:** 896-903, 2001.

Kirby M.L., Gale T.F. and Mattio T.G. Effects of prenatal capsaicin treatment on fetal spontaneous activity, opiate receptor binding, and acid phosphatase in the spinal cord. Ex. Neurol. **76:** 298-308, 1982.

Kwong K., Wu Z.X., Kashon M.L., Krajnak K.M., Wise P.M. and Lee L.Y. Chronic smoking enhances tachykinin synthesis and airway responsiveness in guinea pigs. Am. J. Respir. Cell Mol. Biol. **25:** 299-305, 2001.

LaMotte R.H., Simone D.A., Baumann T.K., Shain C.N. and Alreja M. Hypothesis for novel classes of chemoreceptors mediating chemogenic pain and itch. In: Dubner, R., Gebhart, G.F., Bond, M.R. (Eds.) Pain Research and Clinical Management, Elsevier, Amsterdam. 529-535, 1988.

Lawson T. and Gannett P. The mutagenicity of capsaicin and dihydrocapsaicin in V79 cells. Cancer Lett. **48:** 109-113, 1989.

Lawson S.N., Crepps B.A. and Perl E.R. Relationship of substance P to afferent characteristics of dorsal root ganglion neurons in guinea pig. Physiology **506:** 177-191, 1997.

Lee H.P., Day N.E. and Shammugaratnam K. Trends in cancer incidence in Singapore 1968-82. IARC Publication 91, IARC Lyon, 1988.

Legrin G.Y. Capsaicin and its analogues: properties, preparation and applications. Pharmaceut. J. **30:** 60-68, 1996.

Lopez-Carillo L., Hernandez M. and Dubrow R. Chilli pepper consumption and gastric cancer in Mexico: a case-control study. Am. J. Epidemiol. **139:** 263-271, 1994.

Lundberg J.M., Franco-Cereceda A., Hua X.-Y., Hokfelt T. and Fischer J. Co-existence of substance P and calcitonin gene-related peptide immunoreactivities in sensory nerves in relation to cardiovascular and bronchoconstrictor effects of capsaicin. Eur. J. Parmacol. **108:** 315-319, 1985.

Lundberg J.M. and Saria A. Capsaicin induced desensitization of the airway mucosa to cigarette smoke, mechanical and chemical irritants. Nature **302:** 251-253, 1983.

Marques S., Oliveira N.G., Chaveca T. and Rueff J. Micronuclei and sister chromatid exchanges induced by capsaicin in human lymphocytes. Mutat. Res. **517:** 39-46, 2002.

Martin H.A., Basbaum A.I., Kwiat G.C., Goetzl E.J. and Levine J.D. Leukotriene and prostaglandin sensitization of cutaneous high threshold C- and A-delta mechanonociceptors in the hairy skin of rat hindlimbs. Neurosci. **22:** 651-659, 1987.

Mathew A., Gangadharan P., Varghese C. and Nair M.K. Diet and stomach cancer: a case-control study in South India. Eur. J. Cancer Prev. **9:** 89-97, 2000.

Maxwell D.L., Fuller C.M. and Dixon R.W. Ventilatory effects of inhaled capsaicin in man. Eur. J. Clin. Pharmacol. **31:** 715-717, 1987.

Millqvist E. and Bende-M. Capsaicin cough sensitivity is decreased in smokers. Respir. Med. **95:** 19-21, 2001.

Monsereenusorn Y. Subchronic toxicity studies of capsaicin and capsicum in rats. Res. Commun. Chem. Pathol. Pharmacol. **41:** 95-110, 1983.

Monsereenusorn Y., Kongsamut S. and Pezalla P.D. Capsaicin - a literature survey. Crit. Rev. Toxicol. **10:** 321-329, 1982.

Muralidhara and Narasimhamurthy K. Non-mutagenicity of capsaicin in albino mice. Food Chem. Toxicol. **26:** 955-958, 1988.

Nagabhushahn M. and Bhide S.V. Mutagenicity of chilli extract and capsaicin in short-term tests. Environ. Mut. **7:** 881-888, 1985.

O'Halloran R.L. and Lewman L.V. Restraint asphyxiation in excited delirium. Am. J. Forensic Med. Pathol. **14:** 289-295, 1993.

Ootha K. Significance of ulcer. In: R. Kinikasa, T. Nagayo, T. Tanaka (Eds.) Epidemiological, Experimental and Clinical Studies on Gastric cancer. Gann Monographs **3:** 141-150, 1968.

Pandey M. and Shukla V.K. Diet and gallbladder cancer: a case-control study. Eur. J. Cancer Prev. **11:** 365-368, 2002.

Park K.K., Chun K.S., Yook J.I. and Surh Y.J. Lack of tumor promoting activity of capsaicin, a principal pungent ingredient of red pepper, in mouse skin carcinogenesis. Anticancer Res. **18:** 4201-4205, 1998.

Perfumi M. and Sparapassi L. Rat offspring treated prenatally with capsaicin do not show some of the irreversible effects induced by neonatal treatment with neurotoxin. Pharmacol. Toxicol. **84:** 66-71, 1999.

Pellicer F., Piazo O., Gomez-Tagle B. and Roldan de La O.I. Capsaicin or feeding with red peppers during gestation changes the thermonociceptive response of rat offspring. Physiol. Behav. **60:** 435-438, 1996.

Pindborg J.J., Poulsen H.E. and Zuchariah J. Oral epithelial changes in 30 Indians with oral cancer and submucous fibrosis. Cancer **20:** 1141-1146, 1967.

Reilly D.M. and Green M.R. Eicosanoid and cytokine levels in acute skin irritation in response to tape stripping and capsaicin. Acta Derm. Venereol. **79:** 187-190, 1999.

Rockwell P. and Raw I. A mutagenic screening of various herbs, spices and food additives. In: Nutrition and Cancer, Raven Press, New York. 1-9, 1979.

Ruddick J.A. A toxicological review of capsaicinoid (Oleoresin of Capsicums). Canadian Police Research Center TR-02-93: 1-13, 1993.

Rumsfield J.A. and West D.P. Topical capsaicin in dermatologic and peripheral disorders. DIPC Ann. Pharmacother. **25:** 381-387, 1991.

Saito A. and Yamamoto M. Acute oral toxicity of capsaicin in mice and rats. J. Toxicol. Sci. **21:** 195-200, 1996.

Saria A., Gamse R., Peterman J., Fischer J.A., Theodorsson-Norheim E. and Lundberg J.M. Simultaneous release of several tachykinins and calcitonin gene-related peptide from rat spinal cord slices. Neurosci. Lett. **63:** 310-314, 1986.

Szallasi A. and Blumberg P.M. Vanilloid receptors: new insights enhance potential as a therapeutic target. Pain **68:** 195-208, 1996.

Szallasi A., Farkas-Szallasi T., Tucker J.B., Lundberg J.M., Hokfelt T., Krause J.E. Effects of systemic resiniferatoxin treatment on substance P mRNA in rat dorsal root ganglia and substance P receptor mRNA in the spinal dorsal horn. Brain Res. **815:** 177-184, 1999.

Simone D.A., Baumann T.K. and LaMotte R.H. Dose-dependent pain and mechanical hyperalgesia in humans after intradermal injection of capsaicin. Pain **38:** 99-107, 1989.

Sirgot S.M. and Khanolkar V.R. Submucous fibrosis of the palate and pillars of the fauces. Ind. J. Med. Sci. **16:** 189-97, 1962.

Steffee C.H., Lantz P.E., Flannagan L.M., Thompson R.L. and Jason D.R. Oleoresin capsicum (pepper) spray and "in-custody-deaths". Am. J. Forensic Med. Pathol. **16:** 185-192, 1995.

Such, G., Jancso, G.: Axonal effects of capsaicin: an electrophysiological study. Acta Physiol. Hung. 67 (1986) 53-63

Surh Y.-J. and Lee S.S. Capsaicin, a double-edged sword: toxicity, metabolism and chemoprotective potential. Life Sci. **56:** 1845-1855, 1995.

Surh Y.J., Lee R.C., Park K.K., Mayne S.T., Liem A. and Miller J.A. Chemoprotective effects of capsaicin and diallyl sulfide against mutagenesis or tumorigenesis by vinyl carbamate and N-nitrosodimethylamine. Carcinogenesis **16:** 2467-2471, 1995b.

Szolcsanyi J. Disturbances of thermo regulation induced by capsaicin. J. Thermal Biol. **8:** 207-212, 1983.

Szolcsanyi J. and Jancso-Gabor A. Sensory effects of capsaicin congeners I. Relationship between chemical structure and pain-producing potency of pungent agents. Arzneimittelforschung **25:** 1877-1881, 1975.

Szolcsanyi J. and Jancso-Gabor A. Sensory effects of capsaicin congeners. Part II: Importance of chemical structure and pungency in desensitizing activity of capsaicin-type compounds. Arzneimittelforschung **26:** 33-37, 1976.

Tanaka T., Kohno H., Sakata K., Yamada Y., Hirose Y., Sugie S. and Mori H. Modifying effects of dietary capsaicin and rotenone on 4-nitroquinoline 1-oxide-induced rat tongue carcinogenesis. Carcinogenesis. **23:** 1361-7, 2002.

Technische Richtlinie: Reizstoffspruehgeräte mit Capsaicin ("Pfeffer") des Unterausschusses Fuehrungs- und Einsatzmittel. Unterausschuss Fuehrungs- und

Einsatzmittel (UA FEM) des Arbeitskreises II „Innere Sicherheit" der Arbeitsgemeinschaft der Innenministerien der Laender 1-9, 1999.

Teel R.W. and Huynh H.T. Lack of the inhibitory effect of intragastrically administered capsaicin on NNK-induced lung tumor formation in the A/J mouse. In Vivo **13,** 231-234, 1999.

Toth B. and Gannett P. Carcinogenicity of lifelong administration of capsaicin of hot pepper in mice. In Vivo **6:** 59-64, 1992.

Toth B., Rogan E. and Walker B. Tumorigenicity and mutagenicity studies with capsaicin of hot peppers. Anticancer Res. **4:** 117-119, 1984.

United States Testing Company: Report of Test. Dermal Irritation Test on CAP-STUN® Formulation Conducted for Zarc International Inc., P.O. Box 5800, Bethesda, MD 20824, 1-8, 1992a.

United States Testing Company: Report of Test. Guinea Pig Sensitization Test (Allergenicity) on CAP-STUN® Formulation Conducted for Zarc International Inc., P.O. Box 5800, Bethesda, MD 20824, 1-7, 1992b.

United States Testing Company: Report of Test. Eye Irritation Test on CAP-STUN® Formulation Conducted for Zarc International Inc., P.O. Box 5800, Bethesda, MD 20824, 1-10, 1992c.

United States Testing Company: Report of Test. Acute Inhalation Toxicity Test on CAP-STUN® Formulation Conducted for Zarc International Inc., P.O. Box 5800, Bethesda, MD 20824, 1-9, 1992d.

Valtschanoff J.G., Rustioni A. Guo A., and Hwang S.J. Vanilloid receptor VR1 is both presynaptic and postsynaptic in the superficial laminae of the rat dorsal horn. J. Comp. Neurol. **436:** 225-235, 2001.

Veronesi B., Carter J.D., Devlin R.B., Simon S.A. and Oortgiesen M. Neuropeptides and capsaicin stimulate the release of inflammatory cytokines in a human bronchial epithelial cell line. Neuropeptides **33:** 447-56, 1999.

Vesaluoma M., Muller L., Gallar J., Lambiase A., Moilanen J., Hack T., Belmonte C. and Tervo T. Effects of oleoresin capsicum pepper spray on human corneal morphology and sensitivity. Invest. Ophthalmol. Vis. Sci. **41:** 2138-2147, 2000.

Villasenor I.M. and de Ocampo E.J. Clastogenicity of red pepper (Capsicum frutescens L.) extracts. Mutat. Res. **312:** 151-155, 1994.

Watson W.A., Stremel K.R. and Westdorp E.J. Oleoresin capsicum (CAP-STUN) toxicity from aerosol exposure. Ann. Pharmacother. **30:** 733-735, 1996.

Wetli C.V. Fatal cocaine intoxication. A review. Am. J. Forensic Med. Pathol. **8:** 1-2, 1987.

Winsgrad H.L. Acute croup in an older child—an unusual toxic origin. Clin. Paediatr. **16:** 884-887, 1977.

Zehetner E. Pfefferspray—Einsatzerfahrungen in Oesterreich 1994-1998. Bundespolizei Direktion Salzburg, Amtsaerztlicher Dienst 1998.

Pharmacological Perspectives of Toxic Chemicals and Their Antidotes
Editors: S J S Flora, J A Romano, S I Baskin and K Sekhar

Modalities of Nociception and the Underlying Molecular Mechanisms

Louis S. Premkumar and Jeremy Van Buren
Department of Pharmacology, Southern Illinois University
School of Medicine Springfield, IL 62702 USA

INTRODUCTION

Pain is a perception that alerts us to injury and triggers various compensatory behavioral responses. The sensation of pain is a complex experience that is the culmination of many events processed both centrally and peripherally. Several thalamic nuclei and cortical areas are involved in processing pain sensation (Basbaum & Jessell, 2000). Recently significant advances have been made in understanding the molecular mechanisms underlying pain perception. This process, referred to as nociception, is the means which primary sensory neurons detect painful stimuli. Various physiological and pathological states alter our perception of pain. Under certain circumstances, pain no longer serves as a warning system, instead becoming chronic and debilitating. In other instances, one's ability to sense noxious stimuli can become dulled or completely silenced, thus nullifying an inherent need for an organism to interact with its environment and respond to adverse stimuli. Adaptations in central processing play a role in these changes, but a number of mechanisms responsible occur at the level of primary sensory afferents (Raja et al., 1999).

PRIMARY AFFERENT NOCICEPTORS

Recently, great advances have been made in understanding how primary sensory neurons utilize their molecular machinery to detect pain-producing stimuli. A century ago, Sherrington (1906) brought forth the idea of a nociceptor, a primary sensory neuron that responds to agents responsible for tissue damage. Recently, electrophysiological studies have found sensory neurons excited by noxious heat, intense pressure, or chemical irritants but not innocuous stimuli such as light touch and warmth (Burgess & Perl, 1967). Therefore, a subset of sensory neurons exist whose sole function is to respond to noxious stimuli. These primary sensory fibers, which can be characterized based upon anatomical and functional criteria, arise from cell bodies located in trigeminal (TG) and dorsal root ganglia (DRG).

The fibers originating from TG and DRG innervate head and body respectively. The nerve fibres fall into three major categories; Aδ fibers are myelinated and rapidly conducting, which detect innocuous stimuli. C-fibers are unmyelinated and slow conducting and Aδ fibers are thinly myelinated. C and Aδ fibres carry pain sensation (Raja et al., 1999). Finally, it is hypothesized that these fibers mediate the sensation of 'first' (C) and 'second' (Aδ) pain, namely the rapid, acute, sharp pain and the delayed, more diffuse, dull pain that follows (Raja et al., 1999).

Injury heightens our pain experience by increasing the sensitivity of nociceptors to both mechanical and thermal stimuli (Woolf & Salter, 2000). This pain state, termed hyperalgesia, partially results from the release of chemical mediators of both sensory and non-sensory (i.e. fibroblasts, mast cells, neutrophils, and platelets) origin. These components of inflammation, termed the 'inflammatory soup', can alter sensory neuron excitability by either directly activating ion channels (for example, protons, ATP, serotonin) or through second messengers produced by activation of metabotropic receptors (i.e. bradykinin, NGF and prostaglandins) (Julius & Basbaum, 2001).

The phenomenon known as allodynia, or pain being produced by innocuous stimuli, can occur from either increased responsiveness of spinal cord relay neurons (central sensitization) or lowering of nociceptor activation threshold (peripheral sensitization) (Basbaum & Jessell, 2000). Central sensitization, which is similar to Long Term Potentiation (LTP) in the hippocampus, may result from either pathological conditions or direct spinal trauma leading to increased relay neuron responsiveness to primary inputs. In contrast, peripheral sensitization occurs with exposure to conditions of tissue damage and inflammation, where the inflammatory soup can directly increase primary nociceptor activity.

Glutamate is the predominant excitatory amino acid released from the nerve terminals of nociceptors. However, some of these neurons are also peptidergic, releasing both Substance P and calcitonin gene-related peptide (CGRP) (Snider & McMahon, 1998) as well. Because sensory neurons can release both neurotransmitters and peptides, they are able to modulate the content of inflammatory soup by augmenting or inhibiting the release of factors from neighboring non-neuronal cells and vascular tissue, a process known as neurogenic inflammation.

RESPONSE TO HEAT

When sensory ganglia are dissociated and placed into cell culture, the neurons retain many functional characteristics (Kress and Guenther, 1996). About 45% of small- to medium-diameter neurons exhibit heat-evoked currents with a temperature threshold of ~43°C. This group of fibers presumably contains both C and type II Aδ nociceptors. Another 5-10% of cells respond only to noxious thermal stimuli (no capsaicin response) of a high threshold (~52°C) and are thought to be type I Aδ fibers. Together, these two classes of neurons that respond to heat with either 'moderate' or 'high' temperature thresholds can detect a wide range of ambient temperatures.

What is the molecular mechanism through which specific nociceptors discriminate between different heating intensities? In the case of moderate thermal nociception, a transducer was revealed via cloning and characterization of the vanilloid receptor (VR1) in C and type II Aδ afferents. Using capsaicin, a naturally occuring vanilloid compound (found in hot peppers) known to mimic heat sensation when applied to nerve endings (Szallasi & Blumberg, 1999). Julius and colleagues (Caterina and Julius, 2001) were able to isolate a capsaicin-activated channel, VR1 (TRPV1) in a subset of neurons within DRGs that also responded to heat. VR1 is a plasma membrane protein that is a non-specific cation channel with a heat activation threshold of ~43°C. These characteristics are shared with the currents elicited in native sensory neurons. Moreover, single channel recordings of VR1 and native VRs within patches pulled from sensory neuron cell bodies show that VR1 can be an intrinsically heat-sensitive ion channel.

In order to verify that VR1 is the molecular heat detector found in sensory neurons, studies were performed on mice lacking functional VR1 channels (Caterina and Julius 2001). Cultured DRG neurons from these VR1-null mice showed a marked deficiency in moderate heat-induced responses, whereas responses to high-threshold heat stimuli remained unchanged. Furthermore, $VR1^{-/-}$ mice have fewer heat-sensitive C-fibers and total heat-evoked C-fiber output decreased by >85% compared to wild type. Therefore, VR1 accounts for the majority of moderate-heat response and contributes to the overall thermal coding in the peripheral sensory nervous system (Caterina and Julius, 2001).

With respect to sensing higher temperatures, a different transducer has been identified that has homology to VR1, yet shows unique characteristics in terms of its heat threshold and capsaicin insensitivity (Caterina and Julius, 2001). This receptor, called the vanilloid receptor-like channel (VRL-1, also called TRPV2), shares 50% sequence homology to VR1 and is expressed mostly in medium- to large- diameter sensory neurons in DRGs. It's activation by noxious thermal stimuli above that of VR1 (~52°C) makes it an ideal candidate for mediation of high-threshold heat responses.

VR1 and VRL-1 belong to the larger family of transient receptor potential (TRP) channels (Caterina and Julius, 2001). The core transmembrane structure of the channel resembles that of voltage-gated potassium channels and cyclic nucleotide-gated channels. Recently, two other members of the TRP family have been cloned that respond to temperature. First, a cold and menthol responsive channel (CMR-1, also called TRPM8) has been localized to dorsal root and trigeminal ganglion and is thought to mediate cold sensitivity (McKemy et al., 2002). Secondly, a TRP channel that responds to physiological ranges of temperature (22°C to 40°C) has been identified and called TRPV3 (Xu et al., 2002; Smith et al., 2002). VR1 is distributed in areas that do not experience noxious temperature ranges warrant the identification of a potent endogenous activator. In conclusion, an emerging concept is that TRP channel family members act as molecular thermometers and work together in the transduction of a wide range of temperature stimuli.

Possibly the most striking characteristic of the vanilloid receptor is its ability to be opened by a number of different agonists. In addition to capsaicin and heat, protons, anandamide, and bradykinin have all been shown to activate the receptor. One compound, arachidonoyl dopamine (NADA), has recently been suggested to be the endogenous activator for VR1 (Huang et al., 2002).

DIFFERENT PAIN STATES

Nociceptors have the unique ability to detect a wide array of stimulus modalities, including both physical and chemical inputs. This allows the cell to integrate information and respond to complex changes in the environment. Moreover, different pain states emerge based upon the situation in which a pain stimulus is perceived.

TISSUE INJURY AND INFLAMMATORY PAIN

Injury heightens our pain experience by increasing the sensitivity of nociceptors to both mechanical and thermal stimuli. This phenomenon, termed hyperalgesia, partially results from the release of chemical mediators of both sensory and non-sensory (i.e. fibroblasts, mast cells, neutrophils, and platelets) origin. These components of inflammation can alter sensory neuron excitability by either directly activating ion channels (for example, protons, ATP, serotonin, and prostaglandins) or through second messenger signaling cascades activated by metabotropic receptor binding (i.e. bradykinin and NGF).

A common consequence of inflammation following tissue injury is the burning sensation at the site and the increased heat sensitivity that occurs long after the initial injury. A striking discovery in the VR1 knockout mouse is its failure to develop this increased heat sensitivity following tissue damage. The fact that VR1 is stimulated by some of the aforementioned mediators further supports the notion that the receptor may mediate inflammatory thermal pain (Caterina and Julius, 2001).

CHRONIC OR NEUROPATHIC PAIN

VR1 is not involved in chronic or neuropathic pain suggesting the involvement of receptors other than VR1. In a recent study (Tang et al., 1999), over-expression of NR2B subunit of N-methyl-D-aspartate receptor (NMDAR), a subtype of glutamate receptors in the forebrain region was shown to increase pain sensitivity. Another area that could undergo plastic changes is at the level of spinal cord. Glutamate is the major excitatory neurotransmitter at this level, which after release excites the second order neurons that carry the information to the brain. Neuromodulators such as substance P and CGRP released along with glutamate could modulate synaptic transmission. A phenomenon called 'wind up' seems to take place at the spinal cord level, where the excitability of the neurons is heightened in response to increased or decreased peripheral nerve stimulation. Furthermore, if the sensitivity of peripheral nerve terminal is reduced, an increase in receptor expression in the cell bodies at the sensory ganglia may enhance neuronal excitablility. Opioid receptors also play a role at the spinal cord level. The mechanisms involved in inducing chronic pain conditions also seem to induce morphine tolerance or heightened state of pain after chronic use. This process involves the vanilliod receptor mediated CGRP release and the down stream events induced by activation of CGRP receptors.

SECOND MESSENGERS MODULATING NOCICEPTORS

Figure 1 shows the second-messenger pathways that converge to activate or sensitize the VRs. Metabotropic receptor activation modulates second-messenger cascades regulated by adenylate cyclase (AC), phospholipase-C (PLC), phospholipase-D (PLD), and phospholipase-A2 (PLA_2) (Vassort, 2001; Burnstock, 2001). Activation of receptors inhibits AC via G_i or stimulate it via G_s. Activation of PLC induces hydrolysis of phosphatidyl inositol bisphosphate (PIP_2) to produce diacylglycerol (DAG) and inositol trisphosphate (IP_3). The DAG so produced activates PKC while the IP_3 triggers Ca^{2+} release from intracellular stores. Interestingly, G-protein receptors coupled to PLC invariably stimulate PLD activity as well. PLD catalyses the hydrolysis of N-arachidonyl phosphatidylethanolamine to produce anandamide, which in turn can directly activate VRs.

Activation of phospholipase A_2 (PLA_2) produces arachidonic acid (AA), which is subsequently metabolized via lipoxygenase and cyclooxygenase pathways. The lipoxygenase metabolites (LM) hydroperoxyeicosatetraenoic, hydroxyeicosatetraenoic acids and leukotriene B_4 directly, but weakly, activate VRs (Hwang et al., 2000), whereas cyclooxygenase metabolites (prostaglandin E_2) potentiate the response of capsaicin by increasing cAMP levels (Lopshire and Nicol, 1998). In addition, AA can also directly activate some PKC isoforms. VRs have a number of consensus phosphorylation sites that can be phosphorylated by protein kinase A, C, and G, Ca^{2+} calmodulin kinase, and tyrosine kinase.

Activation of PKA strongly potentiates the VR response in neurons, which can be mimicked by prostaglandin E_2 (Lopshire and Nicol, 1998). PKC-mediated phosphorylation directly activates VRs and strongly potentiates the capsaicin- (Premkumar and Ahern, 2000) or heat- (Cesare et al., 1999) induced responses. Particularly, PKC causes a near 10-fold increase in the membrane current induced by weak agonists, such as anandamide (Premkumar and Ahern, 2000). Moreover, recently it has been found that PKC phosporylates VR1 at two specific sites on the intracellular domains of the protein (Numazaki et al., 2002). Furthermore, the increases in intracellular calcium levels by IP_3 or by other means (membrane depolarization or influx via a Ca^{2+} ionophore) can also potentiate heat-induced responses (likely mediated by VRs) (Kress and Guenther, 1999). Interestingly, ligands that are coupled to both PLC and PLA_2, such as bradykinin, are potent agonists; second messengers from two different pathways synergistically induce maximal VR activity (Premkumar 2001). It is also possible that two different agonists acting on separate signaling pathways can achieve a similar response. It is worth noting that in a recent study, intracellular ATP binding to a Walker-type nucleotide-binding domain sensitized the VR response to capsaicin, suggesting that ATP could function as a co-agonist in a fashion similar to glycine acting on the N-methyl-D-aspartate receptor (Kwak et al., 2000).

In support of the involvement of second-messenger pathways in pain signaling, targeted disruption of the PKC- or cAMP-dependent protein kinase gene has revealed their effects on selective modalities of pain. Deletion of PKC-ε gene reduced thermal and acid-induced hyperalgesia (Khasar et al., 1999) whereas deletion of PKC-δ reduced neuropathic pain but preserved acute pain (Malmberg et al., 1997). Mice lacking type I regulatory subunit of cAMP-dependent protein kinase exhibited diminished inflammatory and nociceptive pain but preserved neuropathic pain (Malmberg et al., 1997).

Finally, another mechanism has been proposed whereby PIP_2 physically interacts with VR (Chuang et al., 2001). At resting state, PIP_2 binds VR and suppresses its activity. However, stimulation of PLC will increase PIP_2 hydrolysis and subsequently enhance VR activity by the removal of PIP_2 inhibition. This final type of modulation, which partly mediates the effects of NGF and bradykinin, renders VR channels active even at room temperatures. Thus, different factors of the extracellular milieu can affect VR functioning through a multiplicity of overlapping mechanisms.

Although VR is a molecular target for noxious stimuli, expression of VR in areas that are not likely to experience noxious heat, especially the central nervous system, raises the possibility of VR functions other than nociception. Recently VR1 has been shown to be expressed in neurons throughout the entire neuroaxis (including the cerebral cortex, limbic system, hypothalamus, substantia nigra, and cerebellum) (Mezey et al., 2000). Recently N-arachidonyl dopamine found in the CNS has been demonstrated to be a potent agonist of VR. In the cardiovascular system (other than as nociceptors in the sensory C-fiber endings of the heart), activation of VRs release a variety of vasodilatory signaling molecules such as, nitric oxide, calcitonin gene-related peptide, bradykinin, and histamine (Szallasi and Blumberg, 1999; Zygmunt et al., 1999). Putative endogenous VR ligands identified so far (anandamide and metabolites of AA) are weak agonists and, therefore, are unlikely to induce a sizable response at the concentrations normally found in tissues (Szolcsanyi, 2000). One way to overcome this would be to sensitize the receptor by phosphorylation. Previous studies have shown that the VR function could be modulated by phosphorylation or dephosphorylation (Docherty et al., 1998; Lopshire and Nicol, 1998; Premkumar and Ahern, 2000). Sensitizaton of vanilloid receptor could be achieved by stimulating metabotropic ATP receptors that leads to phosphorylation of VRs. The temperature sensitivity of VRs is reduced when the receptor is phosphorylated; therefore, when the conditions are appropriate, normal body temperature could act as a primary stimulus. Modulation of VRs by these mechanisms is especially important in the central nervous system, where the role of these receptors is yet to be identified. VRs also have an equal or greater significance in pathological conditions. As these receptors are highly calcium permeable, sensitizing the receptor via the activation of metabotropic receptors poses a danger of calcium overload leading to neuronal cell death, which is a daunting prospect in many neurodegenerative diseases.

In summary, nociception is a complex phenomenon with different modalities. Vanilloid receptors play a critical role and act as an integrator of different stimuli. Increased neuronal excitability caused by changes in neurotransmitter receptors (ionotropic or metabotropic) is likely to underlie long lasting pain conditions.

References

Basbaum A.I., Jessell T.M.. Principles of Neural Science. McGraw-Hill: New York, 2000.

Burnstock G. Trends Pharmacol. Sci. **22:** 182-188, 2001.

Burgess P.R. and Perl E.R.. Journal of Physiology, **190:** 541-562, 1967.

Caterina M.J. and Julius D. Annu Rev Neurosci, **24:** 487-517, 2001.

Cesare P., Moriondo A., Vellani V. and McNaughton P.A. Proc. Natl. Acad. Sci. USA **96:** 7658-7663, 1999.

Chuang H.H., Prescott E.D., Kong H., Shields S., Jordt S.E., Basbaum A.I., Chao M.V. and Julius D. Nature, **411:** 957-62, 2001.

Cockayne D.A., Hamilton S.G., Zhu Q.M., Dunn P.M., Zhong Y., Novakovic S., Malmberg A.B., Cain G., Berson A., Kassotakis L., Hedley L., Lachnit W.G., Burnstock G., McMahon S.B. and Ford A.P. Nature **407:** 1011-1015, 2000.

Docherty R.J., Yeats J.C., Bevan S. and Boddeke H.W. Pflugers Arch. 431: 828-837, 1996.

Fields D.R. and Stevens B. Trends Pharmacol. Sci. **23:** 625-633, 2000.

Hwang S.W., Cho H., Kwak J., Lee S.Y., Kang C.J., Jung J., Cho S., Min K.H., Suh Y.G., Kim D. and Oh U. Proc. Natl. Acad. Sci. USA 97, 6155-6160, 2000.

Julius D. and Basbaum A.I. Nature, **413:** 203-10, 2001.

Khasar S.G., Lin Y.H., Martin A., Dadgar J., McMahon T., Wang D., Hundle B., Aley K.O., Isenberg W., McCarter G., Green P.G., Hodge C.W., Levine J.D. and Messing R.O. Neuron. **24:** 253-260, 1999.

Kress M. and Guenther S. J. Neurophysiol. **81:** 2612-2619, 1999.

Kwak J., Wang M.H., Hwang S.W., Kim T.Y., Lee S.Y. and Oh U. J. Neurosci. **20:** 8298-8304, 2000.

Lopshire J.C. and Nicol G.D. J. Neurosci. **18:** 6081-6092, 1998.

Malmberg A.B., Chen C., Tonegawa S. and Basbaum A.I. Science **278:** 279-283, 1997.

Malmberg A.B., Brandon E.P., Idzerda R.L., Liu H., McKnight G.S. and Basbaum A.I. J. Neurosci. **17:** 7462-7470, 1997.

McCleskey E.W. and Gold M.S. Ann. Rev. Physiol. **61:** 835-856, 1999.

Mezey E., Toth Z.E., Cortright D.N., Arzubi M.K., Krause J.E., Elde R., Guo A., Blumberg P.M. and Szallasi A. Proc Natl Acad Sci U S A, **97:** 3655-60, 2000.

Numazaki M., Tominaga T., Toyooka H. and Tominaga M. J. Biol. Chem. **277:** 13375-13378, 2002.

Premkumar L.S. Proc Natl Acad Sci U S A, **98:** 6537-9, 2001.

Premkumar L.S. and Ahern G.P. Nature, **408:** 985-990, 2000.

Raja S.N., Meyer R.A., Ringkamp M., Campbell J.N. Textbook of Pain. Churchill-Livingstone: Edinburg 1999.

Sherrington C. The Integrative Action of the Nervous System. Scribner: New York 1906.

Snider W.D. and McMahon S.B. Neuron **20:** 629-632, 1998.

Souslova V., Cesare P., Ding Y., Akopian A.N., Stanfa L., Suzuki R., Carpenter K., Dickenson A., Boyce S., Hill R., Nebenuis-Oosthuizen D., Smith A.J., Kidd E.J. and Wood J.N. Nature **407:** 1015-1017, 2000.

Szallasi A. and Blumberg P.M. Pharmacol Rev. **51:** 159-212, 1999.

Szolcsanyi J. Trends Pharmacol. Sci. **21:** 203-204, 2000.

Tang Y.P., Shimizu E., Dube G. R., Rampon C., Kerchner G.A., Zhuo M., Liu G. and Tsien J. Z. Nature **401:** 63-69.

Tominaga M., Wada M. and Masu M. Proc. Natl. Acad. Sci. USA 98 : 6951-6956, 2001.

B.E. Woolf C.J. and Salter M.W. Science, **288:** 1765-9, 2000.

Vassort G. Physiol. Rev. **81:** 767-806, 2001.

Zygmunt P.M., Petersson J., Andersson D.A., Chuang H., Sorgard M., Di Marzo V., Julius D. and Hogestatt E.D. Nature. **400:** 452-457, 1999.

Pharmacological Perspectives of Toxic Chemicals and Their Antidotes
Editors: S J S Flora, J A Romano, S I Baskin and K Sekhar

Chronic Arsenic Poisoning: Target Organ Toxicity, Diagnosis and Treatment

Swaran J.S. Flora and Krishnamurthy Sekhar
Division of Pharmacology and Toxicology
Defence Research and Development Establishment
Jhansi Road, Gwalior 474 002, India

INTRODUCTION

Inorganic arsenic compounds, which are found throughout the environment, can cause acute and chronic toxic effects and arsenic has now been recognized as a human carcinogen. Current uses of arsenic compounds are in the glass industry as a clarifier, as a wood preservative (copper arsenite), in the production of semiconductor (gallium arsenide) as a desiccant and defoliant in agriculture, and as a byproduct of the smelting of non-ferrous metals, particularly gold and copper. Human may encounter arsenic in water from wells drilled into arsenic rich ground strata or in water contaminated by industrial or agro chemical waste (Moncure et al., 1992). Exposure via drinking water has been associated with cancer of the skin and various internal organs as well as hyperkeratosis, pigmentation changes and effects on the circulatory and nervous system. Chronic arsenic toxicity due to drinking of arsenic contaminated water has been reported from many countries. Recently, large populations in West Bengal in India and Bangladesh have reported to be affected with arsenic (Smith et al., 2000, Guha Mazumder et al. 1998, Chowdhury et al., 2000). Exposure to arsenic via drinking water is correlated with a significantly elevated risk of skin and bladder cancer (Rossman, 1998; Moore et al., 2002). About 60-90% of soluble arsenic compounds are absorbed from the gastro intestinal tract following ingestion; inhalation exposure may be similar (ATSDR 1990). Absorption through intact skin is negligible. Absorbed pentavalent arsenic is converted to more toxic and carcinogenic trivalent form (Bertolero et al., 1987; Hall, 2002). Once absorbed arsenic is stored in liver, kidneys, heart and lung, while lower amount are present in muscle and neural tissues. Two to four weeks after ingestion, binding to keratin sulfhydryl groups incorporates it into the nails, hair and skin. Transverse white striae (Mee's lines) in nails are indicative of arsenic exposure. The characteristic Mee's line appear as single, solid, transverse white band of about 1 or 2 mm in width completely crossing the nail of all fingers at the same relative distance from the base. In humans, absorbed inorganic pentavalent arsenic is bio-transformed to trivalent arsenic. Trivalent form of arsenic undergoes methylation to form less toxic compounds that are excreted in urine but some inorganic arsenic is excreted in the urine unchanged (Hall, 2002; Johnson and Farmer, 1991, Hopenhayn Rich et al., 1996). Inhalation exposure to arsenic is associated with an increased risk of lung cancer. Absorption and toxicity of arsenic depends on the form in which it is ingested. Arsenic predominantly exists in two oxidation states arsenic (III) and arsenic (V). Soluble inorganic species are most readily absorbed from the gastro intestinal, with typical absorption rate being 40-100% of the ingested amount (Pontius et al., 1994). There is controversy about the absorption rate for both form of arsenic. Few reports maintain that absorption rate is same for both form of arsenic while; the others suggest arsenic (V) is better absorbed than arsenic (III) (Bernstam and Nriagu, 2000). Arsenic (V) is however, less toxic than arsenic (III). Arsenite (As III), the hydrated form of arsenic trioxide, is harmful as it is, owing to its facile covalent reaction with endogenous thiol groups especially, dithiols. Arsenic (III) is extensively bio-transformed into various methylated metabolites with markedly different toxic potential. Methylation of arsenic has long been regarded as a detoxification process because the pentavalent methylated arsenic metabolites; monomethylarsonic acid (MMAsV) and dimethylarsinic acid (DMAsV) are much less toxic and excreted more readily than As (III). S-adenosylmethionine provides the methyl group and enzymatic methylation occurs in the presence of arsenite methyltransferase or monomethylarsonate methyltransferase. It has been reported that the toxicity of arsenic decreases with the increasing methylation i.e. inorganic arsenic > monomethyl arsonic acid (MMA) > dimethylarsenic acid (DMA) > arsenobetaine (AsBe) > arsenocholine (AsC) (Leonard and Bernard, 1993; Petrick et al., 2001). Most rodent and dogs methylate arsenic more efficiently than humans but mainly to DMA, less than a few percent of the urinary arsenic is present as MMA. Many species like chimpanzee, guinea pigs etc lack the ability to methylate arsenic. This lack of methylation of inorganic arsenic results in increased tissue concentration and a lower rate of excretion of arsenic compared to animal species that do not methylate arsenic. The major route of excretion of most arsenic compounds is via the kidneys. There are marked variation in the relative amounts of the metabolism of inorganic arsenic in urine between individuals but very little is known about the causes (Vahter et al., 2000). It is also not known about the factors, which may influence the metabolism of arsenic in humans. Factors that may influence the methylation of inorganic arsenic include gender, age, ethnicity, dose level, pregnancy, nutrition and genetic polymorphism (Vahter, 2000; Vahter and Concha, 2001). Since toxicity of arsenicals is dictated by their rate of clearance from the body (Klasssen, 1996), methylation was considered for many years to be a biotransformation and detoxification mechanism (Yamamuchi and Fowler, 1994; Moore et al., 1997; Vahter, 2002). Recent evidence however, points to the fact that methylated trivalent arsenic species are more toxic than inorganic arsenic (Ahmad et al., 2000; Mass et al., 2001; Petrick et al., 2000, Thomas et al., 2001). Thus methylation may not solely be a detoxification mechanism for this metalloid but could instead be considered a pathway for its activation (Hughes, 2002; Chung et al., 2002).

TOXICITY

The toxicity of arsenic compounds highly depends on the oxidation state and chemical composition of the arsenicals. Traditionally, inorganic arsenicals have been considered more toxic than organoarsenicals. Gallium arsenide (GaAs) is another inorganic arsenic compound of potential human health concern due to its widespread use in the microelectronic industry. Available information suggests that GaAs is poorly soluble; it undergoes slow dissolution and oxidation to form gallium trioxide and arsenite (Flora 2000; Flora et al., 2002; Flora et al., 1999; Webb et al., 1984). Therefore, its toxic effects are attributable to arsenite plus the additional effect of gallium. Arsenic is primarily absorbed by ingestion, inhalation or percutaneously. Arsenic distributes rapidly into erythrocytes and binds to the globin portion of hemoglobin. Redistribution to the liver, kidneys, spleen, lungs and gastro intestinal tract occurs within 24 h. Arsenic impairs cellular respiration by inhibiting mitochondrial enzymes and uncoupling oxidative phosphorylation via inhibition of sulfhydryl group containing cellular enzymes and substitution of phosphate with arsenate in "high energy" compounds (Malachowski, 1990; Tamaki and Frankenberger, 1992). These arsenate compounds are unstable and are rapidly hydrolyzed, termed arsenolysis. Arsenic also blocks steps in the Krebs cycle. The mechanism of arsine gas toxicity differs. Arsenic becomes localized in the erythrocyte in a faxed nonvolatile state; rapid and often severe hemolysis ensues (Graeme and Pollack., 1998). However, as described above monomethylearsenous (MMA) and dimethylarsenious acids (DMA) are more toxic to a variety of human and animal cell types than inorganic arsenic (Styblo et al., 1999, 2000; Petrick et al., 2000). The absorption of GaAs and indium arsenide (InAs) appears to be accompanied by the formation of arsenic oxide. Numbers of studies confirm that the particles of GaAs and InAs are degraded in vivo to release their constitutive elements, which are then distributed to various organs. Two type of toxicity, acute and chronic, are known. The acute arsenic poisoning requires prompt medical attention and usually occurs through ingestion of contaminated food and drink. The main clinical features after acute arsenic exposure include abdominal pain, vomiting and severe diarrhea. These symptoms are followed by severe shock despite fluid replacement and acute renal tubular necrosis and acute respiratory distress. If the subject survives, protracted severe peripheral neuropathy is common (Hindmarsh, 2002).

Chronic arsenic poisoning is much more insidious in nature, often involving multiple hospital admission before the correct diagnosis is made (Saha et al., 1999). The bone marrow, skin and peripheral may become involved after chronic exposure. Chronic exposure to arsenic (III) causes a wide range of toxic effects and this metalloid is classified as a carcinogen in humans (International Agency for Research on Cancer 1987). Long-term occupational exposure to arsenic has been associated with increased prevalence of cancer of the buckle cavity, pharynx, lung, kidney, bone, large intestine and rectum (Enterline et al., 1995). Numerous other studies have associated chronic exposure to inorganic arsenic in drinking water with increased prevalence of skin, lung, liver, bladder, prostate and kidney cancer (Wu et al., 1989; Bates et al., 1992, Tsuda et al. 199 ; Hopenhayn-Rich et al., 1996a,b; Smith et al., 1998; Hopenhayn Rich et al., 1998). A number of issues are still need to resolve however, particularly concerning the dose response relationship for chronic exposure to arsenic and in the induction of cancer (Thomas et al., 2001). The features of chronic arsenic poisoning are general malaise and weakness, general debility, decreased appetite, and often-profound weight loss (Hindmarsh, 2002). After several months of exposure, the skin usually shows the classical raindrop pigmentation and de-pigmentation; also, commonly present are the characteristic arsenic corns and hyperkeratosis on the hands and feet. Hair loss is common and the nails sometimes demonstrate transverse white striations or Mees' lines. Some of the toxic effects of arsenic on various organs, biochemical and physiological function are given below in brief.

HAEM SYNTHESIS PATHWAY

The haem synthesis pathway plays an important role in all nucleated cells to provide chlorophyll and related structure (Moore et al., 2002). In mammalian and avian tissues the principal product of this pathway is haem ferro-protoporphyrin IX, an essential component of various biological functions including oxygen transport systems, mixed function oxidative reactions and other oxidative metabolic processes. This pathway is known to be highly susceptible to alterations induced by environmental pollutants offering the opportunity to use these changes as indicators of damage caused by many metals or metalloids. About 85% of haem are synthesized in bone marrow where it is required for hemoglobin formation. Remaining 15% is synthesized in liver and other organs where it is required for hemoprotein synthesis. All eight steps of the haem synthesis are catalyzed by enzymes, which require functional sulfhydryl (-SH) group for optimal catalytic activity. Since most metals have strong affinity for nucleophilic ligands, each step of the haem biosynthesis pathway is potentially susceptible to direct inhibition as a result of metal-mercaptide bond formation with the functional sulfhydryl groups. Arsenic exposure is shown to produce dose-related increase in urinary excretion in uroporphyrin and coproporphyrin (Woods and Fowler 1978). Arsenic exposure has also been known to influence the activity of several enzymes of haem biosynthesis. It has been reported that arsenic exposure produces a decrease in ferrochelatase, and decrease in COPRO-OX and increase in hepatic 5-aminolevulinic acid synthetase activity (Cebrian et al., 1988; Woods and Southern, 1989). Sub-chronic exposure to arsenic has been reported to produce alterations in the hepatic and renal activities of porphobilinogen deaminase (PBG-D), uroporphyrinogen III synthetase (UROIII-S), uroporphyrinogen decarboxylase (URO-D) and COPRO-PX in rodents (Hernadez-Zavala et al. 1999). In addition arsenite (III) administration to rats decreases the free heme pool (Cebrian et al., 1988) and increases bilirubin excretion resulting from the degradation of recently synthesized heme (Albores et al., 1989). Sub-chronic exposure to arsenic has also been reported to inhibits ALA-S and ferrochelatase activities, which catalyze limiting steps in the heme synthesis pathway, leading to increased uroporphyrin (URO) and coproporphyrin (Woods and Fowler, 1977) and COPRO urinary excretion (Martinez et al., 1983). In chronically exposed humans, arsenic alters heme metabolism as shown by an inversion of the urinary COPRO/URO ratio (Garcia Vargas et al., 1991). Few recent studies also suggested a significant inhibition of blood δ-aminolevulinic acid dehydratase (ALAD) after sub-chronic and chronic arsenic exposure (Flora et al., 2003; Flora, 1999; Kannan et al., 2001). There have been number of recent animal studies which report the effects of GaAs on porphyrin metabolism. Goering et al. (1988) first reported that GaAs after single intratracheal instillation produced a dose dependent inhibition of ALAD. They reported that the activity decreases to 5% of the control. Urinary ALA excretion was also maximum 3 to 6 days post exposure. These data were later supported by number of studies from our group. We reported that single or repeated oral ingestion of GaAs produced a dose dependent inhibition of blood ALAD accompanied by an increase in blood zinc protoporphyrin and urinary ALA excretion (Flora et al., 1998, 1999, 2002a, b). Hernadez-Zavala et al. (1999) reported that chronic exposure to arsenic alters human haem metabolism since it increases PBG-D and URO-D activities producing uroporphyrinuria III and coprophyrinuria. However, they suggested that severity of the effects appears to depend on characteristics of exposure not yet fully characterized. Although, anemia is often noted in humans exposed to arsenic, red and white blood cell counts are usually normal in workers exposed to inorganic arsenicals by inhalation (Morton and Caron, 1989). Anemia and leukopenias are common effects of poisoning and have been reported from acute, intermediate and chronic exposure. These effects may be due to a direct effect hemolytic or cyto-toxic effect on the blood cells and a suppression of erythropoies. Keeping in view the above there was a proposal that the profile of urinary porphyrins could be used as early biomarkers for arsenic toxicity in humans chronically exposed to arsenic via drinking water.

CENTRAL NERVOUS SYSTEM

In the past there were conflicting reports, which suggest that arsenic (III) did not pass into brain after exposure, however, it was subsequently confirmed that arsenic can reach the brain (Ishizaki, 1980). Acute exposure to arsenic in humans has been shown to result in problem of memory, difficulties in concentration, mental confusion and anxiety (Hall, 2002; Morton and Caron, 1989). In children chronic exposure to arsenic, urine level of arsenic was inversely correlated with verbal IQ scores including verbal comprehension and long-term memory (Calderon et al., 2001; Rodrigues et al., 2003). Arsenic alters cholinergic, glutaminergic and monoaminergic system in adult rodents with the dopaminergic system being the most affected. Although, the increase in brain was less pronounced than in other organs. Neurobehavioral studies suggested that the effect of arsenic might be either stimulation or inhibition of locomotor activity. Itoh et al. (1990) reported changes in brain monoamine metabolism and locomotor activity in mice. They found that arsenic passes into the brain in extremely small amounts exerting an influence on metabolism. Arsenic also influences the synthesis of brain monoamine, as well as stimulating locomotor activity at low dose. Notwithstanding at higher doses, it inhibits locomotor activity. We also confirmed that arsenic could cross blood brain barrier and produces alterations in whole rat brain biogenic amines levels in animals chronically exposed to arsenite (Tripathi et al., 1997). Peripheral neuropathy is commonly seen although symptoms may appear after 1 to 3 weeks after an acute exposure (Malachowski, 1990). Flora et al. (1994) reported a moderate effect of repeated oral GaAs exposure on the steady state level of brain biogenic amines (dopamine, norepinephrine and 5-hydroxytrptamine) but a pronounced effect on brain and blood acetylcholinesterase (AChE) activity. Histopathological observations revealed mild effects on cerebral cortex region. Sub-Chronic and chronic exposure to low-level arsenic exposure also produces symmetrical peripheral neuropathy. Despite few reports of arsenic induced central nervous system (CNS) damage, most studies lack quantitative data on the duration of exposure that caused these effects. At the cellular and molecular levels there has been considerable progress in the identification of signals, enzymes and transcription factors affected by arsenic but in the nervous system our knowledge is still very limited. Therefore, many aspects of arsenic neurotoxicity remains to be investigated from its entrance into the brain, to the cellular and molecular targets that when, altered by arsenic exposure, lead to specific central nervous system dysfunctions (Rodriguez et al., 2003).

CARDIOVASCULAR SYSTEM

Acute arsenic poisoning has lead to variety of electrocardiographic abnormalities including conduction blocks, QT interval prolongation and T-wave changes. Ventricular tachycardia, including torsade de pointers and ventricular fibrillation have also been reported after acute exposure while myocarditis and pericarditis have been reported after chronic exposure. Arsenic causes dilation of blood vessels with endothelial damage, resulting in third spacing of fluid and occasionally severe, rapidly progressive shock (Graeme et al., 1998; Gorby, 1988). A high prevalence of peripheral vascular alterations, ranging from abnormal temperature and acrocyanosis in toes and fingers to gangrene, has been reported in human exposed to arsenical pesticides. It has also been suggested that the combination of arsenic and humic substances found in the well water of Blackfoot disease endemic areas in Taiwan shortens prothrombin time of human plasma and that these substance may play a role in the etiology of this disease (Lu, 1990a). Studies on cultured human umbilical vein endothelial cells suggest that arsenic plays an important role in the pathogenesis of Blackfoot Disease by damaging the endothelial cells (Chen et al., 1990). There is however, no conclusive report available on cardiovascular effects in humans after oral exposure to organic arsenicals.

LIVER

Both arsenite and arsenate demonstrate some degree of hepatotoxicity, each may possess somewhat unique mechanism of toxicity in the liver. While toxicity of arsenite is primarily due to its high reactivity with sulfhydryl groups, arsenate toxicity results from its substitution for phosphate in enzyme-catalyzed reactions (Squibb and Fowler, 1983). Arsenite is rapidly and extensively accumulated in the liver, where it inhibits NAD-linked oxidation of pyruvate or α-ketoglutarate. This occurs by complexation of trivalent arsenic with vicinal thiols necessary for the oxidation of this substrate (Squibb and Fowler, 1983). Webb et al. (1984) reported impaired liver function due to arsenic dissociated from GaAs as indicated by increased urinary excretion of uroporphyrin. We reported changes in some key biochemical variables in the liver of rats exposed to various doses of GaAs, but the changes were moderate (Flora, 1996; Flora et al., 1999). Liver injury characteristic of long term or chronic exposure manifests itself initially in jaundice and may progress to cirrhosis and ascites. Toxic effects on hepatic parenchymal cells results in the elevation of liver enzymes in blood and studies in experimental animals show granules and alterations in the ultrastructure of mitochondria, non specific manifestation of cell injury including loss of glycogen (Goyer, 1996). Cutaneous and hepatic manifestation of arsenic toxicity due to chronic consumption of arsenic contaminated ground water are seen commonly among a large number of population of various districts of West Bengal in India. Another important feature of chronic arsenic toxicity in West Bengal is a form of hepatic fibrosis that causes portal hypertension, but does not progress to cirrohosis (Santra et al., 1999, 2000; Rahman et al., 1999). Clinical examinations reveal liver to be swollen and tender. The analysis of blood revealed elevated levels of hepatic enzymes.

GASTROINTESTINAL SYSTEM

Gastrointestinal symptoms are generally seen after acute exposure but not in chronic poisoning. Dehydration and intense thirst are common. Arsenic (III) is corrosive to the eyes, mouth and mucous membranes and perforation of the nasal septum can occur. The patients complain of metallic taste and garlic odour. Dysphasia, nausea, projectile vomiting, abdominal pain and profuse watery or blood diarrhea may ensue. The efficiency of absorption from the gastrointestinal tract is related to their water solubility. The toxic effects of arsenic on GI mucosal vasculature are vasodilatation, transduction of fluid into the bowel lumen, mucosal vesicle formation and sloughing of tissue fragments. Rupture of vesicles may cause bleeding, profuse watery stool and a protein losing enteropathy. Three deaths in India due to chronic arsenic poisoning by drinking water from tube-wells were reported (Chakraborty and Saha, 1987). The most likely mechanism of gastrointestinal toxicity is damage to the epithelial cells with resulting irritation.

SKIN

Skin cancer has been associated with chronic inorganic exposure (Saha et al., 1999). Skin cancers are mostly monocenteric but sometime multicenteric cases are also found. Dermal changes most frequently reported in arsenic exposed humans include hyper pigmentation, melanosis, hyperkeratosis, warts, and skin cancer (Shannon and Strayer, 1989; Saha, 1984). The lesions generally appear 1 to 6 weeks after the onset of illness. In most cases, a diffuse brawny desquamation develops over the trunk and extremities that is dry, scaling and nonpruritic. Patchy hyperpigmentation—dark brown patches with scattered pale spots sometime described as raindrops on a dusty road—occur particularly on the eyelids, temples, axillae, neck, nipples and groin. Arsenic induced skin cancer occurs mostly in unexposed areas such as the trunk, palms and soles. More than one type of skin cancers are reported and most common are Bowen disease, squamous cell carcinomas, basal cell carcinomas, and combined forms. Hyperkeratosis and hyperpigmentation usually are present in patients with dermal malignancies.

MUTAGENIC EFFECTS

Mutagenic effects of arsenic have been generally negative; they exhibit clastogenic properties in many cell types in vivo and in vitro (Jha et al., 1992; Hartman and Speit, 1996). Arsenic does not induce gene mutation in bacteria and was found to be inactive in inducing reverse mutation and mitotic gene conversion in yeast. Arsenic (III) was also reported to induce DNA damage in human lymphocytes with Comet assay (Schaumloffel and Gebel 1998, Yamanaka et al., 1990). Studies have shown that arsenic compounds are also poor mammalian cells assays. Arsenic compounds induce mutation lymphoma cells but these results are yet to be supported by other studies (Harrington Brock et al., 1999). There is however a study by Hei et al. (1998) which reports that arsenic (III) induces mutation in human-hamster hybrid (A_L) cells. Several other studies have suggested that both arsenic form (III and V) are capable of producing chromosome breaks and chromosomal aberrations in human peripheral lymphocytes and human skin culture. It has also been indicated in few other studies that trivalent forms are more potent and genotoxic than pentavalent form (Nordenson and Bechman 1981; Saha et al., 1999). Sordo et al. (2001) studied the cytotoxic and genotoxic effects of arsenic and its metabolite by evaluating the effects on the proliferation and DNA integrity of both cultured human lyphocytes and leucocytes. The results suggested that both MMA and DMA do not alter the proliferation of human lymphocytes in vitro. In contrast, both metabolites did induce DNA single strand breaks.

CARCINOGENIC EFFECTS

The most striking feature of chronic arsenic poisoning appears to be cancer. Numbers of epidemiological studies have reported a strong correlation between environmental, occupational and medical exposure of man to inorganic arsenic and cancer of skin and lungs (Basu et al., 2001; Saha et al., 1999; Pershagen, 1981; IARC, 1987). Fishbein (1987) suggested that the probability of death in a person with arsenical keratosis from lung cancer is 5 to 10 times higher than expected. IARC (1987) reports too concluded, "There is sufficient evidence that inorganic arsenic compounds are skin and lung carcinogen in humans". The time period however, between initiation of exposure and occurrence of arsenic associated lung cancer has been found to be in the order of 35 to 45 years. Epidemiological studies too have shown that chronic exposure to arsenic can result in an increased incidence of cancer of the lung, skin, bladder and liver (IARC 1987, Chen et al. 1992, Nriagu, 1994;

Waalkes, 1995; Schwartz, 1997; Gonsebatt et al., 1997). For reasons that are still not clear, arsenic is not carcinogenic in rodent models (Wang and Rosseman, 1996). Carcinogenic effects were absent in animals after a life time exposure to inorganic arsenic through oral, drinking water or dietary route (Byron et al. 1967, Byron et al., 1967, Thorgeirsson et al., 1994). Mice exposed to 10-ppm arsenite for 26 weeks in drinking water also did not develop tumors (Rossman et al. 2001). There is however few reported studies where carcinogenic effects have been reported in hamsters after intratracheal instillation of inorganic arsenic (Pershahen and Bjorklund 1985, Yamamoto et al., 1987). After exposure to arsenic trioxide, adenocarcinomas, carcinomas and adenomas developed in the respiratory tract. The absence of animal model, in vitro studies become particularly important in providing information on the carcinogenic mechanism of arsenic.

Skin cancer too has been associated with inorganic arsenic exposure. As explained above skin cancers are mostly monocentric but reports of multicentric cancer to have been found (Saha 1998; Saha et al., 1999). The skin cancers related to arsenic differ from ultraviolet light induced tumors in that they generally occur in areas of the body not exposed to sunlight (e.g. Palms and soles) and they occur in multiple lesions.

Developmental and Reproductive Toxicity

Reproductive and developmental effects of inorganic arsenic on humans and animals species have been reported (Hood et al., 1988, Concha et al., 1998). Limited animal studies suggest that arsenic can produce malformation, intrauterine death, and growth retardation (Beaudoin, 1974; Hood, 1983; Golub et al., 1998). These malformations include neural tube defects, renal and gonadal agenesis, eye defects, and rib malformations. Arsenic readily crosses the placenta. There are few other studies which report reduced litter sizes, intrauterine death and postnatal mortality and growth retardation after chronic oral administration or drinking water during gestation and lactation. Several human studies too suggest the association between arsenic exposure and adverse reproductive outcomes; however, interpretation of these studies was thought to be complicated because study populations were exposed to multiple chemicals (Golub et al., 1998).

Very few studies are available on the effects of arsenic on male or female reproductive system. The available data point to dose dependent effects on growth and viability of the conceptus and offspring but no effects on fertility (Schroeder and Mitchener 1971; Golub et al., 1998). Chronic studies did not report any male reproductive organ pathology. However, in human studies a correlation has been observed between arsenic exposure and incidence of abortion (Tabacova et al., 1994). Higher spontaneous abortions and stillbirths were reported in the high arsenic area (arsenic in drinking water > 0.1 mg/L) compared to the control areas. Ahmad et al. (2001) recently observed adverse pregnancy outcome in women chronically exposed to arsenic through drinking water. The authors also conclude that arsenic contamination is also a threat to healthy and safe pregnancy outcome.

Mechanism of Arsenic Action

One of the hallmarks of chronic arsenic toxicity in humans from oral exposure are skin lesions, which are characterized by hyperpigmentation, hyperkeratosis, and hypopigmentation (Cebrian et al., 1983). In Taiwan, Blackfoot disease, a vasoocculusive disease that leads to gangrene of the extremities, is also observed in individuals chronically exposed to arsenic in drinking water. Although arsenic and its mode of action has been the subject of reviews and symposia, little data exist regarding specific mechanism(s) differentiating its action as a carcinogen to cause cancer and as a chemotherapeutic agent used in the treatment of cancer. The molecular mechanisms of arsenic toxicity are still unknown mainly for the reason that (i) its predilection to undergo a variety of complicated metabolic conversion in vivo (ii) the complex interactions between arsenic metabolites and intra and extracellular macromolecules, (iii) the influence of concomitant exposure to other toxic agents, and (iv) the lack of appropriate animal models for most of the pathologies associated with inorganic arsenic exposure (Wildfang et al 2001; Lee and Ho, 1994). Although, high concentration of arsenic (V) have been shown to substitute for phosphate in enzyme-catalyzed reactions, arsenic toxicity is postulated to be primarily due to the binding of arsenic (III) to sulfhydryl group containing enzymes. Glutathione (GSH) plays a critical role in both the enzymatic and non-enzymatic reduction of pentavalent arsenicals to trivalent and in the complexation of arsenicals to form arsenicothiols during methylation process (Chochane and Snow 2001). The interaction of arsenic with glutathione and its related enzymes by changing their redox status and this may lead to the alterations of their biological function. Inactivation of GSH related enzymes could have deleterious effects on the detoxification processes and other critical cellular processes involving GSH mediated redox regulation.

Arsenic is also known to pyruvate dehydrogenase (PDH), a multi sub unit complex that require a dithiol cofactor α-lipoic acid for enzymatic activity. Recent studies have also indicated that arsenic exerts toxicity by generating reactive oxygen species, but the mechanism is still unclear (Ito et al., 1998; Flora, 1999). These reports suggested that intracellular peroxide level is correlated with arsenic-induced cellular apoptosis. An important protective role of glutathione (GSH) against arsenic-induced oxidative damage has been reported (Ito et al., 1998). Ramos et al. (1995) also described lipid peroxidation as one of the mechanisms of arsenic toxicity in female rats together with a concomitant decrease in cellular GSH concentration, which is inversely correlated with lipid peroxidation in the liver but not in other tissues.

Glutathione related enzymes such as glutathione peroxidase (GPx) and glutathione related reductase function either directly or indirectly as antioxidant and glutathione S transferase (GST) plays an important role in metabolic detoxification. Reactive oxygen species have been implicated in the pathogenesis of cancer. Oxidative stress can be involved in initiation, promotion, or progression (Guyton and Kensler, 1993, Chen et al., 1998, Liu et al., 2000, Lu 1990a,b). Several other studies suggest that the genotoxicity of arsenic may be mediated by ROS. Although, the lung is one of the major organs affected by arsenic, little information is presently available for the production of ROS in arsenic exposed lung cells. Kitchin (2001) provided an interesting explanation for arsenic's ability to cause human cancer at high rates in the lungs. High partial pressures of oxygen are found in the lungs. Human lungs may be an organ responsive to arsenic carcinogenesis because of the high partial pressure of oxygen and the fact that dimethylarsine, a gas, is excreted via the lungs (Yamanaka and Okada 1996). Human bladder may be an organ responsive to arsenic carcinogenesis because of the high concentration of DMA and MMA that is stored in the lumen of the bladder and the amount of DMA (III), dimethylarsine or MMA (III) that might be generated by reductive processes (Kitchen, 2001).

As has been suggested above that the trivalent intermediates in the formation of MMA and DMA may have a role in arsenic toxicity as they are known to react with sulfhydryl groups and are highly toxic (Stylblo et al., 1997). Although, MMA (III) has been detected in bile and urine following exposure to inorganic arsenic, the extent to which MMA (III) is released from the site of the methylation reaction and transported to the tissues is not known. In vitro studies have shown that MMA (III) and DMA (III), like As (III) can form GSH complexes and that these are atleast as toxic as As (III) (Vahter and Concha 2001, Styblo et al. 1997, 2000). Buchet and Lauwerys (1987) have suggested that GSH is required for the monomethylation of arsenite, but not the dimethylation step.

Trivalent arsenic also inhibits pyruvate dehydrogenase (PDH), a multi sub unit complex that needed Lipoic acid as a cofactor for enzymatic activity. It has also been reported recently that MMA (III) is more potent inhibitor of PDH than arsenite. PDH oxidizes pyruvate to acetyl CoA, a precursor to intermediates of the citric acid cycle. The citric acid cycles degrades the intermediate, and this provides reducing equivalents to the electron transport system for ATP production. Inhibition of PDH may ultimately lead to decreased production of ATP. Also intermediate of the citric acid cycle can be used in glucogenesis (Hughes 2002). Inhibition of PDH may explain in part the depletion of carbohydrate observed in rats administered arsenic (Reichl et al. 1988, Hughes 2002).

FACTORS INFLUENCING TOXICITY

Animal studies have shown that toxicity of arsenic may be influenced by the dose level, route of administration form of arsenic administered and nutritional status and diseases (Vahter and Marafante 2000, Vahter 1994, Buchet and Lauwery, 1998). In the population studies, it has been reported that arsenic methylation was influenced by age, exposure level and pregnancy. The relative percentage of urinary DMA was lower in children than in adults which may suggest that children retain more arsenic in their tissues and, therefore, are more sensitive to arsenic exposure (Vahter et al., 2000). It has also been reported that relative amount of DMA in urine at the end of pregnancy, as well as in the newborn infant, indicating induction of arsenic methylation during pregnancy. Studies on the effects of nutrition on arsenic toxicity have been hindered by lack of adequate animal model of chronic toxicity and carcinogenicity. It has generally been seen that populations exhibiting arsenic toxicity have mainly been of low economic status and suffering from some form of malnutrition.

Alcohol ingestion is also recently considered to be one of the important factors determining individual susceptibility in arsenic exposure (Flora et al., 1997). Animals co-exposed to arsenic and ethanol (in drinking water) showed more susceptibility to hepatotoxic effects and the uptake of arsenic concentration in the vital organs.

DIAGNOSIS

Symptoms

The early symptoms of arsenic poisoning in population consuming arsenic contaminated water include headache, dizziness, insomnia, weakness, nightmare, numbness in the extremities, anemia, palpitation and fatigue. A temporal appearance of organ system injury seems to be associated with arsenic poisoning. In few cases after a delay of minutes to hours, severe hemorrhagic gastroenteritis become evident and may be accompanied by cardiovascular collapse or death. Sensorimotor peripheral neuropathy may become apparent several weeks after resolution of the initial signs of intoxication. A metallic taste in the mouth and gastrointestinal disturbances may be present. Mee's lines (white transverse bands in the nails) may be seen. Bone marrow depression with anaemia, leukopenia or pancytopenia is common (Hall 2002). Gangrene of the feet (Blackfoot disease) has been associated with chronic ingestion of arsenic in Taiwan. Arsenic hyperkeratosis appears predominantly on the palms and the plantar aspect of the feet although involvement of the dorsum of the extremities and the trunk have also been described. Yeh (1973) classified arsenical keratosis into two types: a benign type A and a malignant type B consisting of lesions of

Bowen's disease, basal cell carcinoma or squammous cell carcinoma. Spotty raindrop pigmentation of the skin distributed bilaterally and symmetrically over trunk and limbs is the best diagnostic features of arsenic hyperpigmentation.

Biological Indicators

Biological indicators of arsenic exposure are blood, urine, and hair although blood arsenic is only reliable within few days of acute exposure. In case of chronic exposure, urinary arsenic is the best indicator of current or recent exposure. Hair or fingernail concentration of arsenic may be useful in evaluating past exposure. Most investigators have used hair rather than nails arsenic because the former is easier to obtain in sufficient quantities. The diagnosis of chronic arsenic poisoning must rely on the characteristic, clinical features of the typical skin lesions, debility, weight loss and neuropathy, and hair arsenic levels are only supportive of the diagnosis.

There are no specific biochemical parameters that reflect arsenic toxicity, but evaluation of clinical effects must be interpreted with knowledge of exposure history. A number of studies have demonstrated that porphyrins and other constituents of the haem synthesis pathway might serve as sensitive and specific biomarkers of toxic metal exposure in humans. All eight steps of the haem synthesis pathway are catalyzed by enzymes, which require functional sulfhydryl (-SH) groups for optimal catalytic activity, either as part of the active site configuration or to maintain their structural integrity. Since most metals have a strong affinity for nucleophilic ligands, each steps of the haem biosynthesis pathway is therefore potentially susceptible to direct inhibition as a result of metal-mercaptide bond formation with the functional sulfhydryl groups. δ-aminolevulinic acid dehydratase (ALAD) is a sulfhydryl containing enzymes that catalyzes the asymmetric condensation of two molecules of ALA to porphobilinogen. This reaction is fundamental in the biosynthesis of tetrapyrroles (such as heme), the prosthetic group of various protein. Due to its sulfhydryl nature, ALAD activity is highly sensitive to the presence of metals, which possess a high affinity for sulfhydryl group. Chronic exposure to arsenic has recently been shown to inhibit ALAD in blood (Kannan et al. 2001, Flora 1999). Cebrian et al. (1988) found an increase in hepatic δ-aminolevulinic acid synthetase (ALAS) activity in rats exposed for 4 weeks to arsenite and interpreted this to be due to reduction in hepatic free haem pool due to an induction of haem oxygenase.

It is advisable that proper investigation should be carried out to define the various manifestation in chronic arsenicosis and these include routine hematological variable like haemoglobin, total and differential count, RBC morphology, urine and stool examination, chest X-ray, electrocardiogram determination of blood sugar, urea and creatinine. The chronic arsenicosis produces protean manifestation which is evident from the report of the clinical features in 156 cases who had drinking arsenic contaminated water in West Bengal, in India (Guha Majumder et al. 1998).

TREATMENT

Prevention

The primary goal of arsenic screening is the identification and elimination of arsenic sources. Arsenic compounds were used for a long time as pesticide in agriculture, especially in wine growing, fruit plantation and forestry or in paints. Such compounds are still increasingly used as wood and cotton preservative. Arsenic is used in the production of lead and copper alloys, special glasses, enamel, catalysis, industrial purification filters and photo and semiconductor materials (Squibb and Fowler, 1983; Flora, 2000). Gallium arsenide is one such intermetallic semiconductor compound of potential health concerns due to its current widespread use in the microelectronic industry (Flora 2000, Burns et al., 1991, Carter and Sullivan, 1992). Available toxicokinetic data suggest that although gallium arsenide is poorly soluble, it undergo slow dissolution and oxidation to form gallium trioxide and arsenite. Therefore its toxic effects are attributable to arsenic plus the additional effects of gallium (Flora et al. 1997, 1998, 2002).

Precautions in many countries have diminished arsenic pollution. Burning of coal, timber and fossil fuel adds to arsenic levels in the atmosphere and causes contamination of soil and water (EPA 1994). The restoration of these soils may become a problem of the future. There is a need to understand the kinetic by which various chemical species of arsenic are biotransformed in the soil by methylation/demthylation cycle because of better understanding of these processes is oriented to evaluating both potential mobility, bioavailability and toxic potential of this element in terrestrial ecosystem. Dietary supplements may also affect arsenic toxicity and carcinogenicity. Arsenic induced bone marrow chromosomal aberrations were decreased in mice fed crude extract of *Emblica officinalis* fruit (Biswas et al. 1999). This may be because of arsenic affinity for the sulphur moieties in many of the chemical compounds found in garlic extracts. Although there are only very few studies available on the effects of dietary nutrients on arsenic but it is clearly evident that nutritional status may play an important role in the expression of arsenic toxicity.

Nutritional Intervention

Experimentally, excesses or deficiencies of essential trace elements and other dietary nutrients facilitate arsenic absorption. There have been few other reports where exposure to arsenic has lead to the alteration in metabolism of body stores of some essential elements like selenium (Gregus et al. 1998) and copper. Few other reports suggest that arsenic exposure abolish the anticarcinogenic effects of selenium in mice (Hill 1975, Peraza et al. 1998). Selenium can also alleviate arsenic toxicity. Selenate partially prevents the uncoupling of oxidative phosphorylation by arsenate (Peraza et al. 1998). Selenium also decreases the teratogenic toxicity of arsenate in hamsters when both salts were injected simultaneously. Thus adequate or even extra selenium in the diet may alleviate arsenic toxicity whereas a selenium deficiency may aggravate arsenic toxicity. Arsenic can also induce metallothionein (MT) (Flora and Tripathi 1998), a low molecular weight cysteine rich metal binding protein. This implies that arsenite can be detoxified by MT. Dietary antioxidants such as vitamin E; C and vitamin A may also be alleviating arsenic toxicity (Chattopadhya et al. 2001, Kannan and Flora 2003, Nair et al., 1970). Addition of vitamin E could atleast in part prevent the arsenic induced killing of human fibroblast (Peraza et al. 1998). Dietary intake of fruits and vegetables particularly yellow and light green vegetable food high in vitamin A precursor was inversely related with the odds ratio of lung cancer in tin miners exposed to arsenic and other risk factors.

CHELATION THERAPY

The current management of acute and chronic arsenic poisoning relies on supportive care and chelation therapy. The general management of arsenic poisoning begins with the elimination of further exposure to the toxic agent and provision of a patent airway, adequate ventilation and cardiovascular support. Chelating agents, compounds that binds to and enhance the urinary and fecal excretion of toxic metals, form complexes with toxic metals in vivo. The water-soluble complexes, which are formed, get readily excreted in the urine and feces leading to a depleted body arsenic burden. After almost 50 years of research we still do not have an effective antidote for arsine (AsH_3), which is almost 10 times more toxic than inorganic arsenic (Muckter et al 1997). All the three thiol chelators BAL, DMSA, and DMPS (at a dose of 160 mmol/kg) afforded protection against arsine at $2LD_{50}$ dose in experimental animals (Krepple et a., 1990). In the following paragraphs some of the chelating agents are listed for their roles in treating arsenic toxicity

2,3- dimercaprol (British Anti Lewisite; BAL)

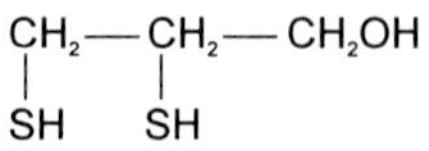

```
CH2—CH2—CH2OH
|    |
SH   SH
```

2,3-Dimercaprol (British Anti Lewisite, or BAL) is the traditional chelating agent that has been used clinically in arsenic poisoning since 1949. It is an oil soluble, clear, colourless liquid with a short half-life. Because of its lipid solubility it is distributed extracellularly and intracellularly. It was originally developed for treating lewisite (dichloro 2-chlorovinyl) arsenate poisoning (Aposhian et al. 1995). In humans and experimental models, the antidotal efficacy of BAL has been shown to be most effective when administered immediately after the exposure. BAL was however toxic at a dose of 40 mmol/kg (Inns et al. 1993). The therapeutic use of BAL is now limited because this chclating agent has got serious disadvantage. Beside rapid mobilisation of arsenic from the body, it causes a significant increase in brain arsenic (Hoover and Aposhian 1983). The other disadvantages include pain at the site of injection and allergic reaction to it (Flora and Tripathi 1998).

Meso 2,3-dimercaptosuccinic acid (DMSA; Succimer)

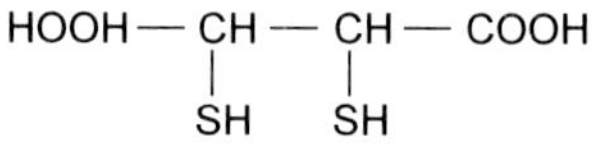

```
HOOH — CH — CH — COOH
       |     |
       SH    SH
```

It has generally been assumed that poisoning by arsine (Inns et al. 1993) and inorganic arsenic (III) in rats (Flora et al. 1995), mice (Kreppel et al. 1993), and guinea pigs (Reichi et al. 1991) all favour treatment with DMSA. DMSA has been tried successfully in animal as well as in few case of human arsenic poisoning (Aposhian and Aposhian 1990, Aposhian 1995, Flora and Tripathi 1998). DMSA has been shown to protect mice due to lethal effects of arsenic. A subcutaneous injection of DMSA provided 80-100% survival of mice injected with s.c. sodium arsenite (Ding and Liang 1991). We also reported a significant depletion of arsenic and a significant recovery in the altered biochemical variables of chronically arsenic exposed rats (Flora et al. 1995, 2003). This drug can be effective if given either oral or i.p. route. Number of other studies appeared in the recent past have recommended that DMSA could be safe and effective for treating arsenic poisoning (Tripathi and Flora 1998). However, in an interesting prospective, double blind, randomised controlled trial study conducted on few selected patients from arsenic affected West Bengal (India) regions with oral administration of DMSA suggested that DMSA was not effective in producing any clinical or biochemical benefits or any histopathological improvements of skin lesions (Guha Mazumder ct al. 1998).

Not many side effects are reported with the use of DMSA, but few reports indicate muscutaneous eruption, nausea and moderate hepatic dysfunction (Grandjean et al. 1991, Flora et al. 1995).

Sodium 2,3-dimercaptopropane 1-sulfonate (DMPS; Unithiol)

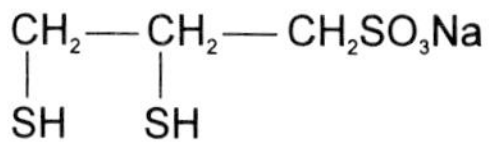

$$\underset{\displaystyle SH}{\underset{|}{CH_2}} — \underset{\displaystyle SH}{\underset{|}{CH_2}} — CH_2SO_3Na$$

DMPS, like DMSA is another analogue of BAL and better known for its antidotal efficacy against mercury, has been reported to be an effective drug for treating arsenic poisoning. In experimental animals, i.p. administration of DMPS increased the lethal dose of sodium arsenite in mice by four folds

(Hauber and Weger 1978, Okonishikova and Nirenberg 1974). We also reported that DMSA and DMPS are equally effective in providing significant depletion of body arsenic burden in chronically arsenite-exposed rats. A quantitative evaluation of three drugs reveals that DMPS is 28 times more effective than BAL in arsenic therapy in mice, while DMSA and DMPS are equally effective. Guha Majumder et al. (1998) recently presented few interesting evidence on the efficacy of treatment of DMPS in a single blind placebo control trials of patients suffering with chronic arsenic poisoning in West Bengal, India. DMPS was given in a dose of 100 mg capsule 4 times a day for a course of 7 days for four courses with one-week drug free period between each course. There was a significant decrease of clinical scores from pre-treatment to post treatment values amongst both DMPS and placebo groups. There was however, no change of skin histology, haematology and liver function test parameters in the patients before and after the therapy with DMPS or placebo. No side effects too were noticed in patients treated with DMPS (Guha Mazumder et al. 2001). DMPS has also been reported to be effective in reducing arsenic induced neuropathy and absence of polyneuropathy after treatment in acute arsenic poisoning (Moore et al. 1994, Paul and Charles 2000).

Although, there have been number of published reports available for the treatment of chronic arsenic exposure in animals. However, for the treatment of subjects exposed to arsenic compounds like GaAs, there is still not much information available. As the toxicology of GaAs has recently been understood, the treatment also remains undefined. Burns et al. (1993) suggested that *in vitro* treatment with meso 2,3-dimercaptosuccinic acid (DMSA; Succimer) could not reverse immunosuppression induced by *in vivo* exposure to GaAs and indicated that an altered form of DMSA (2:1 mixed disulfide) could be useful to bind intracellular arsenic. We compared the efficacy of DMSA and sodium 2,3-dimercaptopropane 1-sulfonate (DMPS) in treating sub-acute GaAs poisoning. The results suggested that both DMSA and DMPS were marginally effective in reversing the altered immunological variables and in reducing tissue arsenic levels (Flora and Kumar 1996).

D-pencillamine

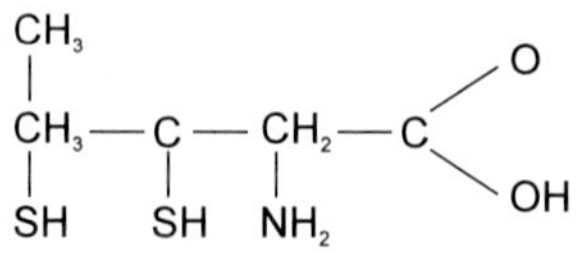

CH_3
|
CH_3 — C — CH_2 — C(=O)OH
| | |
SH SH NH_2

D-Pencillamine (DPA) is another chelator, which has been tried with limited success against chronic arsenicosis either alone or in combination with dimercaprol. However, long-term use of this chelating agent may induce several cutaneous lesions including urticaria, macular or papular lesion, pemphigold lesion etc. The drug has been used for the treatment of long-term exposure to arsenic either alone or in combination with dimercaprol. However, with long-term use, pencillamine induces several side effects like cutaneous lesions including urticaria, macular or popular lesion, pemphigoid lesion etc. Haematological system may also be affected severely causing leucopoenia, anaemia an agranulocytocytosis. Although animal studies point to some encouraging results, few humans studies indicate only limited beneficial effects of DPA (Guha Majumder et 1998, Bansal et al. 1991). There is thus need for detailed animal studies for evaluating the therapeutic effects of DPA either alone or in combination with some other chelator.

Among the four chelating agents described it is clear that BAL is rather toxic by itself and may cause hypertensive episodes and has been shown to redistribute arsenic into organs that are shielded by a blood-brain barrier. There is no oral preparation available and BAL has an unpleasant odour. DMSA and DMPS on the other hand, have been approved in several countries as arsenic antidote in human. Both are less toxic than BAL. DMSA and DMPS have a therapeutic index 14 and 42 times that of BAL respectively (Muckter et al. 1997). Both these drugs have been shown to effectively prevent arsenic from crossing epithelial boundaries and from entering the cells and they enhance the excretion of arsenical from the body more rapidly and completely. There are very few reports available showing the efficacy of these chelators against human arsenic poisoning. However, number of studies is now available demonstrating the efficacy of DMSA and DMPS against chronic lead poisoning in humans it can be assumed that the drug is pretty safe.

Newer Developments in the Treatment of Chronic Arsenic Poisoning

Chelation with DMSA monoesters

DMSA is an effective drug for the treatment of arsenic, cadmium, lead, and mercury intoxication but due to its hydrophilic and lipophobic properties it is unable to pass through cell membranes (Ding and Liang 1991, Bosque et al. 1994). A consequence of the solely extracellular distribution of DMSA is its ineffectiveness in capturing arsenicals intra-cellularly (Planas-Bohne 1981) or in mobilizing intracellular cadmium from metallothionein bound sites (Gale et al. 1993). Monoesters of DMSA, especially the higher analogues, have been reported to be more effective in reducing cadmium (Jones et al. al. 1992, Kostial et al. 1994) and mercury (Kostial et al. 1993), as well as brain lead mobilization (Walker et al. 1992) than the parent drug DMSA. These authors reported that the maximum mobilization of metal was observed with monoisoamyl meso 2,3-dimercaptosuccinic acid (MiADMSA), a C_5-branched chain alkyl monoester of DMSA. The results suggested that DMSA monoesters could be of great medical importance as being more efficient metal mobilizers than DMSA and thus, could be potential drugs for chelation therapy in metal intoxication. We recently reported beneficial effects of MiADMSA in mobilizing arsenic from chronically GaAs pre-exposed and in reducing tissue oxidative stress. However, it was suggested that optimum effects of chelation therapy could be achieved by combined administration of oxalic acid and MiADMSA. Oxalic acid is being an effective gallium chelator (Flora et al. 2002).

Kreppel et al. (1995) also reported the superior efficacy of Monoisoamyl DMSA (MiADMSA) and mono n-amyl DMSA in protecting mice from the lethal effects of arsenic and in reducing body arsenic burden. These studies thus support that MiADMSA could be potential drug to be used in the treatment of chronic arsenic poisoning. Monoisoamyl DMSA (MiADMSA) is thus, a new and one of the most effective of the vicinal class of metal mobilizing agent (Jones et al. 1992, Xu et al. 1995). Although, the compound is more toxic than the parent diacid DMSA (Mehta et al. 2002, Flora and Mehta 2003), its structure features and recent experimental evidence suggest that it might well be effective in chelating arsenic (Flora et al. 1993, Flora et al. 2002a,b 2004). MiADMSA is a monoester of DMSA with a straight and branched chain amyl group thereby increasing the lipophilicity and number of carbon atoms of the compound. Lipophilicity and molecular size of this new drug might be important factors for the removal of arsenic from both intra and extracellular sites possibly leading to better therapeutic efficacy. There could be a possibility of an arsenic redistribution to brain following treatment with the monoester. However, we observed no such effects in a recently conducted study (Flora et al. Unpublished report). It appears plausible that MiADMSA could be decreasing the oxidative stress in tissues either by removing arsenic from the target organs and/or by directly scavenging ROS via its sulfhydryl group. The results also suggest that (i) oral administration being relatively more effective than parenteral administration (ii) a dose of 50 mg/kg produced better results compared to the two other doses.

Combination Therapy

Combination therapy is another new and novel option suggested by us. We have suggested that combined treatment with an antioxidant and a thiol chelator could be a better treatment protocol compared to monotherapy with a chelator. As described earlier, the potential role for oxidative stress in the injury associated with arsenic poisoning suggests that antioxidant may enhance the efficacy of treatment protocol designed to mitigate arsenic induced toxicity. We recently reported that combined administration of n-acetylcysteine and succimer led to a rapid mobilization of arsenic (Flora 1999) and lead (Pande et al. 2001), while, administration of α-lipoic acid and DMSA provided a more pronounced recovery in lead induced altered biochemical variables indicative of oxidative stress (Pande and Flora 2002). In an experimental study conducted by us recently provided an *in vivo* evidence of arsenic induced oxidative stress in number of major organs of arsenic exposed rats and that these effects can be mitigated by pharmacological intervention that encompasses combined treatment with N-acetylcysteine and DMSA (Flora 1999). We also reported that co-administration of naturally occurring vitamins like vitamin E or vitamin C during administration of a thiol chelator like DMSA or MiADMSA may be more beneficial in the restoration of altered biochemical variables

Table 1 Chronic Arsenic Poisoning: Target Organs and common symptoms

Target Organs	*Symptoms*	
	Acute	*Chronic*
Central Nervous System encephalopathy (cortical atrophy)	Delirium, seizures,	Coma and death
Peripheral Nervous System	Fasciculation, painful parenthesias,	Foot and wrist drop
Motor deficit, muscle wasting	Cardiovascular Myocardiatis and prolongation, T-wave change, ventricular tachcardia, ventricular fibrillation, damage	Conduction block, QT interval Pericarditis Dialation of blood vessels, endothelial
Endocrine	None reported	Diabetes
Pulmonary	Pulmonary edema, adult respiratory distress syndrome (ARDS), apnea from phrenic nerve damage	Respiratory cancer
Gastrointestinal	Dehydration, Perforation of nasal Hemorrhagic gastro-septum, mucous membrane irritation, enteritis, Electrolyte metallic taste, garlic odour, and fluid imbalance, Hypovolemic shock,	
Haematopoietic Aplastic anemia,	Pancytopenia, Eosinophilia, Marrow depression, anemia (late acute myelogenous sequela) leukemia, basophilic stippling and rouleaux formation	Bone
Skin (Dermatologic)	Erythema, dilation of cutaneous Conjuctivitis, capillary beds, Hyperpigmentation, Eczematoid, allergic hyperkeratosis, brawny desquamation, Basal cell carcinoma dermatitis, dermatitis, keratosis, cell carcinoma	

Table 2 Possible Diagnostic and Therapeutic Measures for Chronic Arsenic poisoning

Diagnostic Tests	*Therapeutic Measures*
Urinary arsenic levels	Meso 2,3-dimercaptosuccinic acid
Blood arsenic levels	sodium 2,3-dimercaptopropane
Hair or nail arsenic levels	1-sulfonate (DMPS)
2,3-dimercaprol (British Anti Lewisite; BAL)	
Radiographs: KUB for opacities	D-penicillamine (DPA)
	Hemodialysis for renal failure
Peripheral blood smears:	
Basophilic stippling	
Rouleaux formation	Combination therapy: *
	Antioxidant plus DMPS/ DMSA*
	MiADMSA plus DMSA/ DMPS*
δ-aminolevulinic acid dehydratase activity in rbc	

* Based on data from animal studies, Human trials yet to be done

(particularly the effects on haem biosynthesis and oxidative injury) although it has only limited role in depleting arsenic burden. The study reports some new interesting observation particularly the remarkable effects of combined treatment on inhibited blood ALAD activity and in particular its beneficial effect in reducing the arsenic induced oxidative stress. Co-administration of vitamin C and MiADMSA in reducing liver and kidney arsenic burden supports the view that vitamin C acts as detoxifying agent by forming a poorly ionized but soluble complex (Flora 2002).

References

Ahmad S., Kitchin K.T, Cullen W.R. Arsenic species that causes release of iron from ferritin and generation of activated oxygen. Arch. Biochem. Biophys. **382:** 195-202, 2001.

Albores A., Cebrian M.E., Bach P.H., Connell J.C., Hinton R.H. and Bridges J.W. Sodium arsenite induced alterations in bilirubin excretion and heme metabolism. J. Biochem. Toxicol. **4:** 73-78, 1989.

Aposhian H.V., Maiorino R.M., Gonzalez-Ramirez D., Zuniga-Charles M., Xu Z., Hurlbut K.M., Junco-Munoz P., Dart R.C. and Aposhian, M.M. Mobilization of heavy metals by newer, therapeutically useful chelating agents. Toxicology **97:** 23-38, 1995.

Aposhian H.V. and Aposhian M.M. Meso 2,3 dimercaptosuccinic acid: chemical, pharmacological and toxicological properties of an orally effective metal chelating agent. Ann. Rev. Pharmacol. Toxicol. **30:** 279-306, 1990.

ATSDR. ATSDR case studies in environmental medicine. Agency for Toxic Substance and Disease Registry. Atlanta GA, USA, 1990.

Bansal S.K., Haldar N. and Dhank U.K. Phrenic neuropathy in arsenic poisoning. Chest **100:** 878-80, 1991.

Basu A., Mahata J., Gupta S. and Giri A.K. Genetic toxicology of a paradoxical human carcinogen, arsenic: a review, Mutation Research **488:** 171-194, 2001.

Bates M.N., Smith, A.H. and Hopenhayn-Rich C. Arsenic ingestion and internal cancers: A review. Am. J. Epidemiol **134:** 462-476, 1992.

Beaudoin A.R. Teratogenecity of sodium arsenate in rats. Teratology **10:** 153-158, 1974.

Bernstam L. and Nriagu J. Molecular aspects of arsenic stress. J. Toxicol. Environ. Health Part B, **3:** 293-322, 2000.

Bertolero F., Pozzi G., Sabbioni E. and Saffiotti U. Cellular uptake and metabolic reduction of pentavalent to trivalent arsenic as determinants of cytotoxicity and morphological transformation. Carcinogenesis **8:** 803-808, 1987.

Biswas S., Talukder G. and Sharma A. Protection against cytotoxic effects of arsenic by dietary supplementation with crude extract of Emblica officinalis fruit. Phytother. Res. **13:** 513-516, 1999.

Bosque M.A., Domingo J.L, Corbella J., Jones M.M. and Singh P.K. Developmental toxicity evaluation of monoisoamyl meso 2,3-dimercaptosuccinate in mice. J. Toxicol Environ Health **42:** 443-450, 1994.

Buchet J.P. and Lauwerys R. Study of factors influencing the in vivo methylation of inorganic arsenic in rats. Toxicol. Appl. Pharmacol. **91:** 65-74, 1987.

Buchet J.P. and Lauwerys R. Role of thiol in the in vitro methylation of inorganic arsenic by rat liver cytosol. Biochem. Pharmacol. **37:** 3149-3153, 1998.

Burns L.A., Butterworth L.F. and Munson A. E. Reversal of gallium arsenide induced suppression of the in vitro generated antibody response by a mixed disulfide metabolite of meso 2,3-dimercaptosuccinic acid. J. Pharmacol. Exp. Ther. **264:** 695-700, 1993.

Burns L.A., Sikorski E.E., Saady J.J. and Munson A.E. Evidence for arsenic as the immunosuppressive component of gallium arsenide. Toxicol. Appl. Pharmacol. **110:** 157-169, 1991.

Byron W.R., Bierbower G.W., Brouwer, J.B. and Hanse W.H. Pathological changes in rats and dogs from two year feeding of sodium arsenite or sodium arsenate. Toxicol. Appl. Pharmacol. **10:** 132-147, 1967.

Calderon J., Navarro M.E., Jimernez-Capdeville M.E., Santos-Diaz M.A., Golden A., Rodriguez-Leyva I., Borja-Aburto V., Diaz-Barriga F. Exposure to arsenic and lead and neuropsychological development in Mexican children. Environ. Res., **85:** 69-76, 2001.

Carter D.E. and Sullivan J. B. Intermetallic semiconductor and inorganic hydrides. In Hazardous materials Toxicology, Clinical Principles of Environment, Edited by JB Sullivan, GR Kreiger, Williams & Wilkins, MD, USA, pp 916-921, 1992.

Cebrian M.E., Albores A., Aguilar M. and Blakely E. Chronic arsenic poisoning in the north of Mexico. Hum. Toxicol. **2:** 121-133, 1983.

Cebrian M.E., Albores A., Connelly J.C., Bridges J.W. Assessment of arsenic effects on cytosolic heme status using tryptophan pyrrolase as an index. J. Biochem. Toxicol. **3:** 77-86, 1988.

Chakraborty D. and Saha K.C. Arsenical dermatosis from tube well water in West Bengal. Ind. J. Med. Res. **85:** 326-334, 1987.

Chattopadhya S., Ghosh S., Debnath J. and Ghosh D. Protection of sodium arsenite-induced toxicity by co-administration of L-ascorbate (vitamin C) in mature wistar strain rat. Arch. Environ. Contam. Toxicol. **41:** 83-89, 2001.

Chen G.S., Asai T., Suzuki Y., Nishioka K., and Nishiyama S. A possible pathogenesis for Blackfoot disease – effects of trivalent arsenic (As2O3) on cultured human umbilical vein endothelial cells. J. Dermatol. **17:** 599-608, 1990.

Chen C.J., Wu M.M. and Kuo T.L. Cancer potential in lung liver bladder and kidney due to ingested inorganic arsenic in drinking water. Br. J. Cancer **66:** 888-892, 1992.

Chen Y.C., Lin-Shiau S.Y. and Lin J.K. Involvement of reactive oxygen species and caspase 3 activation in arsenite-induced apoptosis. J. Cell Physiol. 177, 324-329, 1998.

Chouchane S. and Snow E.T. In vitro effects of arsenicals compounds on Glutathione-related enzymes. Chem. Res. Toxicol. **14:** 517-522, 2001.

Chowdhury U.K., Biswas B.K., Chowdhuary R.T., Samanta G., Mandal B.K., Basu G.K., Chanda G.R., Lodh D., Saha K.C., Mukherjee S.C., Roy S., Kabir S., Quamruzzaman Q., and Chakraborti D. Groundwater arsenic contamination in Bangladesh and West Bengal, India. Environ. Health Perspect. **108:** 393-397, 2000.

Chung J.S., Kalman D.A., Moore L.E., Kosnett M.J., Arroyo A.P., Boeris M., Mazumder D.N., Hernandez A.L. and Smith A.H. Family correlation of arsenic methylation patterns in children and parents exposed to high concentration of arsenic in drinking water. Environ. Health Perspect. **110:** 729-733, 2002.

Concha G., Nermell B. and Vahter M. Metabolism of inorganic arsenic in children with chronic high arsenic exposure in northern Argentina. Environ. Health Perspect. **106:** 355-359, 1998.

Ding G.S. and Liang Y.Y. Antidotal effects of dimercaptosuccinic acid. J. Appl. Toxicol. 11, 7-14, 1991.

Enterline P.E., Dat R. and Marsh G.M. Cancers related to exposure to arsenic in copper smelter. Occup. Environ. Med. **52:** 28-32, 1995.

EPA. Carcinogenecity Peer Review of Cacodylic Acid. Memorandum from S. Malish and E. Rinde to C.Giles- Parker and J. Ellenberger, U.S. EPA, Office of Pestcides and toxic substances, Washington, D.C. 1994

Flora S.J.S. Alterations in some hepatic biochemical variables following repeated gallium arsenide administration in rats. Int Hepatol Comm **5:** 97-103, 1996.

Flora S.J.S. Arsenic induced oxidative stress and its reversibility following combined administration of n-acetylcysteine and meso 2,3 dimercaptosuccinic acid in rats. Clin. Exp. Pharmacol. Physiol. **26:** 865-869, 1999.

Flora S.J.S. and Kumar P. Biochemical and Immunotoxicological Evaluation of few Metal Chelating Drugs in rats. Drug Invest. **5:** 269-273, 1993.

Flora S.J.S., Dube S.N., Sachan A.S. and Pant S.C. Effect of Multiple Gallium Arsenide Exposure on some Biochemical indices in Rat Brain, Indust. Health **32:** 247-252, 1994.

Flora S.J.S., Dube S.N., Kannan G.M., Arora U. and Malhotra P.R. Therapeutic Potential of Meso 2,3-dimercaptosuccinic acid and 2,3-dimercaptopropane 1-sulfonate against chronic Arsenic Poisoning in Rats. Biometals, **8:** 111-117, 1995.

Flora S.J.S. and Kumar P. Biochemical and Immunotoxicological Alterations Following Repeated Gallium Arsenide Exposure and Their Recoveries by meso 2,3-dimercaptosuccinic acid and 2,3-dimercaptopropane 1-sulfonate administration in rats. Environ. Pharmacol. Toxicol. **2:** 315-320, 1996.

Flora S.J.S., Pant B.P., Tripathi N., Kannan G.M. and Jaiswal D.K. Distribution of Arsenic by Diesters of Meso 2,3- Dimercaptosuccinic Acid during Sub-Chronic Intoxication in rats. J. Occup. Health, **39:** 119-123, 1997.

Flora S.J.S., Kannan G.M., Kumar P. Selenium Effects on Gallium Arsenide induced Biochemical and Immunological Changes in male rats. Chem. Biol. Inter., **122:** 1-13, 1999.

Flora S.J.S., Dube S.N., Arora U., Kannan G.M., Malhotra P.R. Therapeutic potential of meso 2,3 dimercaptosuccinic acid or 2,3, dimercapto propane 1- sulphonate in chronic arsenic intoxication in rats. Biometals **8:** 111-116, 1995.

Flora S.J.S., Pant S.C., Malhotra P.R. and Kannan G.M. Biochemical and histopathological changes in arsenic intoxicated rats co-exposed to ethanol. Alcohol **14:** 563-568, 1997.

Flora S.J.S., Kumar P., Kannan G.M. and Rai G.P. Effects of single oral administration of gallium arsenide on some selected biochemical and immunological variables in male wistar rats. Toxicol. Letters. **94:** 103-113, 1998.

Flora S.J.S and Tripathi N. Hepatic and Renal Metallothionein Induction following Single oral administration of gallium arsenide in rats. Biochem. Molecul. Biol. Inter., **45:** 1121-1127, 1998.

Flora S.J.S. Possible Health hazards associated with the use of toxic metals in semiconductor industries. J. Occupat. Health, **42:** 105-110, 2000.

Flora S.J.S., Dubey R., Kannan G.M., Chauhan R.S., Pant B.P. and Jaiswal D.K. Meso 2,3-dimercaptosuccinic acid (DMSA) and monoisoamyl DMSA effect on gallium arsenide induced pathological liver injury in rats. Toxicol. Lett. **132:** 9-17, 2002.

Flora S.J.S., Kannan G.M, Pant B.P, Jaiswal D.K. Combined administration of oxalic acid, succimer and its analogue in the reversal of gallium arsenide induced oxidative stress in rats. Arch. Toxicol. **76:** 269-276, 2002.

Flora S.J.S., Kannan G.M., Pant B.P and Jaiswal D.K. The efficacy of monoisoamyl ester of dimercaptosuccinic acid in chronic experimental arsenic poisoning in mice. J. Environ. Sci. Health, Part C, **38:** 241-254, 2003.

Flora S.J.S. and Mehta A. Haematological, hepatic and renal alterations after repeated oral or intraperitoneal administration of monoisoamyl DMSA II. Changes in Female Rats. J. Appl. Toxicol. 23, 97-102, 2003.

Flora S.J.S., Mehta A., Rao P.V.L., Kannan G.M., Bhaokar A., Dube S.N., and Pant B.P. Therapeatic potential of monoisoamyl and monomethyl esters of meso 2,3- dimer captosuccinic acid in gallium arsenide intoxicated rats. Toxicology, In Press, 2004.

Gale G.R., Smith A.B., Jones M.M. and Singh P.K. Meso 2,3-dimercaptosuccinic acid monoalkyl esters: effect on mercury levels in mice. Toxicology **81:** 49-56, 1993.

Garcia-Vargas G.G., Garcia-Rangel A., Aguilar-Romo M., Garcia-Salcedo J., Razo L.M., Ostrosky-Wegman P., Nava C.C. and Cebrain M.E. A pilot study on the urinary excretion of porphyrins in human populations chronically exposed to arsenic in Mexico. Hum. Exp. Toxicol. 10, 189-193. 1991.

Gorby M.S. Arsenic poisoning. West J. Med. 149; 308-315, 1988.

Goering P.L., Maronpot R.R. and Fowler, B.A. Effect of intratracheal gallium arsenide administration on d-aminolevulinic acid dehydratase in rats: relationship to urinary excretion of aminolevulinic acid. Toxicol. Appl. Pharmacol. **92:** 179-193, 1988.

Gonsebatt M.E., Vega A.M., Salazar R., Montero P., Guzeman J., Blas L.M., Del Razo G., Garcia-Vargas A., Albores M.E., Cebrian M., Kelsh P. and Ostrosky-Wegman P. Cytogenic effects in human exposure to arsenic. Mut. Res., **386:** 219-228, 1997.

Graeme K.A., Pollack Jr. C.V. Heavy metal toxicity, Part I: arsenic and mercury. J. Emerg. Med. **16:** 45-56, 1998.

Graeme K.A. and Pollack Jr. C.V. Heavy Metal Toxicity, Part II: lead and metal fume fever. J. Emerg. Med. **16:** 171-177, 1998.

Gregus Z., Gyurasics A. and Koszorus L. Interaction between selenium and group Va-metalloid (arsenic, antimony and bismuth) in the biliary excretion. Environ. Toxicol. Pharmacol **5:** 89-99, 1998.

Guha Mazumder D.N., De B.K., Santra A., Ghosh N., Das S., Lahiri S. and Das T., Randomized placebo-controlled trail of 2,3 –dimercaptopropane 1-sulfonate (DMPS) in therapy of chronic arsenicosis due to drinking arsenic exposure contaminated water. Clin. Toxicol. 39, 665-674, 2001.

Guha Mazumder D.N., Das Gupta J. amd Santra A. Chronic arsenic toxicity in West Bengall-The worst calamity in the world. J. Ind. Med. Assoc. **96:** 4-7, 1998.

Guha Mazumder D.N., Ghoshal U.C., Saha J., Santra A., De B.K., Chatterjee A., Dutta S., Angle C.R. and Centeno J.A. Randomized placebo-controlled trial of 2,3-dimercaptosuccinic acid in therapy of chronic arsenicosis due to drinking arsenic-contaminated subsoil water. Clin. Toxicol. 36, 683-690, 1998.

Golub M.S., Macintosh M.S and Baumrind N. Developmental and reproductive toxicity of inorganic arsenic: Animals studies and human concerns. J. Toxicol. Environ Health B. 1; 199-241 1998.

Goyer R.A. Results of lead research: prenatal exposure and neurological consequences. Environ. Health Perspect. **104:** 1050-1054, 1996

Guyton K.Z. and Keusler T.W. Oxidative mechanism in carcinogenesis. Brit. Med. Bull. **49:** 523-544, 1993.

Hall A.H. Chronic arsenic poisoning. Toxicol. Lett. 128, 69-72, 2002.

Harrington-Brock K., Smith T.W., Doerr C.L. and Moore M.R. Effects of arsenic exposure on the frequency of HPRT mutant lymphocytes in a population of copper roaster in Anto Fagasta, Chile: a pilot study. Mutation Res., **431:** 247-257, 1999.

Hartman A. and Speit G. Effects of arsenic and cadmium on the persistence of mutagen-induced DNA lesions in human cells. Environ Mol Mutagen **27:** 98-107, 1996.

Heranandez-Zavala A., Del Razo L.M., Garcia-Vargas G.G., Aguilar C., Borja V.H., Albores A. and Cebrian M.E. Altered activity of heme biosynthesis pathway enzymes in individual chronically exposed to arsenic in Mexico. Arch. Toxicol. **73:** 90-95, 1999.

Hei T.K., Liu S.X. and Waldron C. Mutagenecity of arsenic in mammalian cells; role of reactive oxygen species. Proc. Natl. Acad. Sci. U.S.A. **95:** 8103-8107, 1998.

Hindmarsh J.T. Caveats in hair analysis in chronic arsenic poisoning. Clin. Biochem. **35:** 1-11, 2002.

Hill C.H. Interrelationships of selenium with other trace elements. Fed. Proc 34, 2096, 1975.

Hood R.D. Toxicology if prenatal exposure to arsenic. In Arsenic, Eds, W.H. Lederer and R.J. Fensterheim, pp 134-150, New York, Van Nostrand Reinhold, 1993.

Hopenhayn-Rich C., Biggs M.L., Kalman D.A., Moore L.E. and Smith A.H. Arsenic methylation patterns before and after changing from high to lower concentrations of arsenic in drinking water. Environ. Health Prospect. **104:** 1200-1207, 1996a.

Hopenhayn-Rich C., Biggs ML., Smith AH., Kalman DA., and Moore LE Methylation study of a population environmentally exposed a to arsenic in drinking water. Environ. Health Perspect. **104:** 620-628, 1996b.

Hopenhayn-Rich C., Biggs K.L., Fuchs A., Bergolio R., Tello E.E., Nicolli H. and Smith A.H. Bladder cancer mortality associated with arsenic in the drinking water in Argentina. Epidemiology **7:** 117-124, 1998

Hopenhayn—Rich C., Biggs M.L. and Smith A.H. Lung and kidney cancer mortality associated with arsenic in drinking water in Cordoba, Argentina. Int J Epidemiol **27:** 561-569, 1998.

Hood R.D., Vedel G.C., Zaworotko M.J. and Tatum F.M. Uptake, distribution, and metabolism of trivalent arsenic in the pregnant mouse. J. Toxicol. Environ. Health **25:** 432-434, 1988.

Hoover T.D. and Aposhian H.V. BAL increases the arsenic 74 content of rabbit brain. Toxicol. App. Pharmacol. **70:** 160-162, 1983.

Huges M.F, Arsenic toxicity and potential mechanism of action. Toxicol Lett. **133:** 1-16, 2002.

IARC. Monograph on the Evaluation of Carcinogenic Risks to Humans: Overall Evaluations of Carcinogencity: An updating of IARC Monographs, Vol 1 to 42, IARC, Lyon, 1987.

Inns R.H. and Rice P. Efficacy of dimercapto chelating agents for the treatment of poisoning by percutaneously applied dichloro (2-chlorovinyl) arsine in rabbits. Hum. Exp. Toxicol. **12:** 1241-1246, 1993.

Ishizaki M. Studies on the pharmacokinetics of arsenicals in rats. Jpn J. Hygiene **35:** 584-596, 1980.

Ito H., Okamoto K., Kato K. Enhancement of expression of stress proteins by agents that lower the levels of glutathione in cells. Biochem. Biophys. Acta **1397:** 223-230, 1998.

Itoh T., Zhang Y.F., Murai S., Saito H. The effects of arsenic trioxide on brain monoamine metabolism and locomotor activity of mice. Toxicol. Lett. **54:** 345-353, 1990.

Jacobsen I.A. and Jorgensen P.J. Chronic lead poisoning treated with dimercaptosuccinic acid. Pharmacol Toxicol **68:** 266-269, 1991.

Jha A.N., Nodity M., Nilsson, R. and Natarajan A.T. Genotoxic effects of arsenite on human cells. Mutat. Res. **284:** 215-221, 1992.

Johnson L.R. and Farmer J.G. Use of human metabolic studies and urinary arsenic speciation in assessing arsenic exposure. Bull. Environ. Contam. Toxicol. 46: 53-61, 1991.

Jones M.M., Singh P.K., Gale G.R., Smith A.B. and Atkins L.M. Cadmium mobilization in vivo by intraperitoneal or oral administration of mono alkyl esters of meso 2,3-dimercaptosuccinic acid. Pharmacol. Toxicol. **70:** 336-343, 1992.

Jones M.M., Singh P.K., Gale G.R., Smith A.B. and Atkins L.M. Cadmium mobilization in vivo by intraperitoneal or oral administration of mono alkyl esters of meso 2,3-dimercaptosuccinic acid. Pharmacol. Toxicol. **70:** 336-343, 1992.

Kannan G.M. and Flora S.J.S. Chronic Arsenic Poisoning in Rat: Treatment with Combined Administration of Succimers and an Antioxidant. Ecotoxicol. Environ Safety, In Press, 2003

Kannan G.M., Tripathi N., Dube S.N., Gupta M. and Flora S.J.S. Toxic effects of arsenic (III) on some hematopoietic and central nervous system variables in rats and guinea pigs. Clin. Toxicol. **39:** 675-682, 2001.

Kitchin K.T. Recent advances in arsenic carcinogenesis: Modes of action, animal model systems and methylated arsenic metabolites. Toxicol. Appl. Pharmacol. **172:** 249-261, 2001.

Klaassen C.D. Heavy metals and heavy metal antagonist. In Gilman A.G., Rall TW., Nies A.S., Taylor P. (Eds.) The Pharmacological Basis of Therapeutics, McGraw-Hill, New York, pp 1592-1614, 1996.

Kostial K., Blanusa M., Simonovic I., Jones M. M. and Singh P.K. Decreasing ^{203}Hg retention by intraperitoneal treatment with monoalkyl esters of meso 2,3-dimercaptosuccinic acid in rats. J. Appl. Toxicol. **13:** 321-326, 1993.

Kostial K., Kargacin B., Blanusa M., Piasek M., Jones M.M. and Singh P.K. Monoisoamyl meso 2,3-dimercaptosuccinate as a delayed treatment for mercury removal in rats. Environ. Health Perspect. **102:** 309-312, 1994.

Kreppel H., Reich F.X., Szinicz Fichti B. and Forth W. Efficacy of various dithiol compounds in acute arsenic poisoning in mice. Arch. Toxicol. **64:** 387-392, 1990.

Kreppel H., Paepacke U., Thiermann H., Szinicz I., Reichl F.X., Singh P.K. and Jones M.M. Therapeutic efficacy of new dimercaptosuccinic acid (DMSA) analogues in acute arsenic trioxide poisoning in mice. Arch. Toxicol. **67:** 580-585, 1993.

Kreppel H., Reichl F.X., Kleine A., Szincz L., Singh P.K. and Jones M.M. Antidotal efficacy of newly synthesized dimercaptosuccinic acid (DMSA) monoesters in experimental arsenic poisoning in mice. Fund. Appl. Toxicol. 26, 239-245, 1995.

Lee T.C. and Ho I.C. Differential cytotoxic effects of arsenic on human and animals cell. Environ. Health Perspect. **102** (Suppl 3), 101, 1994.

Liu J., Kaduska M., Liu Y., Qu W., Mason R.P. and Walker M.P. Acute arsenic induced free radical production and oxidative stress related gene expression in mice. Toxicologists **54***:* 280-281, 2000

Lu F.J. Arsenic is a promoter in the effect of humic substances on the plasma prothrombin time in vitro. Thromb Res **58:** 537-541, 1990a

Lu F.J. Fluorescent humic substances and Blackfoot disease in Taiwan. Appl Organometall Chem. **4:** 191-195, 1990b

Maiti A. and Chatterjee A.K. Differential response of cellular antioxidant mechanism of liver and kidney to arsenic exposure and its relation to dietary protein deficiency. Environ. Toxicol. Pharmacol. **8:** 227-235, 2000.

Malachowski M.E. An update on arsenic. Clin Lab Med. **10:** 459-472, 1990.

Martinez G., Cebrian M., Chamorro G. and Jauge P. Urinary uroporphyrin as an indicator of arsenic exposure in rats. Proc West Pharmacol Soc. **26:** 171-174, 1983.

Mass M.J., Tennant A., Roop B.C., Cullen W.R., Styblo M., Thomas D.J. and Kligerman A.D. Methylated trivalent arsenic species are genotoxic. Chem. Res. Toxicol. **14:** 355-361, 2001.

Mehta A., Kannan G.M., Dube S.N., Pant B.P., Pant S.C. and Flora S.J.S. Haematological, hepatic and renal alterations after repeated oral or intraperitoneal administration of monoisoamyl DMSA I. Changes in Male Rats, J. Appl. Toxicol, **22:** 359-369, 2002.

Moncure G., Jankowski P.A. and Drever, J.I. The hydrochemistry of arsenic in reservoir sediments, Miltown, Montana, USA. In water rock-interaction, low temperature environments. Vol. I Eds. (Kharaka, YK, Maest, AS). AA Balkema, Rotterdam, pp. 513-516, 1992.

Moore M.M., Harrington-Brock K. and Doerr C.L. Relative genotoxicity potency of arsenic and its methylated metabolites. Mutat Res **386:** 279-290, 1997.

Moore D.F., O'Callaghan C.A., Berlyne G., Ogg C.S., Davies H.A., House I.M. and Henry J.A. Acute arsenic poisoning: absence of polyneuropathy after treatment with 2,3-dimercaptopropanesulphonate (DMPS). J. Neurol. Neurosurg. Psychiatry **57:** 1133-1135, 1994.

Moore L.E., Lu M. and Smith A.H. Childhood cancer incidence and arsenic exposure in drinking water in Nevada. Arch. Environ. Health **57:** 201-206, 1992.

Morton W.E. and Caron G.A. Encephalopathy; and uncommon manifestation of workplace arsenic poisoning. Am. J. Ind. Med. 287, 698-701, 1998.

Muckter H., Liebl B., Reichi F.X., Hunder G., Walther U. and Fichtl B. Are we ready to replace dimercaprol (BAL) as an arsenic antidote? Hum Exp. Toxicol. 16, 460-465, 1997.

Nair P.P., Murthy H.S. and Grossman N.R. The in vivo effect of vitamin E in experimental prophyria. Biochem. Biophys. Acta **215:** 112-118, 1970.

Nriagu J.O. Arsenic in the environment, Part II, Human health and ecosystem effects. New York; Wiley, 1994.

Nordenson I. and Beckman L. Is the genotoxic effect of arsenic mediated by oxygen free radicals? Hum. Hered. **41:** 71-73, 1991.

Oknonishnivkova I.Y. and Nirember V.L. Absorption, distribution and excretion of ^{35}S-meso dimercaptosuccinic acid (succimer) In: BT Velichkovskli, Editor. Voprosy Dksp Klin Ter Profil Prom Intoksikatsli Sverdkovsk 11-14, 1974.

Pande M., Mehta A., Pant B.P. and Flora S.J.S. Combined Administration of a Chelating Agent and an Antioxidant in the Prevention and Treatment of Acute Lead Intoxication in RatsEnviron. Toxicol. Pharmacol. **9:** 173-184, 2001

Pande M. and Flora S.J.S. Lead induced oxidative damage and its response to combined administration of a-Lipoic acid and succimers in rats. Toxicology, **177:** 187-196, 2002

Paul M.W. and Charles A.T. Recovery from severe arsenic-induced peripheral neuropathy with 2,3 dimercaptopropane 1- sulphonic acid. Clin. Toxicol. 38, 777-780, 2000.

Pershagen G. The carcinogenecity of arsenic. Environ. Health Perspect **40:** 93-100, 1981.

Peraza M.A., Fierro F.A., Barber, D.S., Cesarez, E. and Real, L.T. Effects of micronutrients on metal toxicity. Environ. Health Perspect. **106:** 203-216, 1998.

Pershagen G. Lung cancer mortality among men living near arsenic emitting smelters. Amer J Epidemiol. **122:** 684-94, 1985.

Petrick J.S., Ayala-Fierri F., Cullen W.R., Carter D.E. and Aposhian H.V Monomethylarsenous acid (MMA (III)) is more toxic than arsenite in Chang human hepatocytes. Toxicol Appl Pharmacol **163:** 203-207, 2000.

Petrick J.S., Jagadish B., Mash E.A. and Aposhian H.V. Monomethylarsonous acid (MMA (III)) and arsenite in hamsters and in vitro inhibition of pyruvate dehydrogenase. Chem. Res. Toxicol. **14:** 651-656, 2001.

Planas-Bohne F. The influence of chelating agents on the distribution and biotransformation of methylmercuric chloride in rats. J. Pharmcol. Exp. Ther. **217:** 500-504, 1981.

Pontius F.W., Brown K.G. and Chen C.J. Health implications of arsenic in drinking water. J. Am. Water Works Assoc., **86:** 52-63, 1994.

Rahman M., Tondel M., Ahmad S.A., Chowdhary I.A., Faruquee M.H. and Axelson O. Hypertension and arsenic exposure in Bangladesh. Hypertension. **33:** 74-78, 1999.

Ramos O., Carrizales L., Yanez L., Mejia J., Batres L., Oritz D. and Diaz-Barriga F. Arsenic increase lipid peroxidation rat tissues by a mechanism independent of glutathione levels. Environ Health Perspect **103** (suppl 1): 85-88, 1995.

Reichl F.X., Szinicz L., Kreppel H. and Forth W. effect of arsenic on carbohydrate metabolism after single or repeated injection in guinea pigs. Arch Toxicol. **62:** 473-475. 1988.

Reichl R.X., Kreppel H., Forth W. Pyruvate and lactate metabolism in livers of guinea pigs perfused with chelating agents after repeated treatment with As_2O_3. Arch. Toxicol. **65:** 235-238, 1991.

Rodriguez V.M., Jimenez M.E. and Giordano M. The effects of arsenic on the nervous system. Toxicol. Letters In Press, 2003

Rossman T.G. Arsenic. In Environmental and Occupational Medicine (Rom.W.N. Ed) pp 1011-1019, Lippencott-Raven. Philadelphia, 1998.

Rossman T.G., Uddin A.N., Burns F. and Bosland M.C. Arsenite is a co-carcinogen with solar ultraviolet radiation for mouse skin; an animal model for arsenic carcinogenesis. Toxicol. Appl. Pharmacol. **176:** 64-71, 2001.

Saha J.C., Dikshit A.K., Bandyopadhyay M. and Saha K.C. A review of arsenic poisoning and its effects on human health. Crit. Rev. Environ. Sci. Tech **29:** 281-313, 1999.

Saha K.C. Arsenic poisoning from ground water in West Bengal. Breakthrough **7:** 5-14, 1998.

Saha K.C. Melanokeratosis from arsenic contaminated tubewell water. Ind. J. Dermatol. **29:** 37-46, 1984.

Saha K.C. and Poddar D. Further studies on chronic arsenical dermatosis. Ind. J. Dermatol. **31:** 29-33, 1986.

Santra A., Das Gupta J., De B.K., Roy B. and Guha Mazumder D.N. Hepatic manifestation in chronic arsenic toxicity. Ind. J. Gastroenterol **18:** 152-155, 1988.

Santra A., Maiti A., Das S., Lahiri S., Chakraborty S.K. and Guha Mazumder D.N. Hepatic damage caused by chronic arsenic toxicity in experimental animals. Clin. Toxicol. **38:** 395-405, 2000.

Schaumloffel N. and Gebel T. Heterogeneity of the DNA damage provoked by antimony and arsenic. Mutagenesis **13:** 281-286, 1998.

Schroeder H.A. and Mitchener M. Toxic effects of trace elements on the reproduction of mice and rats. Arch. Environ. Health **23:** 102-106, 1971.

Schwartz R.A. Arsenic and the skin. Int. J. Dermatol. **36:** 241-250, 1997.

Shannon R.L. and Strayer D.S. Arsenic-induced skin toxicity. Hum. Toxicol. **8:** 99-104, 1989.

Smith A.H., Arroyo A.P., Mazumder D.N.G., Kosnett M.J., Hernandez A.L., Beeris M., Smith M.M. and Moore L.E. Arsenic induced skin lesions among Atacameno people Northern Chile despite good nutrition and centuries of Exposure. Environ. Health Perspect. **108:** 617-620, 2000.

Smith A.H., Goycolea M., Haque R. and Biggs M.L. Marked increase in bladder and lung cancer mortality in a region of Northern Chile due to arsenic in drinking water. Am. J Epidemiol **147:** 660-669, 1998.

Sordo M., Herrera L.A., Ostrosky-Wegman P. and Rojas E. Cytotoxic and genotoxic of As, MMA and DMA on leukocytes and stimulated human lymphocytes. Teratogen. Carcinogen. Mutagen. **2:** 249-260, 2001.

Styblo M., Del Raxo L.M., Vega L., Germolec D.R., LeCluyse E.L., Hamilton G.A., Ree W., Wang C., Cullen WR. and Thomas DJ Comparative toxicity of trivalent and pentavalent inorganic and methylated arsenical in rat and human cells. Arch. Toxicol. **74:** 289-299, 2000.

Styblo M., Del Razo L.M., LeCluyse E.L., Hamilton G.A., Wang C., Cullen W.R., and Thomas D.J. Metabolism of arsenic in primary cultures of human and rat hepatocytes. Chem. Res. Toxicol. **12:** 560-565, 1999.

Styblo M., Serves S.V., Cullen W.R., and Thomas D.J. Comparative inhibition of yeast glutathione reductase by arsenicals and arsenothiols. Chem. Res. Toxicol. **10:** 27-33. 1997.

Squibb K.S. and Fowler B.A. The toxicity of arsenic and its compounds. In; Biological and Environmental Effects of Arsenic. Fowler BA., Ed., Elsevier, New York 233-269, 1983.

Tabacova S., Baird D.D., Balabaeva I., Lolova D. and Petrove I. Placental arsenic and cadmium in relation to lipid peroxides and glutathione levels in maternal-infant pairs from a copper smelter area. Placenta **15:** 873-881, 1994.

Tamaki S. and Frankenberger W.T. Environmental biochemistry of arsenic. Rev. Environ. Contam. Toxicol. **124:** 79-110, 1992.

Thomas D.J., Styblo M. and Lin S. The cellular metabolism and systemic toxicity of arsenic. Toxicol. Appl. Pharmcol. **176:** 127-144, 2001.

Thorgeirsson U.P., Daalgard D.W., Reeves J. and Adamson R.H. Tumor incidence in a chemical carcinogenesis study of nonhuman primates. Reg. Toxicol. Pharmacol., **19:** 130-151, 1994.

Tripathi N., Kannan G.M., Pant B.P., Jaiswal D.K., Malhotra P.R. and Flora S.J.S. Arsenic induced changes in certain neurotransmitters levels and their recoveries following chelation in rat whole brain. Toxicology Letters **92:** 201-208, 1997

Tripathi N. and Flora S.J.S. Effects of some thiol chelators on enzymatic activities in blood, liver and kidneys of acute arsenic (III) exposed mice. Biomed. Environ. Sci. **11:** 38-45, 1998.

Tsuda H., Yamasaki H. and Miyayama H. Sebastian platelet syndrome: two Japanese families originally diagnosed with May-Hegglin anomaly. Int. J. Hematol. **70:** 290-293, 1999.

Tsuda T., Nagira T., Yamamoto M., Kurumatani N., Hotta N., Harada M. and Aoyama H. Malignant neoplasm among residents who drank well water contaminated by arsenic from a king's yellow factory. J. UOEH. **11:** Suppl: 289-301, 1989.

Vahter M. Methylation of inorganic arsenic in different mammalian species and population groups. Sci. Prog. **137:** 8-22, 2000.

Vahter M. Genetic polymorphism in the biotransformation of inorganic arsenic and its role in toxicity. Toxicol. Lett. **112 &113:** 209-217, 2002

Vahter M. and Concha G. Role of metabolism in arsenic toxicity. Pharmacol. Toxicol. **89:** 1-5, 2001.

Vahter M. and Marafante E. In vivo methylation and detoxification of arsenic: In: locking F., Craig PJ (Eds). The biological Alkylation of Heavy Elements. Royal Society of Chemistry, London, pp 105-119, 2000.

Waalkes M.P. Target sites of carcinogenetic metals. In: Goyer R.A., Klaassen C.D. and Waalkes M.P. (Eds), Metal Toxicology. Academic press, New York, pp 54-56, 1995.

Walker Jr., E.M., Stone A., Milligan L.B., Gale G.R., Atkins L.M., Smith A.B., Jones M.M., Singh P.K. and Basinger M.A. Mobilization of lead in mice by administration of monoalkyl esters of meso 2.3-dimercaptosuccinic acid. Toxicology **76:** 79-87, 1992.

Wang Z. and Rossman T.G. In: Cheng P, Editor. Toxicology of Metals. Boca Raton, FL; CRC pp 221-229, 1996.

Wang Z. and Rossman T.G. The carcinogenecity of arsenic. In Toxicology of Metals (Chang LW Ed.,) pp 219-227, CRC Press. Boca Raton FL., 1995.

Webb D.R., Sipes I.G. and Carter D.E. In vitro solubility and in vivo toxicity of gallium arsenide. Toxicol Appl Pharmacol. 1984, **76:** 96-104.

Wildfang E., Radabaugh T.R. and Aposhian H.V. Enzymatic methylation of arsenic compounds. IX. Liver arsenite methyltransferase and arsenate reductase activities in primates. Toxicology **168:** 213-221, 2001.

Woods J.S. and Fowler B.A. Effects of chronic arsenic exposure on hematopoetic function in adult mammalian liver. Environ. Health Perspect. **19:** 209-213, 1977.

Woods J.S. and Fowler B.A. Altered regulation of mammalian hepatic heme biosynthesis and urinary porphyrin excretion during prolonged exposure to sodium arsenate. Toxicol. Appl. Pharmacol. **43:** 361-371,1978.

Woods J.S. and Southern M. Studies on the etiology of trace metal-induced porphyria. Effects of porphyrinogenic metals on coproporphyrinogen oxidase in rat liver and kidney. Toxicol. Appl. Pharmacol. **97:** 183-190, 1989.

Wu M.M., Kuo T.L., Hwang Y.H. and Chen C.J. Dose-response relation between arsenic concentration in well water and mortality from cancer and cardiovascular disease. Am. J. Epidemiol. **130:** 1123-1132, 1989.

Xu C., Holscher M.A., Jones M.M. and Singh P.K. Effect of monoisoamyl meso 2,3-dimercaptosuccinate on the pathology of acute cadmium intoxication. J. Toxicol. Environ. Health **45:** 261-277, 1995.

Yamamuchi H. and Fowler B.A. Toxicity and metabolism of inorganic and methylated arsenicals. In Nriagu I.O (Ed.) Arsenic in the Environment. II. Human Health and Ecosystem Effects. Wiley, New York, new York, pp 34-43, 1994.

Yamanaka K., Hoshino M., Okamoto M., Sawamura R., Hasegawa A. and Okada S. Induction of DNA damage by dimethylarsine a metabolite of peroxy radical. Biochem. Biophys. Res. Commun. 168, 58-64, 1990.

Yamanaka K. and Okada S. Induction of lung specific DNA damage by metabolically methylated arsenic via the production of free radicals. Environ. Health Perspect. **102** (Suppl.3): 37-40, 1996.

Yeh S. Skin cancer in chronic arsenism. Human Pathol **4:** 469-485, 1973.

Pharmacological Perspectives of Toxic Chemicals and Their Antidotes
Editors: S J S Flora, J A Romano, S I Baskin and K Sekhar

Mechanisms Involved in the Regulation of Free Radial Generation from Neutrophils

Prashant Sharma and Madhu Dikshit
Division of Pharmacology, Central Drug Research Institute
Lucknow 226 001, India

Polymorphonuclear leukocytes (PMNs) represent 33-75% of the total circulation leukocyte population and constitute the 'first line of defense' against infectious agents or 'nonself' substances that penetrate the body's physical barriers. Migration of PMNs from blood to the specific tissue following a pathological insult is a key feature of the host inflammatory response, which is often localized and protective and helps to eliminate the microorganisms. Their targets include bacteria, fungi, protozoa, viruses, virally infected cells and tumor cells. Once an inflammatory response is initiated, PMNs are the first cells to be recruited to the site of infection or injury. In inflammatory pathologies PMNs unleash their destructive potential against host tissue. More recently, the success of thrombolysis to treat acute coronary obstructions has unmasked the PMNs as an active player in the reperfusion injury and has sparked further research into the regulation of PMNs free radical generation. Thus PMNs have posed a challenge with classical 'friend or foe' dilemma.

The oxidative burst in polymorphonuclear leukocytes (PMNs) plays a fundamental role in defense against microbial and injury associated with inflammation. Multi-protein complex NADPH phagocytic oxidase is responsible for the generation of superoxide anion (Babior, 1984), which initiate a chain of reactions leading to the formation of reactive oxygen species (ROS). Oxygen, which plays a central role in ROS generation, is also a co-substrate for the synthesis of reactive nitrogen species (RNS) such as nitric oxide (NO). The role of NO in regulating several physiological processes became evident in the early 1990s. Dormant neutrophils spontaneously produced NO while activation attenuates this pathway in favor of the oxidative burst. The steady state production of ROS and RNS may dictate the anti/ pro-inflammatory balance. Microbial killing appears to be ROS and RNS dependent in normal neutrophils but nitric oxide may also play a role in modulating ROS generation in PMNs (Seth et al., 1994; Sethi et al., 1999).

NITRIC OXIDE SYNTHESIS

Nitric oxide is formed within the active site of an enzyme commonly known as nitric oxide synthase (NOS) from the oxidation of guanidino nitrogen of L-arginine. NO synthesis is catalyzed in the cells by one of three isoform of enzyme, NOS I (neuronal or nNOS), NOS II (inducible or iNOS) and NOS III (endothelial or eNOS) (Moncada and Higgs 1993). To synthesize NO, all isoforms depend on the substrate L-arginine, cosubstrate oxygen (O_2) as well as on the cofactors/coenzymes nicotinamide adenine dinucleotide phosphate (NADPH), tetrahydrobiopterin (BH_4), flavin adenine dinucleotide (FAD), flavin mononucleotide (FMN), calmodulin (CaM) and protoporphyrin IX (Knowles and Moncada, 1994). The catalytic reaction takes place in two separate mono-oxygenation steps: firstly, L-arginine is hydroxylated by O_2 and NADPH to form N^{ω}- hydroxy-L-arginine, subsequently, N^{ω}-hydroxy-L-arginine is oxidized to yield L-citrulline and NO (Moncada and Higgs, 1993).

All the NOS isoforms are homodimers. NOS homodimers associated with FAD and FMN have been described as tetramers (Alderton et al., 2001). The molecular weight of individual subunit has 130-160 kDa for nNOS and 130-135 kDa for eNOS and iNOS (Park et al., 2000). According to their contribution during NO synthesis, FMN, FAD and calmodulin can be termed as the (carboxy-terminal) "reductase" domain, whereas protoporphyrin IX (heme), BH_4, and the substrate binding site may be termed the (amino-terminal) "oxygenase" domain (Eissa et al., 1998).

Although NOS isoforms have similar NO-synthesizing machinery but synthesize superoxide radical ($O_2^{\bullet -}$) by different mechanisms in absence of vital substrate L-arginine and BH_4, nNOS and eNOS produce $O_2^{\bullet -}$ when the heme center is not occupied by L-arginine. This process requires Ca^{+2}/ calmodulin to facilitate the delivery of electron from the reductase to oxygenase domains (Alderton et al., 2001). In contrast superoxide formation from iNOS was much less sensitive to L-arginine (Abu-Soud and Stuehr, 1993) and cannot be blocked by cyanide, suggesting that the reaction does not occur at the heme. On the other hand, the flavoprotein inhibitor, diphenyleneiodonium (DPI) totally prevented iNOS catalyzed $O_2^{\bullet -}$ formation suggesting that $O_2^{\bullet -}$ synthesis occurs at the flavin binding site of the reductase domain (Xia et al., 1998).

The gene coding for nNOS is located on the long arm of chromosome 12 yielding nNOS mRNA with a size of ~12 kb (Xu et al., 1993). It has 29 exons and 28 introns the translation initiation and termination sites are located at exons 2 and 29 respectively. Transcriptional regulatory sites are present on the 5'-flanking region of the exon 1 of human nNOS. It may be transcriptionally regulated by various transcriptional factors such as activator enhancer binding protein-2 (AP-2), transcription enhancer factor-1 (TEF-1), cyclic AMP response element binding protein (CREB), activating transcription factors (ATF), activator protein complex-1 (AP-1 [cFos+cJun]), nuclear respiratory factor-1 (NRF-1), epithelial specific transcription factor (Ets), neurofibromin type I (NF-1) and nuclear factor kappa B (NF-kB). It suggests that the 5' terminus of the mRNA is tissue specific, though it does not affect the encoded protein. It might be possible thus that the variants facilitate a sophisticated regulatory mechanism for the tissue specific gene expression. Synthesis of NO from nNOS is modulated by changes in the Ca^{+2} levels (Bredt and Snyder, 1990). The enzyme displays binding sites for cofactors /coenzymes such as FAD, FMN, NADPH, calmodulin and protein kinase A (PKA) and protein kinase C (PKC) (Knowles and Moncada, 1994). Neuronal NOS is found in many tissues, including peripheral nerves (Krukoff, 1998), spinal cord (Callsen-Cencic et al., 1999), neutrophils (Greenburg et al., 1998) and human skeletal muscle (Stamler and Meissner, 2001). The eNOS gene has been mapped in the pericentric region of chromosome 7, the corresponding eNOS mRNA is ~4.2-4.4 kb in size (Park et al., 2000). The catalytic activity of eNOS is also dependent on intracellular Ca^{+2}. eNOS is similar in function to nNOS, sharing binding sites for FMN, FAD, PKA and calmodulin, however one major difference in structure is the presence of myristoylation (the addition of a "myristoyl" molecule consisting of 14 carbon atoms) site at the N-terminal part (Pollock et al., 1992) to associate enzyme with membrane (Lamas et al., 1992). Presence of eNOS in the particulate and cytosolic fraction (Knowles and Moncada, 1994), of endothelial cells of arteries (Lamas et al., 1992), and veins (Janssens et al., 1992) as well as in PMNs (Miguel et al., 2002) has been demonstrated. Regulation of eNOS activity is achieved by a complex set of stimuli acting upon transcription of the eNOS gene, mRNA translation and post-translational modification such as myristoylation (Govers and Rabelink, 2001). The iNOS gene is located at chromosome 17 (Xu et al., 1994). The corresponding iNOS mRNA is ~4.2-4.5 kb in size. Calmodulin remains non-covalently bound to the iNOS complex and therefore constitutes an essential subunit of this isoform (Cho et al., 1992). The consensus binding sites for NADPH, FMN and FAD also exist like other isoform of NOS (Knowles and Moncada, 1994). In contrast to the constitutively active isoforms, iNOS exerts its functions independent of Ca^{+2}. iNOS activity has been demonstrated in a wide array of cells and tissues (Stuehr and Griffith, 1992), e.g. macrophages (MacMicking et al., 1997), chondrocytes (Charles et al., 1993), kupffer cells, hepatocytes (Curran et al., 1990; Geller et al., 1993), neutrophils (Sethi and Dikshit, 2000), pulmonary epithelium (Asano et al., 1994) and vasculature (Hickey et al., 2001). iNOS produces NO in response to various stimuli, most prominently endotoxin and endogenous pro-inflammatory mediators. The most prominent cytokines involved in iNOS stimulation are tumor necrosis factor-a (TNF-a), interleukin-1b (IL-1b) and interferon-γ (IFN-γ). Furthermore, lipopolysaccharide (LPS, endotoxic) induced iNOS expression has also been extensively investigated. Regulation of NO production via iNOS necessarily occurs during transcription and translation. One important intracellular signal transduction pathway of these stimuli is the activation of NF-kB. An alternative pathway involves the janus tyrosine kinase (JAK) – signal transducers and activators of transcription (STAT) (Xuan et al., 2001; Kleinert et al., 1998) however, two or more signal transduction pathways are necessary for the maximal induction of iNOS expression. The post-transcriptional regulation of iNOS gene expression predominantly occurs via mechanisms that influence iNOS mRNA stability (Kleinert et al., 1998) and regulation of catalytic activity of the protein (Kunz et al., 1996).

NITRIC OXIDE SYNTHESIS IN PMNS

The ability of PMNs to relax endothelium denuded aortic rings and inhibit platelet aggregation lead to the characterization of a soluble factor released form the PMNs (Rimele et al., 1988; Salvemini et al., 1989; Nicolini and Mehta, 1990; Kadota et al., 1991). Neutrophil derived relaxation factor (NDRF) has a chemical and pharmacological profile similar to that of endothelium derived relaxation factor (EDRF) or NO (Rimele et al., 1988). Molecular biology techniques have shown that both circulating rat (Greenberg et al., 1998) and human PMNs (Wallerath et al., 1997) have the mRNA and express 150 kDa protein for nNOS. Human PMNs contain significantly lower amount of mRNA for NOS than rat–circulating cells. Discrepancies are however, also available in the literature regarding presence of nNOS in human PMNs (Greenburg et al., 1998). Rat circulation PMNs contain more nNOS protein per unit amount of nNOS mRNA than in rat cortex or cerebellum, suggesting that either the rate of translation is greater or the rate of degradation is slower in these cells (Kolb et al., 1994). Rat PMNs express very low level of iNOS mRNA, which is greatly upregulated in glycogen – elicited cells (Miles et al., 1995). NOS (150 kDa) has been purified from the cytosolic fraction of oyster glycogen–elicited rat peritoneal PMNs by (Yui et al., 1991). nNOS in PMNs might also be different from neuronal isozyme and thus may not hybridize with polyclonal anti-nNOS antibody (Wallerath et al., 1997). The activity of NOS was found to be dependent on Ca^{+2}, NADPH and BH_4 having no requirement for calmodulin and thus was different from nNOS (Kolb et al., 1994; Hiki et al., 1991). A calcium independent iNOS activity sensitive to the inhibition by glucocorticoids has also been found in rat peritoneal cells (Padgett, 1995; McCall et al., 1991).

The presence of iNOS mRNA and protein has been shown in unstimulated circulating human PMNs with very low level of activity (Amin et al., 1995). Miles et al., (Miles et al., 1995) were however, unable to detect iNOS mRNA or enzyme activity in the extravasated human cells. The iNOS in human PMNs was though found to be more than 90 % membrane associated (Wheeler et al., 1997). Evans et al., (1996) found that iNOS in human PMNs colocalizes with myeloperoxidase in primary granules. The vacuolar localization of iNOS appears to be an ideal placement for the microbicidal activity. Specific subsets of PMNs, which are in particular stage of differentiation, possibly express iNOS (Amin et al., 1995). It was found that calmodulin (CaM) inhibitors though inhibited NOS activity but CaM was less tightly bound to the enzyme supporting the suggestion made by Yui et al., (1991) that NOS in rodents is CaM independent. The presence of iNOS has been demonstrated in PMNs isolated from oral cavity (Nakahara et al., 1998), septic syndrome (Goode et al., 1995; Tsukahara et al., 1998) and from the PMNs after 16 hrs incubation with a mixture of cytokines (Evans et al., 1996).

NOS purified from PMNs was monomeric (Yui et al., 1991), though the enzyme present in intact cells might actually be a dimer. L-Arginine concentration is an important determinant of NOS activity; however, its levels in PMNs have not been quantitated so far. It has been suggested that normally the intracellular L-arginine levels in PMNs might be sufficient for NO synthesis (Sethi et al., 2001), however, after LPS treatment due to induction in NOS activity and L-arginine transporter, PMNs utilizes more L-arginine to synthesize large amount of NO (Sethi et al., 2001). LPS treatment also augments circulating levels of nitrite (Travares-Murta et al., 1996). L-Arginine transport system in PMNs is independent of the γ-glutamyl cycle and occur through the y^+ system (Riesco et al., 1993). The apparent K_m for cNOS in human PMNs is 1454±160 μM and V_{max} is 0.11±0.02 mM/min/mg of protein (Riesco et al., 1993). For 150 kDa nNOS protein, isolated from rat peritoneal PMNs, the K_m is 22 mM and the V_{max} is 485 nM/min/mg of protein (Yui et al., 1991). The apparent V_{max} values of L-arginine transport for the PMNs isolated from control and LPS treated rat were 5.0 and 9.2 μmol/min/10^7 cells, while, apparent K_m values were 200.4 ± 40 μM and 241.5± 26 μM respectively. PMNs isolated from peritoneal cavity after LPS treatment had the apparent V_{max} 10.7 μmol/min/10^7 and K_m was 30.1± 41 μM, suggesting that peritoneal cells exhibited increase rate of transport of L-arginine and its affinity for NOS in comparison to peripheral PMNs (Sethi et al., 2001).

FORMATION OF NO-DERIVED METABOLITES IN PMNS

Nitrite (NO_2^-), a major end product of NO metabolism, promotes tyrosine nitration through formation of nitryl chloride (NO_2Cl) and nitrogen dioxide ($^\bullet NO_2$) by reaction with the inflammatory mediators hypochlorous acid (HOCl) or myeloperoxidase (MPO) (Riesco et al., 1993; Eiserich et al., 1996, 1998, Van der Vhat et al 1997). Activated PMNs convert NO_2^- in to potent oxidants NO_2Cl and $^\bullet NO_2$ through MPO-dependent pathway (Klebanoff, 1993). These findings reveal that NO_2^- might play an important role in phagocyte-mediated oxidation reaction at the site of inflammation and infections. Colocalization of iNOS and MPO in neutrophils primary granules (Evans et al., 1996) infers that NO_2^- dependent formation of NO_2Cl and $^\bullet NO_2$ in the phagolysosome may represent a mechanism of host defense. This prospect is further strengthened by the observation that NO_2^- can enhance the bactericidal activity of MPO Klebanoff (1993) and phagocytosed bacteria and model particles were nitrated (Evans et al., 1996) and chlorinated (Hazen et al., 1996) on tyrosine residues. Nitrotyrosine levels in peritoneal fluids obtained from MPO-deficient mice infected with *Klebsiella pneumoniae* were decreased relative to the wild type mice-despite the induction of iNOS and comparable accumulation NO_2^- in both the strains (Gaut et al., 2002). By simultaneous measurement of chlorotyrosine and nitrotyrosine in these fluids, the author showed that the formation of nitrotyrosine and chlorotyrosine was comparable in wild type mice, while chlorotyrosine was totally lacking in MPO knockout mice, however theses mice showed detectable nitrotyrosine. It has also been recently reported that ischemia/reperfusion induced cerebral damage was greater in MPO knockout than in wild type and accompanied by increased level of protein tyrosine nitration (Takizawa et al., 2002). Tyrosine nitration thus can be mediated by multiple pathways under different conditions (Beckman, 1996; Simon et al., 1996; Gunttor et al., 1997; Takizawa et al., 2002), and the formation of nitrotyrosine should not be considered a specific marker of $ONOO^-$ formation but a collective indicator of the reactive nitrogen species.

RESPIRATORY BURST

The resting phagocytic cells normally consume a small amount of oxygen, while in activated cells its consumption is increased several folds to produce a large quantity of highly reactive oxygen species (ROS). The respiratory burst response is typical of traditional phagocytes like neutrophils, monocytes and eosinophils. Baldridge and Gerard (1933) were first to observe a burst in oxygen consumption by neutrophils during phagocytosis. It is insensitive to inhibitors of the respiratory chain of mitochondria (cyanide, antimysin) and is associated with an increase in metabolism through the hexose monophosphate shunt. Phagocytosis is mediated by two important processes viz. increase in oxygen consumption due to the activation of membrane bound multi-subunit enzyme complex, NADPH-oxidase and fusion of intracellular granules with the phagosomes to release granular contents into the vesicle/ phagosome.

The enzyme is normally dormant but appropriate stimuli activates it to catalyze reduction of oxygen to superoxide radical (O_2^-) by utilizing NADPH, this phenomenon is responsible for more than 90 % of the total oxygen consumption by the neutrophils (Babior, 1984). The superoxide generated serves as the precursor of immense variety of powerful oxidants such as H_2O_2, oxidized halogens (Thomas and Fishmam, 1986) and chloramines (Thomson et al., 1982). The end products of this process are highly toxic to ingested microorganism and the process has long been recognized as an essential part of host's innate immunity. These oxidants at the same time could also be destructive to the adjacent tissues; activity of the enzyme is therefore very tightly regulated so that the toxic species are formed only under appropriate conditions.

NADPH-OXIDASE

The discovery that neutrophils from certain patients with an autosomal recessive inheritance of chronic granulomatous disease (CGD) failed to mount a respiratory burst led to the identification of NADPH oxidase, and subsequently a significant advancement was made in the research on this enzyme. Multi-protein complex NADPH phagocyte oxidase is responsible for the generation of superoxide anion.

The enzyme is structurally complex, composed of five subunits: $p47^{PHOX}$, $p67^{PHOX}$, $p40^{PHOX}$, $p22^{PHOX}$ and $gp91^{PHOX}$ that are distributed at two location in the resting cells, $p40^{PHOX}$, $p47^{PHOX}$, $p67^{PHOX}$ exists in the cytosol, while $p22^{PHOX}$ and $gp91^{PHOX}$ are localized in the membranes of the specific granules or secondary vesicles (Cross et al., 1981; Bejerrum et al., 1989; Calafat et al., 1993). During cell activation, fusion of these organelles with plasma membrane leads to the assembling of the functional oxidase in the membrane (Calafat et al., 1993; Borregaard et al., 1983).

CYTOSOLIC COMPONENTS

$p47^{PHOX}$: A 47-KD phosphoprotein is mainly responsible for transporting of cytosolic complex to the membrane during activation (Segal et al., 1985). The NCF1 gene responsible for encoding $p47^{PHOX}$ is located on chromosome 7 (Francke et al., 1990). Mutation in this gene causes the A47 type CGD. It acts as the docking site for other cytosolic components of NADPH-oxidase, and the first step in the activation is phosphorylation of this component.

$P67^{PHOX}$: Another cytosolic component of the enzyme NADPH-oxidase is an essential requirement for oxidase activation. It is encoded by NCF2 gene, located on the chromosome 1 (Teahan et al., 1990). The N-terminal half of $p67^{PHOX}$ contains a binding region for Rac. Smith et al., (1996) have reported that this subunit also contained a catalytically essential binding site for NADPH. In the absence of $p47^{PHOX}$, $p67^{PHOX}$ fails to assemble with the membrane subunit of NADPH oxidase, supporting an adapter function for $p47^{PHOX}$ in binding of $p67^{PHOX}$. Corresponding effects of $p67^{PHOX}$ on the binding of $p47^{PHOX}$ have also been demonstrated, indicating mutually facilitated binding of these cytosolic components.

$p40^{PHOX}$: A third cytosolic *PHOX* protein was shown to reside in a complex with $p67^{PHOX}$ in the cytosol of resting neutrophils (Wientjes et al., 1993; Tsunawaki et al., 1994) and proposed to play a role in the stabilization of $p67^{PHOX}$. The function of $p40^{PHOX}$ has recently been examined (Sathyamoorthy et al., 1997), and was found to inhibit activity of the oxidase system. It might also play down-regulatory role to maintain the cytosolic factors in an "off" state. The cDNA of this 40-KD protein has also been cloned with of 339 amino acid sequence.

These cytosolic components appear to reside in a complex of 240-300 KD in the cytosol of resting neutrophils (Park et al., 1992; Iyer et al., 1994). SH3 domain of $p47^{PHOX}$ and $p67^{PHOX}$ are involved in the activation of NADPH oxidase.

MEMBRANE ASSOCIATED COMPONENTS

The heterodimeric $p22^{PHOX}$ and $gp91^{PHOX}$ are designated as cytochrome b_{558} because of optical spectrum with an absorbance peak at 558 nm (Segal, 1987). It is localized in the plasma membrane and in the specific granular vesicles in a ratio of 30:70 or even higher amount (Borregaard et al., 1983; Segal and Jones, 1979). Cytochrome b_{558} contains all the known redox centers of the respiratory burst oxidase it binds FAD and its analogs, and contains two-heme moiety for each molecule of FAD. The large subunit $gp91^{PHOX}$ contains a region of homology with the NADPH-binding region of several other flavoenzymes including cytochrome P_{450} reductase and ferredoxin-$NADP^+$ reductase, and has weaker homology with the FAD binding site of various flavoprotein dehydrogenases.

The small subunit, $p22^{PHOX}$, contains two proline-rich sequences at the C-terminus. This domain binds to one of the SH3 domains of $p47^{PHOX}$. The small subunit has a single histidine, which is conserved across species, and this residue cooperates with the large subunit to form one of the haem binding sites. This arrangement places the proline-rich C-terminus on the cytosolic side of the membrane to interact with $p47^{PHOX}$.

The cDNA's of both subunits of cytochrome b_{558} have been cloned and sequenced. The CYBA gene encoding $p22^{PHOX}$ is located on chromosome 16 (Dinauer et al., 1990). While CYBB gene encoding for $gp91^{PHOX}$ is located on X-chromosomes (Teahan et al., 1987; Dinauer et al., 1987), mutation in this gene account for all cases of X-linked CGD (X91-CGD), which is the most common (about 70% of all patient) form of the disease. The proteins encoded by these two genes has been localized and characterized. The alpha subunit, $p22^{PHOX}$ contain 195 amino acids, with hydrophobic helices in the N-terminal half of the protein that could serve as membrane-spanning domains (Parkos et al., 1988). The β-subunit, $gp91^{PHOX}$ of cytochrome b_{558} contains 570 amino acids with four or five transmembrane domain helices and five N-linked glycosylation site in the amino terminal region (Teahan et al., 1987; Dinauer et al., 1987; Imajoh et al., 1992). Topological studies showed that residues 150-172 are exposed on the outside, and an 18-KD C-terminal fragment is cytosolic (Imajoh et al., 1992). The β-subunit, $gp91^{PHOX}$ appears to be exclusively expressed in cells of the myeloid line (Parkos et al., 1988). The regulation of its restricted expression due to region in the 1.5 kDa 5'-untranslated part of $gp91^{PHOX}$ gene (Skalnik et al., 1991), which contains a duplicate CCAAT box between nucleotide –106 and –124 and a TATA box at the nucleotide –30, the repression of expression involves interaction between the DNA binding protein CDP (CCAAT displacement protein) and the CCAAT box.

GUANINE NUCLEOTIDE BINDING PROTEIN

Small guanine nucleotide (GTP) binding proteins are GTPases that cycle between a GDP-bound inactive state and a GTP-bound active state. These proteins belong to the Ras super family, which are charaterised by low rate of intrinsic cleavage activity of bound GTP into GDP. It is stimulated several fold by GTPase activating proteins (GAP proteins). While, guanine nucleotide exchange factors (GEFs) or GDP-releasing factors (GRFs) catalyze exchange of bound GDP for GTP. In addition to these proteins GDP dissociation (GDIs) also inhibits the exchange of GDP/GTP. To promote membrane interaction and biological functions these proteins undergo prenylation. It is a three step process, addition of isoprenyl tail to the C-terminal cystein in CAAX box; proteolytic cleavage of the AAX peptide and finally methylation of the free carboxyl group (Szaszi, et al 999).

Rac1/Rac2, two small low molecular weight GTP-binding proteins and Rap1A are required for the activation of NADPH oxidase. The Rac proteins are the cytosolic component of the system and share less than 30 % homology with Ras. Rac2 has been identified as an active component in human neutrophils. Rac2 is a hematopoietic-specific Rho family GTPases implicated as an important constituent of the NADPH oxidase complex and shares 92 % amino acid homology with the ubiquitously expressed Rac1. Post translational processing of Rac is required for activation of NADPH oxidase in cell free assay composed of plasma membrane and recombinant $p47^{PHOX}$ and $p67^{PHOX}$ (Heyworth et al., 1993), it is consistent with the ability of inhibitors of isoprenylation to antagonize oxidase activation in the intact cell.

Rac processing is required to interact with a GDP/GTP exchange protein to order to convert Rac to the active state i.e. GTP bound form. Rac exists in phagocyte cytosol as a complex with GDI (guanine nucleotide dissociation inhibitors). Activation of Rac in the cell free system is associated with dissolution of Rac-GDI complex. Several lines of evidence suggest that Rac might exist as a complex with GDI in GTP bound form. The oxidase stimulatory factor, Sigma 1 has been shown to be a dimer of Rac and GDI and it was able to stimulate oxidase activity in absence of GDP (Abo et al., 1991). The formation of primed stage of Rac-GTP was thus, prevented by complexion to GDI, until an activation signal releases GDI and allows Rac to rapidly act before it is converted to GDP due to its high intrinsic rate of GDP hydrolysis (Bokoch and Der, 1993).

Rap1A, a 22-KD low molecular weight GTP binding protein and a member of Rho family of cytoskeleton regulatory G-proteins is located in membrane of resting neutrophils. Rap 1A co-purifies with cytochrome b_{558} (Quinn et al., 1989). The C-terminal region of Rap 1A is a possible site of interaction with Cyt $b_{558.}$ A serine residue (Ser^{180}) at the C-terminal of Rap1A has been identified as a site of phosphorylation by cAMP dependent protein kinase (protein kinase A) (92), phosphorylation at this site disrupted Rap1A-GTP-γ-S binding with cytochrome b_{558} (Heyworth et al., 1989, Quillman et al 1993).

REGULATION OF NADPH-OXIDASE ACTIVATION

When resting cell is exposed to appropriate stimuli, activation of the NADPH oxidase is initiated by the assembly of p47 ^{PHOX}, $p67^{PHOX}$, and Rac with cytochrome b_{558} in a 1:1:1:1 complex in the plasma membrane. $p47^{PHOX}$, $p67^{PHOX}$, and Rac exhibit cooperative binding, since increasing the concentration of any one component lowers the EC_{50} of the other components. The association between cytosolic component and cytochrome b_{558} is necessary to establish a complete electron transport chain between NADPH and oxygen, because NADPH binding site seems to be associated with a cytosolic component while flavin required for oxidase activity is bound to cytochrome b_{558}. Absence of translocation of $p47^{PHOX}$ and $p67^{PHOX}$ following activation in the cells, which lack expression of the cytochrome b (Heyworth et al., 1989) suggest that cytosolic subunits might fall back in to the cytosol due to the deficiency of attachment site on the membrane for these components.

PROTEIN -PROTEIN INTERACTION

Oxidase activation is a multi-step phenomenon; $p47^{PHOX}$ acts as a switch to trigger oxidase assembly, following its phosphorylation a conformational rearrangement exposes cryptic SH3 domains, proline rich regions and a PX domain to mediate interactions both with cytochrome b_{558} and $p67^{PHOX}$.

Various Src homology domains (SH2 and SH3 domain) mediated interactions help in the translocation and assembling of the active membrane bound NADPH oxidase in the whole cells including intermolecular contact within $p47^{PHOX}$ (Leto et al., 1994; DeMendez et al., 1994). $p47^{PHOX}$, $p67^{PHOX}$, and $p40^{PHOX}$ contain both SH2 and SH3 domains. SH2 domains bind to phophotyrosine containing peptides following tyrosine kinase activation. While a proline-rich consensus sequence has been identified as the SH3 domain-binding site. $p47^{PHOX}$, $p67^{PHOX}$, and $p40^{PHOX}$ each of this component contains two SH3 domains. The PX domains of the $p47^{PHOX}$ and $p40^{PHOX}$ bind specifically to phosphoinositides, which might partly facilitate assembly of oxidase at the membrane.

The deduced amino acid sequence (390 residues) of $p47^{PHOX}$ contains a tandem repeat of two SH3 domains near the center of the molecule. The first of these binds to a proline-rich sequence within the C-terminus of $p22^{PHOX}$ in the assembled oxidase complex, while in the non-activated $p47^{PHOX}$ it possibly interacts with the C-terminal proline-rich sequence in the same molecule. Another SH3 domain binds to a proline-rich sequence in $p67^{PHOX}$ near the center of this molecule. The predicted 526 amino acid sequence of $p67^{PHOX}$ contains a central proline-rich region followed by SH3 domain, while a second SH3 domain is present at the C-terminus. $p67^{PHOX}$ binds to $p47^{PHOX}$ via a "tail-to-tail" interaction. The SH3 domain in the C-terminus of the $p67^{PHOX}$ binds to the proline-rich sequence in the C-terminus of the $p47^{PHOX}$. In addition, the second SH3 domain of $p47^{PHOX}$ binds to the proline-rich region of $p67^{PHOX}$.

Interactions of the C terminal proline rich SH3 binding domain of $p47^{PHOX}$ and the SH3 domain of $p47^{PHOX}$ and $p40^{PHOX}$ are observed in the resting complex. The interaction in the $p47^{PHOX}$ keeps the enzyme in inactive state as it prevents its binding to cytochrome b_{558}. Activation initiates occupancy of C-terminal proline rich domain of $p47^{PHOX}$ with C-terminal domain of $p67^{PHOX}$, C terminal SH3 domain of $p47^{PHOX}$ then interacts with N-terminal proline rich domain of the $p67^{PHOX}$, and finally N-terminal SH3 domain of $p47^{PHOX}$ binds to the C-terminal part of the and $p67^{PHOX}$. This complex subsequently translocates to the membrane. The amino acid sequence of $p22^{PHOX}$ has a region of high homology to the proline rich sequence, while similar sequence is not found in $gp91^{PHOX}$. Thus SH3 domains seems to be involved in the direct assembly of the active oxidase as SH3 domain interactions with proline rich region has been demonstrated in $p47^{PHOX}$ and $p67^{PHOX}$; $p47^{PHOX}$ and $p22^{PHOX}$ and $p47^{PHOX}$ and $p47^{PHOX}$ (Babior, 1999; Babior et al., 2002).

ROLE OF PHOSPHORYLATION

Neutrophils exist in one of the three states, quiescent, primed or activated. There is no increase in oxidase activity in the primed state, yet subsequent stimulation provokes a response that is larger than nonprime, activated cells. Priming may result from the crosstalk between Ca^{+2} signaling system, acting through serine-threonine kinases (protein kinase C, PKC), calmodulin dependent kinases and tyrosine kinases such as p55Fgr, p56Lyn, p59Fyn and p59Hck (Hallett et al., 1995).

Generation of superoxide anion by the multi-protein complex NADPH phagocyte oxidase is accompanied by extensive phosphorylation of $p47^{PHOX}$. $p47^{PHOX}$ translocation and its activation require the phosphorylation of one of the serine pairs. The conversion of S303 and S304 or S359 and S370 to alanine equivalents eliminates oxidase activity. PKC seems to be responsible for $p47^{PHOX}$ phosphorylation, although the isoform responsible for this is not yet confirmed. $p47^{PHOX}$ is a substrate for PKC zeta and participates in the signaling cascade between fMLP receptors and NADPH oxidase activation. PKC zeta is an atypical PKC isoform expressed abundantly in human PMNs. A different isoform of PKC such as PKC-β in neutrophils is coupled through Fcg receptor (Dekker et al., 2001). Treatment of dHL-60cells with b-PKC antisense, inhibited phosphorylation of $p47^{PHOX}$, translocation to the membrane and generation of O_2^- in response to fMLP or PMA (Korchak et al., 1998). A potential role of PKC-β isoform is also suggested in the activation of NADPH oxidase (Korchak et al. 2001).

Maximum oxidase activity requires both phosphorylation as well as a group of molecules that act as point of crosstalk between the two systems. At least 8-9 serine phosphorylation site are present at the C-terminus of $P47^{PHOX}$ and up to eight different phosphorylated forms of $p47^{PHOX}$ have been identified (Rotrosen and Leto, 1990). It could be the signal for the oxidase activation as S359 and S370 has to be phosphorylated first during oxidase activation. Later S379 acquire a phosphate to facilitate the translocation of cytosolic complex to cytochrome b_{558}, and finally, S303 and/or S304 are phosphorylated, endowing the oxidase with full catalytic activity (Babior, 1999). It is, therefore, possible that serine kinase forms the basis of crosstalk and the activation of $p47^{PHOX}$ would be achieved by the coordinated effect of the kinases. Staurosporine, a potent PKC antagonist, inhibits the PMA induced respiratory burst at low IC_{50} (25 nM), however it potentiates FMLP and platelet activation factor (PAF) induced respiratory burst at low conc (10 to 200 nM). It suggests towards the possible role of PKC in the regulation of NADPH oxidase activity i.e. a positive regulation in PMA treated cells and a negative regulation in chemoattractant stimulated PMNs (Combadiere et al., 1990). Phosphorylation of $p67^{PHOX}$ and $p47^{PHOX}$ is also reported during the activation of oxidase (Babior, 1999).

Depletion of RACK1 (Receptor for activated C kinase) cytoskeleton and membrane associated anchor molecules, which activated Ca^{+2}/DAG dependent α,β,γ isoform of PKC in β1- PKC null dHL-60 cells exhibited an enhanced O_2^- generation triggered by fMLP or PMA. RACK1 might remove β- PKC from signaling complex, which is otherwise required specifically for the activation of NADPH oxidase. The binding of RACK1 to βII-PKC could thus decrease the $O_2^{\bullet -}$ generation (Davidson-Moncada et al., 2002). PKC isotypes consist of PKC-μ and PKC-ν; PKC-μ is a human homologue of mouse protein kinase D (PKD) on the basis of its regulation by DAG/ phorbol ester (Lint et al., 1995; Johannes et al., 1995). This group, however differ from other PKC family member in its catalytic domain, substrate profile and sensitivity of kinase inhibitors (Lint et al., 1995; Johannes et al., 1995). Davidson-Moncada et al. (2002) have recently shown that PKD is also present in neutrophils which is activated by FCg-receptor stimulation and recruited for particle intake. PKC independent activation of NADPH oxidase shown by using PKD antisense, and mutation or deletion in its regulatory element (pleckstrin homology domain) increase the constitutive activity of kinase (Gay et al., 1997), which might play a crucial role in the regulation of NADPH oxidase. Another protein kinase, Akt, also phosphorylates $p47^{PHOX}$ but cause weak activation of NADPH-oxidase (Babior et al., 2002).

Phosphorylation–dephosphorylation cycle regulates assembly of the oxidase in the resting cells; phosphorylation and dephosphorylation of $p47^{PHOX}$ may exist at equilibrium. Stimulation of PMNs with PMA promotes phosphorylation of $p47^{PHOX}$ and induced its redistribution to the plasma membrane. Addition of H-7, a PKC inhibitor however markedly inhibited the phosphorylation of $p47^{PHOX}$ and decreases the amount of $p47^{PHOX}$ in membrane fraction and NADPH oxidase activity. Addition of calcycllin A, a protein phosphatase inhibitor, in these cells suppressed the effect on $p47^{PHOX}$ and restored the activation of NADPH oxidase and translocation of $p47^{PHOX}$ to the membrane fraction. Activation of NADPH oxidase thus can happen even after the loss of enzyme activity. The activity of NADPH oxidase in the membrane fraction prepared from PMA stimulated neutrophils was inactivated by the addition of the cytosol from the resting neutrophils and this inactivation was suppressed by protein phosphatase inhibitor okadaic acid and calyculin A (Iglesias and Rozengart, 1998). These finding suggested that Type 1 and/or type 2A protein phosphatase inhibitors block the dephosphorylation of $p47^{PHOX}$ and may modulate the O_2^- release. These phosphatases may serve as a potential target for enhancing the oxygen dependent killing mechanism of phagocytic leukocyte (Gay et al., 1997).

Oxidase activation is diminished in neutrophils containing elevated level of cyclic adenosine monophosphate (cAMP) or cAMP analogs. Bengis-Garber et al. (1996) demonstrated that fMLP induced $p47^{PHOX}$ phosphorylation was prevented by cAMP agonist and was blocked by the inhibitor of protein kinase A, however, in PMA treated cells, no effect on $p47^{PHOX}$ phosphorylation was observed. These findings suggest a novel role of protein kinase A in the down regulation of NADPH oxidase activity. Casein kinase 2 (CK-2), a serine threonine kinase is ubiquitously present in the eukaryotes. It is a calcium and cyclic nucleotide independent nucleotide, which utilizes GTP and ATP as phosphate donors. CK-2 has recently been shown to phosphorylate $p47^{PHOX}$ and induce conformation changes that alter the ability of $p47^{PHOX}$ to associate with $p22^{PHOX}$ and reduce interaction between $p47^{PHOX}$ and $p22^{PHOX}$ of cytochrome b_{558} (Park et al., 2001). An additional phosphorylation of membrane associated $p47^{PHOX}$ has been suggested (De Leo et al., 1999) that could mediate the release of $P47^{PHOX}/P67^{PHOX}$ from cytochrome b-enriched phagosomes and termination of oxidase activity. However, the kinase responsible for such an event is not known. PD098059, an inhibitor of mitogen activated protein kinase (MAPK), prevented opsonized zymosan or PMA induced oxidase activation (Hazan et al., 1997). While, SB203580, an inhibitor of p38 MAP kinase antagonized fMLP induced oxidase activation (Rane et al., 1997) and TNF-a and lipopolysaccharide (LPS) prime neutrophils respiratory burst activity and membrane expression of flavocytochrome b_{558} through exocytosis of intracellular granules was blocked by SB203580 (Ward et al., 2000). MAP kinase kinase activates extracellular signal regulated kinase (ERK), suggesting a role of ERK and p38 in the activation of NADPH oxidase. p65 kinase, a mammalian protein kinase related to STE 20 kinase of budding yeast (Laberer et al., 1992) binds specifically to the GTP bound form of Rac. Rac-GTP stimulate autophosphorylation of p65 PAK (p21 mitogen activated protein kinase) and its catalytic activity towards exogenous substrates. Human neutrophils contain a kinase closely related or identical to PAK1 and PAK2 and both kinases autophosphorylate in presence of Rac-GTP. It is confirmed by the activation of both kinases in fMLP stimulated human neutrophils and the rapid translocation of Rac and $O_2^{\bullet -}$ generation observed in response to fMLP. PAK phosphorylates a physiologically relevant site in $p47^{PHOX}$ (ser^{328}), which might play an important role in NADPH oxidase regulation. PAK share common properties of a group of renaturable serine-threonine kinases that participate in NADPH oxidation including inhibition of PAK activation by PI-3 kinase inhibitors (Knaus et al., 1995).

In calcium depleted neutrophils Rossi et al., (1989) had observed normal Con-A mediated phagocytosis and associated respiratory burst events suggesting existence of an alternate transmembrane signaling mechanisms, which is independent of phosphoinoside turnover and rise in $[Ca^{2+}]_{i.}$.

REGULATION MEDIATED BY RAC

Rac mediated regulation of the NADPH oxidase activity in intact cells has been shown by studies using Rac transgenic or antisense to manipulate Rac levels. (Dorseuil et al., 1992; Gabig et al., 1995). Treatment of neutrophils with cytochalasin or botulinum C2 toxin enhanced the chemoattractant activation of respiratory burst but completely inhibited neutrophil migration suggesting a possible cross talk between two member of Rho family, Rac and

Rho. Cell motility is therefore under the control of Rho, while Rac modulates oxidase activation. It indicates existence of a switch mechanism that directs Rac to the oxidase or Rho to the cytoskeleton, thus one of the two processes is activated, and the other is maintained in dormant state. This mechanism may use the property of geranylated Rac and Rho in their GDP bound state to associate with the same GDI (Rho-GDI). The heterodimer, GDI/Rho A (Bourmeyster et al., 1992) and GDI/Rac2 (Kwong et al., 1993) have been isolated in the native form from the cytosol of neutrophils. In cell free system the dissociation of Rac/GDI complex is facilitated by arachidonic acid (Chuang et al., 1993). The first step in the agonist induced activation could be the release of arachidonic acid by phospholipase A_2 (PLA_2) activation, which requires both phosphorylation and an increase in the intracellular free calcium. p42 MAPK (p42 mitogen activated protein kinase), ERK 2 (extracellular signal regulated kinase) has been shown to phosphorylate PLA_2 *in vitro*. Waterman et al., (1996) showed involvement of p38 MAPK in phosphorylation and activation of PLA_2 in tumor necrosis factor-α (TNF-α) stimulated intact human neutrophils. Yamamori et al., (2002) have recently shown that a p38 MAPK inhibitor, SB203580, regulates the Rac activation by preventing translocation and preceding activation leading to partial inhibition of $O_2^{\bullet-}$ generation in OZ (opsonized zymosan) stimulated bovine neutrophils. They also observed that PI3-K (phosphotidylinositol-3-kinase) exists as an upstream regulator of p38 MAPK and suggested a novel alternative-signaling pathway by which PI3-K regulates the activation of Rac via p38 MAPK. Interestingly it has also been shown that PI3-K regulates the activation of GDP/GTP exchange factor vov, via its product PIP3 (phosphatidylinositol 1,4,5 tri phosphate) and the subsequent activation of Rac (Han et al., 1998).

Mutation of amino acids in a region of Rac (residues 26-45), homologous to an effector region in Ras decreases the ability of Rac to support superoxide generation, and these mutations produced an elevated EC_{50} for Rac. These same mutations interfere with the ability of Rac1 to activate PAK, cytoskeleton regulation as well as with the activation of the mitogenic response (Kwong et al., 1995). Rho family small GTPases contain a 12 amino acid region [from 124-135] also involved in regulation of NADPH oxidase, since this region has no counterpart in Ras it is therefore commonly known as insert region. Point mutations and deletion of this region interfere with activation of the NADPH oxidase, but not PAK. These mutations augment the EC_{50} for Rac in supporting superoxide generation. In addition, cytoskeletal regulation by Rac mutants *in vivo* is normal, but blocks the mitogenic action of Rac1. These findings indicate that combinatorial use of the 26-45 effector regions and insert region provides the Rho family GTPases with versatility in their specificity for several downstream pathways (Freeman et al., 1996).

The GTPase activation protein GAPs stimulate the intrinsic rate of GTP hydrolysis and thus accelerate the inactivation of the protein. Exogenous addition of Rac-GAP protein reduces the $O_2^{\bullet-}$ production in a constituted system (Heywarth et al., 1993). Using the semi recombinant and fully purified activation system, Geiszt et al. (2001) demonstrated that Rac-GAP activity was associated with the membrane fraction of resting neutrophils and inversely related to NADPH oxidase activating potency of the Rac-GTP. Thus in intact cell modulation of Rac-GAP activity (by covalent modification or complex formation) can be an important method to decrease (132) or increase Rac-activity to down regulate or up regulate the processes affected by Rac.

REGULATION OF NADPH OXIDASE BY NO

Migration of PMNs from blood to the specific tissue following a pathological insult is a key feature of the host inflammatory response and PMNs might be a potential source of NO at the site of inflammation. PMNs can release upto 10-100 nM NO/5min/10^6cells, which is sufficient to affect the vasoreactivity as well as platelet activation (Salvemini et al., 1989). NO is synthesized by the enzyme nitric oxide synthase (NOS) present in the PMNs cytosol. Work accomplished in Dr. Dikshit's laboratory demonstrated that NO exerts a biphasic regulation on the ROS generation as low concentration augments the free radical release whereas high concentration of ROS inhibits free radical synthesis (Seth et al.,, 1994; Sethi et al., 1999). NO also seems to modulate the increased free radical generation from rat PMNs following hypoxia reoxygenation (Sethi et al., 1999). However, in normal physiological condition an inverse relationship exist between generation of NO and free radicals and PMNs seem to preferentially synthesize more free radicals than NO (Sharma et al., 2002). This finding suggests that changes in the level of nitric oxide in PMNs modulate the activity of NADPH oxidase (Clancy et al., 1992) has shown for the first time that inhibition of $O_2^{\bullet-}$ generation by direct interaction of NO with the membrane component of the oxidase. Fuji et al., (1997) have suggested NO suppresses the $O_2^{\bullet-}$ production by affecting cytosolic or membrane bound proteins of the oxidase while the binding site and redox centers are not impaired. Lee et al., (2000) have observed that direct inhibition of PMNs $O_2^{\bullet-}$ generation only at higher conc. ($\geq$50 μM) of NO), $ONOO^-$ [formed by the reaction of NO and $O_2^{\bullet-}$] (Beckman and Koppenol, 1996) exerts a biphasic modulation of $O_2^{\bullet-}$ production, triggering enhanced $O_2^{\bullet-}$ generation at low physiological concentration but inhibition at higher concentration. The enhancement of $ONOO^-$ mediated $O_2^{\bullet-}$ generation from PMNs is associated with activation of MEK/ERK/MAPK pathway.

The inhibitory effect of $ONOO^-$ and NO at higher concentration could be due to the direct interaction with sulfhydryl groups on enzyme to form S-nitrosothiols. Nitrosothiols (RSNO) has also been shown to inhibit oxidase activation by preventing translocation of cytosolic oxidase component $p47^{PHOX}$ and $p67^{PHOX}$ through an action on membrane (Park, 1996). These findings suggest a critical role of cysteine residue present in the components of oxidase. Furthermore, high concentration of $ONOO^-$ can directly oxidize cysteine residue leading to loss of enzyme activity (Radi et al., 1991). Beckman (1996) suggested that high concentration of peroxynitrite might induce protein tyrosine nitration leading to the inhibition of enzyme function and cellular toxicity (Lee et al., 2000). Agents that affect the NO availability exert modulatory effect on PMNs free radical generation (Kiemer & Vollmar 1998).

17β–Estradiol (E_2) has been shown to attenuate the $O_2^{\bullet -}$ formation in phagocytes due to enhance NO production (Beksi et al., 2000). Increasing the bioavailability of NO by shifting the nitric oxide/superoxide balance in the vessel wall following E_2 treatment might be able to prevent atherosclerosis in the blood vessels. E_2 has been shown to inhibit the expression of $gp91^{PHOX}$ and PMA stimulated superoxide production in human umbilical vein culture endothelial cells (Wagner et al., 2001).

REGULATION OF NADPH OXIDASE BY ION CHANNELS

The transfer of electron from cytosolic NADPH to extracellular oxygen through NADPH oxidase generates a massive depolarization. If uncompensated, it would prevent further electron transfer and associated production of $O_2^{\bullet -}$, because large amount of acid equivalent are released in to the cytosol during the hydrolysis of NADPH and its resynthesis by the hexose monophosphate shunt. The oxidase has been proposed to act as a H^+ channel to preserve electoneutrality and allow extruding of intracellular acid (Henderson and Chappell, 1992). The oxidase is coupled to efflux of H^+ ions through the proton channel. The inhibition of the channel results in inhibition of superoxide generation, thus understanding of the oxidase requires identification of the protein(s) that function as and/or regulate the opening and functioning of the channel. Arachidonic acid activates the H^+ channel with the oxidase activity Henderson and Chappell (1996) suggest that the channel is part of oxidase complex. Several lines of evidence have suggesting that the $gp91^{PHOX}$ component of NADPH oxidase complex might be the proton channel, which is activated during the respiratory burst although contradictory evidences exist regarding its function as proton channel (Nanda et al., 1993; Maturana et al., 2001; DeCoursey et al., 2001). Neutrophils have a high permeability to potassium ions, and its efflux occurs mainly through potassium channel, however the knowledge of functional and physiological significance of K^+ channels in neutrophils is relatively sparse. Neutrophils suspended in high potassium buffer exhibited partial depolarization of the resting cell membrane and a blunted depolarization response to stimulus and produce less superoxide and H_2O_2 (Martin et al., 1988). Reeves et al., (2002) have recently shown influx of K^+ into the phagocytic vacuole during the activation of NADPH oxidase with a concomitant decrease in pH. It supports the optimal activity of elastase and cathepsin G following their release from the anionic proteoglycans matrix to destroy engulfed bacteria.

REGULATION BY REACTIVE OXYGEN SPECIES

Superoxide and probably other oxidants (ROS) serve as messengers at low concentrations, while larger amounts are required for inducing damage. ROS have been traditionally regarded as toxic byproducts of aerobic metabolism. However, ROS can also act as intracellular signaling molecules. The specific response elicited by reactive oxygen intermediaries is determined by their specific intracellular target (s). This, in turn, is dependent on the species of oxidant(s) produced, the source and therefore subcellular localization of the oxidant(s), the kinetics of production, and the quantities produced. NADPH oxidase might play a critical role in the early steps leading to the development of atherosclerosis. $O_2^{\bullet -}$ might act as a small second messenger molecule, while in disease conditions oxidation of low density lipoprotein by $O_2^{\bullet -}$ might be detrimental for the pathology. We thus describe a model for the initiation and development of atherosclerosis that suggests targeted inhibition of NADPH oxidase as a powerful site for prevention and treatment of atherosclerosis.

The concept of ROS as a "signaling molecules" has gained significant recognition over the past several years. In addition to their classical role in bacterial killing endogenous reactive oxygen intermediates (ROI) generated by NADPH oxidase have been known to regulate several intracellular pathways. ROI, acting on endogenous vanadium derivatives to form active compound such as peroxo-vanadyl or vanadyl hydro peroxide which by inactivating protein tyrosine phosphatase leads to the protein tyrosine kinase (PTK) activation and hence promote tyrosine phosphorylation (Trudel et al., 1991; Zor et al., 1993). PTK activation also leads to the activation of the MAP-K cascade and the PLA_2 activation. (Trudel et al., 1991). A series of studies done by Fialkow et al. (1993,1994,1997) have also demonstrated that activation of NADPH oxidase in permeabilized neutrophils by direct stimulation of GTP binding protein also resulted in enhanced tyrosine phosphorylation, while exposure of neutrophils to exogenous oxidants induce both activation of MAPK or ERK and inhibition of protein tyrosine phosphatase CD45, all these events are paralleled by production of $O_2^{\bullet -}$, however in similar conditions neutrophil from a patient of chronic granulomatous disease (CGD) demonstrated no such phosphotyrosine accumulation, activation of kinase and inhibition of CD 45. Oxidant signaling by the PMNs NADPH oxidase is an important determinant of TNF-a induced NF-kappa B activation and ICAM-1 expression in co-cultured endothelial cells (Fan et al., 2002). Gardai et al. (2002) have shown recently that NADPH oxidase derived oxidant regulates neutrophils apoptosis by blocking Akt-derived anti apoptotic pathway to enhance apoptosis. The blockade of Akt activation dependent on reduced phsphoinoside 3,4,5 triphosphate level not on decrease phosphatidyinositol-3-kinase activity implicate the involvement of an inositol phosphatase, SHIP (SH2-containing inositol –5-phosphatase) (Gardai et al., 2002). It is recruited from the cytoplasm to the plasma membrane during activation. SHIP depends on Lyn protein tyrosine kinase activation, which helps in recruitment of SHIP to the plasma membrane. The activation of Lyn kinase in the neutrophils has been shown to depend upon oxidant production. Pretreatment of human neutrophils with an inhibitor of NADPH oxidase inhibit the mycobacterium tuberculosis stimulated activation of caspase-3 and alteration of Bax/Bcl-x (L) expression to prevent apoptosis (Perskvist et al., 2002). These potential pathways for the stimulation of neutrophil apoptosis may be critical for the resolution of inflammation and subsequent clearance, thus protecting tissue from damage and exacerbation of the inflammatory response.

References

Abo A., Pick E., Hall A., Totty N., Teahan C.G. and Segal A.W. Activation of NADPH oxidase involved the small GTP binding protein p21rac1. Nature (London), **353:** 668-670, 1991.

Abu-Soud H.M. and Stuehr D.J. Nitric oxide synthase reveal a role for calmodulin in controlling electron transfer. Proc. Natl. Acad. Sci. USA **90:** 10769-10772, 1993.

Alderton W. K., Cooper C.E. and Knowles R.G. Nitric oxide synthase: structure, function and inhibition. Biochem. J. **357:** 593-615, 2001.

Amin A.R., Attur M., Vyas P., Leszczynska-Piziak J., Levartovsky D., Rediske J., Clancy R.M., Vora K.A. and Abramson S.B.J. Expression of nitric oxide synthase in human peripheral blood mononuclear cells and neutrophils. Inflammation **47:** 190-205, 1995.

Asano K., Chee C.B., Gaston B., Lilly C.M., Gerard C., Drazen J.M. and Stamler J.S. Constitutive and inducible nitric oxide synthase gene expression, regulation and activity in human lung epithelial cells. Proc. Natl. Acad. Sci. USA **91:** 10089-10093, 1994.

Babior B.M. NADPH oxidase: an update. Blood 93, 1464-1476, 1999.

Babior B.M, Lambarth J.D. and Nauseef W. The neutrophil NADPH oxidase. Arch. Biochem. Biophys. **397:** 342-344, 2002.

Babior B.M. The respiratory burst of phagocytes. J. Clin. Invest. **73:** 599-601, 1984.

Baldridge C.W. and Gerard R.W. The extra respiration of phagocytes. Am. J. Physiol. **103:** 235-236, 1933.

Beckman, J.S. Oxidative damage and tyrosine nitration from peroxynitrite. Chem. Res. Toxicol. **9:** 836-844, 1996.

Beckman J.S. and Koppenol W.H. Nitric oxide, superoxide and peroxynitrite: the good, the bad, and the ugly. Am. J. Physiol. Cell Physiol. **271:** C1424-C1437, 1996.

Bejerrum O.W. and Borregaard N. Dual granule localization of the dormant NADPH oxidase and cytochrome b_{559} in human neutrophils. Eur. J. Haematol. **43,** 67-77, 1989.

Beksi G., Kakucs R., Varbiro S., Racz K., Sprintz D., Feher J. and Szekacs B. In vitro effect of different steroid hormones on superoxide anion production of human neutrophils granulocytes. Steroids. **65:** 889-894, 2000.

Bengis-Garber C. and Gruener N. Protein Kinase A down regulates the phosphorylation of p47 phox in human neutrophils: A possible pathway for inhibition of the respiratory burst. Cell Signal. **8:** 291-296, 1996.

Bokoch G.M., Der C.J. Emerging concepts in the Ras super-family of GTP-binding proteins. FASEB J. **7:** 750-759,1993.

Bokoch G.M., Quilliam L.A, Bohl B.P, Jesaitis A.J, Quinn M.T. Inhibition of Rap1A binding to cytochrome b558 of NADPH oxidase by phosphorylation of Rap1A. Science, **254:** 1794-1796, 1991.

Borregaard N., Heiple J.M., Simons E.R. and Clarck R.A. Subcellular localization of the b-cytochrome component of human neutrophils microbicidal oxidase: translocation during activation. J. Cell Biol. **9:** 52-61, 1983.

Bourmeyster N., Stasia M.J., Garin J. Gagnon J., Boquet P., Vignais P.V. Copurification of Rho protein and the Rho-GDP dissociation inhibitor from bovine neutrophils cytosol. Effect of phosphoinsitides on Rho ADP-ribosylation by the C3 exoenzyme of Clostridium botulinum. Biochemistry **31:** 12863-12869, 1992.

Bredt D.S. and Snyder S.H. Isolation of nitric oxide synthetase, a calmodulin requiring enzyme. Proc. Natl. Acad. Sci. USA **87:** 682-685, 1990.

Calafat J., Kuijpers T.W., Janssen H., Borregaard N., Verhoeven A.J. and Ross C. Evidence for small intracellular vesicles in human blood phagocytes containing b_{558} and the adhesion molecule CD11b/CD18. Blood, **81:** 3122-3129, 1993.

Callsen-Cencic P., Hoheisel U., Kaske A., Mense S. and Tenschert S. The controversy about spinal neuronal nitric oxide synthase: under which condition is it up- or downregulated? Cell Tissue Res. **295:** 183-194, 1999.

Charles I.G., Palmer R.M., Hickery M.S., Bayliss M.T., Chubb A.P., Hall V.S., Moss D.W. and Moncada S. Cloning, characterization and expression of a cDNA encoding an inducible nitric oxide synthase from human chondrocytes. Proc. Natl. Acad. Sci. USA **90:** 11419-11423, 1993.

Cho H.J., Xie Q.W., Calaycay J., Mumford R.A., Swiderek K.M., Lee T.D. and Nathan C. Calmodulin is a subunit of nitric oxide synthase from mammals. J. Exp. Med. **176:** 599-604, 1992.

Chuang T.H., Bohl B.P., Bokoch G.M.. Biologically active lipids are regulators of Rac-GDI complexation. J. Biol. Chem. **268:** 26206-26211, 1993.

Clancy R.M., Leszczynska-Piziak, J. and Abramson S.B. Nitric oxide, an endothelial cell relaxation factor inhibits neutrophil superoxide anion production via a direct action on the NADPH oxidase. J. Clin. Invest. **90:** 1116-1121, 1992.

Combadiere C., Hakim J., Giroud J-P. and Peranim A. Staurosporine, a protein kinase inhibitor up-regulates the stimulation of human neutrophils respiratory burst by N-formyl peptide and platelet activating factor. Biochem. Biophys. Res. Comun.**168:** 65-70, 1990.

Cross A.R., Jones T.G., Harper A.M. and Segal A.W. Oxidation-reduction properties of cytochrome b found in the plasma membrane fraction of human neutrophils. Biochem. J. 194, 599-606, 1981.

Curran R.D., Billiar T.R., Stuehr D.J., Ochoa J.B., Harbrecht B.G., Flint S.G. and Simmons R.L. Multiple cytokines are required to induce hepatocytes nitric oxide production and inhibit total protein synthesis. Ann. Surg. **212:** 462-469, 1990.

Davidson-Moncada J.K., Lopez-Lluch G., Segal A.W. and Dekker L.V. Activation of the NADPH oxidase in neutrophils. Biochem. J. **363:** 95-103, 2002.

De Leo F.R., Allen L.A., Apicella M. and Nauseef W.M. NADPH oxidase activation and assembly during phagocytosis. J. Immunol. **163:** 6732-6740, 1999.

De Mendez I., Garrett M.C., Adams A.G. and Leto T.L. Role of p67 phox SH3 domains in assembly of NADPH oxidase system. J. Biol. Chem., **269:** 16326-16332, 1994.

DeCoursey T.E., Chenny V.V., Morgan D., Kartz B.Z. and Dinauer M.C. The gp91 component of NADPH oxidase is not the voltage gated proton channel in phagocytes, but it helps. J. Biol. Chem. **276:** 36063-36066, 2001.

Dekker L.V., Lestges M., Altschuler G., Mistry N., McDermott A., Rose J. and Segal A.W. Protein kinase C-b, contributes to NADPH oxidase activity in neutrophils. Biochem. J. **347:** 285-289, 2001.

Dinauer M.C., Orkin S.H., Brown R., Jesaitis A.J. and Parkos C.A. The glycoprotein encoded by the X-linked chronic granulomatous disease locus is a component of the neutrophil cytochrome b complex. Nature, **327:** 717-720, 1987.

Dinauer M. C., Pierce E.A., Bruns T.G.A., Curnutte J.T. and Orkin S.H. Human neutrophils cytochrome b light chain (p22-phox) gene structure, chromosomal location, and mutation in cytochrome negative chronic granulomatous disease. J. Clin. Invest. **86:** 1729-1737, 1990.

Dorseuil O., Vazquez A., Lang P., Bertoglio J., Gacon G. and Leca G. Inhibition of superoxide production in B-lymphocytes by Rac antisense oligonucleotide. J. Biol. Chem. **267:** 20540-20542, 1992.

Eiserich J. P., Hristova M., Cross C.E. Jones A.D., Freeman B. A., Halliwell B. and Van der Vliet A. Formation of nitric oxide derived inflammatory oxidants by myeloperoxidase in neutrophils. Nature. **391:** 393-397, 1998.

Eiserich J.P., Cross C.E., Jones A.D., Halliwell B. and van der Vliet A. Formation of nitrating and chlorinating species by reaction of nitrite with hypochlorous acid. J. Biol. Chem. **271:** 19199-19208, 1996.

Eissa N.T., Yuan J.W., Haggerty C.M., Choo E.K., Palmer C.D. and Moss J. Cloning and characterization of human inducible nitric oxide synthase splice variants: A domain, encoded by exons 8 and 9 is critical for dimerization. Proc. Natl. Acad. Sci. USA **95:** 7625-7630, 1998.

Evans T.J., Buttery L.D.K., Carpenter A., Sprngall D.R., Polak J.M. and Cohen J. Cytokine- treated human neutrophils contain inducible nitric oxide synthase that produces nitration of ingested bacteria. Proc. Natl. Acad. Sci. USA **93:** 9553-9558, 1996.

Fan J., Frey R.S., Rehman A. and Malik A.B. Role of neutrophil NADPH oxidase in the mechanism of tumor necrosis factor-alpha–induced NF-kappa B activation and intracellular adhesion molecule-1 expression in endothelial cells. J. Biol.Chem. **277:** 3404-3411, 2002.

Fialkow L., Chan C.K., Grinstein S. and Downey G.P. Regulation of tyrosine phosphorylation in neutrophils by NADPH oxidase. J. Biol.Chem. **268:** 17131-17137, 1993.

Fialkow L., Chan C.K., Rotin D., Grinstein S. and Downey G.P. Activation of mitogen activated protein kinase signaling pathway in neutrophils. J. Biol. Chem. **269:** 31234-31242, 1994.

Fialkow L., Chan C.K. and Downey G.P. Inhibition of CD45 during neutrophil activation. J. Immunol. **158:** 5409-5417, 1997.

Francke U., Hsieh C.L., Foellmer B.E., Lomax K.J., Malech H.L. and Leto T.L. Genes for two autosomal recessive forms of chronic granulomatous disease assigned to 1q25 (NCF 2) and 7q11.23 (NCF 1). Am J Hum Genet. **47:** 483-492, 1990.

Freeman J.L., Abo A. and Lambeth J.D. Rac "insert region" is a novel effector region that is implicated in the activation of NADPH oxidase, but not PAK65. J. Biol. Chem. **271:** 19794-19801, 1996.

Fuji H., Ichimore K., Hoshia K. and Nakazawa H. Nitric oxide inactivates NADPH Oxidase in pig neutrophils by inhibiting its assembling process. J. Biol. Chem. **272:** 32773-32778, 1997.

Gabig T.G., Crean C.D., Mantle P.L. and Rosli R. Function of wild type and mutant type Rac 2 and Rap IA GTPases in differentiated HL-60 cells NADPH oxidation. Blood. **85:** 804-811, 1995.

Gardai S., Whitlock B.B., Helgason C., Ambruso D., Fadok V., Bratton D. and Henson P.M. Activation of SHIP by NADPH oxidase-stimulated Lyn leads to enhanced apoptosis in neutrophils. J. Biol. Chem. **277:** 5236-5246, 2002.

Gaut J.P., Byun J., Tran H.D., Lauber W.M., Carroll J.A., Hotchkees R.S., Belaaouaj A. and Heinack J.W. Myeloperoxidase produces nitrating oxidant in vivo. J. Clin. Invest. **109:** 1311-1319, 2002.

Gay J.C., Raddassi K., Truett I.A.P. and Murray J.J. Phosphatase activity regulates superoxide anion generation and intracellular signaling in human neutrophils. Biochem. Biophys. Acta **1326:** 243-253, 1997.

Gciszt M., Dagher M.C., Molnár G., Havasi A., Faure J., Paclet M.H., Morel F., Ligeti E. Characterization of membrane-localized and cytosolic Rac-GTPase-activating proteins in human neutrophil granulocytes: Contribution to the regulation of NADPH oxidase. Biochem. J. **355:** 851-858, 2001.

Geller D.A., Lowenstein C.J., Shapiro R.A., Nussler A.K., Di Silvio M., Wang S.C., Nakayama D.K., Simmons R.L., Snyder S.H. and Billiar T.R. Molecular cloning and expression of inducible nitric oxide synthase from human hepatocytes. Proc. Natl. Acad. Sci. USA **90:** 3491-3495, 1993.

Goode H.F., Morodle P.D., Barry E.W. and Niget R.W. Nitric oxide synthase activity is increased in patients with sepsis syndrome. Clin. Sci. **88:** 131-133, 1995.

Govers R. and Rabelink T.J. Cellular regulation of endothelial nitric oxide synthase. Am. J. Physiol. Renal. Physiol. **280:** F193-F206, 2001.

Greenburg S.S., Ouyang J., Zhao X-F. and Giles J.D. Human and rat neutrophils consitutively express neural nitric oxide synthase mRNA. Nitric Oxide: Biol.Chem. **2:** 203-212, 1998.

Gunttor M.R., His L.C., Curtis J.F., Giersa J.K. Marnett L.J., Eling T.E. and Mason R.P. Nitric oxide trapping of tyrosyl radical of prostaglandin H synthase-2 leads to tyrosine imminoxy radical and nitrotyrosine formation. J. Biol. Chem.**272:** 17086-17090, 1997.

Hallett M.B. and Lloyds D. Neutrophils priming: The cellular signals that say ‘amber’ but not ‘green’. Immunol. Today **16:** 264-268, 1995.

Han J., Luby-Phelps K., Das B., Shu X., Xia Y., Mostller R.D., Krishna U.M., Falck J.R., White M.A. and Broek D. Role of substrate and product of PI-3 kinase in regulation activation of Rac related guanosine triphosphate by vov. Science **279:** 558-560, 1998.

Handerson L. M. and Chappel J. B. NADPH oxidase of neutrophils. Biochem. Biophys. Acta, **1273:** 87-107, 1996.

Hazan I., Dana R., Granot Y. and Levy R. Cytosolic phospholipase A2 and its mode of activation on human neutrophils by opsonized zymosan. Correlation between 42/44 kDa mitogen-activated protein kinase, cytosolic phospholipase A2 and NADPH oxidase. Biochem. J. **326:** 867-876, 1997.

Hazen S.L., Hsu F.F., Mueller D.M., Crowley J.R. and Heinecke J.W. Human neutrophils employ chlorine gas as an oxidant during phagocytosis. J. Clin. Invest. **98:** 1283-1289, 1996.

Henderson L.M. and Chappell J.B. The NADPH oxidase associated H^+ channel is opened by arachidonic acid. Biochem. J. **283:** 171-175, 1992.

Heyworth P.G., Shrimpton C.F. and Segal A.W. Localization of the 47Kda phosphoprotein involved in the respiratory burst NADPH oxidase of phagocytic cells. Biochem. J. **260:** 243-248, 1989.

Heywarth P.G., Knous U.G., Selltleman J., Curnutte J.T. and Bokoch G.M. Regulation of NADPH oxidase activity by Rac-GTPase activation protein(s). Mol. Biol. Cell, **4,** 1217-1223, 1993.

Heywarth P.G., Knous U.G., Xu X., Uhliger D.J., Conroy L., Bokoch G.M. and Curnutte J.T. Requirement of post translational processing of Rac GTP binding proteins for activation of human neutrophils NADPH oxidase. Mol. Biol. Cell, **4:** 261-269, 1993.

Hickey M.J., Granger D.N. and Kubes P. Inducible nitric oxide synthase (iNOS) and regulation of leukocyte endothelial cell interactions: studies in iNOS-deficient mice. Acta Physiol. Scand. **173:** 119-126, 2001.

Hiki K., Yui Y., Hattori R., Eizawa H., Kosuga K. and Kawai C. Three regulation mechanisms of nitric oxide synthase. Eur. J. Pharmacol. **206:** 163-164, 1991.

Iglesias T and Rozengart E. Protein kinase D activation by mutation within its pleckstrin homology domain. J. Biol. Chem. **273:** 410-416, 1998.

Imajoh S., Ohmi K., Ochiai T.H., Nakamura M. and Kanegasaki S.K. Topology of cytochrome b_{558} in neutrophils membrane analyzed by anti-peptide antibodies and proteolysis. J. Biol. Chem., **267:** 180-194, 1992.

Iyer S.S., Pearson D.W., Nauseef W.M. and Clark R.A. Evidence for a readily dissociable complex of p47phox and p67phox in cytosol of unstimulated human neutrophils. J. Biol. Chem. **269:** 22405-22411, 1994.

Janssens S.P., Shimouchi A., Quertermous T., Bloch D.B. and Bloch K.D. Cloning and expression of cDNA encoding human endothelium derived relaxing factor/nitric oxide synthase. J. Biol. Chem. **267:** 14519-14522, 1992.

Johannes F.J. Prestte J., Dieterich S., Oberhangemarn P., Link G. and Pfizanmaier K. Characterization of activator and inhibitor of protein kinase C m. Eur. J. Biochem. **227:** 303-307, 1995.

Kadota K., Yui Y., Hattori R., Uchizumi H. and Kawai C. A new relaxing factor in supernatant of incubated rat peritoneal neutrophils. Am. J. Physiol. 260, H 967-H 972, 1991.

Kiemer A.K. and Vollmar A.M. Autocrine regulation of inducible nitric oxide synthase in macrophages by atrial natriuretic peptide. J. Biol. Chem. **273:** 13444-13451, 1998.

Klebanoff S.J. Reactive nitrogen intermediates and antimicrobial activity: Role of nitrite. Free Radic.Biol.Med.**14:** 351-360, 1993.

Kleinert H., Wallerath T., Fritz G., Ihrig-Biedert I., Rodriguez-Pascual F., Geller D.A. and Forstermann U. Cytokine induction of NO synthase in human DCDY cells: role of the JAK-STAT, AP-1 and NFkB-signaling pathway. Br. J. Pharmacol. **125,** 193-201, 1998.

Knaus U.G., Moris S., Dang H.-J., Chernoff J. and Bokoch, G.M. Regulation of phagocyte oxygen radical production by the GTP-binding protein Rac 2. Science **269,** 221-223, 1995.

Knowles R.G. and Moncada S. Nitric oxide synthase in mammals. Biochem. J. **298:** 249-258, 1994.

Kolb J.P., Paul-Eugene N., Damais C., Yamaoka K., Drapier J. C. and Dugas B. Interleukin 4 stimulate cGMP production by IFN-g activated human monocytes, involvement of nitric oxide synthase pathway J. Biol. Chem. **269,** 9811-9816, 1994.

Korchak H.M. and Kilpatrick L.E. Roles of b-II protein kinase C and RACK1 in positive and negative signaling for superoxide anion generation in differentiated HL60 cells. J. Biol. Chem. **276,** 8910-8917, 2001.

Korchak H.M., Rossi M.W. and Kilpatrick L.E. Selective role of b-protein kinase C in signaling for $O2^{-\cdot}$ Generation but not degranulation or adherence in differentiated HL-60 cells. J. Biol. Chem. **273:** 27292-27299, 1998.

Krukoff T.L. Central regulation of autonomic function: no brakes? Clin. Exp. Pharmacol. Physiol. **25:** 474-478, 1998.

Kunz D., Walker G., Eberhardt W. and P Feilschifter J. Molecular mechanism of dexamethasone inhibition of nitric oxide synthase expression of interleukin-1b stimulated mesangial cells: evidence for the involvement of transcriptional and post transcriptional regulation. Proc. Natl. Acad. Sci. USA **93:** 255-259, 1996.

Kwong C.H., Malech H.L., Rotrosen D. and Leto T.L. Regulation of human neutrophils NADPH oxidase by Rho-related proteins. Biochemistry. **32:** 5711-5717, 1993.

Kwong C.H., Adams A.G. and Leto T.L. Characterization of the effector-specifying domain of Rac involved in NADPH oxidase activation. J. Biol.Chem. **270:** 19868-19872, 1995.

Laberer E., Dignard D., Harcus D., Thomas Y. and Whiteway M. The protein kinase homologue ste20p is required to link the yeast pheromone response G-protein bg subunit to downstream signaling components. The EMBO. J. **11:** 4815-4824, 1992.

Lamas S., Marsden P.A., Li G.K., Tempst P. and Michel T. Endothelial nitric oxide synthase: molecular cloning and characterization of a distinct constitutive enzyme isoform. Proc. Natl. Acad. Sci. USA **89:** 6348-6352, 1992.

Lee C.-I., Miura K., Liu X. and Zweier J.L. Biphasic regulation of leukocytes superoxide generation by nitric oxide and peroxynitrite. J. Biol. Chem. **275:** 38965-38972, 2000.

Leto T.L., Adams A.G. and DeMendez I. Assembly of Phagocytic NADPH oxidase: Binding to Src homology 3 domains to prolein rich targets. Proc. Natl. Acad. Sci. USA. **91:** 10650-10654, 1994.

Lint J.V., Sinnett-smith J. and Razangurt E. Expression and characterization of PKD, a phorbol ester and diacylglycerol–stimulated serine protein kinase. J. Biol. Chem. **270:** 1455-1461, 1995.

Mac Micking J., Xie Q. W. and Nathan C. Nitric oxide and macrophage function. Annu. Rev. Immunol. **15,** 323-350, 1997.

Martin M. A., Nouseef W.M. and Clark R.A. Depolarization blunt the oxidative burst of human neutrophils, paralleled effect of monoclonal antibodies, depolarizing buffer and glycolytic inhibitors. J. Immunol. **140:** 3928-3935, 1988.

Maturana A., Arnaudeau S., Ryser S., Banfi B., Hossle J. P., Schlegel W., Krause K-H., and Demaurex N. Heme histidine ligands within gp91phox modulate proton conductance by phagocyte NADPH oxidase. J. Biol. Chem. **276:** 30277-30284, 2001.

McCall T.B., Palmer R.M.J. and Moncada S. Induction of nitric oxide synthase in rat peritoneal neutrophils and its inhibition by dexamethasone. Eur. J. Immunol. **21:** 2523-2527, 1991.

Miguel L.S., Arriero M.M., Farre J., Jimenez P., Garcia-Mendez A., Frutos, T., Jimenez A., Garcia R., Cabestrero F., Gomez J., Andres R., Monton M., Martin E., Calle-Lombana L.M.D., Rico, L. Romero J., Lopez-Farre A. Nitric oxide production by neutrophils obtained from patients during coronary syndromes expression of the nitric oxide synthase isoforms. J Amer Coll Cardiol. **39:** 818-825, 2002.

Miles A.M., Owens M.W., Mulligan S., Johanson G.G., Fields J.Z., Ing, T.S., Kottapali V., Keshavarzian A. and Griham M.B. Nitric oxide synthase in circulation vs extravasated polymorphonuclear leukocytes. J. Leukoc. Biol. **58:** 616-622, 1995.

Moncada S. and Higgs A. The L-arginine-nitric oxide pathway. N. Engl. J. Med. **329:** 2002-2012, 1993.

Nakahara H., Sato E.F., Ushisaka R., Kanno T., Yoshioka T., Yasuda T., Inove M.S. and Utsumi K. Biochemical properties of human oral polymorphonuclear leukocytes. Free Radic. Res. **28:** 485-495, 1998.

Nanda A., Grinstein S. and Curnutte J.T. Abnormal activation of H^+ conductance in NADPH oxidase defective neutrophils. Proc. Natl. Acad. Sci. USA, **90:** 760-764, 1993.

Nicolini F.A. and Mehta J.L. Inhibitory effect of unstimulated neutrophils on platelet aggregation by release of a factor similar to endothelium-derived relaxing factor (EDRF). Biochem. Pharmacol. **40:** 2265-2269, 1990.

Padgett E.L. and Prunett S.B. Rat, mouse and human stimulated by a variety of activating agents produce much less nitrite than rodent macrophages. Immunology **84:** 135-141, 1995.

Park C.S., Krishna G., Ahn M.S., Kang J.H., Chung W.G., Kim D.J., Hwang H.K., Lee J.N., Paik S.G. and Cha Y.N. Differential and constitutive expression of neuronal, inducible, and endothelial nitric oxide synthase mRNA and proteins in pathophysiologically normal human tissue. Nitric Oxide: Biol. Chem. **4:** 459-471, 2000.

Park H.-S., Lee S.M., Lee J.H., Kim Y.-S., Bae Y.-S. and Park J.-W. Phosphorylation of the leukocyte NADPH oxidase subunit P47 Phox by casein kinase 2: conformation-dependent phosphorylation and modulation of oxidase activity. Biochem. J. **358:** 783-790, 2001.

Park J.W.M., Ma J., Ruedi M., Smith R.M. and Babior B.M. The cytosolic component of respiratory burst oxidase exists as M(r). approximately 240,000 complex that acquires a membrane binding site during activation of the oxidase in a cell-free system. J. Biol. Chem. **267:** 17327-173232, 1992.

Park J.W. Attenuation of p47 Phox p67 Phox membrane translocation as the inhibitory mechanism of S-nitrosothiol on the respiratory burst oxidase in human neutrophils. Biochem. Biophys. Res. Commun. **220:** 31-35, 1996.

Parkos C., Dinauer A.M.C., Walker L.W., Allen R.A., Jesaitis A.J. and Orkin S.H. Primary structure and unique expression of the 22-kilodalton light chain of human neutrophils cytochrome b. Proc. Natl. Acad. Sci. USA, **85:** 3319-3323, 1988.

Perskvist N., Long M., Stendahl O. and Zheng L. Mycobacterium tuberculosis promotes apoptosis in human neutrophils by activation caspase-3 and altering expression of Bax-Bcl-xL via an oxygen–dependent pathway. J. Immunol. **168:** 6358-6365, 2002.

Pollock J.S., Nakane M., Forstermann U. and Murad F.J. Particulate and soluble bovine endothelial nitric oxide synthase are structurally similar proteins yet different from soluble brain nitric oxide synthase. Cardiovasc. Pharmacol. **20,** S50-S53, 1992.

Quilliam L.A., Muller H., Bohl B.P., Prossnitz V., Sklar L.A, Der C.J. and Bokoch G.M. Rap 1A is a substrate for cyclic AMP dependent protein kinase in human neutrophils. J. Immunol. **147:** 1628-1635, 1991.

Quinn M.T., Parkos C.A., Walker L. Orkin S.H., Dinauer M.C. and Jesaitis L. Association of Ras–related protein with cytochrome bof human neutrophils. Nature **342:** 198-200, 1989.

Radi R., Beckman J.S., Bush K.M. and Freeman B.A. Peroxynitrite oxidation of sulfhydryls: The cytotoxic potential of superoxide and nitric oxide. J. Biol. Chem. **266:** 4244-4250, 1991.

Rane M.J., Carrithers S.L., Arthur J.M., Klein J.B. and Mcleish K.R. Formyl peptide receptors are coupled to multiple mitogen activated protein kinase cascade by distinct signal transduction pathways: Role in activation of reduced nicotinamide adenine dinucleotide oxidase. J. Immunol. **159:** 5070-5078, 1997.

Reeves E.M., Lu H., Jacobs H.L., Massina G.M.C., Bolsoner S., Gabella G., Potma E.O., Warley A., Segal A.W. Killing activity of neutrophils is mediated through activation of proteases by K^+ flux. Nature **416:** 291-297, 2002.

Riesco A., Caramelo C., Blum G., Monton M., Gallego M.J., Casado S. and Farre A.L. Nitric oxide-generating system as an autocrine mechanism in human polymorphonuclear leukocytes. Biochem. J. **292:** 791-796, 1993.

Rimele T.J., Sturm R.J. and Adams L.M. Interaction of neutrophils with vascular smooth muscle: identification of a neutrophil-derived relaxing factor. J. Pharmacol. Exp. Ther. **245:** 102-111, 1988.

Rossi F., Bianca D., Grzeskowiak M. and Bazzoni F. Studies on molecular regulation of phagocytosis in neutrophils Con A-mediated ingestion and associated respiratory burst independent of phosphanositide turnover, rise in [Ca^{+2}], and arachidonic acid release. J. Immunol. **142:** 1652-1660, 1989.

Rotrosen D. and Leto T.L. Phosphorylation of neutrophil 47-kDa cytosolic oxidase factor. J. Biol. Chem. **265:** 19910-19915, 1990.

Salvemini D., de Nucci G., Gryglewski R.J. and Vane J.R. Human neutrophils and mononuclear cells inhibit platelet aggregation by releasing a nitric oxide like factor. Proc. Natl. Acad. Sci. USA **86:** 6328-6332, 1989.

Sathyamoorthy C.A., de Mendez I., Adams A.G. and Leto T.L. p40phox down regulates NADPH oxidase activity through interaction with its SH3 domain. J. Biol. Chem. **272:** 9141-9146, 1997.

Segal A.W. and Jones O.T.G. The subcellular distribution and some properties of the cytochrome b compound of the microbicidal oxidase system of human neutrophils. Biochem. J. **182:** 181-188, 1979.

Segal A.W., Heyworth P.G., Cockcroft S. and Barronwman M.M. Stimulated neutrophils from patients with autosomal recessive chronic granulomatous disease fail to phosphorylate a Mr.-44,000 protein. Nature **316:** 547-549, 1985.

Segal A.W. Absence of both cytochrome b_{245} subunit from neutrophils in X-linked chronic granulomatous disease. Nature **326:** 88-91 1987.

Seth P., Kumari R., Dikshit M. and Srimal R.C. Modulation of rat peripheral polymorphonuclear leukocyte response by nitric oxide and arginine. Blood. **84:** 2741-2748, 1994.

Sethi S., Singh M.P. and Dikshit M. Nitric oxide mediated augmentation of polymorphonuclear free radical generation after hypoxia-reoxygenation. Blood **93,** 333-340, 1999.

Sethi S. and Dikshit M. Modulation of polymorphonuclear leukocytes function by nitric oxide. Thromb. Res. **100:** 223-247, 2000.

Sethi S., Sharma P. and Dikshit M. Nitric oxide and oxygen derived free radical generation from control and lipopolysaccharide treated rat polymorphonuclear leukocytes. Nitric Oxide Biol Chem. **5:** 482-493, 2001.

Sharma P., Barthwal M. K. and Dikshit M. NO synthesis and its regulation in arachidonic acid stimulated rat polymorphonuclear leukocytes. Nitric Oxide: Biol. Chem. **7:** 119-126, 2002.

Simon D.I., Mullins M.E., Jia L., Gaston B., Singel D.J. and Stamler J.S. Polynitrosylated proteins, characterization, bioactivity, and functional consequences. Proc. Natl. Acad. Sci. USA, **93,** 4736-4741, 1996.

Skalnik D.G., Strauss E.C and Orkin S.H. CCAAT displacement protein as a repressor of the myelomonocyte specific gp91-phox gene promoter. J. Biol. Chem. **266:** 16736-16744, 1991.

Smith R.M., Conon J.A., Chen L.M. and Babior B.M. The cytosolic subunit p67phox contains an NADPH-binding site that participates in catalysis by the leukocytes NADPH oxidase. J. Clin. Invest. **98:** 977-983, 1996.

Stamler J.S. and Meissner G. Physiology of nitric oxide in skeleton muscle. Physiol. Rev. **81,** 209-237, 2001.

Stuehr D.J. and Griffith O.W. Mammalian nitric oxide synthase. Adv. Enzymol. Relat. Areas Mol. Biol. **65:** 287-346, 1992.

Szaszi K., Korda A., Wolfl J., Paclet M.H., Morel F. and Ligeti E. Possible role of RAC-GTPase-activating protein in the termination of superoxide production in phagocytic cells. Free Radic. Biol. Med. **27:** 764-772, 1999.

Takizawa S., Aratani Y., Fukuyama N., Maeda V., Hirabayashi H., Koyama H., Shinohara Y. and Nakazawa H. Deficiency of myeloperoxidase increases infarct volume and nitrotyrosine formation in mouse brain. J. Cerb. Blood Flow Metab. **22:** 50-54.

Teahan C.G., Totty N.F., Casimir C.M. and Segal A.W. Purification of the 47 Kda phosphoprotein associated with the NADPH oxidase of human neutrophils. Biochem. J. **267:** 485-489, 1990.

Teahan C.G., Rower P., Parker P., Totty N. and Segal A.W. The X-linked chronic granulomatous disease codes for the beta chain of cytochrome b_{245}. Nature **327:** 720-721, 1987, 1987.

Thomas E.L. and Fishmam M. Oxidation of chloride and thiocyanate by isolated leukocytes. J. Biol. Chem., **261:** 9694-9702, 1986.

Thomson E.L., Jefferson M.M., Grisham M. Myeloperoxidase-catalyzed incorporation of amines into proteins: role of hypochlorous acid and dichloramines. Biochemistry, **21:** 6299-6308, 1982.

Travares-Murta B.M., Cunha F.Q. and Ferreira S.H. Differential production of nitric oxide by endotoxin–stimulated rat and mouse neutrophils. Braz. J. Med. Biol. Res. **29:** 381-388, 1996.

Trudel S., Paquet M.R. and Grinstein S. Mechanism of vanadate -induced activation of tyrosine phosphorylation and the respiratory burst in HL-60 cells. Biochem. J. **276:** 611-619, 1991.

Tsukahara Y., Morisaki J., Horita Y., Torisu M. and Tanaka M. Expression of inducible nitric oxide synthase in circulating neutrophils of the systemic inflammatory response syndrome and septic patients. J. Surg. **22:** 771-777, 1998.

Tsunawaki S., Mizunari H., Nagata M., Tatsuzawa O. and Kuratsuji T. A novel cytosolic component , p40 phox, of respiratory burst oxidase associates with p67phox and is absent in patients with chronic granulomatous disease who lack p67phox. Biochem. Biophys Res. Commun. **199,** 1378-1387, 1994.

Van der Vliet A., Eiserich J.P., Halliwell B. and Cross C.E. Formation of reactive nitrogen species during peroxidase-catalyzed oxidation of nitrite. A potential additional mechanism of nitric oxide dependent toxicity. J. Biol. Chem. **272:** 7617-7625, 1997.

Wagner A.H., Schroeter M.R. and Hecker M. 17 beta-estradiol inhibition of NADPH oxidase expression in human endothelial cells FASEB J., **15:** 2121-2130, 2001.

Wallerath T., Gathi I., Aulitzky W.E., Polok, J.S., Kleinert H. and Forstermann U. Identification of nitric oxide synthase isoforms expressed in human neutrophil granulocytes, megakaryocytes and platelets. Thromb. Haemost. **77:** 163-176, 1997.

Ward R.A., Nakamura M., McLeish K.R. Priming of the neutrophil respiratory burst involves p38 mitogen-activated protein kinase-dependent exocytosis of flavocytochrome b558-containg granules. J. Biol. Chem. **275:** 36713-36719, 2000.

Waterman H., Molski T.F., Huang C.K., Adams J.L. and Sha'afi R.I. Tumor necrosis factor a induced the phosphorylation and activation of cytosolic phspholipase A2 are abrogated by an inhibitor of p38 mitogen activated protein kinase cascade in neutrophils. Biochem. J., **319:** 17-20, 1996.

Wheeler M.A., Smith S.D., Cardena G.G., Nathan C.F., Weiss R.M. and Sessa W.J. Bacterial infection induces nitric oxide synthase in human neutrophils. Clin. Invest. **99:** 110-116, 1997.

Wientjes F.B., Hsuan J.J., Totty N.F. and Segal A.W. p40phox, a third cytocolic component of the activation complex of the NADPH oxidase to contain src homology 3 domains. Biochem. J. **296:** 557-561, 1993.

Xia Y., Raman L.J., Mosters S.S. and Zweier J.L. Inducible nitric oxide synthase generates superoxide from the reductase domain. J. Biol. Chem. **273:** 22635-22639, 1998.

Xu W., Charles I.G., Moncada S., Gorman P., Sheer D., Liu L. and Emson P. Mapping of the genes encoding human inducible and endothelial nitric oxide synthase (NOS2 and NOS3) to the pericentric region of chromosome 17 and to chromosome 7, respectively. Genomics **21:** 419-422, 1994.

Xu W., Gorman P., Sheer D., Bates G., Kishimoto J., Lizhi L. and Emson P. Regional localization of the gene coding for human brain nitric oxide synthase (NOS1) to 12q24.2 12q24.31 by fluorescent in situ hybridization. Cytogenet. Cell Genet. **64:** 62-63, 1993.

Xuan Y.T., Guo Y., Han H., Zhu Y. and Bolli R. An essential role of the JAK-STAT pathway in ischemia preconditioning. Proc. Natl. Acad. Sci. USA **98:** 9050-9055, 2001.

Yamaguchi M., Sasaki J.-I., Kuwana M., Sakai M., Okamura N. and Ishibashi S. Cytosolic protein phsphatase may turn off activated NADPH oxidase in guinea pig neutrophils. Arch. Biochem. Biophys. **306:** 209-214, 1993.

Yamamori T., Inanami O., Sumimoto H., Akasaki T., Nagahata H and Kuwabara M. Relationship between p38 mitogen-activated protein kinase and small GTPase Rac for the activation of NADPH oxidase in bovine neutrophils. Biochem. Biophys. Res. Commun. **293:** 1571-1578, 2002.

Yui Y., Hattori R., Kosuga K., Eizawa H., Hiki K. and Kawai C. Purification of nitric oxide synthase from rat macrophages. J. Biol. Chem. **266:** 12544-12547, 1991.

Zor U., Ferber E., Gergely P., Szücs K., Dombrádi V. and Goldman R. Reactive oxygen species mediate phorbol ester-regulated tyrosine phosphorylation and phospholipase A2 activation: potentiation by vanadate. Biochem. J. **295:** 879-888, 1993.

Pharmacological Perspectives of Toxic Chemicals and Their Antidotes
Editors: S J S Flora, J A Romano, S I Baskin and K Sekhar

Toxicological Consequences of Modulation of Brain Cytochrome P450s by Environmental Chemicals

Devendra Parmar, Sanjay Yadav, Ashu Johri, Nidhi Kapoor,[1] Aditya B. Pant, Alok Dhawan and Prahlad K. Seth
Industrial Toxicology Research Centre, PO Box 80,
M.G. Marg, Lucknow 226 001, India
[1]Dept. of Biochemistry, University of Lucknow, Lucknow 226 002, India

Cytochrome P450 (P450) dependent-monooxygenases are the major enzymes of phase I reactions which are involved in the metabolism of a wide variety of drugs and foreign chemicals as well as of endogenous substrates such as steroids and fatty acids etc. P450 dependent metabolism generally determines whether the chemical will be detoxified or metabolically activated to toxic mutagenic and carcinogenic forms. Although much of the P450 mediated endogenous metabolism is highly specific, there appears to be a great degree of overlapping substrate specificity in the metabolism of foreign chemicals. This extraordinarily broad substrate specificity of the enzyme results from the multiplicity of different molecular forms of the P450, which have distinct but overlapping, substrate specificities as evidenced by identification, purification and characterization of different molecular forms of mammalian liver P450s (Lewis 1996 a,b).

Some of the P450s are highly inducible; for example, expression of P450 1A1 can be elevated 100 fold or more in liver and many extrahepatic tissues following exposure to 2,3,7,8-tetrachlorodibenzo-p-dioxin (TCDD), 3-methylcholanthrene (MC) or other polycyclic aromatic hydrocarbons (PAHs) (Whitelock, 1989; Lewis, 1996 a,b). Phenobarbital (PB), as well as a large number of structurally unrelated chemicals termed 'PB-like' inducers, induce the expression of P450 2A, 2B & 2C subfamilies both in the laboratory animals and in the humans (Waxman & Azaroff, 1992). P450 2E1 expression in liver is increased several folds following exposure to ethanol, acetone and other drugs and chemicals while clofibrate and other peroxisome proliferators induce the expression of P450 4A in liver and kidney (Lewis et al, 1996 a,b).

P450s in Brain

Although liver is the major organ involved in the P450 mediated metabolism of endogenous compounds and xenobiotics, several extrahepatic tissues like skin, kidney, lung, intestine, adrenal, ovaries, testis, pancreas, brain, etc. have been shown to be equipped with P450 monooxygenases (Fang & Strobel, 1978; Guengerich, 1977; Gram et al., 1996). Moreover, tissue-specific expression of certain P450s isoenzymes is also reported. The central nervous system (CNS) has been shown to be the target for the toxicity of a variety of environmental chemicals and drugs including drugs of abuse. Several of these xenobiotics, reaching the brain could be biotransformed within the brain and the metabolite(s) so formed are often polar and therefore, cannot move out of the blood-brain barrier. Bioactivation of xenobiotic *in situ* within the brain could cause damage to macromolecules in the brain cells and/or bind at different receptor sites there. Considering the limited regenerative capability of the CNS, such damages for the lifetime could have far reaching consequences in causing neoplastic transformation or disruption of neuronal functions.

PHYSIOLOGICAL FUNCTIONS OF P450s IN BRAIN

Besides being involved in regulating the concentration of the chemicals or their metabolites at the target sites, P450s in brain have been shown to act on a wide variety of endogenous substrates in physiologically significant manner. The cerebral effects of a xenobiotic might be interpreted as being either due to disruption of the normal physiological function of the P450 or by acting as a substrate of P450 thereby leading to the formation of a reactive species (Mensil et al., 1984). The most important role for cerebral P450 demonstrated to date is the transformation of androgens to estrogens by aromatization and of the latter to catechol estrogens by 2-hydroxylation (Mensil et al 1984). Significant quantities of catechol estrogens formed in specific areas of brain, such as hypothalamus and pituitary and the fact that these compounds are formed under normal circumstances strongly suggests that they may have a physiological role within the brain. Fishman et al. (1976) provided evidence for an association in the sites of N-demethylation of morphine and presence of opiate receptor in brain indicating an involvement of receptor-mediated mechanism of opiate action in N-demethylation.

The binding properties, affinities, and drug specificities of sigma receptor ligands were identical in membranes isolated from rat brain and liver microsomes, suggesting that the sigma receptor is, infact, a P450 isoenzyme

(Ross 1990, 1991). The high affinity of proadifen (SKF-525A), a potent inhibitor of some isoforms of P450, for the sigma sites have indicated that the protein binding the sigma ligands is a proadifen-sensitive isoform of P450. Niznik et al. (1990) provided more direct evidence for the involvement of brain P450s in the cerebral neurotransmission process. A major binding protein for several inhibitors of dopamine transporter have been identified in striatum as a P450. These studies have indicated that there is an overlap of substrate specificities between the dopamine transporter and P450 2D1, an isoenzyme of P450 (Tyndale et al. 1991). A lack of good correlation between serum concentrations of neuroleptics and the clinical effects can be attributed in part to central metabolism of these drugs, whereby the actual concentration of the drugs at the CNS receptor may not be related to the concentration of the drug in the serum.

Members of P450 4A subfamily are key enzymes involved in the synthesis and degradation of metabolites of arachidonic acid, which are of physiological importance in the brain (Stromstedt et al., 1993). Junier et al. (1990) also demonstrated that arachidonic acid is metabolised in the hypothalamus by an P450 dependent epoxygenase pathway to metabolites which appear to be involved in the transmembrane signalling mechanism by which activation of D2 dopamine receptors stimulates secretion of the neuropeptide SRIF.

Steroid hormones act on the CNS to produce diverse neuroendocrine and behavioral effects. The sedative and anesthetic effects of gonadal steroid, progesterone and the mineralocorticoid- deoxycorticosterone, as well as several of their metabolites, is due in part to their ability to enhance the inhibitory action of GABA. Changes in mood and sleep/ wakefulness pattern during the stress, pregnancy and the menstrual cycle are thought to be due to the fluctuations in the levels of 3α-OH-DHP and 5α-THDOC (Purdy et al., 1991). The levels of these steroids in the brain are determined both by their rate of synthesis and by their inactivation and / or elimination. Studies of Stromstedt et al. (1993) have shown that P450s expressed in brain are involved in the elimination of these steroids and have indicated that this inactivation pathway in the brain may serve to regulate the levels of GABA receptor active steroids. Interest in brain steroid metabolism has been further aroused by the finding that the brain derived steroids can modulate cognitive function and synaptic plasticity. A novel P450, *hct-1* has been identified in rat and mouse brain hippocampus (Stapleton et al., 1995) and its expression was particularly enriched in the hippocampus though small amounts were detectable in rat liver and kidney.

Expression of P450s in brain

In a tissue as heterogenous as brain, it is often difficult, if not impossible to differentiate the role this enzyme system might play in metabolism of endogenous and foreign compounds. Detailed studies undertaken in our laboratory and elsewhere on the various forms of P450 have provided evidence that brain is capable of handling a wide variety of P450 substrates. Although broadly, the brain enzymes were found to resemble the liver enzyme, some differences were observed in the regulation of cerebral enzymes as compared to the liver P450s. Though expression of multiple forms of P450s has been shown in the mammalian brain, a debate exists as to the precise levels of xenobiotic metabolizing P450s in brain. Strobel et al. (1989) were the first to report that the treatment of rats with PB and tricyclic antidepressants markedly increase the brain RNA hybridising with P450 2B1 cDNA probe. RT-PCR studies from our laboratory have demonstrated significant inducible mRNA expression of P450 1A1 and 2B1/2B2 in whole rat brain isolated from the rats administered repeated doses of MC or PB respectively, as used to induce these P450s in liver (Fig.1). While no significant mRNA expression of P450 1A1 or 2B1/2B2 was observed in control brain, pretreatment with MC or PB resulted in significant increase in the mRNA expression of P450 1A1 or 2B1/2B2 respectively in rat brain. However, the magnitude of induction observed in brain was several fold less when compared with the expression observed in the liver of the induced animals (Fig.1). Using RT-PCR, Hodgson et al. (1993) have also demonstrated the expression of P450 1A1, 2B2, 2D and 2E1 in rat brain, though they used the different combination of inducers and treatment schedule than that used by others. However, cytochrome P450 reductase expression was also detected in the brain samples, giving evidence that the brain contains a competent mixed function oxidase system. Schilter & Omiecinski (1993) reported that expression of P450 1A1, 1A2, 2B1, 2B2 and 3A1 mRNA was region specific. Highly heterogeneous inter-regional profile of P450 1A1 mRNA expression observed in control brain which was altered after β-naphthoflavone (β-NF) treatment. The P450 1A2 mRNA was increased in region specific manner following β-NF treatment. However, mRNA expression of P450 2B1 and 2B2 observed in control animals exhibited a complex pattern of distribution which has been attributed to an acute PB treatment of 18 hr., whereas in most of the other studies animals were exposed chronically to PB or by multiple intra-peritonial administration (Schilter & Omiecinski, 1993). Studies on the different isoforms of P450 and P450 reductase in brain tumors have shown the expression of P450 1A1, 1A2, 2B1, 2B2, 2C7 and P450 reductase in rat glioma C6 cell lines (Geng & Strobel 1993, 1995, 1998). Using RT-PCR & restriction digestion of the PCR products, the induction of P450 1A1 and 2B has been quantified. Ten- and five- fold induction of P450 1A and 2B mRNA after BA or PB treatments respectively, were detected by competitive PCR. Similar results were observed with BA or PB treated rat brain. Control glioma C6 cells were found to possess P450 1A1, 2B1 & 2B2 levels similar to that of control rat brain.

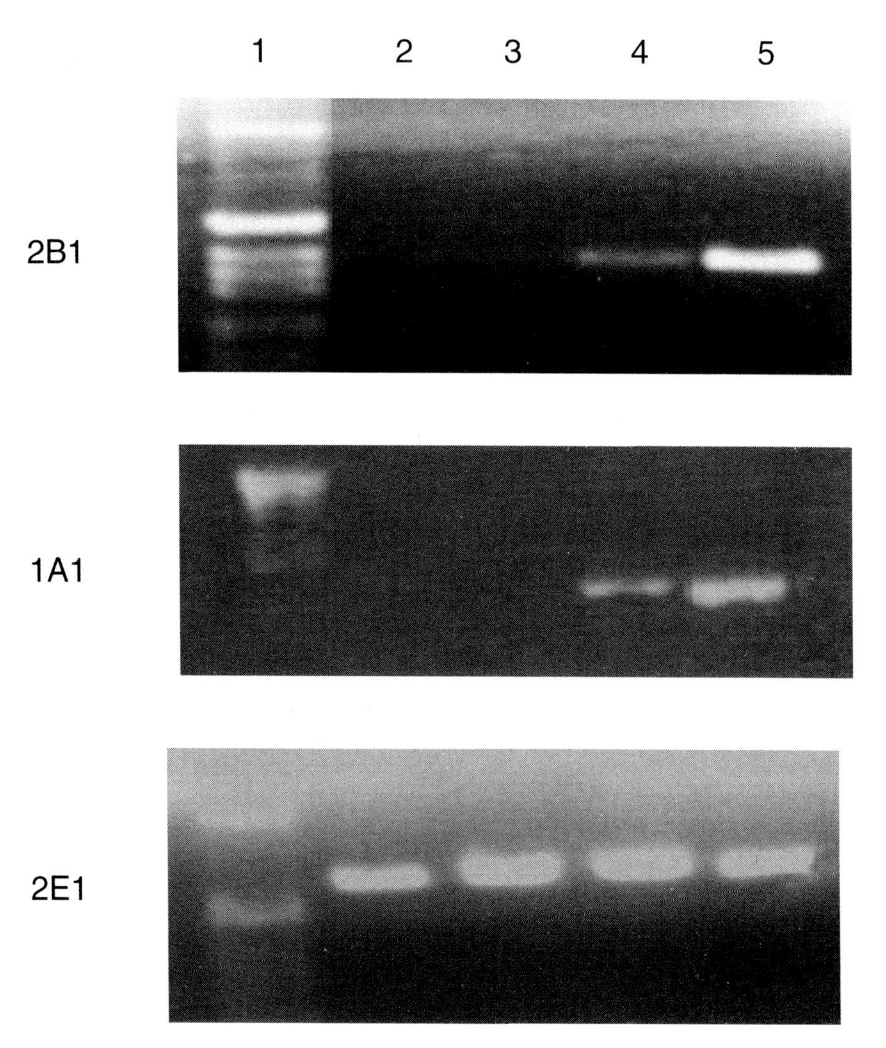

Fig. 1 RT-PCR analysis demonstrating mRNA expression of xenobiotic metabolising P450s (*Lane 1: DNA ladder; 2: RNA for 2B1 and 1A1, con. brain RNA for 2E1, 3: con. brain RNA for 2B1 and 1A1, ethanol brain RNA for 2E1; 4: induced brain RNA for 1A1 and 2B1, ethanol liver RNA for 2E1; 5: induced liver RNA for 1A1 and 2B1, con. Liver RNA for 2E1*)

Immunohistochemical studies have also provided evidence for spatial distribution and neuronal or glial occurrence of P450 1A, 2B, 2C, 2E & 3A isoenzymes in mammalian brain (Kohler et al., 1988). Western blot studies using whole brain microsomes have been inconclusive and failed to differentiate between the constitutive and inducible forms of P450. While no immunoreactivity with anti-P450 1A1/1A2 and anti-P450 2B1/2B2 was observed by many investigators in P450s extracted from brain microsomes of untreated or β-naphthoflavone treated rat brain, Warner et al. (1988) reported immunoreactivity in partially purified preparations with these antiseras only in some brain regions and concluded that most of the P450s in the brain consist of forms other than P450 2B1/2B2 or 1A1/1A2 and remains to be characterised. Studies of Ravindranath and her group (Ravindranath et al., 1989; Ravindranath & Anandatheerthavarada, 1989 & 1990; Anandatheerthavarada et al., 1990) showed that though the P450 levels in the brain were low in comparison to the liver, control brain microsomes isolated from rat, mouse and human brain exhibit significantly high immunoreactivity with antibodies raised against purified rat hepatic and brain P450 2B1/2B2 and 1A1/1A2 isoenzymes.

Consistent with the mRNA expression, studies from our laboratory have demonstrated low, but significant immunoreactivity of anti-P450 1A1/1A2 with MC-pretreated solubilised brain microsomes and anti-P450 2B1/2B2 with PB-pretreated microsomes (Fig. 2). No immunoreactivity was observed with anti-P450 1A1/1A2 and 2B1/2B2 in untreated brain microsomes and could be attributed to extremely low level of expression of these isoenzymes in brain (Parmar et al., 1998a). Densitometric analysis of immunoblots revealed significant increase in immunoreactivity with MC or PB pretreated brain microsomes. As shown in Fig.2, microsomes isolated from the liver of PB or MC pretreated rats exhibited 10-15 fold increase in immunoreactivity with anti-P450 1A1/1A2 or anti-P450 2B1/2B2 as compared to control liver microsomes which was several fold greater than the increase (2-3 fold) observed in brain microsomes isolated from MC or PB pretreated rats (Parmar et al., 1998a). Furthermore, significant regional differences were observed in the expression of P450 2B1 2B2 as indicated by cross-reactivity with anti-P450 2B1/2B2 in the region of 52 kDa which comigrated with immunoreactive band in PB-treated liver microsomes, used as positive control (Fig. 3). Among the control brain regions, corpus striatum exhibited maximum cross-reactivity as evident from visual inspection, followed by olfactory lobe, mid brain and hippocampus. However, no immunoreactivity was observed in other brain regions and whole brain. PB-pretreatment significantly increased the cross-reactivity in olfactory lobe, mid-brain, hippocampus, corpus striatum and hypothalamus (Fig. 3). As observed by us, Morse et al. (1998) also reported low levels of P450 1A2 immunoreactive protein in brain microsomes from the several brain regions isolated from control animals. Tirumalai et al. (1998) also reported distinct regionality in the expression of P450 2B1/2B2 isoenzymes in rat brain. Immunoblots from untreated brain regions showed immunological cross reactivity between certain forms of hepatic P450 and cerebral P450, further suggesting that multiple forms of P450 are constitutively but differentially distributed in rat brain regions. Using antipeptide derived antibody, Farin and Omeicinski (1993) also reported constitutive expression of P450 1A1 in human brain. Most of the neurons and some of the astrocytes from various brain regions displayed significant immunoreactivity with anti-P450 1A1.

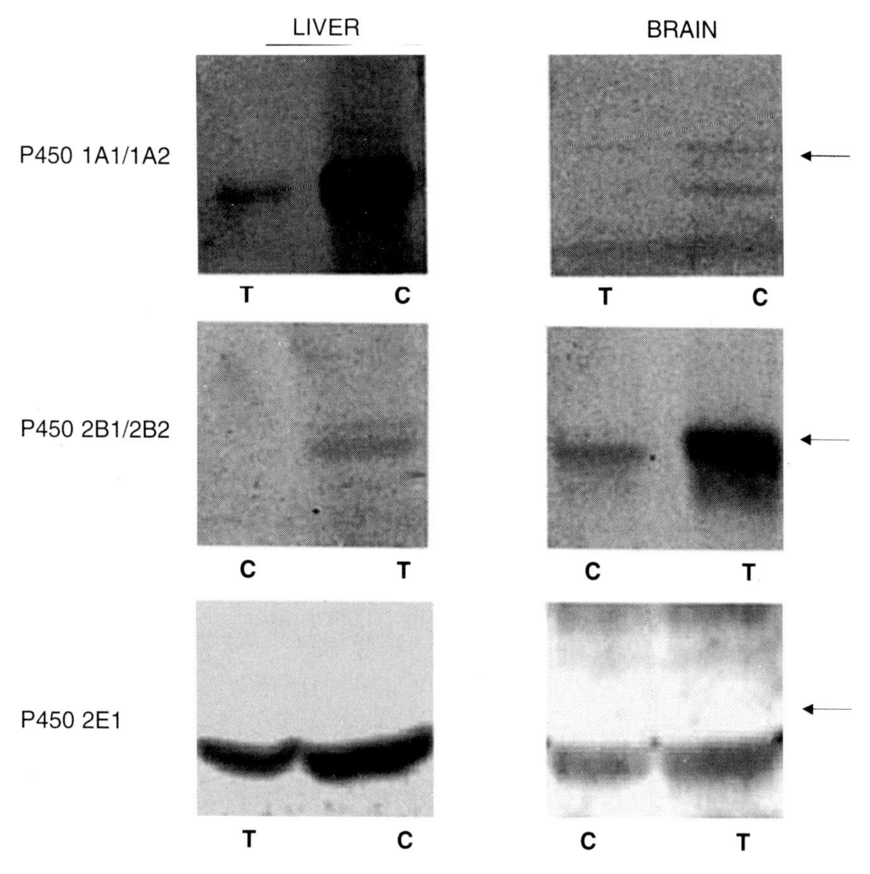

Fig. 2 Western blot analysis of brain P450s *(C-control microsomes; T-microsomes isolated from MC pretreated rats for 1A1/1A2; PB pretreated rats for 2B1/2B2 and ethanol pretreated rats for 2E1).*

Recent studies from our laboratory have also demonstrated the catalytic activity of P450 1A1/1A2 and 2B1/2B2 isoenzymes expressed in rat brain. The specificity of the brain P450 isoenzymes in the O-dealkylation of alkoxyresorufin derivatives and that substituted resorufin derivatives may serve as a useful probe to study the induction of P450 isoforms in the brain has been demonstrated by us (Parmar et al. 1998a,b; Dhawan et al. 1999). Brain microsomes were found to catalyse the O-dealkylation of 7-pentoxyresorufin (PR), and 7-ethoxyresorufin (ER) and this dealkylation was found to be inducer selective (Table 1). While pretreatment with PB resulted in significant induction in the activity of pentoxyresorufin-O-dealkylase (PROD), MC had no effect on the activity of PROD & only a slight effect on that of BROD. MC pretreatment significantly induced the activity of ethoxyresorufin-O-deethylase (EROD) while PB had no effect on it. Kinetic studies have shown that the increase in the activities following pretreatment with P450 inducers

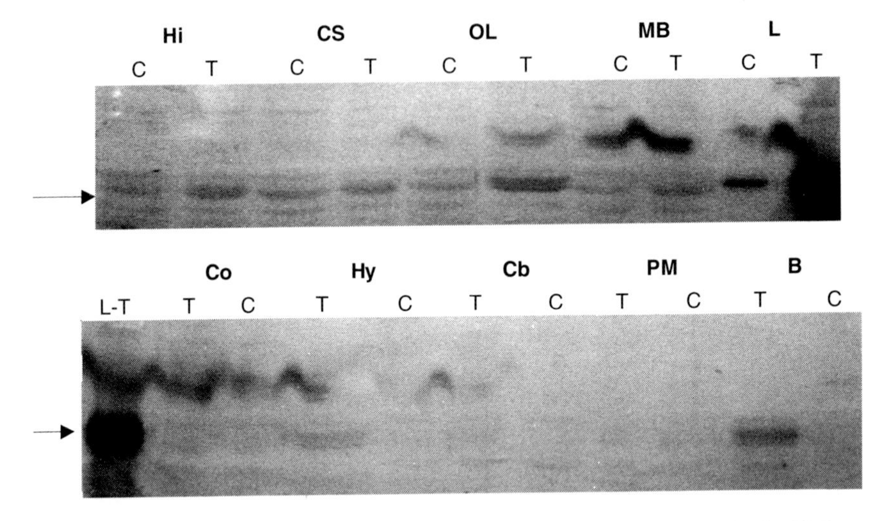

Fig. 3 Region specific expression of P450 2B1/2B2: Western blot analysis (C-Con; T- PB)
CONTROL: Corpus striatum > olfactory lobe > hippocampus > mid brain > cortex = hypothalamus = cerebellum = pons & medulla = brain
PB: Olfactory lobe > brain > hippocampus > Corpus striatum > hypothalamus > mid brain > cortex = cerebellum = pons & medulla

Table 1 Isoenzyme specific P450 monooxygenases in rat brain

	NDMA-d[a]	*EROD*[b]	*PROD*[b]
Control	0.93±0.12	0.84±0.05	0.39±0.02
MC	0.70±0.10	2.60±0.14*	0.30±0.04
PB	1.20±0.19	0.75±0.04	1.20±0.20*
Ethanol	1.86±0.22*	0.83±0.06	0.38±0.05

* $p < 0.05$ when compared to the controls.
a—nmoles HCHO/min/mg protein
b—pmoles resorufin/min/mg protein

was associated with a significant increase in the velocity of the reaction (V*max*) of O-dealkylation. *In vitro* studies using organic inhibitors and antibodies have further provided evidence that the O-dealkylation of alkoxyresorufin is isoenzyme specific. These studies have suggested that as in liver, dealkylation of alkoxyresorufins can be used as a biochemical tool to characterise the xenobiotic metabolising P450s and substrate selectivity of P450 isoenzymes in brain microsomes (Parmar et al. 1998a,b; Dhawan et al. 1999).

In the past decade considerable interest has been centered to investigate the levels of P450 2E1 in mammalian brain. P450 2E1 has been shown to exhibit a broad substrate range, including both endogenous substrates such as acetone, linoleic acid, arachidonic acid and exogenous substrates such as neuroactive chemicals and organic solvents (Tindberg and Ingelman-Sundberg, 1996). While some reports indicated very low levels of expression of P450 2E1 in control rat brain (Hansson et al., 1990; Warner and Gustafsson, 1994, Tindberg and Ingelman-Sundberg, 1996), other reports suggest P450 2E1 levels to be 25% of liver levels (Anandatheerthavarada et al., 1993; Bhagwat et al. 1995). RT-PCR studies from our laboratory, using primers specific for rat hepatic P450 2E1 exhibited significant mRNA expression of P450 2E1 in control rat brain which was significantly induced following pretreatment of ethanol (Fig. 1). Preliminary data from our laboratory have also demonstrated constitutive mRNA expression of P450 2E1 in cultured neuronal and glial cells (Fig. 4). Although earlier studies have reported mRNA expression of P450 2E1 in cultured glial cells, interestingly, the neuronal cells were also found to express significantly higher expression of P450 2E1. The presence of cytochrome P450 2E1 and its induction by ethanol was also demonstrated in primary cultures of astrocytes using immunofluorescence technique, confocal microscopy and dot blot assay by Montoliu et al. (1995). Interestingly, though significant mRNA expression of constitutive and inducible P450 2E1 was observed in rat brain, western blot studies reported low levels of constitutive and inducible expression of P450 2E1 at the protein level (Fig. 2). Olfactory lobes, however exhibited distinct expression of P450 2E1 (Fig. 5). Pretreatment of ethanol was found to result in significant expression of P450 2E1 in hippocampus and produced an increase in the expression in olfactory lobes (Fig. 5). Warner & Gustafsson (1994) have also reported significant increase in the P450 contents in the brain regions after single dose of ethanol (0.8 ml/kg), of which 10 to 20 fold increase being in the olfactory lobes and hypothalamic preoptic area. Region specific distribution of P450 2E1 was also shown by Hansson et al. (1990) in brain using polyclonal antibodies to P450 2E1 of rat. Immunocytochemical studies revealed the preferential localization of P450 2E1 in the neuronal cells of hippocampus, cortex, basal ganglia, hypothalamic nuclei and reticular nuclei in the brain stem (Anandatheerthavarada et al., 1993; Bhagwat et al., 1995). P450 2E1 expressed in brain was also found to catalyse the activity of N-nitrosodimethylamine demethylase (NDMA-d). Like in liver, ethanol pretreatment significantly induced the enzyme activity which was associated with an increase in the Vmax and affinity (Km) of the substrate towards the brain enzyme (Table 1). Significant differences were also observed in the distribution of NDMA-d activity in the different brain regions, with olfactory lobe exhibiting the maximum enzyme activity.

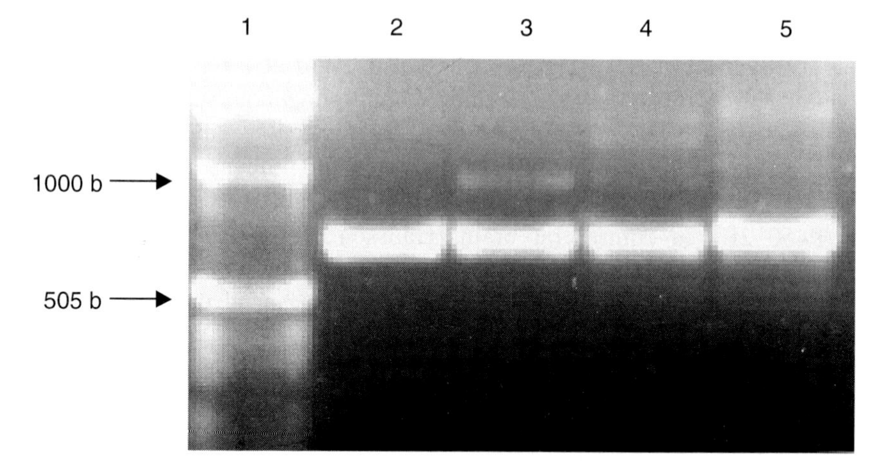

Fig. 4 Constitutive expression of CYP2E1 mRNA in cultured rat cells. *Lane 1 contain 1.2 Kb ladder of Life Technologies . Lane 2&3 contain 5 μl of RT-PCR product of RNA isolated from neuronal and glial cells respectively. Lane 4 contain 5 μl of RT-PCR product of RNA isolated from control rat brain. Lane 5 contains RT-PCR product obtained from RNA of ethanol pretreated rat liver*

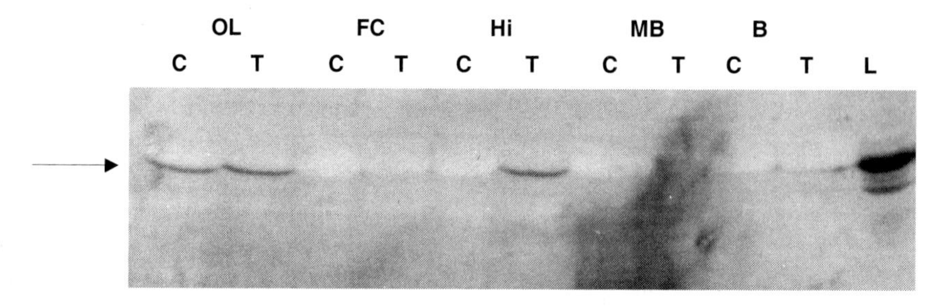

Fig. 5 Region specific expression of P450 2E1: Western blot analysis *(C-Con: T-ethanol treated; OL-olfactory lobes; FC- frontal cortex;Hi-hippocampus; MB-midbrain; B-Brain; L-Liver)*

The mRNA for P450 2D subfamily has also been demonstrated in mammalian brain. Interestingly, P450 2D4 was found to be much more abundant in rat brain than those for P450 2D1 & 2D5, the major hepatic forms. P450 2D2 & 2D3 mRNA which are also abundant forms in liver, were not detectable in brain. Using northern and southern blots and RT-PCR analysis of rat brain mRNA, Komori (1993) also reported expression of only P450 2D4 amongst the 2D P450s in the rat brain. A novel P450 belonging to the 2D family has recently been cloned from the control and imipramine treated rat brain cDNA library (Kawashima and Strobel, 1995a) which was not expressed in the liver. Expression of P450 2D and P450 reductase in the CNS and effect of a number of xenobiotics and hormones upon their levels in the brain has been reported (Bergh & Strobel, 1996).

Cloning studies have led to the identification of novel P450s expressed in mammalian brain (Table 2). Full length cDNA (P450 3A9) from untreated male rat brain cDNA library having 68-76% homology with reported P450 3A sequences has been cloned (Wang et al., 1996, Wang & Strobel, 1997). Region specific expression of members of P450 4A subfamily involved in the synthesis of metabolites of arachidonic acid was demonstrated in rat brain by PCR studies (Stromstedt et al. 1994) and mRNA for P450 4A2 & 4A3. A very weak signal for P450 4A8 was seen by PCR only in the cerebral cortex, while no signal was detected by northern blot analysis since brain content of P450 4A was 0.1% of that in liver as demonstrated by immunoblotting. N-terminal sequencing of protein in purified P450 fractions have also revealed the presence of P450 4A8 and 4A3 in brain. Using P450 4A5 cDNA as a probe, Kawashinma & Strobel (1995b) isolated the cDNA clones belonging to P450 4F subfamily from untreated rat brain cDNA library. The expression levels of these forms of P450s in the brain were somewhat low with relatively high level of expression of similar forms of subfamily 4F P450 in liver and kidney.

Table 2 cDNA cloning of novel rat brain P450s

cDNA	*Source*	*Probe*	*Reference*
2d-29, 2d-30 (P450 2D subfamily)	Control & imipramine treated rat brain cDNA libraries	PCR product of P450 2D	Kawashima & Strobel, 1995a
4f-8, 4f-44, 4f-41 (P450 4F4, 4F5, 4F-6)	Control rat brain cDNA library	P450 4A5 cDNA	Kawashima & Strobel, 1995b
3aH15 (P450 3A9)	Control rat brain cDNA library	PCR generated cDNA fragment of P450 3A2	Wang et al., 1996
hct-1 (P450 7)	Control rat hippocampus cDNA library	Radiolabelled cDNA probes.	Stapleton et al., 1995

Kainu et al. (1995) reported the localisation of aryl hydrocarbon receptor (Ahr) and aryl hydrocarbon receptor nuclear translocator (Arnt) in the rat brain through *in situ* mRNA hybridization. While the distribution of these related transcription factors, involved in the regulation of expression of CYP1A1, were partly similar, differences in their localisation also existed, with Arnt being more abundantly and more widely expressed. Thus, Arnt may have a further role in the brain, in addition to, being the heterodimerization partner of Ahr.

Toxicological Consequences of Expression of P450s in Brain

Role of P450 in the etiopathogenesis of several of the neurodegenerative disorders induced by drugs and environmental toxins have been indicated in recent years. There is a growing evidence suggestive of an interaction between P450 and 1-methyl-4-phenyl-1,2,3,6-tetrahydropyridine (MPTP), an agent known to induce highly selective and irreversible neurotoxicity, which results in clinical syndromes indistinguishable from idiopathic Parkinson's disease in man. MPTP and the environmental toxins structurally similar to MPTP has been shown to be potent inhibitors of rat cerebral P450 2D isoenzymes catalysing debrisoquine 4-hydroxylation, which is one of the best studied example of genetically determined polymorphism of drug oxidation (Langston et al., 1983; Fonne Pfister et al., 1987; Ohta et al., 1990). The presence of P450 2D6 in humans and similar isoform 2D1 in rat brain indicate that this P450 form may be a factor responsible in susceptibility to MPTP neuronal toxicity (Fonne-Pfister et al., 1987; Gilham et al., 1990). P450 2D in brain has also been shown to detoxify tetrahydroisoquinolone, a naturally occurring compound causing Parkinsonism in monkeys (Ohta et al., 1990; Yoshida et al., 1990). It is quite possible that hereditary deficiency of this P450 isoenzyme in brain would result in greater likelihood of bioactivation, thus making individuals more susceptible to the effect of such toxins. In this context, a specific mutation in P450 2D6 coding region was shown to be closely related with Parkinson's disease (Kurth and Kurth, 1993; Tsuneoka et al., 1993; Akhmedova et al., 1995). Thus interindividual genetic variations in this isoenzyme may have role for neurological disorders and toxicity as well as in therapeutic drug response. Moreover, presence of polymorphic enzymes i.e. P450 2D6 also affects the metabolism of neuroleptic and other psychoactive drugs, which are the substrates of these enzymes in brain (Masimirombwa and Hasler, 1997).

Brain P450s have also been shown to be involved in the activation of precarcinogens to carcinogens and mutagens (Das *et al.* 1985). Rouet et al. (1981) reported the qualitative and quantitative differences in metabolites of benzo(a)pyrene (BP) formed in brain microsomes during perinatal development indicating the greater susceptibility of younger ones to these carcinogens. Das et al. (1985) have also shown that BP binds with cerebral DNA which may explain the process of malignant tumor formation in brain by polycyclic aromatic hydrocarbons, PAHs (Napalkov and Alexandrov, 1974). Likewise, treatment of pregnant rats with polycyclic aromatic hydrocarbons yielded a high percentage of brain tumors in the newborns. Species differences in the extent of activity may also be of significance in this context since it appears that mouse fetal brain tissue is very reactive in metabolizing PAHs to mutagenic metabolites in comparison to rat tissue (Mensil et al. 1984). Since the brain P450s are also inducible and altered by the administration of the other xenobiotics, there is some evidence to suggest that this may be a mechanism of importance in determining the carcinogenicity in brain. Interestingly, P450 1A1 inducers, β-NF and cigarette smoke have been reported to protect mice against the toxicity of MPTP (Shahi et al., 1991).

Furthermore, epidemiological studies have shown that Parkinson's disease occurs less frequently in smokers than in non-smokers (Godwin-Austen et al., 1982).

Ethanol has been shown to increase the P450 of rat brain and specifically induce the levels of 2C, 2E & 4A P450s in several brain regions. The induction of brain P450 2E1 by ethanol is also of toxicological interest for several reasons. Besides the toxicological consequences of increased production of the acetaldehyde in the brain, the increase in the levels of P450 2E1 in the brain could lead to i) an increase in the release of active oxygen and increase in lipid peroxidation and destruction of membranes; ii) metabolism of many organic solvents such as n-hexane, benzene and toluene, which easily enter the brain, and are neurotoxins. The possibility that inhalation of organic solvents (e.g. acetone, isopropanol or ethylacetate) that are common in workplace and induce P450 2E1 in the liver, may also induce this enzyme in the brain and increase the risk of neuronal damage further adds to the toxicological consequences related to induction of this enzyme in the brain. P4502E1 was found to be selectively localized in the nigral dopamine containing cells. The expression of P450 2E1, a potent generator of the free radicals, has led them to conclude that P450 2E1 may have a role in neurodegeneration or in the detoxication of environmental neurotoxin similar to the metabolism of MPTP carried out by P450 2D6 and other enzymes. Electron leakage from P450 redox reactions in the substantia nigra might be expected to induce oxidative stress under certain conditions. Such an increase in the formation of reactive species from P450 2E1 in nigral dopaminergic neurons could be a potential cause of the oxidative stress which has been detected in substantia nigra in Parkinson's disease. Alternatively P450 2E1 activity in the nigra might be of importance in the activation of pro-neurotoxins in a manner similar to the activation of MPTP by monoamine oxidase B. There is also increasing evidence that exposure to n-hexane, as a component of petroleum spirit and of other solvents is linked to the development of Parkinsonian syndrome. Sincc hexane is neurotoxic only following P450 2E1 induced activation to 2,5-hexanedione, the presence of P450 2E1 in the nigra may contribute in part to the neurotoxicity of hexane (Watts et al. 1998). Likewise, members of the P450 2C family, the expression of which is known to be induced by ethanol, inactivate steroids and activate several of the procarcinogens to carcinogens. In addition, P450 2C7, which is specifically induced by ethanol in the brain, metabolizes retinol and retinoic acid (Leo et al., 1986). Though the role of P450 2C7 has not been investigated in brain nor there is any information available on the levels of retinoic acid after ethanol administration, induction in the levels of P450 2C7 could be of significance as ethanol consumption has been shown to deplete Vitamin A in liver (Ryle et al., 1986). Another catalytic activity of some members of P450 2C family that may be relevant in brain is the arachidonic acid epoxygenase activity and suggests role of P450 2C in intracellular signalling.

Cerebral P450s have been reported to be involved in Manganese (Mn) neurotoxicity (Qato and Maines 1985; Liccione and Maines, 1989). Moreover, Mn was found to increase P450 dependent hydroxylation activity in both the mitochondrial and microsomal fractions of rat striatum. The effects were more pronounced in the mitochondrial fraction where P450 hydroxylation activities were increased by 2-3 fold. Mn induced selective dopamine depletion in striatum was found to be associated with specific induction of P450 dependent metabolism of amphetamine to p-hydroxyamphetamine, an inhibitor of neural dopamine uptake, particularly in the mitochondrial fractions of striatum which may influence indirect dopaminergic action of amphetamine. In addition, the superoxide anion radicals produced by P450s can lead to the formation of reactive electrophiles of dopamine and other catechols. Thus, the Mn mediated alterations in brain P450 dependent drug metabolism reactions, along with possible concomitant enhanced formation of active oxygen species, might partially explain Mn neurotoxicity of dopamine pathways.

Studies from our laboratory have shown that the neurotoxicity of acrylamide, a monomer used in plastics & polymer industry could be attributed to its metabolite formed during P450 mediated metabolism (Srivastava et al., 1985). Pretreatment of rats with PB, trans-stilbene oxide or dichloro diphenyl trichloroethane (DDT) resulted in earlier onset and subsequent development of acrylamide induced hind limb paralysis that was observed only in animals treated with acrylamide. Pretreatment of cobalt chloride, an inhibitor of P450s, caused a significant delay in the onset and development of hind limb paralysis. Our hypothesis that a "presumed" intermediate of acrylamide formed by the P450 system is responsible for the toxicity of acrylamide is supported by the earlier appearance of neurotoxicity in rats treated with the inducers of P450. The delayed development of acrylamide toxicity in $CoCl_2$ pretreated animals further supports the concept that the observed toxicity of acrylamide is, at least in part, mediated by an intermediate formed by the P450 system (Srivastava et al., 1985).

Recent studies from our laboratory have shown the involvement of P450 mediated metabolism in the neurobehavioural toxicity of deltamethrin, a widely used type II pyrethroid insecticide (Dayal et al., 1999). Deltamethrin has been shown to act on multiple site in the CNS. Its marked neurotoxicity in laboratory animals following acute and chronic exposure has been attributed to its effect on nerve membrane sodium channels and interaction with γ-aminobutyric acid (GABA) receptor ionophore complex. Studies on the levels of insecticide in different tissues of rats have indicated that pharmacokinetics plays an important role in the neurotoxicity of deltamethrin (Rickard & Brodie, 1985; Sheets et al., 1994). Our studies have shown that deltamethrin produces a marked dose- and time dependent increase in the expression and activity of P450 monooxygenases (7-pentoxyresorufin-O-dealkylase, PROD and 7-ethoxyresorufin-O-deethylase, EROD) in rat brain & liver and that these alterations in P450s could be related to the amounts of pyrethroids and its metabolites reaching and accumulating in the brain (Fig. 6). These studies have shown that the increase in the activity of P450 monoxygenases in brain after deltamethrin exposure is due to the increase in the expression of P450 2B1/2B2 & 1A1/1A2 isoenzymes (Dayal et al., 1999). Interestingly, the induction in the activity of cerebral P450 enzymes was associated with an increase in the spontaneous locomotor activity indicating accumulation of deltamethrin or its metabolites in the brain (Table 3). These studies further demonstrated that the metabolism of deltamethrin, possibly formed by P450 enzymes, is involved in the neurobehavioural toxicity of deltamethrin and have also provided evidence that the modulation of individual P450 isoenzymes significantly influence the neurobehavioural toxicity of deltamethrin (Dayal et al., 2003). Our data revealed that pretreatment with 3-methylcholanthrene (MC) or phenobarbital (PB), inducer of P450 1A1/1A2 and 2B1/2B2 isoenzymes respectively, potentiated the deltamethrin neurobehavioral toxicity as reflected by significant increase in the incidence and onset of tremors in PB or MC pretreated animals. Half of the animals pretreated with MC prior to exposure to deltamethrin even exhibited choreoathetosis, the ultimate toxicity symptom of deltamethrin toxicity (Table 4). In contrast to the pretreatment with inducers, animals pretreated with cobalt chloride, a blocker of P450 catalysed reactions did not exhibited any signs of neurobehavioral toxicity indicating that reactive metabolite of deltamethrin is formed by P450 catalysed reactions which is involved in the neurobehavioral toxicity of deltamethrin (Table 4). Deltamethrin was further found to produce region specific differences in the induction of P450 2B1/2B2 & 1A1/1A2 enzymes in rat brain with significant induction occuring in the activity of P450 1A1/1A2 dependent EROD in cerebellum, hippocampus, hypothalamus and medulla-pons and that of P450 2B1/2B2 mediated PROD in hippocampus, hypothalamus, corpus striatum and mid-brain (Table 5). These differences in the induction of individual P450 enzymes in diverse brain regions could play a role in regulating the response of brain to the pyrethroid insecticides by modulating their concentration per se or their metabolites at the target site(s) (Dayal et al. 2001).

Likewise, our studies have shown that the neurotoxicity of lindane is metabolism dependent & isoenzyme specific (Parmar et al., 2003). Oral administration of lindane was found to produce dose- and time- dependent increase in the activity of P450 dependent EROD, PROD and N-nitrosodimethylamine demethylase (NDMA-d) in rat brain and liver (Fig. 7). *In vitro* studies using brain and liver microsomes with organic inhibitors and antibodies specific for individual P450s and Western blotting studies have indicated that the induction in the activity of P450 enzymes is due to the

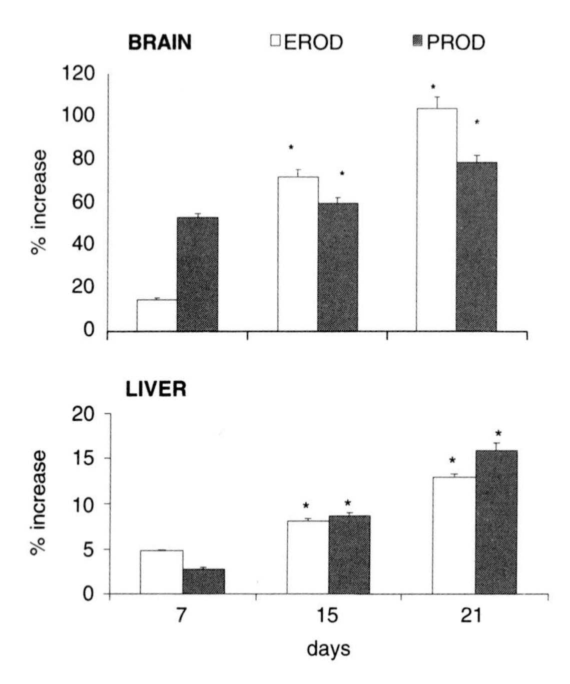

Fig. 6 Time dependent induction of hepatic and brain P450 enzymes by deltamethrin (5mg/kg; po)

increase in the expression of P450 1A1/1A2, 2B1/2B2 & 2E1 isoenzymes respectively. This has also been demonstrated by RT-PCR studies indicating significant increase in the mRNA expression of P450 1A1, 2B1 and 2E1 in rat brain following exposure to lindane (Fig. 8). Our data have further indicated that induction of individual P450 isoenzymes could significantly influence the neurotoxicity of lindane (Parmar et al., 2003). While administration of lindane in rats pretreated with MC, an inducer of P450 1A1/1A2, failed to produce any increase in convulsions when compared to animals treated with lindane alone, pretreatment of PB, an inducer of P450 2B1/2B2 isoenzymes or ethanol, an inducer of P450 2E1, significantly induced the incidence of convulsions in all the animals. Similarly, when the P450 mediated metabolism of lindane was blocked by cobalt chloride, incidence of lindane convulsions were significantly increased in the animals treated with lindane indicating that lindane *per se* or its metabolites formed by PB inducible P450 isoenzymes are involved in its neurotoxicity (Table 6). Though further studies are needed to identify the role of individual P450 isoenzymes in the formation of reactive species of lindane or deltamethrin, the data have assumed significance as the effects observed at doses used in the present study could be possibly reached during occupational exposures (IARC monographs, 1974) and humans are exposed to several of these chemicals through their environment which have the potential to modulate the expression of brain and liver P450 isoenzymes.

Table 3 Time dependent effect of deltamethrin (5mg/kg) on spontaneous locomotor activity of rats

Days of treatment	*Distance travelled*		*Resting time*		*Ambulatory time*		*Horizontal count*	
	Control	*Treated*	*Control*	*Treated*	*Control*	*Treated*	*Control*	*Treated*
1	695.5 ± 70.5	845.4 ± 73.4	177.6 ± 15.2	138.1 ± 14.5	35.1 ± 4.2	42.5 ± 5.0	492.2 ± 65.2	618.5 ± 63.0
7	689.6 ± 92.8	836.2 ± 99.8	179.8 ± 15.1	140.3 ± 17.9	33.9 ± 4.4	41.6 ± 4.6	490.4 ± 66.4	616.8 ± 61.8
15	688.1 ± 73.2	942.9 ± 19.8*	172.4 ± 4.5	131.9 ± 2.5*	34.2 ± 3.2	43.5 ± 1.4*	489.6 ± 33.7	647.8 ± 38.0*
21	709.0 ± 54.7	1127.4 ± 100*	166.2 ± 11.2	114.8 ± 8.6*	33.8 ± 2.2	54.5 ± 4.3*	586.5 ± 60.8	907.0 ± 77.3*

Value represents mean ± S.E. of ten animals

*$p < 0.05$ when compared to control

Table 4 Effect of pretreatment of P450 modifiers on incidence of deltamethrin-induced neurobehavioral symptoms

	Tremors	*Choreoathetosis*	*Latency of tremors (minutes)*
Control	0(10)	0(10)	ND
Deltamethrin	2(10)	0(10)	315.0 ± 15.2
MC + Deltamethrin	6(10)	5(10)*	240.0 ± 30.0
MC	0(10)	0(10)	N.D.
PB + Deltamethrin	6(10)	0(10)	292.5 ± 22.5
PB	0(10)	0(10)	ND
$CoCl_2$ + Deltamethrin	0(10)	0(10)	ND
$CoCl_2$	0(10)	0(10)	ND

*One rat died during choreoathetosis.

Values in parenthesis indicate number of animals in each group.

N.D.—Not detected.

Doses: Deltamethrin (DM) - 10 mg/kg ; MC - 30 mg/kg; i.p. × 5 d; MC+DM-MC, 30 mg/kg; i.p. × 5 d + DM , 10 mg/kg; PB - 80 mg/kg; i.p. × 5 d; PB + DM-PB, 80 mg/kg; i.p. × 5 d + DM, 10 mg/kg; $CoCl_2$ - 23 mg/kg; s.c. x 2 d; $CoCl_2$+DM-$CoCl_2$, 23 mg/kg; s.c. x 2 d + DM, 10 mg/kg

Table 5 Effect of deltamethrin (5 mg/kg X 21 days) on the activity of P450 enzymes in different brain regions

	EROD		*PROD*	
	Control	*Treated*	*Control*	*Treated*
	(pmoles resorufin formed/ min/ mg protein)			
Cerebellum	0.54 ± 0.06	1.79 ± 0.20*	0.23 ± 0.02	0.34 ± 0.04
Medulla and Pons	0.68 ± 0.07	1.46 ± 0.20*	0.20 ± 0.03	0.36 ± 0.03
Hippocampus	0.18 ± 0.01	0.45 ± 0.05*	0.27 ± 0.03	0.54 ± 0.05*
Hypothalamus	0.14 ± 0.01	0.22 ± 0.02*	0.22 ± 0.02	0.45 ± 0.05*
Corpus Striatum	0.54 ± 0.06	0.67 ± 0.08	0.41 ± 0.05	0.74 ± 0.09*
Midbrain	0.72 ± 0.09	1.08 ± 0.20	0.27 ± 0.03	0.67 ± 0.07*
Cortex	0.67 ± 0.07	0.90 ± 0.09	0.18 ± 0.02	0.19 ± 0.13

Values are mean + S.E. of three experiments.

Specific activity in pmoles resorufin/min./ mg protein.

*$p < 0.05$ when compared to control.

Conclusion

P450s in mammalian brain have been identified as the functional enzymes enabling the CNS to metabolize a variety of substrates of both exogenous and endogenous origin. Though levels of P450s are low in the brain (1-5% of the liver enzyme), they are of significance due to protective nature of the organ and its role as target for a variety of drugs and foreign chemicals. Using RT-PCR, immunocytochemistry and cloning approaches, multiple forms of xenobiotic metabolizing P450s have been shown to be expressed in mammalian brain. Enzymatic studies have demonstrated the substrate

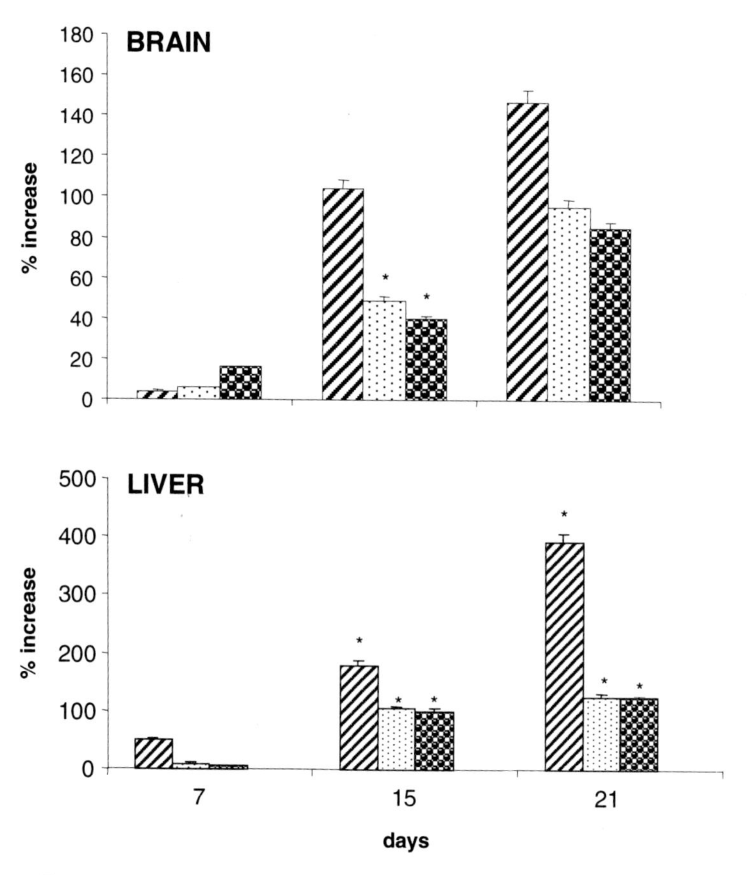

Fig. 7 Time dependent induction of hepatic and brain enzymes by lindane (2.5 mg/kg; po)

specificity and selectivity of the P450s expressed in brain and have provided evidence that the substituted alkoxyresorufins may serve as a biochemical tool to characterize the brain P450 activities. Though overlapping substrate specificity of the xenobiotics with endogenous substrates have made it exceedingly difficult to distinguish the xenobiotic metabolizing activity and the endogenous functions of the P450s in brain, copurification of the brain P450s with dopamine transporter and sigma receptors have indicated the involvement of P450s in cerebral neurotransmission. Several of the neurotoxins including drugs and drug contaminants such as antidepressants, psychotoxins, neuroleptics, N-methyl-4-phenyl-2,5,6-tetrahydro-pyridine (MPTP) and environmental chemicals such as pesticides (synthetic pyrethroids), solvents etc. may interact with the brain P450s as they are known to be metabolized by these enzymes in liver. By acting as specific inducers and inhibitors of the brain P450s, these chemicals may interfere with

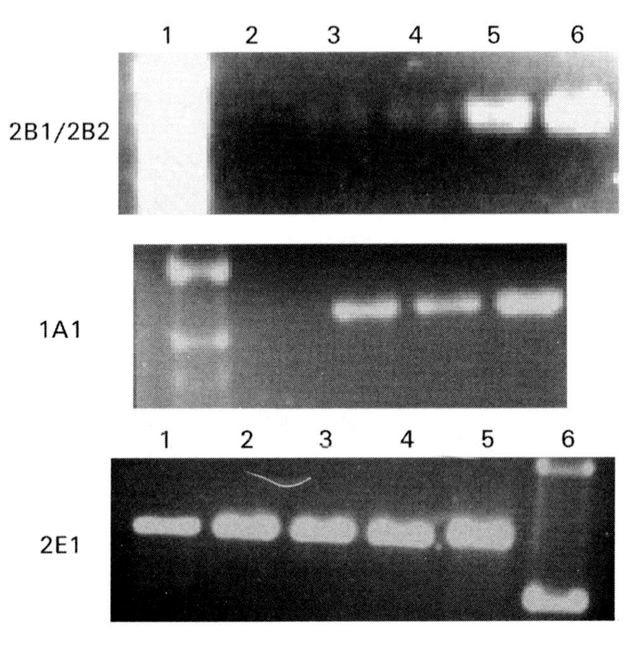

Fig. 8 RT-PCR analysis demonstrating time dependent increase in mRNA expression by lindane *(For 2B1/2B2 and1A1: lane1-DNA ladder, lane2-control;lane3 -5, RNA from 5,15 and 21 days of lindane (2.5 mg/kg, po) exposure; lane 6 RNA from PB treated liver. For 2E1: lane1-control; lane2-4, RNA from 5,15 and 21 day of lindane (2.5 mg/kg, po) exposure; lane 6-ladder).*

Table 6 Effect of pretreatment of P450 modifiers on lindane induced convulsions

	Onset of convulsions (min)	*Incidence*
Control	ND	0/10
Lindane	94.0 ± 2.45	4/10
PB	ND	0/10
PB + Lindane	51.7±8.5	10/10
MC	ND	0/10
MC + Lindane	62.5 ± 3.5	3/10
Ethanol	ND	0/10
Ethanol + Lindane	55.4 ± 7.0	7/10
$CoCl_2$	ND	0/10
$CoCl_2$ + Lindane	36.8 ± 3.0	7/10

Values in parenthesis indicate number of animals in each group.

ND - Not detected.

Doses: Lindane—35 mg/kg, p.o ; MC—30 mg/kg, i.p. x 5 d; PB—80 mg/kg, i.p. x 5 d; Ethanol—15% (v/v) solution in drinking water X 3 d; MC+Lindane-MC,30 mg/kg, i.p. x 5 d + Lindane , 35 mg/kg, p.o; PB + Lindane- PB, 80 mg/kg, i.p. x 5 d + Lindane, 35 mg/kg, p.o; $CoCl_2$—23 mg/kg, s.c. x 2 d; $CoCl_2$ + Lindane—$CoCl_2$, 23 mg/kg, s.c. x 2 d + Lindane, 35 mg/kg; Ethanol+Lindane-15% (v/v) in drinking water X 3 d + Lindane, 35 mg/kg, p.o the functioning and regulation of the cerebral processes or be activated by this system to produce tissue specific toxicity. Because of the low levels in brain, P450s are unlikely to influence the overall pharmacokinetics of the neurotoxins, nevertheless the *in situ* bioactivation could lead to the damage of macromolecules at the target site and may play a critical role in toxication/ detoxication processes for environmental or endogenous toxins involved in the neurodegenerative diseases.

References

Akhmedova S.N., Pushnova E.A., Yakimovsky A.F., Avtonomov V.V. and Schwartz E.I. Frequency of a specific cytochrome P450 2D6 B (CYP 2D6 B) mutant allele in clinically differentiated groups of patients with Parkinson's disease; Biochem. Molec. Med. **54:** 88-90, 1995.

Anandatheerthavarada H.K., Shankar S.K. and Ravindranath V. Rat brain cytochrome P-450: catalytic, immunochemical properties and inducibility of multiple forms; Brain Res **536:** 339-343, 1990.

Anandatheerthavarada H.K., Shankar S.K., Bhamre S., Boyd M.R., Song B.J. and Ravindranath V. Induction of brain cytochrome P-450 IIE1 by chronic ethanol treatment; Brain Res **601:** 279-285, 1993.

Bergh A.F. and Strobel H.W. Anatomical distribution of NADPH-cytochrome P450 reductase and cytochrome P450 2D forms in rat brain: effects of xenobiotics and sex steroids; Mol. Cell Biochem **162:** 31-41, 1996.

Bhagwat S.V., Boyd M.R. and Ravindranath V. Brain mitochondrial cytochromes P-450: Xenobiotic metabolism, presence of multiple forms and their selective inducibility; Arch. Biochem. Biophys **320:** 73-83, 1995.

Das M., Srivastava S.P., Seth P.K. and Mukhtar H. Brain microsomal enzyme mediated covalent binding of benzo(a)pyrene to DNA; Cancer Lett. **25:** 343-350, 1985.

Dayal M., Parmar D., Dhawan A., Dwivedi U.N., Doehmer J. and Seth P.K. Induction of rat brain and liver cytochrome P450 1A1/1A2 and 2B1/2B2 isoenzymes by deltamethrin; Environ. Toxicol. Pharmacol **7:** 169-178, 1999.

Dayal M., Parmar D., Ali M., Dhawan A., Dwivedi U.N. and Seth P.K. Induction of rat brain cytochrome P450s (P450s) by deltamethrin: Regional specificity and correlation with neurobehavioural toxicity; Neurotoxicity Res, **3:** 351-357, 2001.

Dayal M., Parmar D., Dhawan A., Ali M., Dwivedi U.N. and Seth P.K. Effect of pretreatment of cytochrome P450 modifiers on neurobehavioural toxicity induced by deltamethrin; Fd Chem Toxicol. **41:** 431-437, 2003.

Dhawan A., Parmar D., Dayal M. and Seth P.K. Cytochrome P450 (P450) isoenzyme specific dealkylation of alkoxyresorufin in rat brain microsomes; Mol. Cell. Biochem. **200:** 169-176, 1999.

Fang W.F. and Strobel H.W. The drug and carcinogen metabolism system of rat colon microsomes; Arch. Biochem. Biophys. **186:** 128-138, 1978.

Farin F.M. and Omiecinski C.J. Regiospecific expression of cytochrome P450 and epoxide hydrolase in human brain tissue; J. Toxicol. Environ. Hlth. **40:** 323-341, 1993.

Fishman J., Hahn F.E. and Norton B.I. N-demethylation of morphine in rat brain is localized in sites with high opiate receptor content; Nature **261:** 64-65, 1976.

Fonne-Pfister R., Bargetzi M.J. and Meyer U.A. MPTP, the neurotoxin inducing Parkinson's disease is a potent competitive inhibitor of human and rat cytochrome P450 isoenzymes (P450 buf I, P450 db1) catalyzing debrisoquine 4-hydroxylation; Biochem. Biopyhs. Res. Commun. **148:** 1144-1150, 1987.

Geng J. and Strobel H.W. Identification of cytochromes P450 1A2, 2A1, 2C7, 2E1 in rat glioma C6 cell line by RT-PCR and specific restriction enzyme digestion; Biochem. Biophys. Res. Commun. **197:** 1179-1184, 1993.

Geng J. and Strobel H.W. Identification of inducible mixed function oxidase system in rat glioma C6 cell line; J. Neurochem. **65:** 554-563, 1995.

Geng J. and Strobel H.W. Expression, induction and regulation of the cytochrome P450 monooxygenase system in the rat glioma C6 cell line; Brain Res. **784:** 276-83, 1998

Gilham D.E., Cairns W., Paine M.J., Modi S., Poulsom-R., Roberts G.C. and Wolf C. Metabolism of MPTP by cytochrome P450 2D6 and the demonstration of 2D6 mRNA in human foetal and adult brain by in situ hybridization; Xenobiotica **27:** 111-125, 1997.

Godwin-Austen R.B., Lee P.N., Marmot M.G. and Stern G.M. Smoking and Parkinson's disease; J Neurol. Neurosurg Psychiatry **45:** 577-581, 1982.

Gram T.E., Okine L.A. and Gram R.A. The metabolism of xenobiotics by certain extrahepatic organs and its relation to toxicity; Ann. Rev. Pharmacol. **26:** 259-291, 1996.

Guengrich F.P. Preparation and properties of highly purified cytochrome P450 and NADPH cytochrome P450 reductase from pulmonary microsomes of untreated rabbits; Mol. Pharmacol **13:** 911-923, 1977.

Hansson T., Tindberg N. and Kohler C. Regional distribution of ethanol- inducible cytochrome P-450 IIEI in the rat central nervous system; Neurosci. **34:** 451-463, 1990.

Hodgson A.V., White T B, White J.W. and Strobel H.W. Expression analysis of the mixed function oxidase system in rat brain by the polymerase chain reaction; Mol. Cell. Biochem. **120:** 171-179, 1993.

IARC Monographs on the evaluation of carcinogenic risk of chemicals to man. **5:** 47-74, 1974.

Junier M.P., Dray F., Blair I., Capdevila J., Dishman E., Falck J.R. and Ojeda S.R. Epoxygenase products of arachidonic acid are endogenous constituents of the hypothalamus involved in D2 receptor-mediated, dopamine induced release of somatostatin; Endocrinol. **126,** 1534-1540, 1990.

Kainu T., Gustafsson J.A. and Pelto-Huikko M. The dioxin receptor and its nuclear translocator (Arnt) in the brain; Neuroreport **6:** 2557-2560, 1995.

Kawashima H. and Strobel H.W. cDNA cloning of a novel rat brain cytochrome P450 belonging to the CYP 2D subfamily; Biochem. Biophys. Res. Commun. **209,** 535-540, 1995a.

Kawashima H. and Strobel H.W. cDNA cloning of three new forms of rat brain cytochrome P450 belonging to the CYP 4F subfamily; Biochem. Biophys. Res. Commun. **217:** 1137-1144, 1995b.

Kohler C., Ericksson L.G., Hanssan T., Warner M. and Gustafsson J.-A. Immunohistochemical localisation of cytochrome P450 in the rat brain; Neurosci. Lett. **84:** 109-114, 1988.

Komori M. A novel P450 expressed at high level in rat brain; Biochem. Biophys. Res. Commun. **196:** 721-728, 1993.

Kurth M.C. and Kurth J.H. Variant cytochrome P450 CYP 2D6 allelic frequencies in Parkinson's disease; Amer J. Med. Genet. **48:** 166-168, 1993.

Langston J.W., Ballard P., Tetrud J.W. and Irwin I. Chronic Parkinsonism in humans due to a meperidine analog synthesis; Science **219:** 979-980, 1983.

Leo M.A., Lasker J.M., Raucy J.L., Kim C.I., Black M. and Lieber C.S. Metabolism of retinol, retinoic acid by human liver cytochrome P450 IIC8. Arch. Biochem. Biophys; **269:** 305-312, 1989.

Lewis D.F.V. (Ed.) In "Cytochromes P450 : Structure, function and mechanism"; TJ Press, Great Britain, pp. 169-208, 1996a.

Lewis D.F.V. (Ed.) In "Cytochrome P-450 : Structure, function and Mechanism", TJ Press, Great Britain pp. 115-1167, 1996b.

Licione J.J. and Maines M.D. Manganese-mediated increase in the rat brain mitochondrial cytochrome P450 and drug metabolism activity: susceptibility of striatum; J. Pharmacol. Exp. Ther. **248:** 222-248, 1989.

Masimirombwa C.M. and Hasler J.A. Genetic polymorphism of drug metabolizing enzymes in African populations: implications for the use of neuroleptics and anti-depressants; Brain Res. Bull. **44:** 561-571, 1907.

Mensil M., Testa B. and Jenner P. Xenobiotic metabolism by brain monoxygenases and other cerebral enzymes; Adv. Drug Res. **13:** 95-207, 1984.

Montoliu C., Sancho-Tello M., Azorin I., Burgal M., Vafles S., Renau-Piqueras J. and Gueni C. Ethanol increases cytchrome-P450 2EI and induces oxidative stress in astrocytes; J. Neurochem. **65:** 2561-2570, 1995.

Morse D.C., Stein A.P., Thomas P.E. and Lowndes H.E. Distribution and induction of cytochrome P450 1A1 and 1A2 in rat brain; Toxicol. Appl. Pharmacol. **152:** 232-239, 1998

Napalkov N.P. and Alexandrov V.A. Hemotropic effect of 7,12-dimethyl benz(a)anthracene in transplacental carcinogens J. Natl. Cancer Inst. **52:** 1365-1366, 1974.

Niznik H.B., Tyndale R.F., Sallee F.R., Gonzalez F.J., Hardwick J.P., Inaba T. and Kalow W. The dopamine transporter and cytochrome P450 IIDI (debrisoquine 4-hydroxylase) in brain: resolution of two distinct [^{3}H]GBR-12935 binding protiens; Arch. Biochem. Biophys. **276:** 424-432, 1990.

Ohta S., Tachikawa O., Maldno Y., Tdsaki Y. and Hirobe M. Metabolism and brain accumulation of tetrahydroisoquinoline (TIQ), a possible parkinsonism- inducing substance, in an animal model of a poor debrisoquine metabolizer; Life Sci. **46:** 599-60, 1990.

Parmar D., Dhawan A., Dayal M. and Seth P.K. Immunochemical and biochemical evidence for expression of phenobarbital and 3- methylcholanthrene inducible isoenzymes of cytochrome P450 in rat brain; Int. J. Toxicol. **171:** 1-12, 1998a.

Parmar D., Dhawan A, & Seth P.K. Evidence for O-dealkylation of 7-pentoxyresorufin by cytochrome P450 2B1/2B2 isoenzymes in brain; Mol. Cell. Biochem. **189:** 201-205, 1998b.

Parmar D, Yadav S, Dayal M, Johri A, Dhawan A and Seth PK 2003 Effect of lindane on hepatic and brain cytochrome P450s and influence of P450 modulation in lindane induced neurotoxicity; Fd. Chem. Toxicol. (**41:** 1077-87, 2003)

Purdy R.H., Morrow A.L., Moore P.H. and Paul S.M. Stress induced elevations of g-aminobutyric acid type A receptor active steroids in the rat brain; Neurobiol. **88:** 4553-4557, 1991.

Qato M.K. and Maines M.D. Regulation of heme and drug metabolism in the brain by manganese; Biochem. Biophys. Res. Commun. **128:** 18-24, 1985.

Ravindranath V. and Anandatheerthavarada H.K. High activity of cytochrome P-450 linked aminopyrene N-demethylase in mouse brain microsomes and associated sex related difference; Biochem. J. **261:** 769-773, 1989.

Ravindranath V., Anandatheerthavarada H.K. and Shankar S.K. Xenobiofic metabolism in human brain-presence of cytochrome P450 and associated monooxygenases; Brain Res. **496:** 331-335, 1989.

Ravindranath V. and Anandatheerthavarada H.K. Preparation of brain microsomes with cytochrome P450 activity using calcium aggregation; Anal. Biochem. **187:** 310-313, 1990.

Rickard J. and Brodie M.E. Correlation of blood and brain levels of the neurotoxic pyrethiroid deltamethrin with the onset of symptoms in rats; Pestic. Biochem. Physiol. **23:** 143-156, 1985.

Ross S.B. Is the sigma opiate receptor a proadifen-sensitive sub-form of cytochrome P450?; Pharmacol. Toxicol. **67:** 93-94, 1990.

Ross S.B. Heterogeneous binding of sigma radioligands in the rat brain and liver: possible relationship to subforms of cytochrome P-450; Pharmacol. Toxicol. **68:** 293-301, 1991.

Rouet P., Alexandrov K., Markovits P., Frayssinet C. and Dansette P.M. Metabolism of benzo(a)pyrene by brain microsomes of fetal and adult rats and mice. Induction by 5,6 benzoflavone, comparison with liver and lung microsomal activities; Carcinogenesis **2:** 919-926, 1981.

Ryle P.R., Hodges A. and Thomas A.D. Inhibition of ethanol induced hepatic vitamin A depletion by administration of N,N'-diphenyl-p-phenylene diamine (DPPD); Life Sci. **38:** 695-702, 1986.

Schilter B. and Omiecinski C.J. Regional distribution and expression modulation of cytochrome P-450 and epoxide hydrolase mRNAs in the rat brain; Mol. Pharmacol. **45:** 990-996, 1993.

Shahi G.S., Das N.P. and Moochhala S.M. 1-methyl-4-phenyl-1,2,5,6- tetrahydropyidine-induced neurotoxicity: partial protection against striato-nigral dopamine depletion in C57 BL/6J mice by cigarette smoke exposure and β-naphthoflavone- pretreatment; Neuroscience Lett. **127:** 247-250, 1991.

Sheets L.P., Doherty J.D., Law M.W., Reiter L.W. and Crofton K.M. Age-dependent differences in the susceptibility of rats to deltamethrin; Toxicol. Appl. Phramacol. **126:** 186-190, 1994.

Srivastava S.P., Seth P.K., Das M. and Mukhtar H. Effects of mixed-function oxidase modifiers on neurotoxicity of acrylamide in rats; Biochem. Pharmacol. **34:** 1099-1102, 1985.

Stapleton G., Steel M., Richardson M., Mason J.O., Rose K.A., Morris R.G.M. and Lathe R. A novel cytochrome P450 expressed primarily in brain; J. Neurochem. **270:** 29739-29745, 1995.

Strobel H.W., Cattaneo E., Adesnik M. and Maggi A. Brain cytochromes P-450 are responsive to phenobarbital and tricyclic amines; Pharmacol. Res. **21:** 169-175, 1989.

Stromstedt M., Warner M. and Gustafsson J.A. Cytochrome P450 of the 4A subfamily in the brain; J. Neurochem. **63:** 671-676, 1994.

Tindberg N. and Ingelman-Sundberg M. Expression, catalytic activity, and inducibility of cytochrome P450 2EI (CYP 2EI) in the rat central nervous system; J. Neurochem. **67:** 2066-2073, 1996.

Tirumalai P.S., Bhamre S., Upadhya S.C., Boyd M.R. and Ravindranath V. Expression of multiple forms of cytochrome P450 and associated mono- oxygenase activities in rat brain regions; Biochem. Pharmacol. **56:** 371-375, 1998.

Tsuneoka Y., Matsuo Y., lwahashi K., Takeuchi H. and Ichikawa Y. A novel Cytochrome P450 2D6 mutant gene associated with Parkinson's disease; J. Biochem. **114:** 263-266, 1993.

Tyndale R.F., Sunahara R., Inaba T., Kalow W., Gonzalez F.J. and Niznik B.B. Neuronal cytochrome P450 IIDI (debrisoquine/sparteine-type): Potent inhibition of activity by (-)-cocaine and nucleotide sequence identity to human hepatic P450 gene CYP2D6; Mol. Pharmacol. **40:** 63-68, 1991.

Wang H., Kawashima H. and Strobel H.W. cDNA cloning of a novel Cyp 3A from rat brain; Biochem. Biophys. Res. Commun. **221:** 157-162, 1996.

Wang H. and Strobel H.W. Regulation of CYP 3A9 gene expression by estrogen and catalytic studies using cytochrome P450 3A9 expressed in Escherichia coli; Arch. Biochem. Biophys. **344:** 365-72, 1997.

Warner M., Kohler C., Hansson T. and Gustafsson J.A. Regional distribution of cytochrome P-450 in the rat brain: spectral quantification and contribution of P-450 b,e and P-450 c,d; J. Neurochem. **50:** 1057-1065, 1988.

Warner M. and Gustafsson J.A. Effect of ethanol inducible cytochrome P450 in the rat brain; Proc. Natl. Acad. Sci. USA **91:** 1019-1023, 1994.

Watts P.M., Riedl A.G., Douek D.C., Edwards R.J., Boobis A.R., Jenner P. and Marsden C.D. Colocalization of P450 enzymes in the rat substantia nigra with tyrosine hydroxylase; Neurosci. **86:** 511-519, 1998.

Waxman D.J. and Azaroff L. Phenobarbital induction of cytochrome P-450 gene expression; Biochem. J. **281:** 577-592, 1992.

Whitelock J.P. The control of Cytochrome P450 gene expression by dioxin; Trends Pharmacol. Sci. **10:** 285-288, 1989.

Yoshida M., Niwa T. and Nagatsu T. Parkinsonism in monkeys produced by chronic administration of an endogenous substance of the brain, tetrahydroisoquinoline: the behavioral and biochemical changes; Neurosci. Lett. **119:** 109-113, 1990.

Section C
Pharmacology of CBW Antidotes

Pharmacological Perspectives of Toxic Chemicals and Their Antidotes
Editors: S J S Flora, J A Romano, S I Baskin and K Sekhar

Development of Treatments for Intoxication by Botulinum Neurotoxin

Michael Adler[a], Robert E. Sheridan[a], Heather A. Manley[a1], Adler Manley[a1], James Apland[a], Sharad S. Deshpande[b] and James Romano[c]

[a]Neurotoxicology Branch, Pharmacology Division,
US Army Medical Research Institute of Chemical Defense,
Aberdeen Proving Ground, Maryland 21010, USA
[b]Battelle Eastern Science & Technology Center, Aberdeen, MD 21001, USA
[c]Office of the Commander, US Army Medical Research
Institute of Chemical Defese,
Aberdeen Proving Ground, Maryland 21010-5400 USA
[1]National Research Council Postdoctoral Fellow

INTRODUCTION

The botulinum neurotoxins (BoNTs) consist of seven immunologically distinct protein toxins secreted by the anaerobic bacteria *Clostridium botulinum, Clostridium baratti* and *Clostridium butyricum* (Simpson, 1981; 1989; Sellin, 1981). The median human lethal dose has been estimated to be 1 ng/kg, making BoNT the most lethal substance known to mankind (Schantz and Sugiyama, 1974; Franz, 1997). Intoxication by BoNT leads to flaccid paralysis that is bilateral and descending, involving primarily skeletal muscle but also structures innervated by autonomic fibers (Habermann and Dreyer, 1986; Simpson, 1989; Shapiro et al., 1998). Botulism becomes fatal when the diaphragm and intercostal muscles become sufficiently compromised to

---

The opinions or assertions contained herein are the private views of the authors' and are not to be construed as official or as reflecting the views of the Army or the Department of Defense.

In conducting the original research described in this report, the investigators complied with the regulations and standards of the Animal Welfare Act and adhered to the principles of the Guide for the Care and Use of Laboratory Animals (NRC 1996).

impair ventilation (Franz et al., 1993), or when patients succumb to secondary infections following long periods of intensive care (Hatheway et al., 1984; Shapiro et al., 1998). Human intoxication is generally caused by exposure to serotypes A, B, E and to a much lesser extent to serotype F, and is manifested as foodborne, wound or infant botulism (Habermann and Dreyer, 1986; Simpson, 1981; 1989).

The BoNTs belong to the family of A-B toxins where one component mediates binding (B) and the second contains the active (A) moiety. For BoNT, the binding component is the 100 kDa heavy chain (HC), and the active component is the 50 kDa light chain (LC). In addition to the BoNTs, this group also includes diphtheria toxin, cholera toxin, tetanus toxin (TeNT), *Shigella* toxin as well as ricin. Common traits include synthesis as an inactive protoxin and a subunit structure that segregates function to distinct structural entities. Characteristics that differentiate BoNT from other members of the A-B toxin family include a more discriminating target selectivity and considerably higher potency. Selectivity of BoNT entails both the exclusive targeting of peripheral cholinergic nerve terminals by the HC during the initial binding step (Schmitt et al., 1981; Simpson, 1981) as well as cleavage of specific nerve terminal and synaptic vesicle proteins during the LC-mediated catalytic step (Montecucco and Schiavo, 1993).

Receptors for BoNT (ectoacceptors) are thought to consist of protein-ganglioside complexes, which appear to be unique for each serotype (Marxen et al., 1989; Daniels-Holgate and Dolly, 1996). The cholinergic symptomatology for botulism can be attributed to the abundance of these functional ectoacceptors on cholinergic nerve terminals (Daniels-Holgate and Dolly, 1996). In high concentrations, relatively insensitive systems such as CNS cells and even tumor cell lines can be effectively poisoned by clostridial neurotoxins (Wellhöner and Neville, 1987; Ray et al., 1993; Trudeau et al., 1998).

The LCs cleave one of three neuronal proteins that are involved in the release of neurotransmitter, collectively designated as SNARE proteins (Söllner et al., 1993). The SNARE family consists of synaptobrevin, a small integral membrane protein associated with the surface of small synaptic vesicles, and two proteins associated with the cytoplasmic surface of the nerve terminal, SNAP-25 and syntaxin (Hanson et al., 1997). Although only a single peptide bond is cleaved by each BoNT LC, this action is sufficient to cause total inhibition of neurotransmitter release. Similar SNARE proteins are required for evoked release in essentially all synapses (Söllner et al., 1993; Hanson et al., 1997), therefore inhibition of transmitter release by BoNT LC will occur in most systems if concentration and exposure times are appropriately adjusted.

Two members of the A-B toxin family, namely BoNT and ricin, are considered to be potential biological warfare agents (Franz, 1997). This is based on factors such as potency, ease of concealment and history of prior weaponization or stockpiling by hostile powers (Franz, 1997; Arnon et al., 2001). Other potent toxins, such as tetanus toxin, are generally precluded from this role due to widespread immunization programs that exist in most developed countries. Ironically, BoNT is also used as a therapeutic agent and is the treatment of choice for a number of focal muscle spasms and movement disorders (Brin, 1997; Johnson, 1999). Clinical use involves the same action of BoNT as that producing systemic botulism, but the effect is confined to specific muscle groups due to the local nature of the injection and the minute quantities and volumes involved (Johnson, 1999).

The mechanism of action of BoNT is not completely understood in spite of the enormous progress made during the last decade in identifying targets for the LC and in elucidating the complexities of neurotransmitter exocytosis (Montecucco and Schiavo, 1993; Schiavo et al., 1993; Trudeau et al., 1998). Based on analogy to other A-B toxins, Simpson (1981) proposed a multi-step process for BoNT that forms the framework for our current understanding. Accordingly, selective binding of the HC to a cell surface receptor on the presynaptic motor nerve ending is followed by internalization of the LC via endocytotic vesicles. Acidification of these vesicles produces a conformational change in the toxin leading to delivery of the LC into the cytoplasm. The LC causes selective proteolysis of a SNARE protein and consequent block of nerve-evoked transmitter release.

Translocation of the LC from endosome to cytosol has long been thought to involve transmembrane ion channels contributed by the N-terminal portion of the HC (Donovan and Middlebrook, 1986; Rauch et al., 1990; Sheridan, 1998), and recently Koriazova and Montal (2003) provided convincing evidence for the validity of this proposal. Using artificial lipid bilayers, these authors demonstrated that channels formed by incorporation of HC were able to translocate sufficient LC molecules to produce cleavage of SNAP-25 in the trans compartment.

TREATMENT OF BOTULISM

Currently, there is no approved pharmacological treatment for botulism, and therapy consists of supportive care and infusion of a trivalent equine antitoxin. In addition a pentavalent vaccine (serotypes A-E) is available for personnel considered to be at high risk such as laboratory workers and military personnel.

Immunology

Botulinum toxins are classified into seven serotypes: A, B, C, D, E, F and G based on immunoreactivity of the C-terminus of the heavy chain that is responsible for binding to the cell surface receptor (Simpson and DasGupta,

1983; Mullaney et al., 2001). Within these serotype classes, there are smaller structural differences in the toxins from various bacterial strains that may also produce variations in immunogenicity (Shimizu et al., 1974). Not all antibodies binding to BoNT are capable of neutralizing the toxin's activity, and effective neutralization of the toxin appears to require binding to more than one epitope (Amersdorfer et al., 2002; Nowakowski et al., 2002).

The toxin neutralization process can occur after binding to the cell surface receptor (Simpson and DasGupta, 1983) but must take place before the toxin is internalized into a protected endosome. The latter restricts the duration of the therapeutic window between exposure and treatment and is the greatest limitation of the immunological approach to botulinum therapy.

Vaccination

Currently, the most effective treatment for botulism is immunization prior to exposure. Due to the relatively few cases of human botulism, the efficacy of the vaccine cannot be absolutely determined (Flock et al., 1963). However, the value of vaccination can be readily established in animal models, including primates, and the circulating titer against botulinum toxin can be used as a surrogate marker for human efficacy (Gelzleichter et al., 1999).

The currently available vaccine for botulinum toxin in the US is a pentavalent mixture of formalin-inactivated type A, B, C, D and E toxins. Although in use since 1966, the vaccine is still classified as an Investigational New Drug (IND) due to limited human efficacy data. The vaccine is available for restricted use under protocols from the Centers for Disease Control and Prevention (CDC) and the US Department of Defense (DoD). The current protocol for the vaccination series consists of injections at 0, 2, and 12 weeks with a yearly booster and/or titer check.

Development of effective titers can take months. In one study, antibody to type A toxoid was detectable by ELISA four weeks after the initial vaccination. However, despite continued increases in the ELISA titer, effective neutralization of toxin in the mouse bioassay could not be detected for another 15 weeks (Dezfulian et al., 1987). Approximately half of individuals required the one-year booster to reach acceptable titers (Siegel, 1988). Once successfully evoked, the antibody titer can remain effective for years without additional annual boosting.

At this time, vaccination for botulinum toxins is not considered practical for civilian use except for laboratory investigators who may be exposed to lethal levels of BoNT (Franz, 1997). A further disadvantage of vaccination for the general population is the increasing use of BoNT in clinical practice for producing relatively long-term muscle relaxation (Brin, 1997; Johnson, 1999), an application precluded by prior vaccination.

Antitoxin

In the absence of prior vaccination, clinical cases of acute BoNT intoxication require prompt treatment upon diagnosis. Current recommendations are for treatment with antitoxin and supportive care to include assisted ventilation if needed (Arnon et al., 2001). It is important to note that for maximal efficacy, antitoxin should ideally be administered prior to onset of symptoms since internalization of toxin precedes the development of frank paralysis (Franz et al., 1993). In practice, antitoxin is generally administered after development of paralysis to prevent further internalization of toxin and has a plasma half-life of 5 to 7 days (Hatheway et al., 1984). BoNT already located in presynaptic terminals will not be affected and can continue its catalytic cleavage of presynaptic proteins for weeks (Keller et al., 1999; Adler et al., 2001, Meunier et al., 2003).

The only available licensed therapeutic for botulinum toxin is a trivalent (A, B, E) equine antitoxin available from the CDC. This product has been associated with hypersensitivity reactions and in one study of 268 humans receiving the antitoxin, 9% developed either acute or delayed reactions (Black and Gunn, 1980). Serum sickness was correlated with the dose of antitoxin administered and was noted to be higher than in other equine-derived antisera (Black and Gunn, 1980). In general, a substantial excess of antitoxin is administered, in the range of 10-1000 times that required for toxin neutralization (Hatheway et al., 1984). It would be inadvisable to lower these doses, however, due to potentially deleterious interactions between antitoxin and toxin at low ratios (Sheridan et al., 2001). Other approaches to reduce the reactions to antitoxin have involved production of a heptavalent (A, B, C, D, E, F, G) despeciated Fab equine product, available from DoD as an IND since 1994, and the use of human-derived immune globulin from immunized volunteers (Metzger and Lewis, 1979).

Although the treatment of acute botulinum poisoning by administration of antitoxin is the only currently accepted therapy, it has several shortcomings including induction of hypersensitivity reactions and serum sickness, the necessity to be administered as soon as symptoms appear and the inability to restore function in already paralyzed muscle. For these reasons, despite the availability of effective immunological approaches to BoNT poisoning, work continues on the development of pharmacological treatments for botulism.

PHARMACOLOGICAL TREATMENTS

Potassium Blockers

The first attempt to treat human botulism with a specific therapeutic agent was reported over 30 years ago by Cherington and Ryan (1968). The patient was a 57-year old woman who became critically ill after consumption of home canned string beans. Detection of obvious spoilage by the victim restricted ingestion to a single specimen. The following day, the patient experienced vertigo, diplopia, blurred vision and slight nausea. Symptoms progressed to dysphagia, dyspnea, cyanosis and respiratory arrest over the subsequent week requiring mechanical ventilation. The patient was treated with guanidine hydrochloride by a nasogastric tube intermittently for 20 days starting at day 10. Guanidine was selected because it blocks K^+ channels, resulting in an increase in Ca^{2+} influx during the nerve terminal spike and an enhancement of transmitter release (Hirsch et al., 1979). Guanidine was able to increase the size of action potentials evoked by electrical stimulation, relieve ptosis and extraocular palsies and increase strength in the extremities. Treatment by guanidine did not promote the recovery of spontaneous ventilation, however, and the patient died shortly after failure of the mechanical respirator.

During the next decade, guanidine was administered to a total of 52 patients with 75% receiving some benefit from the drug (Puggiari and Cherington, 1978). As with the first patient, improvements were noted in the eye and limb muscles but not in vital capacity or in pulmonary function. However, even this limited efficacy of guanidine has been questioned. In a double-blind crossover study involving 6 individuals intoxicated by BoNT/A from the same source, Kaplan et al. (1979) found no improvement in muscle strength or duration of hospitalization by guanidine treatment.

More recent studies have failed to resolve the uncertainty over the efficacy of K^+ blockers for treatment of human botulism. In one report, a 31-year old patient was given a similar but more selective agent, 3,4-diaminopyridine (DAP), for severe intoxication by BoNT/A (Davis et al., 1992). Little or no beneficial actions were observed. On the other hand, a 69 year-old patient was treated with DAP for intoxication by BoNT/B and showed progressive improvement (Doc et al., 2002). The latter result is unexpected since numerous animal experiments have shown that DAP is not effective against poisoning by serotype B, and is, in fact, only effective in serotype A intoxication (Simpson, 1986; Siegel et al., 1986; Adler et al., 1996). The lack of consistent results with K^+ channel blockers and the inability to achieve a better clinical outcome stems, in part, from the inability of patients to tolerate these drugs at doses necessary to achieve appropriate restoration of muscle function. The outcome may also be limited by the condition of the patient prior to onset of therapy. In addition, the serotype classification may not have been accurate, especially in some of the older studies (Hatheway et al., 1984).

DAP has thus far proved to be the most promising candidate, but it has several shortcomings that have prevented further progress in its use as a post-exposure treatment for botulism. First, it is rapidly eliminated while the action of BoNT/A is highly persistent, often lasting for many months (Siegel et al., 1986; Adler et al., 1996). This problem can be addressed in principle by use of an infusion system for delivery of DAP. Using 7-day osmotic minipumps for continuous drug delivery, Adler et al. (2000) demonstrated that DAP can antagonize the paralysis caused by BoNT/A for a one week period with no loss of efficacy.

A more serious problem is the systemic toxicity of DAP, especially the production of seizures. DAP is a tertiary amine and is able to cross the blood-brain-barrier where it can augment transmitter release in the CNS. Since BoNT is unable to gain access to the CNS, the central action of DAP is unopposed and results in seizures (Molgo et al., 1980). An attempt to solve this problem by use of a quaternary analog of DAP was unsuccessful, since the binding site for aminopyridines resides on the cytoplasmic surface of the K^+ channel, and the quaternary analog of DAP was unable to gain access to the nerve terminal cytoplasm (Adler et al., 1995).

Although difficulties with the use of currently-available K^+ blockers have thus far prevented their use for treatment of botulism, discovery of new potassium blockers with lower CNS toxicity should allow for development of this class of drugs for treatment of the highly persistent serotype A.

Concept of a Universal Inhibitor

Since there are at least three BoNT serotypes that are capable of causing human disease (A, B and E), it would be useful to have a single inhibitor with efficacy against all three serotypes. A single therapeutic agent would simplify treatment of intoxicated personnel since serotype identification is time consuming and could unduly delay treatment when early intervention may produce a more favorable outcome. To develop universal inhibitors, one needs to target common pathways that are involved in the BoNT intoxication process. The most obvious candidates are the ectoacceptor present at the nerve terminal surface, components of the translocation/internalization pathway and Zn^{2+}, which is required for proteolytic cleavage of SNARE proteins.

A schematic illustrating BoNT interactions with the nerve terminal and sites for therapeutic intervention is presented in Fig. 1.

Blockade of BoNT Ectoacceptor

Gangliosides were proposed as components of the BoNT ectoacceptor, based in part on earlier work implicating gangliosides as receptors for cholera, pertussis and tetanus toxins (Schiavo et al., 1993). BoNTs bind to complex gangliosides such as GQ1b and GT1b, which are most abundant on presynaptic nerve terminals (Kitamura et al., 1999). The importance of this binding to toxicity was recently confirmed in mice lacking complex gangliosides due to a deletion of the gene coding for β1,4 GalNAc-transferase (Bullens et al., 2002). These mice were insensitive to the toxic actions of BoNT, as demonstrated by an absence of neuromuscular impairment following exposure to a paralytic dose of BoNT/A.

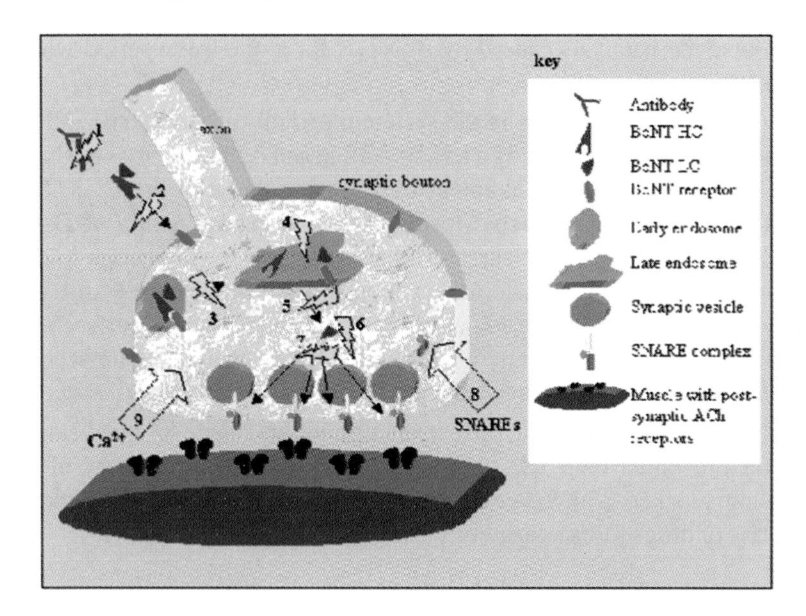

Fig. 1 Internalization pathway for BoNT and strategies for therapeutic intervention. Neutralization of BoNT with anti-BoNT antibodies (1) (e.g., antitoxin or vaccine) is a successful strategy up to and including binding to ectoacceptors. Binding of BoNT to gangliosides can be inhibited by use of lectins (e.g., *Triticum vulgaris* lectin) (2). After intoxication, BoNT is processed through the endosomal pathway. Potential therapeutic strategies may be targeted to prevention of internalization (3), of endosomal acidification (4) and of translocation of LC into the cytoplasm after dissociation from the HC (5). Other areas to target include elimination of cofactors necessary for LC action (i.e., chelation of Zn^{2+}) (6), prevention of SNARE cleavage by protease inhibitors (7), incorporation of non-cleavable SNARE proteins (8) and administration of K^{+} channel blockers to enhance Ca^{2+} influx (9).

Inhibition of ganglioside binding by a potential therapeutic agent was first reported by Bakry et al. (1991), who demonstrated that the time to block of twitch tension was delayed by a factor of 2 when isolated diaphragm muscles were pretreated by *Tritiucum vulgaris* lectin (TVL) prior to exposure to BoNT. Importantly, TVL was effective against the actions of all BoNT serotypes examined, although to different degrees.

Since isolated muscles have a limited viability *in vitro*, they are generally evaluated under conditions of high toxin concentration, far in excess of what may be encountered in actual outbreaks. To test potential inhibitors under more realistic conditions, we examined the actions of TVL in *extensor digitorum longus* (EDL) muscle exposed to either 0.63 U of BoNT/A or to 40 U of BoNT/E by local injection. The doses of toxin were selected to produce $\geq$ 95% inhibition of the nerve evoked muscle twitch in 48 hr. TVL was injected locally in the EDL muscle 30 min before or 6 hr after toxin. The results for the 30 min pretreatment data are shown in Table. 1.

Table 1 Pretreatment by TVL protects EDL muscle from paralysis following local injection of BoNT/A or /E

BoNT serotype (dose)	*Pretreatment*	*TVL dose (mg)*	*Number of muscles*	*Twitch tension (g)*
None	None		5	69.4 ± 3.4
BoNT/A (0.63 U)	None		4	2.0 ± 0.9
None	TVL	0.375	3	61.2 ± 10.0
		0.75	4	41.7 ± 3.4
BoNT/A (0.63 U)	TVL	0.375	3	9.7 ± 3.1
		0.75	6	23.8 ± 6.4
None	None		4	64.7 ± 8.0
BoNT/E (40 U)	None		4	9.8 ± 3.7
None	TVL	0.375	3	57.7 ± 3.3
BoNT/E	TVL	0.375	3	53.3 ± 7.3

The doses of BoNT/A and /E were selected to produce equivalent levels of muscle paralysis 48 hr after injection. TVL was injected 30 min before BoNT/A or /E. Tensions were tested at different times (7 days after BoNT/A and 4 days after BoNT/E) since paralysis by the two serotypes recover at different rates. Values are expressed as mean ± SEM. BoNT dose is expressed in terms of mouse i.p. LD50 units (U). Rats were anesthetized during tension recordings.

These data clearly demonstrated that pretreatment by TVL protects muscles from the toxic actions of both BoNT/A and /E. TVL was more effective in protecting BoNT/E-intoxicated muscles from paralysis as demonstrated by the finding of a near normal twitch tension in muscles pretreated with TVL and challenged with BoNT/E. Twitch tension in muscles pretreated with TVL and challenged by BoNT/A was only a fraction of its normal value. The above results extend those of Bakry et al. (1991) by demonstrating that TVL pretreatment can provide sustained protection of muscle tension with doses of BoNT that mimic actual exposures. In the study of Bakry et al. (1991), where high concentrations of BoNT were applied in an *in vitro* twitch bath, TVL pretreatment was only able to delay the time-to-block, but could not prevent complete paralysis.

In contrast to its efficacy as a pretreatment, TVL was unable to protect muscles when administered 6 hr after a local injection of BoNT/A or /E. This is consistent with the finding that TVL inhibits binding of BoNT to its nerve terminal ectoacceptor (Bakry et al., 1991). Since binding of toxin is rapid, a 6 hr delay in TVL administration renders any strategy based on protection of receptors ineffective.

Identity of the protein component of the BoNT ectoacceptor is less certain, but several lines of evidence point to synaptotagmin as a possible candidate for BoNT/A /B and /E binding (Nishiki et al., 1994; Li and Singh, 1998; Dong et al., 2003). A common protein receptor appears to be at variance with reports that there is little competition for binding among different serotypes (Lalli et al., 1999). These apparently conflicting results may be reconciled if the ganglioside and protein ectoacceptor components have distinct geometric patterns or 3-dimensional conformations that are unique to each serotype. To date, there have been no attempts to exploit the protein component of the BoNT ectoacceptor for production of BoNT antagonists.

Interference with BoNT Internalization

A second common step in BoNT intoxication involves the internalization of the toxin and its translocation into the cytosol. Internalization is currently the least understood step in the intoxication by BoNT; however, some general conclusions may be drawn about the process. These conclusions are based on inhibition of BoNT activity by treatments that interfere with specific cellular processes. Thus, a better understanding of these processes and development of selective inhibitors offer promising avenues for therapeutic intervention.

In the first phase of internalization, BoNT that is attached to the ectoacceptor accumulates in endosomes. This process can be visualized with labeled BoNT at early stages of intoxication, with accumulation of toxin in discrete regions of the cytosol (Black and Dolly, 1986; Goodnough et al., 2002). There is currently little data available on the possibility of inhibiting endosome formation as a means of antagonizing BoNT action.

In the second phase, a pH drop in the endosome occurs, which is required for the internalization of the toxin. This step is confirmed by a number of experiments where the acidification of the endosome is prevented, with a concomitant reduction in toxin activity (Simpson, 1983; Simpson et al., 1994; Williamson and Neale, 1994; Sheridan 1996). Acidification of endosomes is a common requirement for activation of all BoNT serotypes. This suggests that inhibitors of endosomal acidification may be appropriate candidates for development of a universal antagonist. Potential BoNT therapeutics that have been shown to antagonize acidification and BoNT activity include pH buffers such as ammonium chloride and methylamine hydrochloride, protonophores such as nigericin and monensin, and the vesicular ATPase inhibitor bafilomycin A1.

In the final phase of internalization/translocation, the toxin, or at least the LC, is released from the acidified endosome into the cytosol. Although not a confirmed mechanism, this is commonly thought to involve transport of LC through an ion channel formed by the N-terminal segment of the toxin HC (Koriazova and Montal, 2003). The formation of ion channels by toxin or its HC is a common feature of both tetanus toxin and all BoNTs that have been examined (Hoch et al., 1985; Sheridan, 1998). Thus antagonists of the BoNT ion channel also offer the possibility of a universal BoNT antagonist. The formation of BoNT channels is facilitated by a pH gradient across the membrane where the channels are formed, with optimal channel formation occurring with acidic pH on the cis (toxin) side and neutral pH on the trans (cytosol) side (Donovan and Middlebrook, 1986; Hoch et al., 1985; Bouquet and Duflot, 1982; Rauch et al., 1990). This raises the possibility that the pH dependence of toxin internalization may be governed in part by the formation of BoNT channels.

Aminoquinolines have been investigated as both potential BoNT channel-blocking drugs and as inhibitors of endosomal acidification. Aminoquinolines were originally considered as BoNT antagonists due to their ability to inhibit pH changes in lysosomes as part of their antimalarial action (Simpson, 1982). However, chloroquine was reported to block the channels produced by botulinum C2 toxin in artificial lipid bilayers (Schmid et al., 1994), and the structure-activity relationships of a number of aminoquinoline derivatives indicate that potency against BoNT is not related to antimalarial activity. Instead, efficacy is more consistent with direct actions on a putative BoNT-mediated ion channel (Sheridan et al., 1997).

Further investigation of translocation antagonists reveals a critical weakness in the strategy of targeting a transient state in BoNT intoxication. Aminoquinolines or similar drugs must be present when BoNT is translocated into the cytosol. If the duration of BoNT internalization exceeds the clearance of the drug at the neuromuscular junction, the efficacy of the inhibitor is dramatically reduced (Simpson, 1982; Deshpande et al., 1997). In attempting to antagonize persistent serotypes such as BoNT/A, aminoquinolines have little efficacy, even in the pretreatment mode, as illustrated in Fig. 2.

Inhibition of Catalytic Activity

The third broad target for pharmacological intervention is the Zn^{2+} metalloprotease activity of the LC, which is common to all BoNT serotypes (Schiavo et al., 1992b; Montecucco and Schiavo, 1993). In principle, one can approach this problem by developing specific inhibitors for each LC or by using chelators with a high affinity for transition state metals that would inhibit the activity of all serotypes. Historically, the latter approach was attempted first since chelators with the desired characteristics already existed, whereas no specific lead compounds were identified until recently.

Zinc Chelators

Since Zn^{2+} is required for the protease activity of BoNT LC, and catalysis of SNARE proteins takes place inside the nerve terminal cytoplasm, an effective chelator would need to have a high affinity for Zn^{2+} as well the ability to penetrate the plasma membrane to gain access to the nerve terminal cytosol. In addition, it would be desirable to have a chelator with low affinity for Ca^{2+} and Mg^{2+}, since these cations are involved in evoked release of neurotransmitter, muscle contraction and energy production. The ideal candidate appeared to be N,N,N'N'-tetrakis(2-pyridylmethyl) ethylenediamine (TPEN), a chelator which is lipid soluble, has a high affinity for zinc (association constant = $10^{15.6}$) and a low affinity for Ca^{2+} and Mg^{2+} (Anderegg et al., 1977).

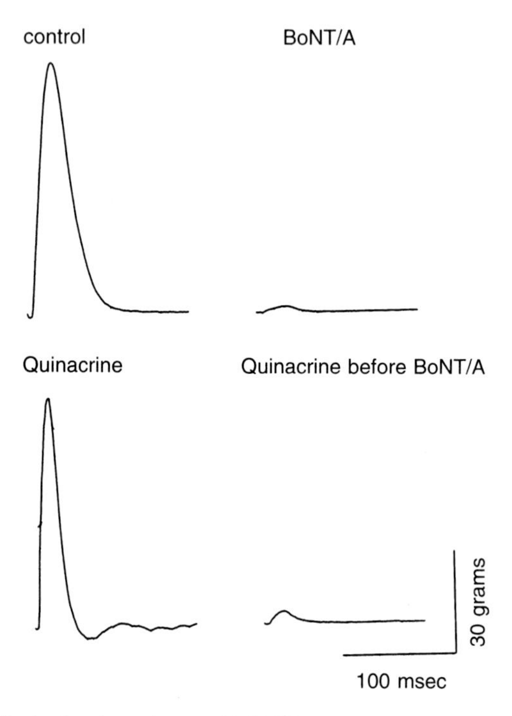

Fig. 2 Lack of effect of quinacrine on BoNT/A paralysis. Paired rat EDL muscles were locally injected with either saline or 1.6 μg of quinacrine in 20 μl of saline. Thirty minutes later, one muscle in each pair was injected with 0.63 units of BoNT/A. After seven days, the muscles were isolated and stimulated via the peroneal nerve to produce the twitch tensions shown. Pretreatment with quinacrine had little or no effect on inhibition of tension in the BoNT/A-injected muscles. Rats were anesthetized during tension recordingss

Initial studies with TPEN on isolated BoNT-intoxicated muscles were encouraging. Pretreatment with TPEN or co-administration of TPEN with BoNT resulted in a pronounced reduction in the time-to-block of neurally-elicited muscle tensions in preparations challenged by paralytic doses of toxin (Simpson et al., 1993; Sheridan and Deshpande, 1995). Furthermore, TPEN was effective against all BoNT sertoypes (Simpson et al., 1993), making it potentially useful as a universal therapeutic.

Subsequent studies, however, revealed significant problems with TPEN usage. First, the chelator was not effective in the post-exposure mode. Sheridan and Deshpande (1995) determined that TPEN became progressively less effective when applied during the course of BoNT-induced paralysis, and nerve-muscle preparations that were completely paralyzed failed to improve by addition of TPEN. Second, TPEN was found to cytotoxic, and concentrations above 1 μM produced apoptosis and necrosis (Adler et al., 1997; Sheridan and Deshpande, 1998; Adler et al., 1999b). Third, the efficacy of TPEN *in vivo* was considerably less than its apparent efficacy *in vitro*. In a survival study where TPEN was administered 30 min before and in 4 divided doses after BoNT, the mean increase in survival time was only 3.3 hr for mice challenged with 20 units of serotype A and 2.1 hr by those challenged with 20 units of serotype B. Importantly, all of the TPEN-treated mice succumbed to a dose of BoNT that was survived by animals pretreated with antitoxin (Sheridan et al., 1999).

The combined TPEN dose in the above study was 25 mg/kg, administered over a 6.5 hr period, which was the highest sign-free dose that mice were able to tolerate. A single TPEN dose of 20 mg/kg produced ataxia and loss of coordination and 30 mg/kg TPEN was fatal in all animals examined. To evaluate TPEN under essentially *in vivo* conditions but without the limitation of its systemic toxicity, we injected both TPEN and BoNT locally in the EDL muscle and examined the protective efficacy of the chelator by measuring nerve-evoked muscle tension *in situ*. The results are illustrated in Fig. 3. In EDL muscles paralyzed by injection of 0.63 units of BoNT/A, TPEN locally injected in the same muscle 30 min before BoNT was unable to prevent paralysis or to restore tension during the 3-day period of observation.

With serotype E, the findings were more encouraging. Muscles injected locally with TPEN 30 min prior to BoNT exhibited tension that was 30-40% of control, which was sustained for the 4-day period of observation.

Metalloprotease Inhibitors

The Zn^{2+} metalloproteases with the greatest structural similarity to the BoNT/A catalytic domain are thermolysin and leishmanolysin, with the similarity being confined to the helix containing the HEXXH Zn^{2+}-binding motif and to a four-stranded β-sheet near the helix. The rest of the catalytic domain of BoNT/A has a substantially different secondary structure from the known metalloproteases (Lacy et al., 1998). For BoNT/B, the catalytic domain contains two separate substrate recognition sites that are highly specific: the domain binds to the SNARE secondary recognition sequence (SSR) on the substrate only if it is of the proper sequence, possesses a minimum length, and is the correct distance from the bond to be hydrolyzed (Hanson and Stevens, 2000).

The first approach to finding an effective BoNT inhibitor was to test available metalloprotease inhibitors on BoNT light chain activity. Some potent inhibitors of Zn^{2+} metalloproteases, including phosphoramidon, peptide hydroxamate, thiorphan and captopril, were reported to have weak inhibitory activity on TeNT and BoNTs (Schiavo et al., 1992a; 1992b; Cornille et al., 1997; Deshpande et al., 1995; Foran et al., 1994; Adler et al., 1999a; Martin et al., 1999). However, none were found to be potent inhibitors.

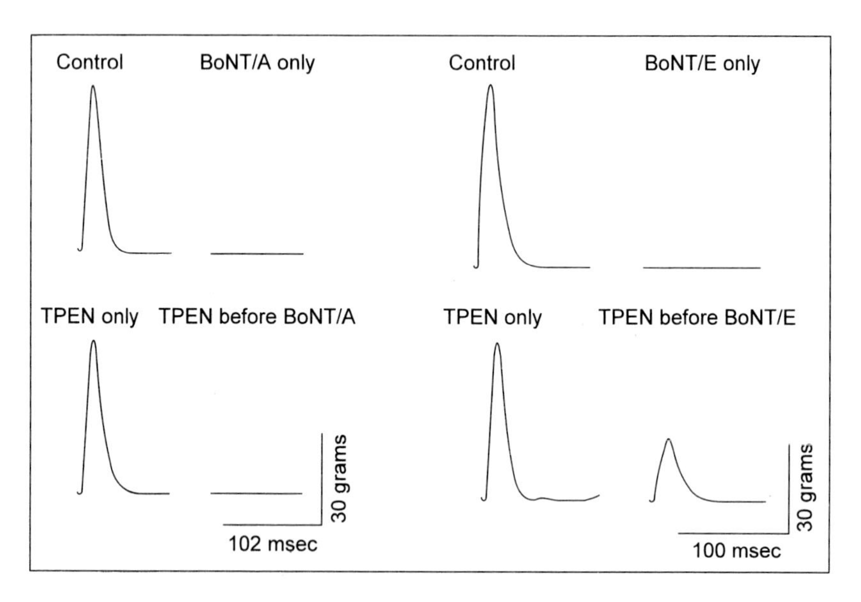

Fig. 3. Twitch tensions elicited *in situ* from EDL muscles under the indicated conditions. BoNT/A records were obtained 3 days after injection of BoNT/A (0.63 units in 15 μl) or TPEN (128 mg in 30 μl) 30 min prior to BoNT/A. Note that TPEN was unable to protect muscles from the paralytic action of BoNT/A. BoNT/E records were obtained 4 days after injection of BoNT/E (40 units in 15 μl) or TPEN (106 mg in 50 μl) 30 min prior to BoNT/E. In this case, TPEN was able to protect a significant fraction of muscle tension from BoNT/E. Rats were anesthetized during tension recordings.

Rational Design of Inhibitors

Small organic molecules

The low potency of phosphoramidon on BoNT/B was suggested to be due to its inability to fit optimally into the highly specific catalytic site of the BoNT/B LC. To optimize binding to the LC catalytic site, Adler et al. (1999a) designed other small dipeptide metalloprotease inhibitors based upon the structure of phosphoramidon but these were modified to mimic the residues at the scissile bond of the substrate, synaptobrevin. Of the resulting analogues, only one (ICD 2821) was more potent than phosphoramidon. However, each of these compounds, including phosphoramidon, required high concentrations (mM range) to be effective. The first organic inhibitor for BoNT/B effective in the sub-mM range was 7-N-phenylcarbamoylamino-4-chloro-3-propyloxyisocoumarin (ICD 1578), originally designed as a matrix metalloprotease inhibitor (Starks et al., 1996; Adler et al., 1998).

Combinatorial libraries

In the attempt to synthesize a more potent inhibitor, Roques and colleagues synthesized a combinatorial library of pseudotripeptide compounds mimicking the substrate cleavage site (Gln76-Phe77). The most potent compounds from this study inhibited BoNT/B in the mM range (Martin et al., 1999; Roques et al., 2000). To improve upon the potency of these compounds, another combinatorial library was synthesized using the pseudotripeptides as parent compounds, modifying them by incorporation of non-natural aromatic and aliphatic amino acids. The most potent BoNT/B antagonist from this series had a K_i of 250 nM. Although further optimization is needed, the combinatorial approach has the potential to yield potent BoNT antagonists (Anne et al., 2003).

Investigators at US Army Medical Research Institute of Infectious Diseases followed a similar approach to generate inhibitors of BoNT/A by producing analogues of SNAP-25 (Schmidt and Bostian, 1997; Schmidt et al., 1998; Schmidt and Stafford, 2002). Schmidt et al. (1998) found that short peptides were able to bind to BoNT/A with high affinity and that substitution of two residues near the BoNT/A cleavage site was critical for formation of potent inhibitors: 1) a D- or L-cysteine at the P1 site (N-terminal to the cleavage site) to contribute an SH group for binding to the catalytic $Zn^{2+}$of BoNT/A; 2) replacement of L-arginine with its D enantiomer at the P1' site (C-terminal to the cleavage site) to bind to BoNT/A. With the exception of the P1' site, other subsites of BoNT/A exhibited a marked tolerance for alteration of amino acid side chains.

The most potent inhibitor of BoNT that Schmidt and Stafford (2002) synthesized was the peptide N-acetyl-CRATKML-amide, which has a Ki of 330 nM. The important modifications were 1) replacing the P1 cysteine with other SH-containing compounds to orient the sulfhydryl in a more favorable position to bind Zn^{2+} and 2) adding a phenyl ring on the β carbon to reduce the freedom of rotation, thereby increasing the binding affinity of the inhibitor to BoNT/A.

Rational design of inhibitors based on BoNT crystal structure: BABIMs

Optimal binding of synaptobrevin to the active site of BoNT/B requires a length of 51 amino acids around the scissile bond (Gln76-Phe77), which approaches the Zn^{2+} binding domain of BoNT/B. Bis(5-amidino-2-benzimidazolyl) methane, or BABIM, a metalloprotease inhibitor which acts by chelating Zn^{2+}, was designed to interact with this site (Eswaramoorthy et al., 2002). When two molecules of BABIM bind to BoNT/B, partially occupying the synaptobrevin binding site, the BoNT/B active site rearranges, resulting in a loss of the Zn^{2+}. The Zn^{2+} is reportedly transported to another site within the light chain, which may be responsible for the observed loss of catalytic activity.

Other Approaches

SNARE replacements

Cleavage of SNAREs by BoNT LC results in inhibition of neurotransmitter release. In theory, replacement of SNAREs in the affected nerve terminals could result in restoration of synaptic function. However, these SNAREs would also be cleaved by BoNT LC, which is thought to reside in the nerve terminal for varying amounts of time, depending upon serotype (Keller et al., 1999). One solution would be introduction of SNAREs with a mutation rendering them non-cleavable and thus resistant to the effect of BoNT LC. The major limitation in utilizing non-cleavable SNAREs would be the difficulty in their introduction into the nerve terminal cytoplasm.

Targeted delivery vehicle

The available BoNT inhibitor drug candidates suffer from relatively poor efficacy (Adler et al., 1999a) and high toxicity (Adler et al., 1997; Deshpande et al., 1997). In addition to developing more suitable BoNT inhibitors, a method to selectively deliver the drugs to affected peripheral cholinergic neurons would help circumvent some of the toxicity resulting from CNS effects and would also theoretically deliver sufficiently high concentrations to be effective. Zhou et al. (1995) and Goodnough et al. (2002) have addressed the problem of targeted delivery by exploiting the selectivity of the heavy chain of botulinum toxin. Zhou et al. (1995) demonstrated that catalytically inactive BoNT holotoxin was able to transport tetanus toxin LC.

Goodnough et al. (2002) designed a prototype drug delivery vehicle which is a protein chimera consisting of the HC of BoNT covalently attached to an inert carrier of potential antagonists. They found that the delivery vehicle was specifically taken up by neurons, and the uptake was inhibited in a concentration-dependent manner by BoNT holotoxin.

DISCUSSION

Although cases of botulism from food or wound contamination and infant infection occur each year, intoxication by BoNT is a rare event. Due to their extreme toxicity, BoNTs are considered to be potential weapons of mass destruction (Arnon et al., 2001; Starks et al., 1996; Franz, 1997) with the potential of producing large numbers of critically ill casualties. At the current time there is no effective pharmacologic treatment for botulism and intoxication of more than a handful of people would easily overwhelm the capacity of medical care facilities.

Early efforts to treat human botulism focused on the use of K^+ channel blockers. Increasing synaptic function with agents such as guanidine and DAP has produced mixed results (Cherington et al., 1978; Puggiari et al.,

1978; Molgo et al., 1980; Siegel et al., 1986; Adler et al., 1995; 1996; 2000). However, this is a promising avenue of study given the efficacy of DAP in animal models (Adler et al., 2000). Development of a K^+ channel blocker that stays in the periphery (i.e., does not cross the blood-brain barrier) would address some of the toxicity issues, and would likely be effective for the BoNT/A serotype. Efficacy for other serotypes has not been conclusively demonstrated even *in vitro*.

Other efforts to develop pharmacological treatments for botulism have targeted the binding, internalization and catalytic activity of BoNT rather than reversing its effects after exposure (Fig. 1). This strategy has been moderately successful using TVL pretreatment in rat EDL studies with BoNT/E (Table 1). However, for these to be effective strategies, therapy has to precede the full paralysis produced by BoNT. A limiting factor for many of these approaches is that the toxin often relies upon normal cellular processes such as endosome formation and vesicle acidification for its action, and these processes are also vital components of normal neuronal function.

The most promising avenue of research into development of BoNT antagonists is inhibition of catalytic activity. Efforts using Zn^{2+} chelators and classical metalloprotease inhibitors (Simpson et al., 1993; Deshpande et al, 1995; Sheridan et al., 1995; Adler et al., 1997; 1999a), have met with limited success due in part to the long duration of toxin residence in the terminal relative to the inhibitor employed (Fig. 3) and the inherent toxicity of non-selective inhibitors of zinc metalloproteases. The rational design of molecules to bind specifically to the catalytic site of the LC of BoNT offer the possibility of both reducing side effects of these therapeutics and also of reducing the duration of paralysis from the more persistent toxin serotypes, such as BoNT/A (Keller et al., 1999; Adler et al., 2001).

A further development to be used in conjunction with efforts to produce a less toxic BoNT catalytic inhibitor is the targeted delivery vehicle. By concentrating therapeutic agent within the sensitive nerve population, generalized toxicity can be reduced. Two research groups have demonstrated that this is a feasible approach. Zhou et al., (1995) and Goodnough et al. (2002) have designed prototypic delivery vehicles consisting of the heavy chain of BoNT. Goodnough et al. (2002) also demonstrated that it is possible to attach the heavy chain to carriers of potential BoNT antagonists. More work is needed to optimize such targeted drug carrier complexes.

The problem of developing therapies against botulinum intoxication is a difficult one. There are many obstacles to overcome, including binding affinity, toxicity of compounds, access of drugs to the nerve terminal, and targeting of drugs to the appropriate neurons to minimize side effects. Despite the difficult nature of this problem, given the current advances in molecular biology and rational drug design, it is likely that therapeutic agents will be within reach.

References

Adler M., Capacio B. and Deshpande S.S. Antagonism of botulinum toxin A-mediated muscle paralysis by 3,4-diaminopyridine delivered via osmotic minipumps. Toxicon **38:** 1381-1388, 2000.

Adler M., Dinterman R.E. and Wannemacher R.W. Protection by the heavy metal chelator N,N,N',N'-tetrakis(2-pyridylmethyl)ethylenediamine (TPEN) against the lethal action of botulinum neurotoxin. Toxicon **35:** 1089-1100, 1997.

Adler M., Keller J.E., Sheridan R.E. and Deshpande S.S. Persistence of botulinum neurotoxin A demonstrated by sequential administration of serotypes A and E in rat EDL muscle. Toxicon **39:** 233-243, 2001.

Adler M., Macdonald D.A., Sellin L.C. and Parker G.W. Effect of 3,4-diaminopyridine on rat extensor digitorum longus muscle paralyzed by local injection of botulinum neurotoxin. Toxicon **34:** 237-249, 1996.

Adler M., Nicholson J.D., Cornille R. and Hackley B.E. Efficacy of a novel metalloprotease inhibitor on botulinum neurotoxin B activity. FEBS Lett. **429:** 234-238, 1998.

Adler M., Nicholson J.D., Starks D.F., Kane C.T., Cornille F. and Hackley B.E. Evaluation of phosphoramidon and three synthetic phosphonates for inhibition of botulinum neurotoxin B catalytic activity. J. Appl. Toxicol. **19:** S5-S11, 1999a.

Adler M., Scovill J., Parker G., Lebeda F.J. Piotrowski J. and Deshpande S.S. Antagonism of botulinum toxin-induced muscle weakness by 3,4-diaminopyridine in rat phrenic nerve-hemidiaphragm preparations. Toxicon **33:** 527-537, 1995.

Adler M., Shafer H., Hamilton T. and Petrali J.P. Cytotoxic actions of heavy metal chelator TPEN on NG108-15 neuroblastoma-glioma cells. NeuroToxicology **20:** 571-582, 1999b.

Amersdorfer P., Wong C., Smith T.J., Chen S., Deshpande S.S., Sheridan R.E. and Marks J.D. Genetic and immunological comparison of anti-botulinum type A antibodies from immune and non-immune human phage libraries. Vaccine **20:** 1640-1648, 2002.

Anderegg G., Hubmann E., Podder N.G., Wenk F. Die thermodynamik der metallkomplexbildung mit bis-tris und tetrakis [(2-pyridyl)methyl]-aminen. Helv. Chim. Acta **60:** 123-140, 1977.

Anne C., Turcaud S., Quancard J., Teffo F., Meudal H., Fournie-Zaluski M.C. and Roques B.P. Development of potent inhibitors of botulinum neurotoxin type B.J. Med. Chem. **46(22):** 4648-56, 2003.

Arnon S.S. et al. Botulinum toxin as a biological weapon. JAMA **285:** 1059-1070, 2001.

Bakry N., Kamata Y. and Simpson L.L. Lectins from *Triticum vulgaris* and *Limax flavus* are universal antagonists of botulinum neurotoxin and tetanus toxin. J. Pharmacol. Exp. Ther. **258:** 830-836, 1991.

Black J.D. and Dolly J.O. Interaction of ^{125}I-labeled botulinum neurotoxins with nerve terminals. J. Cell Biol. **103:** 535-544, 1986.

Black R.E. and Gunn R.A. Hypersensitivity reactions associated with botulinal antitoxin. Am. J. Med. **69:** 567-570, 1980.

Bouquet P. and Duflot E. Tetanus toxin fragment forms channels in lipid vesicles at low pH. Proc. Natl. Acad. Sci. USA **79:** 7614-7618, 1982.

Brin M.F. Botulinum toxin: chemistry, pharmacology, toxicity, and immunology. Muscle Nerve **6:** S146-S168, 1997.

Bullens R.W.M., O'Hanlon G.M., Wagner E., Molenaar P.C., Furukawa K., Furukawa K., Plomp J.J. and Willison H.J. Complex gangliosides at the neuromuscular junction are membrane receptors for autoantibodies and botulinum neurotoxin but redundant for normal synaptic function. J. Neurosci. **22:** 6876-6884, 2002.

Cherington M. and Ryan D.W. Botulism and guanidine. N. Engl. J. Med. **278:** 931-933, 1968.

Cornille F., Martin L., Lenoir C., Cussac D., Roques B.P. and Fournie-Zaluski M.-C. Cooperative exocite-dependent cleavage of synaptobrevin by tetanus toxin light chain. J. Biol. Chem. **272:** 3459-3464, 1997.

Daniels-Holgate P.U. and Dolly J.O. Productive and non-productive binding of botulinum neurotoxin A to motor nerve endings are distinguished by its heavy chain. J. Neurosci. Res. **44:** 263-271, 1996.

Davis L.E., Johnson J.K., Bicknell J.M., Levy H. and McEvoy K.M. Human type A botulism and treatment with 3,4-diaminopyridine. Electromyogr. Clin. Neurophysiol. **32:** 379-383, 1992.

Deshpande S.S., Sheridan R.E. and Adler M.A study of zinc-dependent metalloendopeptidase inhibitors as pharmacological antagonists in botulinum neurotoxin poisoning. Toxicon **33:** 551-557, 1995.

Deshpande, S.S., Sheridan, R.E. and Adler, M. Efficacy of certain quinolines as pharmacological antagonists of botulinum neurotoxin poisoning. Toxicon **35:** 433-445, 1997.

Dezfulian M., Bitar R.A. and Bartlett J.G. Kinetics study of immunological response to *Clostridium botulinum* toxin. J. Clin. Microbiol. **25:** 1336-1337, 1987.

Doc M., Ben Ali A., Karras A., Misset B., Garrouste-Orgeas M., Deletie E., Goldstein F. and Carlet J. Treatment of severe botulism with 3,4-diaminopyridine. Presse Med. **31:** 601-602, 2002.

Dong, M., Richards, D.A., Goodnough, M.C., Tepp, W.H., Johnson, E.A. and Chapman, E.R. Synaptotagmins I and II mediate entry of botulinum neurotoxin B into cells. J. Cell. Biol. **162:** 1293-1303, 2003.

Donovan J.J. and Middlebrook J.L. Ion-conducting channels produced by botulinum toxin in planar lipid membranes. Biochemistry **25:** 2872-2876, 1986.

Eswaramoorthy S., Kumaran D. and Swaminathan S. A novel mechanism for *Clostridium botulinum* neurotoxin inhibition. Biochemistry **41:** 9795-9802, 2002.

Flock M.A., Carnella M.A. and Gearinger N.F. Studies on immunity to toxins of *Clostridium botulinum*. J. Immunol. **90:** 697-702, 1963.

Foran P., Shone C.C. and Dolly J.O. Differences in the protease activities of tetanus and botulinum B toxins revealed by the cleavage of vesicle-associated membrane protein and various sized fragments. Biochemistry **33:** 15365-15374, 1994.

Franz D.R. Defense against toxin weapons, in *Textbook of Military Medicine: Medical Aspects of Chemical and Biological Warfare*, Part I, Zajtchuk, R. ed., Office of the Surgeon General at TMM Publications Borden Institute, Washington, 1997, chap. 30.

Franz D.R., Pitt L.M., Clayton M.A., Hanes M.A. Efficacy of prophylactic and therapeutic administration of antitoxin for inhalation botulism. In: Botulinum and Tetanus Neurotoxins, DasGupta, B.R. ed., Plenum Press, NY, 1993, 473-476.

Gelzleichter T.R., Myers M.A., Menton R.G., Niemuth N.A., Matthews M.C., Langford M.J. Protection against botulinum toxins provided by passive immunization with botulinum human immune globulin: Evaluation using an inhalation model. J. Appl. Toxicol. **19:** S35-S38, 1999.

Goodnough M.C., Oyler G., Fishman P.S., Johnson E.A., Neale E.A., Keller J.E., Tepp W.H., Clark C., Hartz S. and Adler M. Development of a delivery vehicle for intracellular transport of botulinum neurotoxin antagonists. FEBS Lett. **513:** 163-168, 2002.

Habermann E. and Dreyer F. Clostridial neurotoxins: handling and action at the cellular and molecular level, Current Top. Microbiol. Immunol. **129:** 93-179, 1986.

Hanson M.A. and Stevens R.C. Cocrystal structure of synaptobrevin-II bound to botulinum neurotoxin type B at 2.0 Å resolution. Nat. Struct. Biol. **7:** 687-692, 2000.

Hanson P.I., Heuser, J.E. and Jahn, R. Neurotransmitter release-four years of SNARE complexes. Curr. Opin. Neurobiol. **7:** 310-315, 1997.

Hatheway C.H., Snyder J.D., Seals J.E., Edell T.A. and Lewis G.E. Antitoxin levels in botulism patients treated with trivalent equine botulism antitoxin to toxin type A, B, and E. J. Infect. Dis. **150:** 407-412, 1984.

Hirsch J., Kirpekar S.M. and Prat J.C. Effect of guanidine on release of noradrenaline from the perfused spleen of the cat. Br. J. Pharmacol. **66:** 537-546, 1979.

Hoch D.H., Romero-Mira M., Ehrlich B.E., Finkelstein A., DasGupta B.R. and Simpson L.L. Channels formed by botulinum, tetanus, and diphtheria toxins in planar lipid bilayers: Relevance to translocation of proteins across membranes. Proc. Natl. Acad. Sci. USA **82:** 1692-1696, 1985.

Johnson, E.A. Clostridial toxins as therapeutic agents: benefit of nature's most toxic protein. Annu. Rev. Microbiol. **53:** 551-575, 1999.

Kaplan J.E., Davis L.E., Narayan V., Koster J. and Katzenstein D. Botulism, type A, and treatment with guanidine. Ann. Neurol. **6:** 69-71, 1979.

Keller J.E., Neale E.A., Oyler G. and Adler M. Persistence of botulinum neurotoxin action in cultured spinal cord cells. FEBS Lett. **456:** 137-142, 1999.

Kitamura M., Takamiya K., Aizawa S., Furukawa K. and Furukawa K. Gangliosides are the binding substances in neural cell for tetanus and botulinum toxins in mice. Biochim. Biophys. Acta. **1441:** 1-3, 1999.

Koriazova L.K. and Montal M. Translocation of botulinum neurotoxin light chain protease through the heavy chain channel. Nat. Struct. Biol. **10:** 13-18, 2003.

Lacy D.B., Tepp W., Cohen A.C., DasGupta B.R. and Stevens R.C. Crystal structure of botulinum neurotoxin type A and implications for toxicity. Nat. Struct.l Biol. **5:** 898-902, 1998.

Lalli G., Herreros J., Osborne S.L., Montecucco C., Rossetto O. and Schiavo G. Functional characterization of tetanus and botulinum neurotoxins binding domains. J. Cell. Sci. **112:** 2715-2724, 1999.

Li L. and Singh B.R. Isolation of synaptotagmin as a receptor for types A and E botulinum neurotoxin and analysis of their comparative binding using a new microtiter plate assay. J. Nat. Toxins **7:** 215-226, 1998.

Martin L., Cornille F., Turcaud S., Meudal H., Roques B.P. and Fournie-Zaluski M.-C. Metallopeptidase inhibitors of tetanus toxin: A combinatorial approach. J. Med. Chem. **42:** 515-525, 1999.

Marxen P., Fuhrmann U. and Bigalke H. Gangliosides mediate inhibitory effects of tetanus and botulinum A neurotoxins on exocytosis in chromaffin cells. Toxicon **27:** 849-859, 1989.

Metzger J.F. and Lewis G.E. Human-derived immune globulins for the treatment of botulism. Rev. Infect. Dis. **1:** 689-692, 1979.

Meunier F.A., Lisk G., Sesardic D. and Dolly J.O. Dynamics of motor nerve terminal remodeling unveiled using SNARE-cleaving botulinum toxins: the extent and duration are dictated by the sites of SNAP-25 truncation. Mol. Cell. Neurosci. **22:** 454-466, 2003.

Molgo J., Lundh H. and Thesleff S. Potency of 3,4 diaminopyridine and 4-aminopyridine on mammalian neuromuscular transmission and the effect of pH changes. Eur. J. Pharmacol. **61:** 25-34, 1980.

Montecucco C. and Schiavo G. Tetanus and botulism neurotoxins: a new group of zinc proteases. Trends Biochem. Sci. **18:** 324-327, 1993.

Mullaney B.P., Pallavicini M.G. and Marks J.D. Epitope mapping of neutralizing botulinum neurotoxin A antibodies by phage display. Infect. Immun. **69:** 6511-6514, 2001.

Nishiki T., Kamata Y., Nemoto Y., Omori A., Ito T., Takahashi M. and Kozaki S. Identification of protein receptor for *Clostridium botulinum* type B neurotoxin in rat brain synaptosomes. J. Biol. Chem. **269:** 10498-10503, 1994.

Nowakowski A., Wang C., Powers D.B., Amersdorfer P., Smith T.J., Montgomery V.A., Sheridan R.E., Blake R., Smith L.A. and Marks J.D. Potent neutralization of botulinum neurotoxin by recombinant oligoclonal antibody. Proc. Natl. Acad. Sci. USA **99:** 11346-11350, 2002.

Puggiari M. and Cherington M. Botulism and guanidine ten years later. JAMA **240:** 2276-2277, 1978.

Rauch G., Gambale F. and Montal M. Tetanus toxin channel in phosphatidylserine planar bilayers: conductance states and pH dependence. Eur. Biophys. J. **18:** 79-83, 1990.

Ray P., Berman J.D., Middleton W. and Brendle J. Botulinum toxin inhibits arachidonic acid release associated with acetylcholine release from PC12 cells. J. Biol. Chem. **268:** 11057-11064, 1993.

Roques B.P., Anne C., Turcaud S. and Fournie-Zaluski M.-C. Mechanism of action of clostridial neurotoxins and rational inhibitor design. Biol. Cell. **92:** 445-447, 2000.

Schantz E.J. and Sugiyama H. Toxic proteins produced by *Clostridium botulinum* toxins. J. Agric. Food Chem. **22:** 26-30, 1974.

Schiavo G., Benfenati F., Poulain B., Rossetto O., Polverino de Laureto P., DasGupta B.R. and Montecucco C. Tetanus and botulinum-B neurotoxins block neurotransmitter release by proteolytic cleavage of synaptobrevin. Nature **359:** 832-835, 1992a.

Schiavo G., Poulain B., Rossetto O., Benefanti F., Tauc L. and Montecucco C. Tetanus toxin is a zinc protein and its inhibition of neurotransmitter release and protease activity depend on zinc. EMBO J. **11:** 3577-3583, 1992b.

Schiavo G., Rossetto O., Catsicas S., Polverino de Laureto P., DasGupta B.R., Benfenati F. and Montecucco C. Identification of the nerve terminal targets of botulinum neurotoxin serotypes A, D, and E. J. Biol. Chem. **268:** 23784-23787, 1993.

Schmid A., Benz R., Just I. and Aktories K. Interaction of *Clostridium botulinum* C2 toxin with lipid bilayer membranes. J. Biol. Chem. **269:** 16706-16711, 1994.

Schmidt, J.J. and Bostian, K.A. Endoproteinase activity of type A botulinum neurotoxin: substrate requirements and activation by serum albumin. J. Prot. Chem. **16(1):** 19-26, 1997.

Schmidt, J.J. and Stafford, R.G. A high-affinity competitive inhibitor of type A botulinum neurotoxin protease activity. FEBS Lett. **532(3):** 423-426, 2002.

Schmidt J.J., Stafford R.G. and Bostian K.A. Type A botulinum neurotoxin proteolytic activity: development of competitive inhibitors and implications for substrate specificity at the S1' binding subsite. FEBS Lett. **435:** 61-64, 1998.

Schmitt A., Dreyer F. and John C. At least three sequential steps are involved in the tetanus toxin-induced block of neuromuscular transmission. N.-S. Arch. Pharmacol. **317:** 326-330, 1981.

Sellin L.C. The action of botulinum toxin at the neuromuscular junction. Med. Biol. **59:** 11-20, 1981.

Shapiro R.L., Hatheway C. and Swerdlow D.L. Botulism in the United States: a clinical and epidemiological review. Ann. Intern. Med. **129:** 221-228, 1998.

Sheridan R.E. Protonophore antagonism of botulinum toxin in mouse muscle. Toxicon **34:** 849-855, 1996.

Sheridan R.E. Gating and permeability of ion channels produced by botulinum toxin types A and E in PC12 cell membranes. Toxicon **36:** 703-717, 1998.

Sheridan R.E. and Deshpande S.S. Interactions between heavy metal chelators and botulinum neurotoxin at the neuromuscular junction. Toxicon **33:** 539-549, 1995.

Sheridan R.E. and Deshpande S.S. Cytotoxicity induced by intracellular zinc chelation in rat cortical neurons. In vitro & Molec. Toxicol. **11:** 161-169, 1998.

Sheridan R.E., Deshpande S.S., Amersdorfer P., Marks J.D. and Smith T.J. Anomalous enhancement of botulinum toxin type A neurotoxicity in the presence of antitoxin. Toxicon **39:** 651-657, 2001.

Sheridan R.E., Deshpande S.S., Nicholson J.D. and Adler M. Structural features of aminoquinolines necessary for antagonist activity against botulinum neurotoxin. Toxicon **35:** 1439-1451, 1997.

Sheridan R.E., Deshpande S.S. and Smith T. Comparison of in vivo and in vitro mouse bioassays for botulinum toxin antagonists. J. Appl. Toxicol. **19:** S29-S33, 1999.

Shimizu T., Kondo H. and Sakaguchi G. Immunological heterogeneity of *Clostridium botulinum* type B toxins. Jpn. J. Med. Sci. Biol. **27:** 99-100, 1974.

Siegel L.S. Human immune response to botulinum pentavalent toxoid determined by a neutralization test and by ELISA. J. Clin. Microbiol. **26:** 2351-2356, 1988.

Siegel L.S., Johnson-Winegar A.D. and Sellin L.C. Effect of 3,4-diaminopyridine on the survival of mice injected with botulinum neurotoxin type A, B, E, or F. Toxicol. Appl. Pharmacol. **84:** 255-263, 1986.

Simpson L.L. The origin, structure, and pharmacological activity of botulinum toxin. Pharmacol. Rev. **33:** 155-188, 1981.

Simpson L.L. The interaction between aminoquinolines and presynaptically acting neurotoxins. J. Pharmacol. Exp. Ther. **222:** 43-48, 1982.

Simpson L.L. Ammonium chloride and methylamine hydrochloride antagonize clostridial neurotoxins. J. Pharmacol. Exp. Ther. **225:** 546-552, 1983.

Simpson L.L. A preclinical evaluation of aminopyridines as putative therapeutic agents in the treatment of botulism. Infect. Immun. **52:** 858-862, 1986.

Simpson L.L. Peripheral actions of the botulinum toxins, in *Botulinum Neurotoxins and Tetanus Toxin*, Simpson L.L. ed., Academic Press, New York, 153-187, 1989.

Simpson L.L., Coffield J.A. and Bakry N. Chelation of zinc antagonizes the neuromuscular blocking properties of the seven serotypes of botulinum neurotoxin as well as tetanus toxin. J. Pharmacol. Exp. Ther. **267:** 720-727, 1993.

Simpson L.L., Coffield J.A. and Bakry N. Inhibition of vacuolar adenosine triphophatases antagonizes the effects of clostridial neurotoxins but not phospholipase A2 neurotoxins. J. Pharmacol. Exp. Ther. **269:** 256-262, 1994.

Simpson L.L. and DasGupta B.R. Botulinum neurotoxin type E: studies on mechanism of action and on structure-activity relationships. J. Pharmacol. Exp. Ther. **224:** 135-140, 1983.

Söllner T., Bennett M.K., Whiteheart S.W., Scheller R.H. and Rothman J.E. A protein assembly-disassembly pathway in vitro that may correspond to sequential steps of synaptic vesicle docking, activation, and fusion. Cell **75:** 409-418, 1993.

Starks D.F., Kane C.T., Nicholson J.D., Hackley B.E. and Adler M. Proceedings of USMRAMC Medical Defense Bioscience Review **3:** 1608-1615, 1996.

Trudeau L-E., Fang Y. and Haydon P.G. Modulation of an early step in the secretory machinery in hippocampal nerve terminals. Proc. Natl. Acad. Sci. USA **95:** 7163-7168, 1998.

Wellhöner H.H. and Neville Jr., D.M. Tetanus toxin binds with high affinity to neuroblastoma X glioma hybrid cells NG108-15 and impairs their stimulated acetylcholine release. J. Biol. Chem. **262:** 17374-17378, 1987.

Williamson L.C. and Neale E.A. Bafilomycin A1 inhibits the action of tetanus toxin in spinal cord neurons in cell culture. J. Neurochem. **63:** 2342-2345, 1994.

Zhou, L., de Paiva, A., Liu, D., Aoki, R. and Dolly, J.O. Expression and purification of the light chain of botulinum neurotoxiin A: a single mutation abolishes its cleavage of SNAP-25 and neurotoxicity after reconstitution with the heavy chain. Biochem. **34(46):** 15175-15181, 1995.

Pharmacological Perspectives of Toxic Chemicals and Their Antidotes
Editors: S J S Flora, J A Romano, S I Baskin and K Sekhar

Human Butyrylcholinesterase : A Future Generation Antidote for Protection Against Organophosphate Agents

Ashima Saxena[a], C. Luo[a], R. Bansal[a], W. Sun[a], M.G. Clark[a], Y. Ashani[b], and B.P. Doctor[a]

[a]Divisions of Biochemistry and Neurosciences, Walter Reed Army Institute of Research, Silver Spring, MD 20910-7500, USA
[b]Israel Institute for Biological Research, P. O. Box 19, Ness-Ziona, Israel

INTRODUCTION

Exposure to organophosphorus compounds (OPs) in the form of nerve agents and pesticides poses an ever increasing military and civilian threat. The serious medical challenges posed by chemical warfare agents to the military healthcare system were first demonstrated in World War I for HD. The sixth and seventh exposures in the 20th century were verified in the Iran-Iraq conflict and the 1995 Tokyo subway incident, respectively. The acute toxicity of OPs is usually attributed to their irreversible inhibition of acetylcholinesterase (AChE). The resultant increase in acetylcholine concentration manifests at the cholinergic synapses of both the peripheral and central nervous system by over-stimulation at the neuromuscular junctions. This precipitates a cholinergic crisis characterized by miosis, increased tracheobronchial and salivary secretions, broncho-constriction, bradycardia, fasciculation, behavioral incapacitiation, muscular weakness, and convulsions, culminating in death by respiratory failure. Current antidotal regimens for OP poisoning consist of a combination of pretreatment with a spontaneously reactivating reversible AChE inhibitor such as pyridostigmine bromide, and post-exposure therapy with anticholinergic drugs such as atropine sulfate and oximes such as 2-PAM chloride (Gray, 1984). Although these antidotal regimens are effective in preventing lethality of animals from OP poisoning, they do not prevent post-exposure incapacitation, convulsions, performance deficits or in many cases, permanent brain damage (Dirnhuber, 1979; McLeod, 1985; Dunn and Sidell, 1989). These problems stimulated the development of enzyme bioscavengers as a pretreatment to sequester highly toxic OPs before they reach their physiological targets and prevent the *in vivo* toxicity of OPs and post exposure incapacitation (Dunn and Sidell, 1989).

Among the enzymes examined as potential scavengers of highly toxic OP nerve agents, significant advances have been made using cholinesterases (ChEs). Exogenous administration of plasma-derived ChEs such as AChE from fetal bovine serum (FBS) and butyrylcholinesterase (BChE) from human and equine serum (Eq), in both rodent and non-human primate models, has been successfully used as a safe and efficacious prophylactic treatment to prevent poisoning by OPs (Doctor et al., 2001). Of the ChEs evaluated so far, Hu BChE has several advantages as an exogenously administered prophylactic for human use (Ashani, 2000). First, it reacts rapidly with all highly toxic OPs, offering a broad range of protection for nerve agents including, soman, sarin, tabun, and VX. Studies in mice (Raveh et al., 1993), rats (Brandeis et al., 1993), guinea pigs (Allon et al., 1998) and rhesus monkeys (Raveh et al., 1997) clearly demonstrated that Hu BChE could function as a universal prophylactic antidote. These studies also showed that pretreatment with Hu BChE was effective in preventing mortality as well as development of behavioral deficits without the need for additional post-exposure therapy (Brandeis et al., 1993; Raveh et al., 1997). Second, it possesses a long retention time in human circulation and is readily absorbed from sites of injection. Although the reported values of half-life of exogenously administered Hu BChE in humans vary from 3.4 to 11 days (Garry et al., 1974; Ostergaard et al., 1988), they suggest that the circulatory stability of the enzyme is sufficient for its use as a pretreatment drug. The extended stability of Hu BChE following i.m. and i.v. injections, was also demonstrated in mice and rats (Raveh et al., 1993), guinea pigs (Allon et al., 1998) and rhesus monkeys (Raveh et al., 1997). These results suggest that a single injection of Hu BChE will provide long-lasting protection if used as a prophylactic treatment. Third, since the enzyme is from a human source, it should not produce any adverse immunological responses upon repeated administration into humans. The lack of reports indicating untoward side-effects in humans following plasma transfusions and i.v. injections of partially purified Hu BChE (Cascio et al., 1988; Naguib et al., 1996) support our contention. Similarly, the exogenous administration of 13-20 mg/Kg doses of Hu BChE did not seem to affect gross behavior in mice (Raveh et al., 1993), rats (Brandeis et al., 1993), or guinea pigs (Allon et al., 1998) and no behavior alterations were reported in rhesus monkeys treated with 13-34 mg of Hu BChE (Raveh et al., 1997). The utility of homologous BChE as an effective and safe scavenger was also demonstrated by the long mean residence times of 225 ± 19 h and the absence of significant adverse effects following administration of homologous macaque BChE into macaques (Rosenberg et al., 2002). Extrapolation of data obtained from prophylaxis experiments with Hu BChE in four species, suggests that a dose of 200 mg of Hu BChE can protect humans from exposure of up to 2 × LD_{50} of nerve agents (Ashani, 2000). Smaller doses of 50 mg of enzyme would be sufficient to provide protection against low-level exposure to nerve agents. In addition to its use as a pretreatment for a variety of wartime scenarios, including covert actions, it also has potential use for first responders (civilians) reacting to intentional/ accidental nerve gas release or pesticide overexposure. In addition, since Hu BChE catalyzes the hydrolysis of cocaine and short-acting muscle relaxants succinylcholine and mivacurium, it could be an effective treatment for cocaine intoxication, as well as succinylcholine-and mivacurium-induced apnea (Ashani, 2000).

The foremost requirement to advance Hu BChE as a bioscavenger for human use was to obtain sufficient amounts of purified enzyme for conducting animal and clinical studies. Although a procedure for the purification of Hu BChE from human plasma, which contains ~2 mg of enzyme per liter of plasma, was described, this source is not suitable for producing gm quantities of Hu BChE (Grunwald et al., 1997). A richer source of Hu BChE was identified as Cohn Fraction IV-4 paste, which is a by-product of human plasma generated during the production of human proteins, such as γ-globulin. This paste contains ~150 mg of enzyme per Kg, which is much higher than human plasma and contains much lesser quantities of other plasma proteins due to the fractionation steps deployed in the production process. A procedure for the large-scale purification of Hu BChE from Cohn Fraction IV-4 paste was developed, which uses batch procainamide affinity chromatography (De La Hoz et al., 1986) followed by anion exchange chromatography. Approximately, 6 gm of purified enzyme was obtained from 120 Kg of Cohn fraction IV-4 and its characterization with respect to purity, enzymatic activity, thermal stability, safety and efficacy in rodents were accomplished.

MATERIALS AND METHODS

Purification of Hu BChE from Cohn Fraction IV-4

Cohn Fraction IV-4 paste (120 Kg; MedImmune Inc., Frederick, MD) was resuspended in 1200 L of 25 mM sodium phosphate, pH 8.0, containing 1 mM EDTA. The suspension was filtered, the filtrate was concentrated to a volume of 320 L, and the fat was separated by decantation. Eighty liters of this extract were mixed with 4 L of procainamide-Sepharose affinity gel (Sigma Chemical Co., St. Louis, MO) and gently stirred overnight at 4°C. The gel was allowed to settle and most of the clear supernatant was decanted. The gel was rinsed with 25 mM sodium phosphate, pH 8.0, packed in a column, and washed with 50 mM NaCl in 25 mM sodium phosphate, pH 8.0, until the absorbance of the effluant at 280 nm was <0.05. The enzyme was eluted with 0.1 M procainamide in 25 mM sodium phosphate, pH 8.0. Fractions were assayed for BChE activity (Ellman et al., 1961), pooled, and precipitated with ammonium sulfate (75% saturation). The precipitate was collected by centrifugation, dissolved in 25 mM sodium phosphate, pH 8.0, and dialyzed against milli-Q water.

The pH of the enzyme solution was adjusted to 4.0 with 0.1 M acetic acid, the precipitated proteins removed by centrifugation, and the enzyme was loaded on a 6 L DEAE Sepharose column pre-equilibrated with 20 mM sodium acetate, pH 4.0. The column was washed with 50 mM NaCl in 25 mM sodium phosphate, pH 8.0, until the absorbance of the effluant at 280 nm was < 0.02. Hu BChE was eluted with 200 mM NaCl in 25 mM sodium phosphate, pH 8.0. Fractions containing activity were pooled and precipitated with 75% ammonium sulfate. The precipitate was collected by centrifugation, dissolved in 25 mM sodium phosphate, pH 8.0, and dialyzed against the same buffer.

The enzyme was further purified by affinity chromatography on a procainamide-Sepharose column. The column was washed with 50 mM NaCl in 25 mM sodium phosphate, pH 8.0, until the absorbance of effluant at 280 nm was < 0.02. The enzyme was eluted using 100 mM procainamide in 25 mM sodium phosphate buffer, pH 8.0. Fractions containing Hu BChE activity were pooled, precipitated with 75% ammonium sulfate, and dialyzed against 25 mM sodium phosphate, pH 8.0.

THERMAL STABILITY OF HuBChE

Aliquots of enzyme were stored in lyophilized (1 mg) or liquid form (10 mg/ml in 50 mM sodium phosphate, pH 8.0 + 10% glycerol + 1 mM EDTA) at 4°, 25°, 37° or 45°C. The thermal stability of the enzyme was monitored by measuring enzyme activity (Ellman et al., 1961), at various time intervals.

CIRCULATORY STABILITY AND TOXICITY OF HuBChE IN MICE

Research was conducted in compliance with the Animal Welfare Act and other federal statutes and regulations relating to animals and experiments involving animals and adheres to principles stated in the *Guide for the Care and Use of Laboratory Animals*. Five doses of Hu BChE (0.1-10X of the therapeutic human dose of 200 mg) were administered into CD-1 mice by i.m. or i.p. injections. Ten µl of blood was drawn from the tail vein at various time intervals for the determination of enzyme activity (Ellman et al., 1961). The mice were euthanized after 14 days and blood samples were collected by cardiac puncture for serum chemistry, hematology, immunologic and enzyme determination. Following blood collection and euthanasia, a complete necropsy was performed and a full set of tissues, including muscle injection sites, were examined for histologic changes.

BEHAVIORAL TOXICITY OF HuBChE AND ITS EFFICACY AGAINST PHYSOSTIGMINE IN MICE

The subjects were 12-week-old male C57BL/6J mice. Separate groups of mice were utilized for monitoring behavior and *in vivo* BChE activity. The amplitude and latency of the acoustic startle reflex (ASR) was recorded using a Startle Monitor System (Hamilton Kinder, Poway, CA). Acoustic startle responses were measured daily during the light phase of the light/dark cycle. Mice were placed in the Plexiglas restraint in the dark sound-attenuating chamber. Each test session began with a 3-min chamber adaptation period during which the background noise level was set to 60 dB SPL. Acoustic startle pulses consisting of 0, 70, or 120 dB white noise bursts, sometimes preceded by 70-dB prepulses, were presented through a loudspeaker. The sound level was verified using a modified Realistic sound level meter (Hamilton Kinder, Poway, CA) with the microphone placed in the position of the subject's head. The stimulus trials were presented ten times each using a randomized block design. The trial types were a 50 ms, 120-dB noise bursts alone or with a prepulse, a prepulse alone or a no stimulus trial.

Subjects were placed in startle chambers and habituated to the apparatus for 6 days prior to drug testing. Mean startle amplitudes were computed using the startle amplitude to 100 and 120 dB noise bursts across the last three pretest days for each subject. Subjects were matched based on mean startle amplitude, and divided into four experimental groups (saline-saline, saline-physostigmine, Hu BChE-physostigmine, and Hu BChE-saline). Prior to the next test session, the subjects were exposed to their respective treatment as follows: 2 h before testing, the mice were pretreated with saline or Hu BChE (2000 U, i.p.). Fifteen minutes before testing, subjects were exposed to either saline or physostigmine hemisulfate (0.4 mg/kg, sc). Physostigmine was used as a surrogate ChE inhibitor in lieu of an OP nerve agent. Following each stimulus presentation, the movement of each animal was measured for a period of 200 ms. Peak startle amplitude (V_{max}) was recorded as the highest observed force occurring during the 200 ms window following stimulus presentation. Latency to peak startle amplitude (T_{max}) was recorded as the time in ms after stimulus presentation at which the peak startle amplitude occurred. The amount of prepulse inhibition (PPI) produced upon drug administration, was calculated following behavioral testing and represents the difference between the pulse alone and the prepulse plus pulse trials, divided by the pulse alone and multiplied by 100.

RESULTS AND DISCUSSION

Purification of Hu BChE from Cohn fraction IV-4

A simple procedure for the large-scale purification of Hu BChE from Cohn fraction IV-4 was developed. The procedure is a modification of the previously published methods used for the purification of FBS AChE (De la Hoz et al., 1986) and Hu BChE from plasma (Grunwald et al., 1997), and involved three steps, which are summarized in Table 1. The procedure yielded ~6 g of purified Hu BChE from 120 Kg of Cohn Fraction IV-4, with an overall yield of 30 %. The purified enzyme had a specific activity of ~750 U/mg and migrated as a single band on Sodium dodecyl sulfate-polyacrylamide gel electrophoresis (SDS-PAGE; Figure 1). These results show that a relatively simple procedure using a combination of procainamide affinity and DEAE anion-exchange chromatography can be used to purify gram quantities of Hu BChE from Cohn fraction IV-4.

Table 1 Purification of Hu BChE from Cohn Fraction IV-4

Procedure	*BChE Activity (U/ml)*	*Protein (mg/ml)*	*Specific Activity (U/mg)*
Resuspension of Cohn fraction IV (1: 3.2)	46	174	0.26
Batch extraction Using procainamide affinity gel	1210	8.1	149
DEAE chromatography	3620	13	278
Analytical procainamide affinity chromatography	10,500	14	750

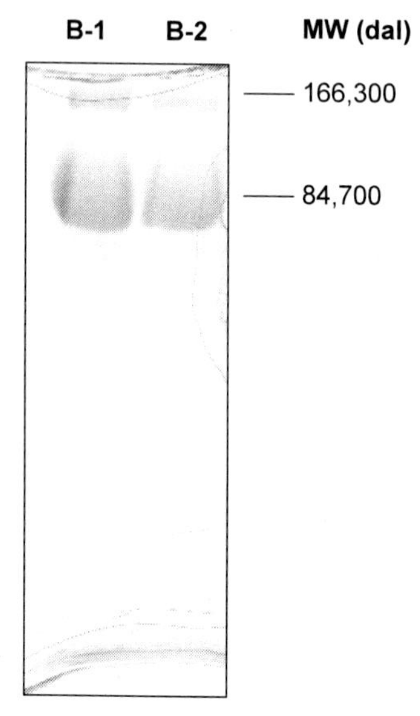

Fig. 1 10% SDS-PAGE of Hu BChE batches 1 and 2 (B-1, B-2), purified from Cohn fraction IV-4.

THERMAL STABILITY OF HuBChE

The thermal stability of the enzyme purified from Cohn fraction IV-4, in lyophilized and liquid form is shown in Figure 2A and 2 B, respectively. The enzyme activity was stable when stored in lyophilized form at 4°, 25°, 37° or 45° and in liquid form at 4° and 25° to date (214 days).

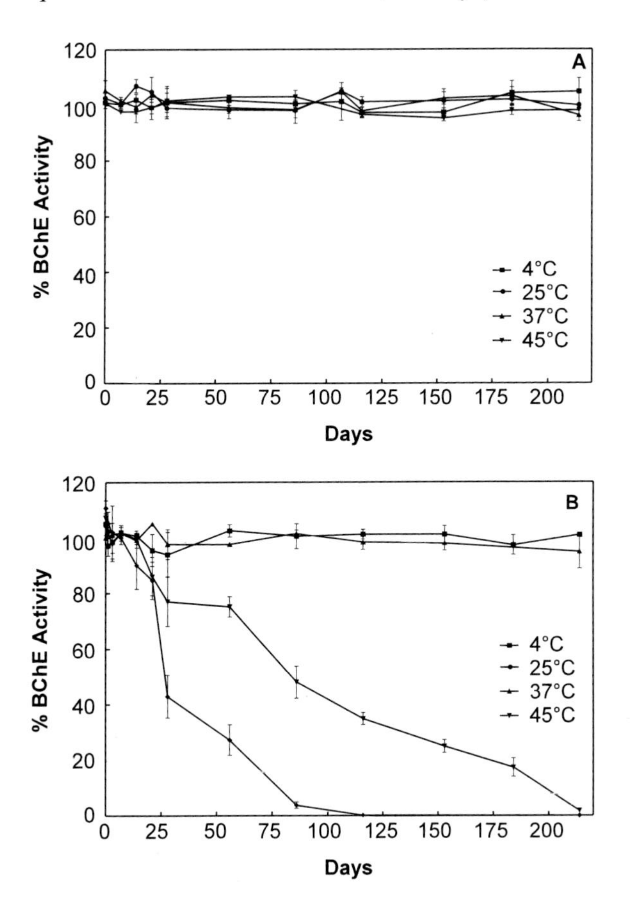

Fig. 2 Thermal stability of Hu BChE in lyophilized (panel A) and liquid form (panel B)

CIRCULATORY STABILITY AND TOXICITY OF HuBChE IN MICE

The time course of Hu BChE purified from Cohn fraction IV-4 was assessed in the circulation of mice by two different routes of administration e.g., i.m. and i.p. As shown in Figure 3, enzyme purified from Cohn fraction IV exhibited circulatory stability profiles similar to those observed previously for enzyme purified from human plasma in rats and mice (Raveh et al., 1993), guinea pigs (Allon et al., 1998), and rhesus monkeys (Raveh et al., 1997). Mice that were administered 7-700 U of Hu BChE by i.p. injection showed a rapid increase in BChE activity, which reached peak levels between 2 and 4 h. On the other hand, when the same doses of enzyme were delivered by i.m. injections, peak levels of activity were attained ~10-12 h. General observation of mice with circulating levels of BChE as high as 100 U/ml, did not display any signs of toxicity. Similarly, necropsy performed on animals after 14 days following euthanasia, did not reveal any toxicity in mice as measured by, serum chemistry, complete blood count, as well as gross and histologic tissue changes.

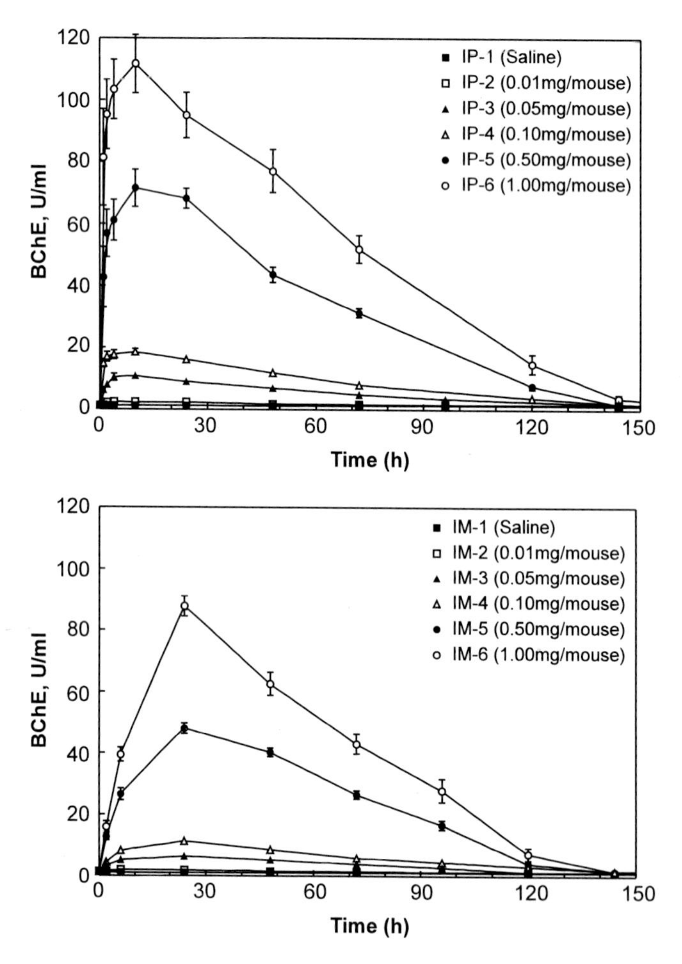

Fig. 3 Average enzyme levels in blood of mice following i.p. (top panel) and i.m. (bottom panel) injections of various doses of Hu BChE

BEHAVIORAL TOXICITY OF HuBChE AND ITS EFFICACY AGAINST PHYSOSTIGMINE IN MICE

The behavioral toxicity of Hu BChE and its efficacy against physostigmine, a carbamate cholinesterase inhibitor, were assessed in mice. Control mice not exposed to Hu BChE, displayed a constant blood BChE level of ~1 U/ml. Consistent with data shown in Figure 3 (top panel), mice that were administered 2000 U of Hu BChE showed a rapid increase in BChE activity, which reached peak levels (~600 U/ml) between 2 and 4 h. In the Hu BChE-physostigmine group, a 40% reduction in circulating BChE level was observed following the physostigmine challenge. This result is consistent with the molar ratio of Hu BChE to physostigmine of 1.6 and shows that there was sufficient enzyme to neutralize the dose of physostigmine.

The effects of Hu BChE and physostigmine on Vmax, Tmax and PPI are shown in Figure 4. As expected, physostigmine significantly decreased the amplitude of the ASR. Furthermore, significant increases in the Tmax as well as the amount of PPI of the acoustic startle reflex were also observed. When physostigmine was presented as a pharmacological challenge to subjects that had received Hu BChE, protection was provided against the performance altering effects of physostigmine. Despite the presence of excess Hu BChE relative to physostigmine, a partial protection to physostigmine was observed, which is probably due to the reversible nature of the inhibitor.

With regard to the behavioral toxicity of Hu BChE, we observed little if any change in performance following exposure to Hu BChE. Similar results indicating that, exogenously administered Eq and Hu BChE did not cause toxic neurobehavioral effects at doses lower than those used in this study, were reported previously in rhesus monkeys (Raveh et al., 1997; Matzke et al., 1999) and rats (Brandeis et al., 1993; Genovese and Doctor, 1996). The observations made in the present study are particularly compelling in light of the fact that the subjects had a mean circulating BChE level 600-fold higher than the normal physiological level. Moreover, the 2000 U dose of Hu BChE is equivalent to 30 times the dose proposed to be necessary for protection against 2 x LD_{50} of nerve agent in humans.

Conclusion

Taken together, these results demonstrate that Cohn fraction IV-4 is a suitable source for purifying gram quantities of Hu BChE. The purified enzyme displays remarkable *in vitro* stability in lyophilized form and high bioavailability and extended circulatory stability *in vivo*. Also, animals exposed to much higher doses than the proposed therapeutic dose did not display any signs of histopathological and behavioral toxicity. Work is currently underway to transition this procedure to isolate tens of gm quantities and eventually Kg amounts of BChE from Cohn fraction IV under GMP conditions and evaluate its use as a bioscavenger for safety and efficacy in humans. The completion of this concept exploration phase should provide the data required for filing an investigational new drug application.

OP nerve agents represent a very real threat to both the warfighter and to the public in covert terrorist events. Nerve agents have already been used

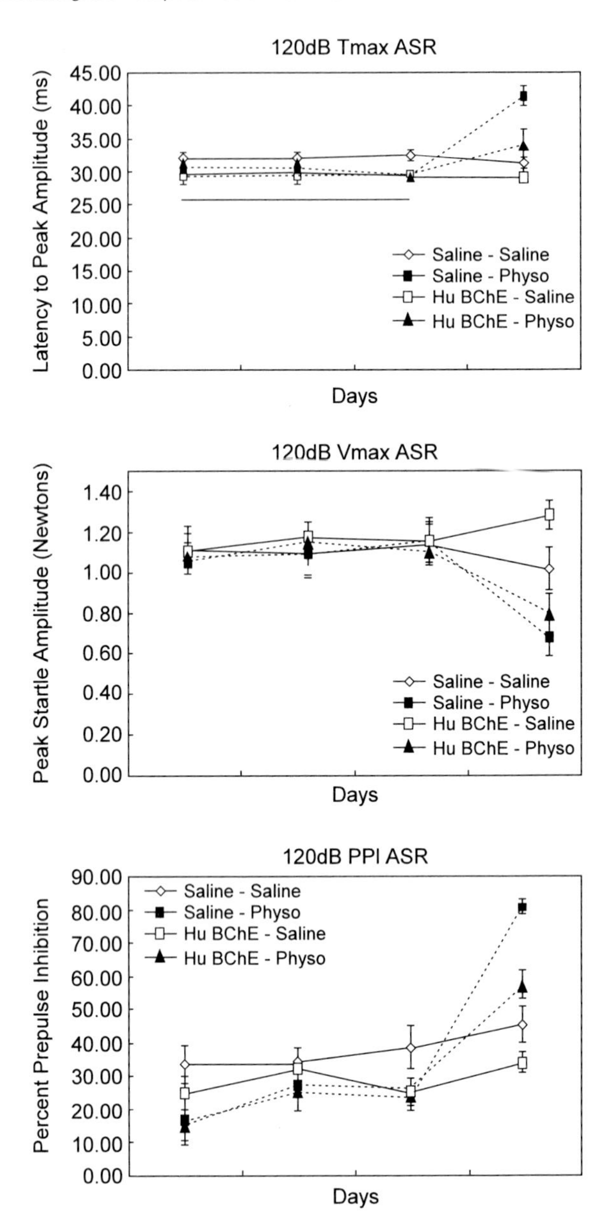

Fig. 4 The effects of Hu BChE and physostigmine on acoustic startle reflex in mice

against civilians and protection presents medical challenges that we are addressing with the use of the bioscavenger, Hu BChE. As described above, it will be the first successful pretreatment/treatment, offering protection against a wide variety of nerve agents. It can afford protection against not only mortality, but also against the adverse physiological and behavioral effects of nerve agent exposure. It can be administered prophylactically, precluding the need for immediate and extensive post-exposure therapy. The use of Hu BChE as a defense against OP intoxication in humans has many advantages and few apparent disadvantages. Our initial results provide convincing data that Hu BChE is a safe and effective bioscavenger that should continue to be assessed for future inclusion into the protective regimen against chemical warfare nerve agents.

References

Allon N., Raveh L., Gilat E., Cohen E., Grunwald J. and Ashani Y. Prophylaxis against soman inhalation toxicity in guinea pigs by pretreatment alone with human serum butyrylcholinesterase, Toxicol. Sci. **43:** 121-128, 1998.

Ashani Y. Prospective of human butyrylcholinesterase as a detoxifying antidote and potential regulator of controlled-release drugs, Drug Development Res. **50:** 298-308, 2000.

Brandeis R., Raveh L., Grunwald J., Cohen E. and Ashani Y. Prevention of soman-induced cognitive deficits by pretreatment with human butyrylcholinesterase in Rats, Pharmacol. Biochem. Behav. **46:** 889-896, 1993.

Cascio C., Comite C., Ghira M., Lanza G. and Ponchione A. The use of serum cholinesterase in severe phosphorous poisoning. personal experience, Minerv. Anestesiol. **54:** 337-338, 1988.

De La Hoz D., Doctor B.P., Ralston J.S. and Wolfe A.D. A simplified procedure for the purification of large quantities of fetal bovine serum acetylcholinesterase, Life Sci. **39:** 195-199, 1986.

Dirnhuber P., French M.C., Green D.M., Leadbeater I. and Stratton J.A. The protection of primates against soman poisoning by pretreatment with pyridostigmine, J. Pharm. Pharmacol. **31:** 295-299, 1979.

Doctor B.P., Maxwell D.M., Ashani Y., Saxena A. and Gordon R.K. New approaches to medical protection against chemical warfare nerve agents, in Chemical Warfare Agents: Toxicity at low levels, Somani, S.M. and Romano, J.A. Jr., Eds., CRC Press, New York, pp 191-214, 2001.

Dunn M.A. and Sidell F.R. Progress in medical defense against nerve agents, J. Am. Med. Assoc. **262:** 649-652, 1989.

Ellman G.L., Courtney K.D., Andres V. and Featherstone R.M. A new and rapid colorimetric determination of acetylcholinesterase activity, Biochem. Pharmacol. **7:** 88-95, 1961.

Garry P.J., Prince L.C. and Notari R.E. Half-life of human serum cholinesterase following blood transfusion, Res. Commun. Chem. Pathol. Pharmacol. **8:** 371-380, 1974.

Genovese R.F. and Doctor B.P. Behavioral and pharmacological assessment of butyrylcholinesterase in rats, Pharmacol. Biochem. Behav. **51:** 647-654, 1995.

Gray A.P. Design and structure-activity relationships of antidotes to organophosphorus anticholinesterase agents, Drug Metab. Rev., **15:** 557-589, 1984.

Grunwald J., Marcus D., Papier Y., Raveh L., Pittel Z. and Ashani Y. Large-scale purification and long-term stability of human butyrylcholinesterase: a potential bioscavenger drug, J. Biochem. Biophys. Methods. **34:** 123-135, 1997.

Matzke S.M., Oubre J.L., Caranto G.R., Gentry M.K. and Galbicka G. Behavioral and immunological effects of exogenous butyrylcholinesterase in rhesus monkeys, Pharmacol. Biochem. Behav. **62:** 523-530, 1999.

McLeod C.G. Pathology of nerve agents: perspectives on medical management, Fundam. Appl. Toxicol. **5:** S10-S16, 1985.

Naguib M., Selim M., Bakhamees H., Samarkandi A. and Turkistani A. Enzymatic versus pharmacokinetic antagonism of profound mivacurium-induced neuromuscular blockade, Anesthesiology. **84:** 1051-1059, 1996.

Ostergaard D., Viby-Mogensen J., Hanel H.K. and Skovgaard L.T. Half-life of plasma cholinesterase, Acta Anaesth. Scand. **32:** 266-269, 1988.

Raveh L., Grunwald J., Marcus D., Papier Y., Cohen E. and Ashani Y. Human butyrylcholinesterase as general prophylactic antidote for nerve agent toxicity, Biochem. Pharmacol. **45:** 2465-2474, 1993.

Raveh L., Grauer E., Grunwald J., Cohen E. and Ashani Y. The stoichiometry of protection against soman and VX toxicity, Toxicol. Appl. Pharmacol. **145:** 43-53, 1997.

Rosenberg Y., Luo C., Ashani Y., Doctor B.P., Fischer R., Wolfe G. and Saxena A. Pharmacokinetics and Immunologic Consequences of Exposing Macaques to Purified Homologous Butyrylcholinesterase, Life Sci. **72:** 125-134, 2002.

Pharmacological Perspectives of Toxic Chemicals and Their Antidotes
Editors: S J S Flora, J A Romano, S I Baskin and K Sekhar

Management of Commonly Encountered Drug Poisonings

S.K. Gupta, Amita Srivastava, Thomas Kaleekal and Sharda Shah Peshin
National Poisons Information Centre, Department of Pharmacology,
All India Institute of Medical Sciences, New Delhi 110 029, India

MANAGEMENT OF COMMONLY ENCOUNTERED DRUG POISONINGS

The current age is witnessing a plethora of chemical preparations. There is a multitude of products available for various purposes, and drugs form an important component. Though the uses of drugs are incalculable yet they have the potential to cause poisoning if misused, abused or overused. Drugs normally considered safe can cause severe morbidity or mortality if taken in overdose. Patients suffering from depression, elderly and children are especially at risk.

One of the leading causes of poisoning world over is drugs. The management of toxicology emergencies is an overwhelming problem and efficient management of such patients depends on provision of good supportive care, diagnosis and antidotal therapy.

Our experience at the National Poisons Information Centre (NPIC), in the Department of Pharmacology, at the All India Institute of Medical Sciences, New Delhi, highlights that next to household products, incidence of poisoning due to drugs is high. The drugs commonly consumed include antihistamines, barbiturates, benzodiazepines, carbamazepine, lithium, opioids, paracetamol and tricyclic antidepressants.

ANTIHISTAMINES

Histamine (H_1) receptor antagonists are commonly found in over-the-counter and prescription medications used for motion sickness, control of allergy-related itching, cough and cold palliation and as sleep aids. Antihistamines block certain histamine-induced effects. The response of histamine is mediated via three distinct histamine receptors H_1, H_2 and H_3. H_1 receptors are present in the smooth muscles of bronchi, blood vessels, heart, CNS, autonomic ganglia and afferent nerve fibers and mediate the allergic and inflammatory response. H_2 receptors are present on gastric parietal cells, vascular smooth muscles, heart, CNS and T lymphocytes. They are mainly responsible for gastric acid secretion. H_3 receptors are present presynaptically on nerve terminals in the brain and periphery. They mediate the feedback control of histamine synthesis and release.

H_1 antihistamines (particularly of first generation) are the ingredient of several common cold, cough, sleep, preanaesthetic medication etc. They are used in the treatment of allergic disorders, anaphylactic shock, motion sickness, vestibular disease, insect bites, common cold, parkinson's disease, vomiting, asthma and insomnia.

H_1 antihistamines are structurally related to histamine and antagonize the effects of histamine on H_1 receptor sites. They possess anticholinergic effects except the "nonsedating" agents like terfenadine, astemizole and loratadine. They may also stimulate or depress the central nervous system. Some agents (diphenhydramine) have local anaesthetic and membrane-depressant effects in large doses.

H_2 antihistamines are used in the treatment of duodenal ulcers, gastric ulcers, Zollinger Ellison syndrome, systemic mastocytosis, basophilic leukemia, reflux esophagitis, stress ulcers and hiatus hernia. The important drugs in this group include cimetidine, ranitidine and famotidine. They are selective and do not block H_1 receptors. These antihistamines also inhibit the secretion elicited by gastric as well as to lesser extent by muscarinic agonists.

The toxic dose of first generation H_1 antihistamines is high. The estimated fatal oral dose of diphenhydramine is 20-40 mg/kg. In general, toxicity occurs after ingestion of 3-5 times the usual daily dose. Accidental poisoning is more likely in children than adults and children are more susceptible to toxicity.

Antihistamines sedate at therapeutic doses. Overdose may result in CNS depression and/or stimulation and may resemble anticholinergic overdose. Effects may include fixed and dilated pupils, flushed face, dry mouth, excitation, hallucinations and tonic-clonic seizures. In young children CNS stimulation is predominant. Adult overdose usually causes CNS depression with drowsiness or coma followed by excitement, seizures, and postictal depression. Severe toxicity in children and adults may result in deep coma, cardiovascular and respiratory collapse, or death. Onset of symptoms may occur within 30 minutes to 2 hours after ingestion. Death may occur several days after onset of toxic symptoms. Toxicity of second generation antihistamines like astemizole and terfenadine can lead to QT prolongation and torsades de pointes especially if the patient is taking erythromycin, ketoconazole or other cytochrome P-450 inhibitors.

Acute overdose of the specific H_2 receptor antagonists typically causes only minor toxic effects such as drowsiness and mild bradycardia. More serious effects such as hypotension or bradycardia are likely to occur with intravenous overdoses and not with ingestion of the drug. Other effects rarely reported include hypersensitivity hepatitis, bone marrow suppression, and renal failure with long-term therapeutic doses.

Common over-the-counter antihistamines are usually detectable in general urine toxicology screening. The laboratory parameters such as electrolytes, glucose, arterial blood gases or pulse oximetry and ECG monitoring are indicated in diphenhydramine, terfenadine and astemizole poisoning/overdose.

Management of poisoning with antihistamines is supportive and symptomatic. Ipecac induced emesis is not recommended because of the potential for CNS depression and seizures. Seizure control is mandatory prior to gastric lavage which should be performed within 1 hour of ingestion. Administration of activated charcoal and a cathartic is indicated. Repeat doses of activated charcoal are not effective. Physical cooling is recommended in hyperthermia. Antipyretics (salicylates) are not indicated to control hyperthermia. Anticonvulsants are given to treat seizures and agitation. Recurrent seizures are controlled with phenytoin. Physostigmine is used to treat severe central and peripheral anticholinergic symptoms (narrow QRS complex, refractory superaventricular tachycardia with either haemodynamic instability or ischaemic chest pain) not responding to other agents, as antihistamines carry the risk of development of seizures in overdose. Torsades de pointes is managed with IV magnesium sulphate. Myocardial depression and QRS interval prolongation is treated with IV sodium bicarbonate. Hypotension is managed with IV fluids and vasopressors. Epinephrine should be avoided. Forced diuresis is not effective. The role of hemodialysis, hemoperfusion, peritoneal dialysis and exchange transfusion is questionable. Cardiac status should be monitored for 18-24 hours after ingestion of terfenadine and astemizole. There is no specific antidote for poisoning due to antihistamines.

BARBITURATES

Barbiturates are a heterogeneous group of compounds used in the treatment of insomnia, anxiety, agitation, convulsions, or withdrawal syndromes from alcohols. Depending upon the onset of action, barbiturates are divided into ultrashort, short, intermediate, and long acting. It is, also, based on their elimination half-lives. The long acting barbiturates have elimination half-lives in humans greater than or equal to 48 hours. The short acting agents are highly lipid soluble and rapidly penetrate the CNS. They induces toxic effects rapidly at lower concentrations than longer acting barbiturates.

The CNS depressant effect is believed to be due to their actions in the reticular activating system through GABA (gamma-aminobutyric acid) inhibitory neurotransmitter system. Barbiturates binding to the $GABA_A$ receptor leads to GABA dependent chloride channels which remain open for a longer duration and increase the inhibitory synaptic transmission. Barbiturates activate inhibitory $GABA_A$ receptors and inhibit excitatory AMPA receptors which explain the CNS depressant effects. They are metabolized either partially or completely by the liver microsomal enzyme system.

In sensitive individuals, 8 mg/kg phenobarbital (long acting) may produce toxicity. Addicts may tolerate up to 1,000 mg/day. Phenobarbital concentrations of 80 mcg/ml(8 mg/dL, 355 mcmol/L) are generally associated with significant CNS depression. Concentrations above 30 to 40 mcg/ml (3 to 4 mg/dL, 130 to 180 mcmol/L) are usually associated with lethargy and ataxia in non-tolerant individuals.

Common clinical features of overdose include nystagmus, dysarthria, ataxia, drowsiness, and mild somnolence to coma, loss of corneal reflexes, and respiratory depression. Other effects include hypotension, cardiovascular collapse, and hypothermia. Phenobarbital poisoning may be associated with unusual findings such as bullous skin lesions (at pressure areas and also between toes and fingers) and focal neurological signs. Hepatotoxicity secondary to idiosyncratic or hypersensitivity reactions has been reported. Crystalluria has been reported following intoxication with primidone. Renal failure secondary to hypotension or rhabdomyolysis may occur. Following high-dose intravenous barbiturate administration, thrombophlebosis has been reported. Erythematous or hemorrhagic skin blisters may occur. Discontinuation of chronic barbiturate therapy may produce withdrawal symptoms.

The laboratory monitoring parameters include plasma barbiturate levels, CBC, electrolytes, glucose, blood urea nitrogen and creatinine. Arterial blood gases in patients with respiratory depression should be obtained. ECG should be monitored in patients with signs and symptoms of cardiovascular compromise.

All patients with a history of phenobarbital ingestion should have an initial serum level done and have continuous cardiac and respiratory monitoring. Due to slow and possibly delayed absorption, asymptomatic patients should be monitored for at least 6 hours. The onset of toxic effects is usually within 2 hours but peak toxicity may not occur for 18 hours or more. Barbiturate elimination can be increased by hemodialysis or charcoal hemoperfusion, but these techniques should be reserved for patients with hemodynamic compromise refractory to aggressive supportive care.

Gastric lavage should be done preferably within the first four hours of ingestion. In some cases, the ingested drug may form a mass leading to delayed absorption and prolonged toxicity. Activated charcoal is most effective, if administered, within one hour of exposure. Hypotension is managed with isotonic intravenous fluids. Rapid infusion of isotonic intravenous fluids should be followed by vasopressors as needed. Withdrawal symptoms can be treated by reinstitution of phenobarbital and a gradual reduction over three weeks as desired. A tapering schedule of 10% every 3 days has been used.

Renal excretion of long acting barbiturates is increased with urinary alkalanization but this does not correlate well with improved clinical outcome and it should be considered only in severe barbiturate toxicity with life-threatening signs and symptoms. Sodium bicarbonate, IV should be administered at a rate sufficient to produce a urinary pH of at least 7.5. Additional sodium bicarbonate (1 to 2 meq/kg) and potassium chloride (20 to 40 meq/L) may be needed to achieve an alkaline urine. Hourly intake/output and urine pH should be obtained. Adequate hydration and renal function are to be assured prior to alkalinization. Potassium should not be given to an oliguric or anuric patient. Fluid and electrolyte balance should be carefully monitored.

Indications for hemodialysis are stage four coma, and blood levels >10mg/dl for long acting barbiturates and >5 mg/dl for short acting barbiturates. Exchange transfusion has been successful in decreasing plasma phenobarbital levels. There is no specific antidote and the mainstay of treatment remains symptomatic and supportive care.

BENZODIAZEPINES

Benzodiazepines constitute the most widely prescribed group of psychoactive drugs in therapeutic use because of efficacy, safety, fewer side effects and relatively low addiction potential. They have largely replaced other sedative-hypnotic drugs and have a widespread use in anxiety, insomnia, panic disorder syndromes, acute mania in bipolar disorder, neuroleptic-induced dystonias, catactonia and jet-lag. They have also been found useful in treatment of cancer patients for the relief of anticipatory anxiety and nausea, chemotherapy-induced emesis, neuralgias and psychiatric disorder secondary to use of high-dose steroids.

These drugs are subject to misuse because of widespread availability and are common ingestants or co-ingestants in drug overdose. The potency, specificity, duration of effect, presence and absence of effect and clinical use of individual agents varies widely.

Benzodiazepines act on central nervous system and the effects appear to be mediated through the inhibitory neurotransmitter GABA with the opening of chloride channels. GABA acts at two receptors $GABA_A$ and $GABA_B$ and former is the site of action of most benzodiazepines.

A remarkable property of benzodiazepines is their relative safety after overdose, because they are generally of low order toxicity unless ingested with other CNS or respiratory depressants. The toxic-therapeutic ratio is very high. However, they are frequently encountered in acute poisoning. Ingestion of 500-2000mg of diazepam has resulted in minor toxicity. Though, in general, deaths from overdose ingestions alone are rare, few fatalities have been clearly documented due to oral benzodiazepine overdose alone. When taken in combination with other agents like ethanol, analgesics, other sedative-hypnotics, tricyclic antidepressants, phenothiazines or barbiturates, there has been a substantial increase in potential for serious toxicity.

The common clinical signs after overdose are drowsiness, ataxia, slurred speech with midposition and reactive pupils. The patient may present in low grade coma without focal neurologic abnormalities. Respiratory depression is the primary clinical concern. Overdose may depress respiratory rate and tidal volume and airway protective reflexes. Severity depends on amount ingested and absorbed, type of benzodiazepine ingested and coingestants. Nystagmus and disconjugate gaze may be present. Bowel sounds though usually present may be slightly diminished and muscle tone tends to be decreased. Hypothermia is reported. Concomitant ingestion of other drugs may lead to profound coma, significant hypotension, respiratory depression or hypothermia.

Unusual cases of prolonged deep coma, prolonged cyclic coma with focal neurologic signs after overdose have been reported. Rhabdomyolysis may develop after prolonged coma. Severe complications are noted more likely with newer shortacting agents or when other depressant drugs have been ingested. Most obtunded patients become arousable within 12-36 hours.

Treatment mainly involves supportive care including care of airway, breathing and circulatory status. Respiratory support and mechanical ventilation should be provided if necessary. Activated charcoal instillation may be required for gastric decontamination. Enhanced elimination methods like forced diuresis has no proven value. Hemoperfusion, hemodialysis and peritoneal dialysis are relatively ineffective in removing clinically significant amount of benzodiazepines.

Flumazenil is the specific benzodiazepine antagonist, but should not replace the supportive care. It may be given in the doses of 0.2 mg over 30 sec intravenously. The dose may be repeated at an interval of 20 min. No more than 3mg is indicated in an hour except in profound coma. Though it reverses the CNS and respiratory depressant effects of benzodiazepines, it is not routinely recommended in overdose. It precipitates convulsions in epileptics, and is contraindicated in patients on cyclic antidepressants. It may also precipitate withdrawal symptoms in benzodiazepine dependent individuals. In the presence of myoclonic jerking, altered vital signs or EKG conduction abnormalities the drug is contraindicated as the risk of seizures outweighs the potential benefits.

CARBAMAZEPINE

Carbamazepine is an anticonvulsant, belonging to the class of iminostilbenes. It was first employed for trigeminal neuralgia but it is now used as an anticonvulsant for generalized tonic-clonic and simple and complex partial seizures. It is structurally and pharmacologically similar to the tricyclic antidepressants such as imipramine. It has also been used in the treatment of bipolar-affective disorders, dyscontrol syndrome, alcohol withdrawal syndrome, resistant schizophrenia, pain syndrome and restless leg syndrome. Although overdose may cause apnea, seizures and coma, fatalities are relatively uncommon.

Carbamazepine (sometime referred to as CBZ) is a lipophilic compound that is relatively insoluble in aqueous media and erratically absorbed from the gastrointestinal tract. Absorption is more rapid on a full stomach. After therapeutic doses, 72% is absorbed and 28% is excreted in the feces. Peak levels may be delayed for 6-24 hours, but they may occur as late as 72 hours after overdose. The apparent volume of distribution ranges from 0.8 to 2.0 L/kg in adults on therapeutic doses. Carbamazepine is 75 to 78% bound to plasma protein in adults. The presence of the drug in the feces may be due to incomplete absorption or biliary excretion of metabolites. At therapeutic doses carbamazepine, like phenytoin, blocks sodium channels. It has anticholinergic, antidepressant, antidiuretic, muscle relaxant, and antidysrhythmic activities in higher doses.

Overdoses cause varying levels of CNS depression with or without concomitant respiratory depression. Common clinical untoward effects include drowsiness, ataxia, nystagmus, slurred speech, dyskinesias, vomiting and tachycardia. Life threatening effects include coma, seizures, and respiratory arrest. CNS effects are a major component of the clinical picture and are common at concentrations greater than 9 μg/ml. At therapeutic concentrations, it causes nystagmus, tremors, and dysarthria. Agitation and hallucinations, may occur followed by mental status depression and coma. Prolonged coma may result in bullous skin lesions, rhabdomyolysis and renal failure. Pupillary effects are not consistent, and both mydriasis and miosis as well as a variable response to light have been reported. Anticholinergic syndrome which is common mainly consists of tachycardia, mydriasis, confusion, decreased bowel activity and dry, flushed skin. Pediatric carbamazepine intoxication generally produces a picture similar to that in adults.

Carbamazepine is structurally similar to the tricyclic antidepressants and may produce similar signs and symptoms. Another tricyclic compound, cyclobenzapine (also sometimes referred to as "CBZ") can produce similar anticholinergic and cardiac findings. Based upon its structural similarity to tricyclic antidepressants, carbamazepine might potentially cause QRS and QT interval prolongation. After an acute overdose, manifestations of intoxication may be delayed for several hours because of erratic absorption. Cyclic coma and rebound release of symptoms may be caused by continued absorption from a tablet mass as well as enteroheptic circulation of the drug.

Carbamazepine is highly protein bound, and both free and total drug concentration must be measured. Free drug therapeutic levels range from 1.6 to 2.4 μg/mL. The active metabolite of carbamazepine (CBZ-E) concentration has a reference range of 0.8 to 3.2 μg/mL. However, patients may be relatively symptom free with carbamazepine levels over 30 μg/mL.

The initial treatment of carbamazepine poisoning should be directed towards assessment, stabilization, and general supportive care. Gastric lavage may be considered in patients with significant ingestion, even hours after ingestion as due to its anticholinergic properties and poor aqueous solubility it remains in the gastrointestinal tract for an extended period of time.

Activated charcoal adsorbs carbamazepine. It should be given in a dose of 1g/kg orally. A cathartic is recommended in view of the anticholinergic effects of carbamazepine on bowel motility. Multiple doses of activated charcoal may be indicated in cases of large ingestions, especially if there is evidence of ongoing absorption. Seizures should be treated with benzodiazepines, such as intravenous diazepam. The second drug of choice is phenobarbital. There is no specific antidote for carbamazepine intoxication.

LITHIUM CARBONATE

Lithium carbonate is a fine white powder. It is used in the treatment of a variety of psychiatric disorders. Most of the lithium intoxications may develop during prolonged therapy and may not be presented with dramatic features of poisoning. Gastrointestinal absorption is rapid and complete after ingestion. Peak serum levels are reached within 1 to 2 hours. Sustained release preparations may take 24 hours for the drug to cross the blood brain barrier effectively. Onset of effects may be delayed up to sixteen hours after an acute ingestion. The ingested dose is not a reliable indicator of toxicity due to variable absorption.

Serum concentration of 0.6 to 1.2 mEq/L is the therapeutic range. In mild to moderate toxicity 1.5 to 2.5 mEq/L, in severe cases 2.5 to 3.5 mEq/L and levels >3.5 mEq/L may be fatal, especially in patients on chronic lithium therapy. In patients not chronically taking lithium, levels as high as 3 to 6 mEq/L have been noted in asymptomatic patients after an acute overdose.

The toxic mechanism of lithium is poorly understood. Like sodium, lithium also is filtered and reabsorbed in the renal proximal tubule. Lithium, a monovalent cation acts competitively against other monovalent and divalent cations (sodium, potassium, calcium, magnesium) at the level of cell membrane and neuronal synapses. It has the ability to pass through sodium channels and can block potassium channels at higher concentrations. It's ability to inhibit adenine cyclase decreases the level of intracellular cAMP. Chronic toxicity from therapeutic use may occur if renal resorption of sodium is increased by dehydration or drug interactions. Calcium channel blockers increase the risk of lithium neurotoxicity. Drugs that decrease lithium excretion include diuretics, NSAIDS, ACE-inhibitors, and metronidazole.

Acute overdose results in gastrointestinal symptoms and higher lithium levels may be tolerated without developing neurologic effects. Mild to moderate intoxication may produce nausea, vomiting, tremors, muscle twitching, hyperreflexia, agitation, lethargy, tinnitus, ataxia, CNS depression, slurred speech, muscle rigidity and hypertonia. Patients with severe toxicity may develop seizures, myoclonus, hypotension and coma.

In chronic toxicity from therapeutic use, CNS effects predominate. They are often precipitated by dehydration, renal insufficiency, hyponatremia, low sodium diet, metabolic stress, drug interactions or infection at levels that are even within the normal range.

With lithium poisoning nausea, vomiting and diarrhea are common; anorexia, abdominal discomfort and dry mouth are also reported. Usually, these GI effects are mild and reversible. Excessive tearing, burning, and nystagmus may develop. Photophobia has been reported following chronic lithium intoxication. Transient blurred vision and blindness may occur. T wave flattening or inversion, bundle branch blocks, bradycardia, junctional rhythms, and hypotension may develop following severe, usually chronic intoxication. ARDS and respiratory failure have been reported with severe intoxication. Agitation, tremor, hyperreflexia, ataxia, slurred speech, lethargy, confusion, and cogwheel rigidity are reported. Seizures and coma may develop with severe poisoning. Diabetes insipidus may develop. Therapeutic use is associated with hypothyroidism, goiter and even papilledema.

Chronic poisoning/therapeutic use may result in glomerular and tubular abnormalities. It increases BUN and creatinine, and oliguria has been reported. If renal water concentrating ability is significantly impaired, diabetes insipidus may occur. Dehydration is common in patients with chronic toxicity. Severe hypernatremia may develop in patients with diabetes insipidus. Decreased anion gap is noticed and leucocytosis is generally observed. Lithium overdose may lead to elevated neutrophil and eosinophil counts. Dermatological effects like psoriasis, dermatitis, are also observed.

Serum electrolytes and serial lithium concentrations should be determined. Urianalysis and determination of serum creatinine to assess renal function should be carried out. ECG, in symptomatic patients, should be monitored.

Management of acute lithium overdose includes GI decontamination. Gastric lavage should be done, if it can be performed soon after ingestion (generally within 1 hour). Whole bowel irrigation with polyethylene glycol (PEG) may be performed in large ingestions or sustained release preparations. Rate of administration is 2 L/hr in adults and 500ml/hr in children until the rectal effluent is clear. Induction of emesis is controversial. Lithium is not adsorbed to charcoal. Airway is protected by placement in Trendelenburg and left lateral decubitus position or by endotracheal intubation. Seizure control is mandatory. Normal saline should be administered intravenously to maintain adequate urine output and to enhance renal clearance of lithium. Thiazides and spironolactone should be avoided as they impair lithium excretion.

Hemodialysis is indicated for moderate to severe intoxications. It is most effective in acute poisoning as further distribution of lithium to CNS is prevented. Hemodialysis should be continued till the serum lithium level is less than 0.5 mEq/L. Lithium levels rebound after hemodialysis due to it's redistribution from peripheral sites. Hemodialysis should be repeated if severe symptoms persist and or lithium level is elevated. Furosemide slightly enhances lithium elimination, but is not routinely recommended. It may be used in case of volume overload from hydration therapy. Intravenous magnesium sulfate has been used to treat ventricular tachycardia where other pharmacologic measures are not effective. There is no specific antidote.

OPIOIDS

Opioids are a broad class of alkaloids having opium/morphine-like actions. They are used as analgesics, preanaesthetic medications, antitussive and antidiarrheal agents. Apart from their clinical uses their addictive potential is very high. The toxicity varies depending on the compound involved, it's potency, elimination $t_{1/2,}$ rate and route of exposure, and concomitant use of other CNS depressants. Some of the newer opioids have agonist and antagonist properties with unpredictable effects in overdose.

Ingestion of more than 5 mg/kg of codeine has caused respiratory arrest. Ingestion of greater than 1 mg/kg of codeine may produce toxic effects in children. The estimated lethal dose of codeine in adults is 7 to 14 mg/kg. Infants and children may demonstrate unusual sensitivity and habituated adults may have extreme tolerance to opioids. 2.5 mg of hydrocodone has been lethal in infants.

The opioids stimulate a number of specific opiate receptors in the central nervous system. The three main opioid receptors are classified as mu (μ), kappa (κ) and delta (δ). Activation of these receptors results in inhibition of synaptic transmission in both the central nervous system and the neural plexus in the periphery. ACTH inhibition may occur with toxic opioid doses. After absorption, opioids are rapidly cleared from blood and deposited primarily in kidney, liver, brain, lung, spleen, skeletal muscle and placental tissue. Excretion of opioids is primarily through kidneys by glomerular filtration.

The classic features of opioid overdose include central nervous system and respiratory depression and miosis. Other common features are hypotension, shock, gastric hypomotility with ileus, and non-cardiogenic pulmonary edema. Hypotension and hyperthermia or hypothermia may occur. Pupils are usually pinpoint but severe acidosis, hypoxia, or respiratory depression may lead to dilation. Cardiac effects include arrhythmias, pulmonary hypertension, and cyanosis. Respiratory depression leading to respiratory arrest, pulmonary edema, hypoxia, bronchospasm, acute asthma, bullous pulmonary damage, and pneumonitis can occur. Coma, seizures, myoclonic reactions, and spongiform encephalopathy and myelopathy have been reported in abusers of opioids. Heroin abuse may lead to stroke, liver toxicity, acute tubular necrosis secondary to rhabdomyolysis and myoglobinuria, glomerulonephritis, glomerulosclerosis, renal amyloidosis, and renal failure. Hypomotility with ileus may occur following an overdose. Rashes may develop during therapeutic use of opioid narcotics. Leukocytosis has been reported in heroin abusers. Scleroderma following heroin abuse has been reported and may be linked to talc mixed with heroin. Seizures may be precipitated with opioid overdose. Muscle rigidity and spasms have also been noticed. Hypoglycemia has been noticed following a heroin overdose. Effects on sensorium may range from euphoria to dysphoria and sedation to coma. Withdrawal symptoms include nausea, vomiting and abdominal cramps.

Hypotension, dysrhythmias, and respiratory depression should be monitored and need for endotracheal intubation must be evaluated. Hypoglycemia, electrolyte disturbances, or hypoxia, if present, should be corrected. Ventilation and oxygenation need to be maintained and evaluated with frequent arterial blood gas or pulse oximetry monitoring. Early use of PEEP and mechanical ventilation may be needed. Plasma opioid levels are not clinically useful. Chest x-ray for patients with significant exposure is required.

Naloxone is the specific antidote for opioid intoxication. Ipecac-induced emesis is not recommended because of the potential for CNS depression and seizures. Consider gastric lavage if it can be performed within 1 hour. Airway is protected by placement in Trendelenburg and left lateral decubitus position or by endotracheal intubation. Control of seizures is mandatory. Charcoal is administered as a slurry.

Hypotension is treated with IV fluids and if unresponsive to these measures, dopamine or norepinephrine should be administered and titrated as needed to desired response. In both adults and children administer naloxone at the rate of 0.4 to 2.0 mg intravenously and repeat as needed to reverse signs and symptoms of toxicity. If no response is observed after 10 mg of naloxone have been administered, the diagnosis of opioid-induced toxicity should be questioned. Continuous IV infusion may be beneficial following overdosage with long-acting narcotics. Two-third of the initial naloxone bolus is administered on an hourly basis. Add ten times this dose to each liter of diluent and infuse at a rate of 100 mL/hr. Infusion rate and concentration should be adjusted to achieve the desired antagonistic effect to avoid fluid overload. Mixtures should be used within 24 hours. Infusion rate should be increased and/or additional IV bolus doses of naloxone administered as needed to assure adequate ventilation. Respiratory rate should be carefully monitored during infusion. Management of withdrawal should initially be achieved by stopping the naloxone infusion until withdrawal abates. Naloxone should then be reinstated at a rate capable of maintaining ventilation without withdrawal. Later withdrawal may be accomplished with the administration of clonidine, by the substitution of methadone, or with the reintroduction of the original addicting agent. Opioid induced pulmonary edema is treated with naloxone, oxygen and appropriate ventilatory support.

PARACETAMOL

Paracetamol, also known as acetaminophen (N-acetyl-p-aminophenol), is a popular and widely available over-the-counter analgesic/antipyretic. It is a nonsteroidal anti-inflammatory drug with weak anti-inflammatory property at therapeutic doses which makes it unsuitable for use in inflammatory conditions. It is currently available either alone or in combination with other pharmaceuticals in more than 100 preparations. Since its inception, morbidity and mortality from overdose have continued to climb steadily. Paracetamol currently is one of the most frequent causes of poisoning due to a pharmaceutic agent worldwide.

Paracetamol is metabolized almost exclusively in the liver. More than 90 percent is directly converted to nontoxic glucuronide and sulfate conjugates and less than 5 percent is excreted unchanged in the urine. The rest (approximately 5 percent) is oxidized by various cytochrome P-450 enzymes, including P4502E1, P4501A2 and P4503A4. Metabolism via these enzymes produces highly reactive electrophile, N-acetyl-p-benzoquinoneimine (NAPQI). In acetaminophen overdose the glucuronidation and sulfation pathways become saturated and metabolism by P-450 increases. This increases the production of highly reactive metabolite, N-acetyl-p-benzoquinoneimine. It reacts directly with hepatic macromolecules, causing liver injury. Renal damage may occur by the same mechanism owing to renal metabolism. Overdose during pregnancy has been associated with fetal death and spontaneous abortion.

The minimal single acute toxic dose is 7.5 g for an adult and 150 mg/kg for a child. Liver toxicity is likely to occur with oral ingestion of 140 mg/kg in adults. A single acute dose of 10-15 g of the drug is potentially fatal. The risk of toxicity increases in chronic alcoholics and in patients chronically taking isoniazid, rifampicin or both presumably because of induction of liver microsomal enzymes and impairment of glutathione synthesis and consequently increased formation of toxic metabolites of paracetamol. Children younger than 5 years appear to be more resistant to the toxic effects of paracetamol. Very few deaths have been reported, and the incidence of hepatotoxicity is much lower than that observed in general population. Alcoholism and Gilbert's disease are also risk factors for paracetamol poisoning.

Early recognition of acute paracetamol overdose is essential, as the prognosis is best when antidotal treatment is initiated within 8 hours of overdose. Early signs of toxicity may include malaise, nausea and vomiting, with few findings on physical examination. Many patients with toxic paracetamol levels and a significant potential for hepatotoxicity are initially asymptomatic following acute ingestion (phase I). Signs of liver injury, such as abdominal pain, persistent vomiting, icterus and right upper quadrant tenderness, only become apparent 24 to 48 hours after an acute ingestion (phase II).

Serum transaminases begin to rise as early as 16 hours following a significant ingestion and are always elevated at the time of initial manifestations of hepatotoxcity.

During phase III, which occurs 3 to 4 days after ingestion, the full extent of hepatic injury and the prognosis can be estimated. Hepatotoxicity can range from mild signs and symptoms accompanied by elevations in serum hepatic transaminases (AST >1000 IU/L) to fulminant hepatic failure with abdominal pain, right upper quadrant tenderness, jaundice, hypoglycemia, dramatic serum hepatic transaminase elevation (>10000 IU/L), acidosis, coagulopathy and encephalopathy.

Fatalities occur during this stage from complications such as adult respiratory distress syndrome, cerebral edema, uncontrollable hemorrhage, infection or multiorgan failure. Patients without preexisting liver disease who survive this stage generally recover completely (phase IV).

In a patient who presents within 8 hours of acute paracetamol ingestion, measurement of a postingestion paracetamol level at 4 hours or later is the only laboratory test needed. Because of over-the-counter availability of paracetamol and lack of early symptoms after overdose, serum paracetamol concentrations should be measured in all suicidal overdose patients. Baseline AST level, ALT level, bilirubin level, prothrombin time, creatinine level, pregnancy test for women of child bearing age should be done. A salicylate level should also be done in the patients who present with overdose of paracetamol.

The treatment of acute overdose includes gastrointestinal decontamination followed by activated charcoal. Although rapid drug absorption limits its effectiveness more than 2 hours after ingestion, paracetamol is well adsorbed by activated charcoal. Oxygenation and ventilatory support should be taken care of before decontamination.

With acute overdose, patients who have serum paracetamol levels above the nomogram line should receive the first dose of NAC (N-Acetyl cysteine) an effective antidote preferrably within 8 hours of the ingestion.

The oral preparation of NAC (Mucomyst) is available as 20% formulation. It has a foul sulfurous odour and should be diluted to a 5% solution with either fruit juice or a carbonated beverage. An oral loading dose of 140mg/kg should be administered, followed every 4 hours by 17 maintenance doses of 70 mg/kg. If a patient vomits within 1 hour of receiving NAC, that dose must be repeated. Antiemetic therapy is critical, when the oral preparation of NAC is used because emesis of the antidote prevents effective therapy. In some cases, retention of oral NAC requires intravenous administration of antiemetics, such as intravenous metoclopramide. Intravenous administration of the oral solution is an option when other attempts to administer the antidote have failed.

TRICYCLIC ANTIDEPRESSANTS

The tricyclic antidepressants (TCA's) which form an essential class of drugs are used to treat a wide range of disorders like depression, panic disorder, social phobia, bulimia, narcolepsy, attention deficit disorder, obsessive compulsive disorder, childhood enuresis and chronic pain syndromes. Toxicity due to TCA's has increased over the past years around the world because of their widespread availability to patients of depression who have a high risk for suicide. Further, the toxicity is compounded due to rapid absorption, tight binding to plasma proteins and tissues, enterohepatic recycling and low therapeutic margin. They are one of leading causes of death worldwide due to prescription drug overdose.

The tricyclic antidepressants are classified as secondary or tertiary amines and have usually three ringed structures with the exception of amoxapine and maprotiline which have additional rings. They have a narrow therapeutic index and doses less than 10 times the therapeutic daily dose may produce severe intoxication. Variations are observed in the adult toxic dose. Most antidepressants result in moderate to serious toxicity after an ingestion 10-20 mg/kg. Fatality in children may be caused by as little as 3.5 mg/kg.

The toxicity affects, primarily, the cardiovascular and central nervous systems and is due to blockade of cardiac sodium and potassium channels, blockade of α-adrenergic, cholinergic muscarinic and histaminic receptors and inhibition of neuronal catecholamine reuptake.

The most distinctive feature of serious overdose is QRS interval prolongation which also serves as a marker for risk of adverse cardiac or central nervous system events like ventricular tachycardia or seizures. Though QT interval is mildly prolonged with therapeutic doses, it is more prolonged with overdose. Nonspecific changes in ST-T wave morphology are common. Sinus tachycardia is most common and may be aggravated by concurrent hypoxia, hypotension, hyperthermia or the use of β-adrenergic agonists. Ventricular tachycardia is commonly seen in patients with marked QRS prolongation or hypotension and may be precipitated by seizures. Torsades de pointes though reported is rare. Vasodilation or impaired cardiac contractility lead to hypotension. Hyperthermia and tachydysrhythmias may aggravate hypotension. Very fast or slow heart rates, intravenous volume depletion, hypoxia, hyperthermia, acidosis, seizures or coingestion of other cardiodepressant or vasodilating drugs may also contribute to hypotension. Anticholinergic syndrome caused due to cholinergic muscarinic receptor blockade may include sinus tachycardia, delirium, coma, mydriasis, impaired gut motility, urinary retention, impaired sweating and dry mucosa. However, patients need not have all of these findings and the absence of anticholinergic effects does not rule out an overdose. Excessive heat production due to agitation, myoclonus or seizures along with reduced heat dissipation from impaired sweating may contribute to hyperthermia. Pulmonary edema is common probably due to aspiration and hypotension rather than to a specific effect of tricyclic antidepressants. Generalized and brief seizures develop within first few hours of ingestion. In patients with coexisting cardiovascular toxicity, seizures may lead to rapid and marked worsening of hypotension or dysrhythmias. Myoclonus is less common than seizures. Death is usually due to hypotension or ventricular dysrhythmias.

All patients with known or suspected overdose should be closely observed with intravenous access cardiac, monitoring and frequent measurement of temperature. Gastric lavage is given within one hour of presentation followed by activated charcoal. Induction of emesis is contraindicated. The mainstay of treatment for overdose consists of supportive treatment with attention to the adequacy of airway, breathing and circulatory status. Respiratory support and mechanical ventilation should be provided as necessary.

The cardiotoxic manifestations are treated with hypertonic sodium bicarbonate (1M or 1m Eq/ml) in the doses of 1-2 mEq/kg intravenously and repeated as needed to maintain the arterial pH between 7.45-7.55. Ventricular tachycardia has been experimentally managed with hypertonic sodium bicarbonate, hyperventilation and lidocaine which should be used in case sodium bicarbonate is ineffective. It's administration should be slow to avoid precipitating seizures. Hyperventilation to raise blood pH has been used in patients with anecdotal benefit, but may be useful in cases of pulmonary edema where sodium bicarbonate is relatively contraindicated. Though phenytoin has been used to reduce QRS duration, but efficacy data is lacking. It may increase the incidence of ventricular tachycardia. Hypotension is managed with crystalloids. Hypertonic sodium bicarbonate and sympathomimetic agents may be given for hypotension unresponsive to fluids.

Diazepam may be effective for treating seizures. Brief neuromuscular blockade may be considered for refractory seizures avoiding prolonged paralysis. Use of physostigmine is contraindicated as it may cause bradyarrhythmias or seizures. Hyperthermia may be managed with rapid cooling through evaporation or ice water gastric lavage. Temporary neuromuscular blockade may be useful if seizures or agitation persist during hyperthermia. Use of flumazenil is contraindicated because it may precipitate seizures.

Enhanced elimination procedures like dialysis and hemoperfusion are not effective. Efficacy of repeat dose activated charcoal is unproven, though it has been reported to accelerate elimination.

Generally the plasma levels of TCA's are not required because the QRS interval and the toxidrome of overdose are reliable and more readily available indicators of toxicity.

All asymptomatic patients who have been given activated charcoal should be monitored for temperature, vital signs and ECG for 6 hours and in case of signs of toxicity, the patients should be admitted to an intensive care setting for at least 24 hours.

Use of class IA or IC anti dysrhythmic drugs like quinidine, procainamide, flecanide, encaimide is contraindicated as they may aggravate toxicity. Seizures should be treated promptly to prevent hypoxia and lactic acidosis which may predispose to the development of life-threatening ventricular dysrhythmias.

Further Reading

Ellerhorn M.J. Ellerhorn's Medical Toxicology: Diagnosis and Treatment of Human Poisoning. 2nd ed., Williams & Wilkins, Baltimore, 1996.

Haddad L.M. Clinical Management of Poisoning and Drug Overdose. 3rd ed., W.B. Saunders Company Philadelphia, 1998.

Olson K.R. Poisoning and Drug Overdose. 2nd ed., Appleton & Lange, Norwalk, 1994.

Dart, RC. The 5 Minute Toxicology Consult. Lippincott, Williams & Wilkins, Philadelphia, 2000.

Ford M.D., Delaney K.A., Ling L.J. and Erickson T. Clinical Toxicology. 1st ed., W.B. Saunders Company, Philadelphia, 2001.

Goldfrank L.R., Flomenbaum N.E., Lewis N.A. Weisman R.S., Howland M.A. and Hoffman RS. Goldfrank's Toxicologic Emergencies, 5th ed. Appleton & Lange, Connecticut 1994.

Pharmacological Perspectives of Toxic Chemicals and Their Antidotes
Editors: S J S Flora, J A Romano, S I Baskin and K Sekhar

α-ketoglutarate: A Promising Antidote to Cyanide Poisoning

R. Bhattacharya
Division of Pharmacology and Toxicology, Defence Research and Development Establishment, Jhansi Road, Gwalior 474 002, India

INTRODUCTION

Cyanide is considered as a potent suicidal, homicidal, genocidal and chemical warfare agent (Way, 1983; 1984; Ballantyne, 1987; Borowitz et al., 1992; Marrs et al., 1996; Ryan, 1998; Bhattacharya, 2000; Borowitz et al., 2001). Occupational exposure to cyanide in industry (Blanc et al., 1985; Peden et al., 1986; Ryan, 1998) and ingestion of cyanide-containing foods have caused serious toxic problems (Osuntokun, 1980; Rosling, 1989; Suchard et al., 1998). Administration of certain drugs like sodium nitroprusside (Vesey and Cole, 1985) and laetrile (Kalyanaraman et al., 1983) have also been shown to generate toxic levels of cyanide in the body. Elevated levels of blood cyanide have been recorded in victims and survivors of fire smoke inhalation (Clark et al., 1981; Barillo et al., 1994) and concomitant inhalation of hydrogen cyanide (HCN) and carbon monoxide is largely responsible for the toxicity of fire smoke (Baud et al., 1991).

The toxic effect of cyanide has been attributed to its production of a histotoxic anoxia by the inhibition of cytochrome oxidase, the terminal oxidase of the mitochondrial respiratory chain (Isom and Way, 1974; Isom et al., 1975; Solomonson, 1981; Way 1984; Ballantyne, 1987). Cyanide produces a rapid onset of toxicity which warrants immediate treatment to prevent morbidity and mortality. The currently approved treatment for cyanide poisoning involves administration of amyl nitrite and sodium nitrite (SN) to induce methemoglobin, which reversibly binds with cyanide (Chen and Rose, 1952; Friedberg, 1968), and administration of sodium thiosulfate (STS) which converts cyanide to thiocyanate, mediated by mitochondrial enzyme rhodanese (van Heijst and Meredith, 1990; Baskin et al., 1992; Marrs et al., 1996). However, in many instances of cyanide poisoning, therapeutic problems have been associated with the use of nitrites (Graham et al., 1977; van Heijst et al., 1987; van Heijst and Meredith, 1990) and thiosulfate (Baumeister et al., 1975). Nitrites are slow methemoglobin formers (Kruszyna et al., 1982) and cause hypotension which can aggravate the condition of victims of smoke inhalation because the cellular respiration is critically impaired due to carboxylation of the hemoglobin by carbon monoxide (Moore et al., 1987; Hall et al., 1989; Kulig, 1991). Therefore, methemoglobin formers cannot be used against smoke inhalation toxicity involving combination of cyanide and carbon monoxide (Norris et al., 1986; Cohen and Guzzardi, 1988; Hall et al., 1989; Kulig, 1991). In addition, the inhibitory effects of carbon monoxide and cyanide on cellular metabolism in the brain may be additive or even synergistic (Pitt et al.,1979). Because of this, other cyanide antidotes not involving methemoglobin formation are being sought (Niknahad et al., 1994).

Cyanide, being a reactive nucleophile, is known to interact with carbonyl moieties like aldehydes and ketones to form cyanohydrin intermediates (Morrison and Boyd, 1976). Previous studies have shown that sodium pyruvate and a-ketoglutarate (α-KG) conferred significant protection against acute cyanide poisoning in rodents (Schwartz et al., 1979; Moore et al., 1986). Subsequent studies from our laboratory showed that on the basis of protective index and plasma cyanide levels, parenteral administration of α-KG alone or in combination with SN and/or STS afforded significant protection in mice exposed to cyanide through parenteral or inhalation routes (Bhattacharya and Vijayaraghavan, 1991). Although, it is over fifteen years since α-KG was first shown to antagonize cyanide poisoning, unfortunately for reasons unknown it could not be developed as an antidote.

An oral form of a-KG is sold as an over-the-counter nutritional supplement or sports medicine in USA (Klaire Laboratories, San Marcos, CA). Additionally, various salts of a-KG such as calcium, sodium and ornithine have been evaluated as nutritional support (Cynober, 1999). Considering the efficacy and safety of a-KG, there was a resurgence of interest to develop it as an effective treatment against cyanide poisoning. Therefore, an elaborate study on the pharmacology and toxicology of α-KG was undertaken. Our study principally focused on the application of a-KG through oral route. Its utility is particularly envisaged for fire fighters, chronic occupational exposures, accidental or deliberate oral ingestion of cyanide or as an oral pretreatment for personnel engaged in evacuation/decontamination, apprehending cyanide exposure. In circumstances where nitrites may not be suitable, α-KG with STS is anticipated to offer significant protection while in cases where nitrites can be administered, an augmented protection with α-KG and STS is anticipated. This chapter addresses the present state of cyanide antidotes with particular reference to the work carried out on α-KG in our laboratory and the research efforts made elsewhere.

TOXICOLOGY OF CYANIDE

Lethal Dose of Cyanide

On a molar basis, cyanide appears innocuous when compared to some more commonly ingested essential vitamins in human diet (Solomonson, 1981). However, at toxic levels it is known to produce death within a few minutes. It is not easy to determine what are the lethal doses of cyanide to man but taken orally the fatal dose of HCN to adults is estimated at 50-100 mg, and for potassium cyanide (KCN), about 150-250 mg (Ballantyne, 1974). However, victims ingesting as much as 3 g of KCN have been saved with immediate therapy (van Heijst et al., 1987). A blood cyanide concentration of 2.5-3.0 μg/ml is usually found to be lethal. The severity of the poisoning however, depends on the form of cyanide and the dose, and route of exposure (Ballantyne, 1974).

Mechanism of Toxicity

Cyanide toxicity may not be attributed solely to a single biochemical lesion but a complex phenomenon. Mechanism of acute and chronic cyanide toxicity has been largely discussed elsewhere (Way, 1984; Ballantyne, 1987; Maduh, 1989; Borowitz et al., 1992; 2001). Briefly, the cascade of events occurring after cyanide intoxication is primarily attributed to histotoxic hypoxia consequent to inhibition of cytochrome oxidase, the terminal enzyme in mitochondrial electron transport chain (Isom and Way, 1975; Solomonson, 1981). This enzyme contains two heme A moieties and two copper ions. Cyanide has a special affinity for the heme iron and the reaction of cyanide with the multimeric iron enzyme complex is facilitated by first penetration of cyanide to protein crevices, with initial binding of cyanide to the protein followed by binding of cyanide to heme iron (Way, 1984). Thereby, a cyanide-heme cytochrome oxidase complex is formed which renders the enzyme incapable of utilizing the oxygen. This reaction is reversible but forms a relatively stable complex (Solomonson, 1981; Way, 1984). Other enzymes critical to cyanide poisoning are nitrate reductase, myoglobin, ribulose diphosphate carboxylase, catalase, superoxide dismutase, and glutathione peroxidase, etc. (Way, 1984; Ardelt et al., 1989). Cyanide is also potent stimulator of neurotransmitter release both in the CNS and peripheral nervous system (Kanthasamy et al., 1991; 1994). Another important mechanism of cyanide toxicity may be attributed to its affinity to Schiff base intermediates, e.g. ribulose diphosphate carboxylase and 2-keto-4-hydroxy glutarate aldolase involving formation of cyanohydrin intermediate (Ballantyne, 1987; Borowitz et al., 1992). Formation of cyanohydrin is important in biological systems because of the existence of several biologically active carbonyl compounds and perhaps for this reason many nontoxic carbonyl compounds have been exploited as cyanide antidotes (Cittadini et al., 1971; 1972; Schwartz et al., 1979; Moore et al., 1986). Also, increased rate of disappearance of free blood cyanide in the presence of carbonyl compounds *in vitro* has been demonstrated using gas chromatography (Aldous et al., 1984). The present chapter discusses the role of similar cyanohydrin former namely α-KG, as a promising antidote to cyanide poisoning.

Clinical Presentations

The clinical picture of acute cyanide poisoning varies in both time and intensity depending upon the magnitude of exposure. Various non-specific signs and symptoms like headache, dizziness, nausea, vomiting, confusion, coma and incontenence of feces and urine occur (Ballantyne, 1974; 1987; Borowitz et al., 1992). Physiologically a series of events like dyspnea, incoordination of movement, cardiac irregularities, convulsive seizures, coma and respiratory failure may occur leading to death (Ballantyne, 1974; Way, 1983; 1984; Baskin et al., 1992). Classically, cyanide has been said to produce an odor of bitter almonds. This finding, although helpful if present, cannot be relied on. Studies have shown that 20% to 40% of the population are not capable recognizing this odor. An important clue in cyanide poisoning is the funduscopic appearance of the arteries and veins (Ryan, 1998). With cyanide poisoning, the veins have a red color in contrast to the usual blue tint, and are difficult to distinguish from the arteries. This change is due to the decreased tissue extraction of oxygen, resulting in high venous oxygen saturation. However, reliability of this finding has not been evaluated in clinical studies (van Heijst et al., 1987; Ryan, 1998). Pathologically no particular lesions can delineate cyanide toxicity, albeit animal experiments indicate that the lesions are principally in the central nervous system, predominantly necrosis in the white matter (Ballantyne, 1974; Way, 1984). Subacute or chronic cyanide poisoning is characterized by prolonged energy deficit, loss of ionic homeostasis and oxidative stress leading to CNS pathology (Kanthasamy et al., 1994). Probably the most wide-spread pathologic condition attributed to chronic cyanide poisoning is tropic ataxic neuropathy following cassava consumption (Osuntokun, 1980).

PRESENT CYANIDE ANTIDOTES

The rapid intracellular action of cyanide is due to its good lipid and water solubility, causing severe hypoxia in vital organs resulting in death. Therefore, it warrants immediate and vigorous treatment to prevent the toxicity. There are numerous cyanide antidotes and no unanimity of opinion on which is the most effective regimen (Way, 1984; van Heijst et al., 1987; Borowitz et al., 1992). A number of reviews have discussed the efficacy of various mechanistic based antidotes against experimental cyanide poisoning but their clinical safety and correct indication for use have not been determined so far (Way, 1983; 1984; Isom and Borowitz., 1995; Bhattacharya, 2000). This section will briefly focus only on the cyanide antidotes which are clinically used in various countries.

NITRITE-THIOSULFATE

In the United States, the only Food and Drug Administration- approved cyanide antidote kit which is popular in many other countries as well including India, contains three antidotes: amyl nitrite, SN and STS. For many years, nitrites were thought to cause removal of cyanide from tissues by the preferential binding of cyanide to the methemoglobin induced by these agents. However, other protective mechanisms of nitrites have been proposed in the recent past. Nitrites also are likely to generate nitric oxide, which is an effective cyanide antidote independent of methemoglobin formation (Sun et al., 1995). Inhalation of amyl nitrite as a first aid measure to cyanide poisoning is known for many years (Way, 1983; 1984; Way et al., 1984; van Heijst et al., 1987; van Heijst and Meredith, 1990; Marrs et al., 1996). However, its efficacy as methemoglobin inducer remained disputed due to its inability to generate methemoglobin greater than 6% (Jandorf and Bodansky, 1946; Bastian et al., 1959), while about 15% is required to challenge one LD_{50} of cyanide (van Heijst et al., 1987). Protective effect of amyl nitrite is now attributed to its vasodilatory properties (van Heijst et al., 1987; van Heijst and Meredith, 1990). Artificial ventilation with amyl nitrite broken into ambu bags is a life saving therapy for comatose patients. The drug of choice for cyanide poisoning is SN (Chen and Rose, 1952; Way, 1983; 1984; Way et al., 1984; van Heijst et al., 1987; van Heijst and Meredith, 1990; Marrs et al., 1996). When given intravenously (i.v.) it takes about 12 min to generate approximately 40% of methemoglobin (van Heijst et al., 1987). In spite of this delay in inducing a significant level of methemoglobinemia, immediate protection is offered by SN (Kruszyna et al., 1982). The cyanide antagonism of nitrites has also been shown to persist despite administration of methylene blue to prevent formation of methemoglobin (Way et al., 1984). Perhaps for these reasons SN is thought to protect by its vasodilatory action or some other mechanism (van Heijst and Meredith, 1990). A serious drawback with SN is that i.v. administration may be accompanied by serious cardiovascular embarrassment, particularly in children, for whom an adjusted dose is recommended (Berlin, 1970). SN is also contraindicated in many instances of cyanide poisoning (Graham et al., 1977; Way, 1983; Moore et al., 1987; van Heijst et al., 1987; Hall et al., 1989; van Heijst and Meredith, 1990; Kulig, 1991). Another important antidote for cyanide poisoning is STS which provides sulfur for enzymatic conversion of cyanide to thiocyanate, which is non toxic and can be excreted by the kidneys (Isom and Johnson, 1987). The exact site of action of STS and involvement of rhodanese in this reaction remains controversial, as rhodanese is a mitochondrial enzyme and STS has difficulty getting into the mitochondria. Some researchers have suggested that the transformation to thiocyanate may involve albumin and a sulfur donor, with minimal involvement of rhodanese (Ryan, 1998). Like nitrites, therapeutic problems are also associated with STS (Baumeister et al., 1975) and it is contraindicated in patients with renal insufficiency as the thiocyanate formed may cause toxicity (van Heijst and Meredith, 1990).

Dicobalt Edetate (Kelocyanor)

The usual preparation of dicobalt edetate is Kelocyanor (SERB, Paris). The 20-ml ampoules contain a solution of 0.196-0.240 g/100 ml free cobalt and 1.35-1.65 g/100 ml dicobalt edetate, as well as 4 g of glucose per ampoule (Marrs et al., 1996). Kelocyanor has been widely adopted in Europe and a number of case reports testify its efficacy in the treatment of cyanide poisoning in man. It is the current treatment of choice in the United Kingdom provided that cyanide toxicity is definitely present (van Heijst and Meredith, 1990). Serious side effects like vomiting, urticaria, anaphylactoid shock, hypotension, ventricular arrythmias and gastrointestinal problems have been reported in patients receiving kelocyanor (van Heijst and Meredith, 1990; Salkowski and Penney, 1994).

Hydroxocobalamin (Vitamin B 12a)

This agent is perhaps the most promising cyanide antidote used in human toxicology (Hall and Rumack, 1987; van Heijst et al., 1987). With the exchange of hydroxy group of hydroxocobalamin for cyanide, non toxic cyanocobalamin (Vitamin B12) is formed. However, use of this antidote remained limited because of the large dose required to challenge cyanide poisoning. In France, a formulation is available which contains 4 g of hydroxocobalamin powder that has to be dissolved in 80 ml of 10% STS solution prior to use (van Heijst and Meredith, 1990). Some authors have recorded reduced antidotal effect as a result of mixing these two antidotes (Friedberg and Shukla, 1975; van Heijst and Meredith, 1990). In France and Germany, desirable antidotal effects were also obtained with an injectable preparation of 5 g hydroxocobalamin in water (van Heijst and Meredith, 1990). Recorded side effects of hydroxocobalamin includes anaphylactoid reactions and acne (Linnell, 1987; van Heijst and Meredith, 1990; Salkowski and Penney, 1994).

4-Dimethylaminophenol

The relatively slow rate of methemoglobin formation by SN prompted the development of rapid methemoglobin formers like aminophenols. An i.v. dose of 3.25 mg/kg of 4-dimethylaminophenol (DMAP) was reported to produce methemoglobin level of 30% within 10 min and 15% methemoglobinemia was attained within one minute without any immediate effect on cardiovascular system (Kiese and Weger, 1969; van Dijk et al., 1987; van Heijst and Meredith, 1990). DMAP has been successfully instituted in many case of severe cyanide poisoning. However, there are differences in individual susceptibility to DMAP which may result in an undesirable levels of methemoglobinemia even after normal therapeutic doses (van Dijk et al., 1987). Intramuscular injection of DMAP results in local abscess, fever and other toxicological implications like nephrotoxicity (Weger, 1983). Because of many unfavourable clinical experiences this drug has not been widely adopted outside Germany (van Dijk et al., 1987; van Heijst and Meredith, 1990; Marrs et al., 1996).

Oxygen

Oxygen appears to be a physiological antagonist and at 1 ATM is found to potentiate the antidotal efficacy of nitrite-thiosulfate treatment (Salkowski and Penney, 1994). Although, several explanations have been offered on the action of oxygen in cyanide detoxification, its exact mechanism is still not clearly elucidated (Litovitz, 1987).

CARBONYL COMPOUNDS AS CYANIDE ANTIDOTE

Cyanide being a reactive nucleophile is known to interact with various a-keto acids such as pyruvate and α-KG, which are carbohydrate/amino acid intermediary metabolites to yield cyanohydrin derivatives (Way, 1983; 1984; Way et al., 1984; Marrs, 1987). Perhaps, for the first time Nosek et al. (1957) demonstrated the antidotal property of pyruvate against experimental cyanide poisoning. More elaborate studies were conducted by Cittadini et al (1972) and they showed that sodium pyruvate rapidly reversed the cyanide-inhibited respiration of Ehrlich ascites tumor cells. Subsequently, it was found to offer significant, but small degree of protection to mice poisoned with cyanide when administered at 250 mg/kg, i.v. 30 sec later. Even greater protection was afforded by 500 mg/kg. Schwartz et al. (1979), also using mice, found that sodium pyruvate, 1 g/kg given by the intraperitoneal (i.p.) route 10 min before subcutaneous (s.c.) potassium cyanide (KCN), gave minimal protection but, it did however, potentiate the antidotal effects of STS, and of SN and STS in combination. The reasons for investigating sodium pyruvate are that it has many apparent theoretical advantages over other cyanide antagonists like SN. Cyanide reacts directly with pyruvate to form cyanohydrin derivatives. Also, pyruvate is more likely to be distributed to sites of cyanide localization, as there is specific carrier for active transport of pyruvate (Schwartz et al., 1979; Way, 1983). The value of sodium pyruvate as a potential cyanide antagonist is that it provides a different approach to the development of cyanide antidotes. As a possible supplement to the nitrite-thiosulfate combination, it provides a reason for decreasing the dose of SN, as this antagonist has caused fatalities in susceptible individuals (Way, 1984). In addition, no convulsions from cyanide were observed when sodium pyruvate was present (Schwartz et al., 1979). Sodium pyruvate or its analogs also have potential advantage over cobalt compounds considering its minimal toxicity and therefore, greater dose can be tolerated (Way, 1983). However, due to limited efficacy, sodium pyruvate does not appear to have been used clinically on humans. (Marrs, 1987), but it prompted research in similar lines with newer cyanohydrin formers. Similarly, another a-ketocarboxylic acid viz. α-KG although not used in humans as cyanide antidote has shown very encouraging results in animals (Moore et al., 1986; Norris et al., 1990). Niknahad et al (1994) compared various α-keto acid and carbohydrate cellular nutrients as antidotes to cyanide-induced cytotoxicity in isolated rat hepatocytes. They discovered that dihydroxyacetone and glyceraldehyde were more effective than pyruvate and α-KG as cyanide antidote. Their cytoprotective mechanism involved trapping cyanide and restoring respiration and ATP levels. Further, *in vivo* studies showed that dihydroxyacetone in combination with STS provided better protection than combination of STS with pyruvate or α-KG (Niknahad and O'Brien, 1996). Authors proposed that dihydroxyacetone could therefore prove an effective antidote to cyanide poisoning particularly in the cases of smoke inhalation lethality attributed to cyanide in the presence of carbon monoxide. However, the most promising approach to medical management of cyanide poisoning may be the use of the combination of STS and α-KG (Litovitz et al., 1983). This treatment protected mice against 15 LD_{50} doses of cyanide (Moore et al., 1986). These two substances are very non-toxic and may provide the best available treatment for cyanide intoxication (Borowitz et al., 1992; 2001).

α-KETOGLUTARATE AS CYANIDE ANTIDOTE

For the first time structure activity relationship of keto acids in antagonizing cyanide-induced lethality was demonstrated by Norris and Hume (1986). Soon after, detailed *in vivo* studies on the efficacy of α-KG in the antagonism of cyanide intoxication was shown in mice by Moore et al. (1986). On the basis of potency ratio, a pre-treatment of α-KG was found to increase the LD_{50} value of cyanide by a factor of five, a value statistically equivalent to that ascertained in mice pre-treated with SN and STS. The combination of α-KG and STS increased the LD_{50} value of cyanide by fifteen fold and addition of SN further increased it to ninteen fold. Unlike SN, no induction of methemoglobin was observed with a-KG pre-treatment and its protective efficacy was attributed to its ability to react with or bind cyanide. On the basis of animal mortality, α-KG in conjunction with STS was considered a better treatment regimen than the combination of SN and STS. Yamamoto in 1989 showed that administration of cyanide produced hyperammonemia, increased brain neutral and aromatic amino acid levels, and encephalopathy in mice. α-KG completely blocked the development of loss of consciousness and hyperammonemia and also significantly inhibited the increase of the neutral and aromatic amino acid levels in brain. Authors concluded that hyperammonemia and the increase of neutral and aromatic amino acids may play an important role in the development of loss of consciousness by cyanide. Most of the research work on the efficacy of α-KG against experimental cyanide poisoning were evidenced in the 1990s. Protection against cyanide-induced convulsions was observed after treatment with α-KG, either alone or in combination with STS (Yamamoto, 1990). However, STS alone did not protect against cyanide induced convulsions. Cyanide treated mice showed convulsions accompanied by depleted levels of brain α-aminobutyric acid (GABA). Combination of α-KG and STS completely abolished the decrease in GABA levels induced by cyanide. Furthermore, STS alone also completely abolished the decrease in GABA levels. The results suggested that depletion of brain GABA levels may not be directly related to cyanide-induced convulsions. On the other hand, cyanide significantly increased the calcium levels in brain crude mitochondrial fractions in mice with convulsions. The elevated levels of calcium were completely abolished by the combined administration of α-KG and STS, but not affected by STS alone. The above studies indicated that hyperammonemia, elevated levels of brain neutral and aromatic amino acid levels, and calcium were associated with cyanide-induced convulsions. Effect of α-KG on such neurochemical mechanism was due to cyanide binding with the ketone group adjacent to the carboxylic group (Norris and Hume, 1986). Further, Norris et al., in 1990 elucidated the mechanism of antagonizing cyanide-induced lethality by α-KG. The diagram below illustrates the mechanism of this binding.

$$HO-\overset{\overset{\displaystyle O}{\|}}{C}-CH_2-CH_2-\overset{\overset{\displaystyle O}{\|}}{C}-\overset{\overset{\displaystyle O}{\|}}{C}-OH + CN^- \rightleftharpoons HO-\overset{\overset{\displaystyle O}{\|}}{C}-CH_2-CH_2-\overset{\overset{\displaystyle ^-O}{|}}{\underset{\underset{\displaystyle CN}{|}}{C}}-\overset{\overset{\displaystyle O}{\|}}{C}-OH$$

Several investigative approaches were undertaken to determine the existence of this binding. Authors injected various molar ratios of α-KG : cyanide into high pressure liquid chromatogram and demonstrated that addition of cyanide reduced the peak area of α-KG at a molar ratio greater than 1:5. Second, blood from naive mice was spiked with α-KG and cyanide, and α-KG was found to reduce the peak area of hydrogen cyanide released into the head space at molar ratios of >1:2.5. Effect of cyanide on the ultraviolet spectrum of α-KG was also determined as an indication of binding at 316 nm. Further, cyanide-induced inhibition of brain cytochrome oxidase activity was found to be prevented by pre-treatment of α-KG and this protection was ascribed to binding of α-KG with cyanide. This was perhaps the first substantial evidence of interaction of α-KG and cyanide *in vivo* in favour of its mechanism of antagonizing cyanide-induced lethality. The reaction was thought to be similar to that produced by another α-ketocarboxylic acid, pyruvic acid, bonded to cyanide as the mechanism for reducing cyanide-induced lethality (Green and Williamson, 1937).

Most of the protection studies were carried out in rodent models, until Dalvi et al., in 1990 reported that antidotal efficacy of α-KG was also observed in dogs intoxicated with cyanide. Although, Moore et al. (1986) reported significant protection by α-KG against cyanide poisoning, an *in situ* binding of these agents could not be ruled out as both α-KG and cyanide were given through the i.p. route. To address this we evaluated the efficacy of α-KG given i.p. against cyanide administered through s.c., i.p or inhalation route (Bhattacharya and Vijayaraghavan, 1991). For the first time we showed that α-KG given parenterally afforded significant protection against HCN administered through inhalation route. Combination of α-KG and STS were equipotent to conventional treatment of SN and STS combination, and plasma cyanide levels were also significantly decreased by α-KG in mice receiving lethal doses of cyanide. The salient finding was that the efficacy of α-KG remained undeterred irrespective of the route of cyanide intoxication, while the magnitude of protection varied. Efficacy of orally administered α-KG, alone or in combination with N-acetylcysteine, in reducing the lethal effects of cyanide was examined in mice by Dulaney Jr., et al., (1991). α-KG significantly reduced the lethality of cyanide in a dose related manner. The protective efficacy of α-KG was observed if given between 10 and 30 min prior to cyanide exposure. Based on lethality and a behavioral scoring system, the protective efficacy of α-KG was found to be enhanced by concomitant administration of N-acetylcysteine.

The effects of various glycolytic substrates and keto acid metabolites including α-KG on the cytotoxic effects of cyanide were studied with isolated rat hepatocytes (Niknahad et al, 1994). Cyanide (2 mM) immediately inhibited cellular respiration followed by ATP depletion. Disruption of plasma membrane occurred when 85-90% ATP levels were depleted and post-treatment of α-KG restored the changes. It was further reported that combination of high oxygen concentration and the presence of either pyruvate or α-KG was necessary to effectively protect cytochrome oxidase against cyanide poisoning *in vitro* (Delhumeau et al., 1994). The results suggested that oxygen displaced cyanide from the enzyme and the poison was then trapped by the keto acids to form the respective cyanohydrins. The antidotal effectiveness of α-KG may be based, in part, upon its ability to bind cyanide nucleophiles in the vascular system, thus decreasing cyanide distribution to the vital tissues. Such decreased distribution would result in a diminution of the consequential histotoxic hypoxia caused by cyanide intoxication. The distribution of cyanide to the brain stem and heart was significantly reduced by pre-treatment of α-KG which was also found to be more effective than the therapeutic doses of cobalt edetate and sodium pyruvate (Hume et al., 1996).

Although, clinical trials on α-KG as cyanide antidote has not yet been conducted in humans, based on its promising results as evidenced above, it is presently envisaged as a potential antidote for cyanide poisoning, particularly in combination with STS (Borowitz et al., 1992; 2001). In view of this, we have initiated extensive work on the pharmacology and toxicology of α-KG to develop it as a promising antidote against cyanide poisoning. Its enormous protective efficacy, safety, availability as a nutritional supplement and its scope to fit into a treatment regimen where other cyanide antidotes are likely to fail, were the pre-disposing factors in our research efforts on α-KG. We focused on the efficacy of α-KG through an oral route. Our initial studies in rodents revealed that regardless of the route of cyanide poisoning, efficacy of α-KG pre-treatment through oral route was appreciable and the efficacy was both dose and time dependent (our unpublished work). However, the efficacy of oral α-KG against cyanide administered parenterally was less as compared to previous studies where both α-KG and cyanide were given parenterally (Moore et al., 1986; Bhattacharya and Vijayaraghavan, 1991). Although, transport of α-KG into the cells have been well defined (Meier et al., 1990; Aussel et al, 1996), its absorption through oral route and then dissipation to interact with cyanide administered parenterally is expected to consume more time. Now, our interest was to see the efficacy of α-KG given orally against cyanide poisoning through the same route. Accidental or deliberate exposure to cyanide through oral route may produce rapid onset of symptoms because many human overdoses involve doses far greater than the minimal lethal dose. Animal studies have shown that absorption of cyanide decreases with more alkaline stomach and that normally most cyanide is absorbed within 2 to 3 hr of ingestion, suggesting that the time course of an oral ingestion should be fairly rapid and should peak within several hours (Ryan, 1998). When cyanides are given orally, the gastric acid environment favors formation of the unionized form of HCN which facilitates absorption (Ballantyne, 1987). The disodium salt of α-KG has an alkaline pH and besides binding with cyanide it is also likely to minimize its absorption when both are administered orally. Absorption of α-KG given orally is equally fast and may not be a deterrent factor for challenging cyanide exposure through other routes, particularly through inhalation. Oral administration of α-KG is anticipated to afford maximum protection when cyanide poisoning occurs through the same route. Also, its efficacy is not likely to be compromised when cyanide poisoning occurs by other routes (Dulaney Jr et al., 1991).

Isolated thymocytes were earlier used to characterize cyanide-induced cytotoxicity and DNA damage, and also their pharmacological interventions *in vitro* (Bhattacharya and Lakshmana Rao, 1997; 2001). Using this cell type, we recently demonstrated the efficacy of α-KG pre-treatment (30 min), simultaneous treatment (0 min) or post-treatment (5 min and 30 min) against KCN induced cytotoxicity and DNA damage *in vitro* (Bhattacharya et al, 2002). Pre-treatment or simultaneous treatment of α-KG minimised the cytotoxicity but could not prevent the DNA damage and mitochondrial dysfunction. Also the intracellular GSH (reduced glutathione) which is implicated in DNA damage remained depleted (Slater et al., 1995). Our previous studies have shown that the dose of cyanide required to produce DNA damage was far less as compared to that needed to cause cytotoxicity, and DNA damage was preceded by mitochondrial dysfunction (Bhattacharya and Lakshmana Rao, 1997; 2001). Perhaps for this reason any unbound cyanide in the presence of a-KG was sufficient to cause DNA damage and mitochondrial dysfunction, although the cytotoxicity was prevented. The gel electrophoresis showed that cyanide caused internucleosomal cleavage of DNA, and this could not be prevented by any of the treatments of α-KG, except 30 min pre-treatment where the laddering of DNA, characteristic of apoptotic type of cell death was minimal (Bhattacharya et al., 2002). The study also showed that pre-treatment of α-KG improved cell viability but surface blebbing and irregular margin of cell membrane was still evident when visualized on cell smears. In this study, protective efficacy of graded doses of α-KG (p.o.) as pre-treatment or simultaneous treatment was also evaluated against acute p.o cyanide poisoning against male mice. The most interesting observation was that a 60 min oral prophylaxis with α-KG, particularly at 0.5 to 2.0 g/kg conferred significant protection against acute cyanide poisoning by the same route. However, considering the short half-life of α-KG, previous reports suggested a prophylactic window of 10 to 15 min only (Moore et al., 1986, Dulaney et al., 1991). This difference in protection could be because of different routes of administration for α-KG and cyanide, adopted by different workers. Besides, in addition to complexing cyanide, other protective mechanisms of α-KG could also be considered (Yamamoto, 1990; Niknahad et al., 1994). An 18-fold and 22-fold protection by a-KG and STS as simultaneous treatment or 10 min pre-treatment respectively and further increase in protection to 26-fold by adjunction of SN with 10 min pre-treatment of α-KG are by any standard enormous protection against cyanide poisoning. Similar to our studies in mice, pre-treatment with α-KG (0.125-2.0 g/kg) in female rats exhibited dose and time dependent effects and was found to be effective even when given 60 min prior to cyanide intoxication (Bhattacharya and Vijayaraghavan, 2002). Addition of STS significantly enhanced the protective efficacy of α-KG at all the doses and time intervals. A 10 min pre-treatment of α-KG increased the LD_{50} of KCN by 7-fold, which was further increased 28-fold by addition of both SN and STS. Simultaneous treatment with α-KG (2.0 g/kg) increased the LD_{50} of KCN by 7-fold, which was doubled by addition of STS. However, addition of SN did not confer any additional protection. This could be due to compounded cardiovascular effects of SN in the presence of α-KG as observed in the case of mice (Bhattacharya et al., 2002). Cyanide is a rapid poison and delayed therapeutic interventions cannot be successful against lethal doses of experimental cyanide poisoning. Our recent unpublished work showed that even after the onset of toxic signs and symptoms following 2.0 LD_{50} KCN, post-treatment of α-KG and STS reduced the mortality of rats by 50% as compared to unprotected animals but did not significantly extend the survival time. Similar experiments were also carried out in mice to evaluate the efficacy of post-treatment of α-KG. Treatment of α-KG marginally improved the mean survival time of mice receiving 2.0 LD_{50} of KCN and also reduced the mortality by 30%. Addition of STS further improved the protection. The difference in the protective efficacy of pre-treatment, simultaneous-treatment or post-treatment of α-KG was also validated by the fact that post-treatment of α-KG could not significantly protect the cyanide-inhibited cytochrome oxidase as compared to the pre-treatment or simultaneous treatments (Bhattacharya and Vijayaraghavan, 2002). In our separate unpublished work, therapeutic efficacy of α-KG was also evaluated in rats administered 0.75 LD_{50} KCN. Here, various biochemical parameters viz. cytochrome oxidase, GSH, GSSG (oxidized glutathione), GPx (glutathione peroxidase) and SOD (superoxide dismutase) were measured in brain 30 or 60 min post exposure. KCN significantly reduced the activity of cytochrome oxidase and GSH after 30 min, and GPx after 60 min. Post-treatment of α-KG alone or with STS significantly attenuated these effects. This shows that post-treatment of α-KG could challenge the effects of 0.75LD_{50} KCN on various biochemical parameters but it could not protect the cytochrome oxidase inhibited by 2.0 LD_{50} KCN. Cyanide is known to cause oxidative stress and treatment with antioxidant are reported to be beneficial (Ardelt et al., 1989;Yamamoto and Tang, 1996; Bhattacharya et al., 1999). Although, α-KG is not known to have antioxidant properties, α-keto and aldehytic metabolites of carbohydrates and amino acids are reported to protect cells from cyanide by more than one mechanism (Niknahad et al., 1994). In another study, we have recently evaluate the effects of pre-treatment or post-treatment of α-KG on various physiological parameters, following administration of 0.75 LD_{50} KCN in rats (unpublished work). Pre-treatment of α-KG significantly protected the decrease in MAP (mean arterial pressure), HR (heart rate), RR (respiratory rate) and NMT (neuromuscular transmission) following KCN intoxication. However, post-treatment of α-KG did not produce any significant protective effects. The above experiments reveal that pre-treatment of α-KG is certainly superior to simultaneous treatment or post-treatment.

In order to establish the safety of α-KG, various hematological, biochemical, physiological and histological parameters were studied following p.o. administration of 2.0 or 4.0 g/kg α-KG in rats (Bhattacharya et al., 2001).

The p.o. LD_{50} (14 days) of α-KG in both male and female rats was found to be >5.0 g/kg and considering the effective dose of α-KG (0.5-2.0 g/kg), a safety margin of 3 to 10-fold was evident. Animals receiving 2.0 g/kg α-KG did not show any significant change in organ-body weight index of vital organs, hematology, biochemistry, physiology and histology. Increase in plasma alkaline phosphatase (ALP) and urea levels after 1 hr and decrease in inorganic phosphorus levels after 7 days of treatment were only transient in nature and were not of any clinical significance. Although, MAP and NMT were decreased at 4.0 g/kg α-KG, it is unlikely to cause any clinical implications as the dose was 2 to 8-fold higher than the proposed doses. In brief, the 2.0 g/kg dose of α-KG (p.o.) which offers maximum antidotal efficacy against cyanide poisoning, was found to be non-toxic (Bhattacharya et al., 2001).

α-KETOGLUTARATE AS NUTRITIONAL SUPPLEMENT

It is now recognized that besides their fundamental role in protein synthesis and energy supply, some amino acids are also important metabolic regulators capable of counteracting functional and metabolic disorders induced by trauma. Supplementation of artificial nutrition with such amino acids are considered as pharmacologic nutrition or nutritional supplements. Two of the best known representatives are glutamine and arginine (Cynober, 1991). α-KG is the natural, ubiquitous collector of amino group ($-NH_2$) in body tissue; upon adding an amino group it becomes glutamic acid. In the CNS there is an additional route for nitrogen transport; glutamic acid adds an additional ammonia as an amide group to form glutamine. Glutamine may then transport the nitrogen out of CNS (Bionostics, Inc., Illinois, 1990). Literature suggests that glutamine synthesis accounts only for marginal part of the disposal of exogenously supplied α-KG and administered α-KG has a potent sparing effect on endogenous glutamine pool (Cynober, 1999). Ornithine-α-ketoglutarate (OKG), a salt formed by two molecules of ornithine and one molecule of α-KG has been successfully used by the enteral and parenteral routes in burn, traumatized, and surgical patients and chronically malnourished subjects. Presently, the role of OKG is being widely studied in clinical nutrition and metabolic care (Cynober, 1991; Le Boucher and Cynober, 1998; Cynober, 1999,). OKG is considered as a potent nutritional modulator characterized by an anticatabolic activity, anabolic activity, or both. In addition, OKG is also found to be an efficient immunomodulator and a key promoter of wound healing (Le Boucher and Cynober, 1998). The mechanism of action of OKG is not fully understood, but the secretion of anabolic hormones (insulin, human growth hormone) and the synthesis of metabolites (glutamine, polyamines, arginine, α-ketoacids) may be involved (Cynober, 1991). Besides, salt of ornithine, calcium and sodium salts of α-KG have also been evaluated in nutritional support (Cynober, 1999). Jon Pangborn of Bionostic Inc., Illinois (USA) first introduced α-KG to the nutritional supplement industry in 1981. He had extensively worked on the citric acid cycle and on nitrogen detoxification. α-KG functions as a collector or scavenger of amino groups ($-NH_2$) in the body tissue preventing hyperammonemia which may occur in many pathological conditions. Excessive ammonia interferes with phosphorylation and oxidation in the citric acid cycle (Krebs cycle). Another important function of α-KG occurs in the formation of carnitine which participates in the metabolism of fat. Because of its chemical structure, α-KG is considered as a potent natural detoxifying agent. In addition to its ability in detoxification of hyperammonemia, hyperaminoacidurias, it is suggested to counteract exposures to toxic nitrogen chemicals such as: cyanide, ammonia, ammonium compounds, amines, hydrazines, etc. (Bionostics, Inc., Illinois, 1990). An oral form of α-KG is sold as an over-the counter nutritional supplement by Klaire Laboratories, San Marcos, CA since early 1980s. It is used in sports medicine to aid athletes and body builders on high-protein diets, and it is used medically by doctors to lower blood and tissue ammonia levels in individuals suffering with hyperammonemia. However, there have been no controlled human studies for detoxification of amines or cyanide. As nutritional supplement the therapeutic dose range of α-KG is approximately 500 to about 2500 mg per day but for detoxification of amines and cyanide higher doses may be required.

Conclusion

On the basis of experimental evidence it is concluded that in circumstances where SN is contraindicated, treatment of α-KG and STS is likely to be a better alternative and instances where treatment of both SN and STS is indicated, an augmented protection could be anticipated by the adjunction of α-KG. In animal studies, the effective dose of α-KG was found to vary between 0.5 to 2.0 g/kg, depending upon the severity of cyanide poisoning. If this dose has to be extrapolated to human, it is certainly high as compared other cyanide antidotes. However, this could be offset by the magnitude of protection it is likely to confer. Further, at this dose level no toxicity was observed. This, however needs introspection and full expression of antidotal potency of α-KG will be best realized after its clinical safety trials.

Acknowledgement

The author is grateful to Dr. R.V. Swamy, Chief Controller, Materials and Life Sciences, Defence Research and Development Organisation, New Delhi, Mr. K. Sekhar, Director and Dr. R. Vijayaraghavan, Head, Pharmacology and Toxicology Division, Defence Research and Development Establishment, Gwalior, for their keen interest and critical suggestions in preparation of this manuscript.

References

Aldous C., Norris J., Balter A., Wilson R., Ho I. and Hume A. Formation of cyanohydrins and antagonism of cyanide toxicity by some carbonyl compounds. Pharmacologist, **26:** 230, 1984.

Ardelt B.K., Borowitz J.L. and Isom G.E. Brain lipid peroxidation and antioxidant defense mechanisms following acute cyanide intoxication. Toxicology. **56:** 147-54,1989.

Aussel C., Coudray-Lucas C., Lasnier E., Cynober L. and Ekindjian O.G., α-ketoglutarate uptake in human fibroblasts. Cell Biol. Intern., **20:** 359-363, 1996.

Ballantyne B. The forensic diagnosis of acute cyanide poisoning. In Forensic Toxicology, Ballantyne B, ed., Wright Pub., Bristol, England. 99-113, 1974.

Ballantyne B. Toxicology of cyanides. In *Clinical and Experimental Toxicology of Cyanides*, Ballantyne B, Marrs TC, eds., Wright Pub. Bristol, England. 41-126, 1987.

Barillo D.J., Goode R. and Esh V. Cyanide poisoning in victims of fire: Analysis of 364 cases and review of literature. J. Burn Care Rehabil. **15:** 46-57, 1994.

Baskin S.I., Horowitz A.M. and Nealley E.W. The antidotal action of sodium nitrite and sodium thiosulfate against cyanide poisoning. J. Clin. Pharmacol. **32:** 368-75, 1992.

Bastian G. and Mercker R.H. Zur Frage der Zweckmässigkeit der inhalation von Amylnitrit in der Behandlung der Cyanidvergiftung. Naunyn. Schmiedeberg's Arch. Exp. Pathol. Pharmacol. **237:** 285-295, 1959.

Baud F.J., Barriot P., Toffis V., Riou B., Vicaut E., Lecarpentier Y., Bourdon R., Astier, A. and Bismuth C. Elevated blood cyanide concentrations in victims of smoke inhalation. New. Eng. J. Med. **325:** 1761-1766, 1991.

Baumeister R.G.H., Schievelbein H. and Zickgraf-Rudel G. Toxicological and clinical aspects of cyanide metabolism. Drug. Res. **25:** 1056-1063, 1975.

Berlin C.M. The treatment of cyanide poisoning in children. Pediatrics, **46:** 193-196, 1970.

Bhattacharya R. Antidotes to cyanide poisoning: Present status. Indian J. Pharmacol. **32:** 94-101, 2000.

Bhattacharya R. and Lakshmana Rao P.V. Cyanide induced DNA fragmentation in mammalian cell cultures. Toxicology. **123:** 207-215,1997.

Bhattacharya R. and Lakshmana Rao P.V. Pharmacological interventions of cyanide-induced cytotoxicity and DNA damage in isolated rat thymocytes and their protective efficacy *in vivo*. Toxicol. Lett. **119:** 59-70, 2001.

Bhattacharya R., Lakshmana Rao P.V. and Vijayaraghavan R. *In vitro* and *in vivo* attenuation of experimental cyanide poisoning by a-ketoglutarate. Toxicol. Letts. **128:** 185-195, 2002.

Bhattacharya R., Lakshmana Rao P.V., Parida M.M. and Jana A.M. Antidotal efficacy of antioxidants against cyanide poisoning *in vitro*. Def. Sci. J. **49:** 55-63, 1999.

Bhattacharya R., Kumar D., Sugendran K., Pant S.C., Tulsawani R.K. and Vijayaraghavan R. Acute toxicity studies of α-ketoglutarate: A promising antidote for cyanide poisoning. J. Appl. Toxicol. **21:** 495-499, 2001.

Bhattacharya R., and Vijayaraghavan R. Cyanide intoxication in mice through different routes and its prophylaxis by a-ketoglutarate. Biomed. Environ. Sci. **4:** 452-60, 1991.

Bhattacharya R. and Vijayaraghavan R., Promising role of a-ketoglutarate in protecting the lethal effects of cyanide. Hum. Exp. Toxicol. **21:** 297-303, 2002.

Bionostics Inc., Lisle, Illinois, (USA), A monograph on alpha-ketoglutaric acid (eds. Pangborn, J.B., Hicks, J.T., Chambers, J.R. and Shahani, K.M. 1-3, 1990.

Blanc P., Hogan M., Malin K., Hryhorezuk D., Hessel S. and Bernard B. Cyanide intoxication among silver reclaiming workers. J. Am. Med. Assoc., **253:** 367-371, 1985.

Borowitz J.L., Isom G.E. and Baskin S.I. Acute and chronic cyanide toxicity. In *Chemical warfare agents: Toxicity at low levels*, Somani S.M. and Romano J.A., (Jr.), eds., CRC Press LLC, U.S.A. 301-319, 2001.

Borowitz J.L., Kanthasamy A.G. and Isom G.E. Toxicodynamics of cyanide. In *Chemical Warfare Agents*, Somani, S.M., ed. Academic Press, San Diego, CA, U.S.A. 209-36, 1992.

Chen K.K. and Rose C.L. Nitrite and thiosulphate therapy in cyanide poisoning. *JAMA*.**149:** 113-119, 1952.

Cittadini A., Galeoti T. and Terranova T. The effect of pyruvate on cyanide-inhibited respiration in intact ascites tumor cells. Experientia. **27:** 633-635, 1971.

Cittadini A., Caprino L. and Terranova T. Effect of pyruvate on the acute cyanide poisoning in mice. Experientia. **28:** 943-944, 1972.

Clark C.J., Campbell D. and Reid W.H. Blood carboxyhaemoglobin and cyanide levels in fire survivors. Lancet, **1:** 1332-1335, 1981.

Cohen M.A. and Guzzardi L.J. The smoke inhalation and cyanide poisoning. Am. J. Emerg. Med, **6:** 203-204, 1988.

Cynober L. Ornithine α-ketoglutarate in nutritional support. Nutrition, **7:** 313-322, 1991.

Cynober L. The use of α-ketoglutarate salts in clinical nutrition and metabolic care. Curr. Opin. Clin. Nutr. Metab. Care, **2:** 33-37, 1999.

Dalvi R.R., Sawant S.G. and Terse P.S. Efficacy of α-ketoglutaric acid as an effective antidote in cyanide poisoning in dogs. Vet. Res. Commun. **14:** 411-414, 1990.

Delhumeau G., Cruz-Mendoza A.M. and Lojero C.G. Protection of cytochrome C oxidase against cyanide inhibition by pyruvate and α-ketoglutarate. Effect of aeration *in vitro*. Toxicol. Appl. Pharmacol., **126:** 345-351,1994.

Dulaney (Jr) M.D., Brumley M., Willis J.T. and Hume A.S. Protection against cyanide toxicity by oral alpha-ketoglutaric acid. Vet. Hum. Toxicol. **33:** 571-575, 1991.

Friedberg K.D. Antidotes bei blausaurevergiftungen. Arch. Toxicol. **24:** 41-48, 1968.

Friedberg K.D. and Shukla U.R. The efficiency of cyanocobalamine as an antidote in cyanide poisoning when given alone or combined with sodium thiosulfate. Arch. Toxicol. **33:** 103-113, 1975.

Graham D.L., Laman D., Theodore J. and Robin E.D. Acute cyanide poisoning complicated by lactic acidosis and pulmonary edema. Arch. Intern. Med. **137:** 1051-1055, 1977.

Green E.E. and Williamson S. Pyruvic and oxaloacetic cyanohydrins. Biochem. J. **31:** 617, 1937.

Hall A.H. and Rumack B.H. Hydroxycobalamin/ sodium thiosulfate as cyanide antidote. J. Emer. Med. **5:** 115-121, 1987.

Hall A.H., Kulig K.W. and Rumack B.H. Suspected cyanide poisoning in smoke inhalation: complications of sodium nitrite therapy. J. Toxicol. Clin. Exp. **9:** 3-9,1989.

Hume A.S., Moore S.J. and Hume A.T. Effects of α-ketoglutaric acid on the distribution of cyanide and acidosis associated with cyanide intoxication. Toxicologist. **3:** 98, 1996.

Isom G.E. and Borowitz J.L. Modification of cyanide toxicodynamics: Mechanistic based antidote development. Toxicol. Lett. **82/83:** 795-799,1995.

Isom G.E. and Johnson J.D. Sulphur donors in cyanide intoxication. In *Clinical and Experimental Toxicology of Cyanides* , Ballantyne B, Marrs TC, eds., Wright Pub., Bristol, England, 1987, 413-426.

Isom G.E., Liu D.H.W. and Way J.L. Effect of sublethal dose of cyanide on glucose catabolism. *Biochem. Pharmacol.* **24:** 871-875, 1975.

Isom G.E. and Way J.L. Effect of oxygen on cyanide intoxication VI. Reactivation of cyanide inhibited glucose catabolism. J. Pharmacol. Exp. Ther. **189:** 235-243, 1974.

Jandorf B.J. and Bodansky O. Therapeutic and prophylactic effect of methemoglobinemia in inhalation poisoning by hydrogen cyanide and cyanogen chloride. J. Ind. Hyg. Toxicol. **28:** 124-132, 1946.

Kalyanaraman U.P., Kalyanaraman K. and Cullinan S.A. Neuropathy of cyanide intoxication due to laetrile (amygdalin). Cancer. **51:** 2126-2133, 1983.

Kanthasamy A.G., Borowitz J.L. and Isom G.E. Cyanide induced increases in plasma catecholamines: relationship to acute toxicity. Neurotoxicol. **12:** 777-784, 1991.

Kanthasamy A.G., Borowitz J.L., Pavlakovic G. and Isom G.E. Dopaminergic neurotoxicity of cyanide: Neurochemical, histological and behavioral characterization. Toxicol. Appl. Pharmacol. **126:** 156-163, 1994.

Kiese M. and Weger N. Formation of ferrihaemoglobin with aminophenols in the human for the treatment of cyanide poisoning. Europ. J. Pharmacol. **7:** 97-105, 1969.

Kruszyna R., Kruszyna H. and Smith R.P. Comparison of hydroxylamine, 4-dimethylaminophenol and nitrite protection against cyanide poisoning in mice. Arch. Toxicol. **49:** 191-202, 1982.

Kulig K.W. Cyanide antidotes and fire toxicology. N. Eng. J. Med. 325: 1801-1802, 1991.

Le Boucher J. and Cynober L.A. Ornithine α-ketoglutarate: The puzzle. Nutrition. **14:** 870-873, 1998.

Litovitz T.L., Larkin R.F. and Myers R.A.M. Cyanide poisoning treated with hyperbaric oxygen. Am. J. Emerg. Med. **1:** 94-101, 1983.

Litovitz T. The use of oxygen in the treatment of acute cyanide poisoning. In *Clinical and Experimental Toxicology of Cyanides*, Ballantyne, B., and Marrs, T.C., eds., Wright Pub., Bristol, England, 1987, 467-472.

Linnell J.L. The role of cobalamins in cyanide detoxification. In *Clinical and Experimental Toxicology of Cyanides*, Ballantyne, B., and Marrs, T.C., eds., Wright Pub., Bristol, England, 1987, 427-439.

Maduh EU. Mechanism of cyanide neurotoxicity. Ph.D Dissertation, submitted to Purdue University, Indiana, U.S.A., UMI Dissertation Services 1989, 1-199.

Marrs T.C. The choice of cyanide antidotes. In *Clinical and Experimental Toxicology of Cyanides*, Ballantyne B. and Marrs, T.C., eds., Wright Pub., Bristol, England, 1987, 383-401.

Marrs T.C., Maynard R.L. and Sidell F.R. Cyanides. In *Chemical Warfare Agents. Toxicology and Treatments*, Marrs, T.C., Maynard, R.L., and Sidell, F.R., eds., John Wiley, England, 1996, 203-219.

Meier P.J., Zimmerli B. and O'Neil L., Rat hepatocytes exhibit basolateral Na^+/ dicarboxylate cotransport and dicarboxylate organic anion exchange. Biol Hoppe-Seyler. **371:** 301, 1990.

Moore S.J., Norris J.C., Ho I.K. and Hume A.S. The efficacy of α-ketoglutaric acid in the antagonism of cyanide intoxication. Toxicol. Appl. Pharmacol. **82:** 40-44, 1986.

Moore S.J., Norris J.C., Walsh D.A. and Hume A.S. Antidotal use of methemoglobin forming cyanide antagonists in concurrent carbon monoxide/cyanide intoxication. J. Pharmacol. Exp. Ther. **242:** 70-73, 1987.

Morrison R.T. and Boyd R.N. Organic Chemistry, Allyn and Bacon, Inc., MA 1976, 637-639.

Niknahad H., Khan S., Sood C. and O' Brien P. Prevention of cyanide-induced cytotoxicity by nutrients in isolated rat hepatocytes. Toxicol. Appl. Pharmacol. **128:** 271-79,1994.

Niknahad H. and O'Brien P.J. Antidotal effect of dihydroxyacetone against cyanide toxicity in vivo. Toxicol. Appl. Pharmacol. **138:** 186-191, 1996.

Norris J.C. and Hume A.S. Structure activity relationship of keto acids in antagonizing cyanide-induced lethality. Fed. Proc. **45:** 916 (Abstract), 1986.

Norris J.C., Moore S.J. and Hume A.S. Synergistic lethality induced by combination of carbon monoxide and cyanide. Toxicology. **40:** 121-129,1986.

Norris J.C., Utley W.A. and Hume A.S. Mechanism of antagonising cyanide induced lethality by α-ketoglutaric acid. Toxicology. **64:** 275-283,1990.

Nosek J., Chmelar V. and Ledvina M. Antidotes for cyanide poisoning: The influence of ascorbic acid, c-ferronate, glucose, dihydroxyacetone and pyruvic acid on the course of experimental poisoning. Ceskoslov. Physio. **6:** 87-94, 1957.

Osuntokun B.O. A degenerative neuropathy wth blindness and chronic cyanide intoxication of dietary origin. The evidence in Nigerians. In *Toxicology in the Tropics*, Smith R.L. and Bababunmi E.A. eds. Taylor & Francis, London, 16-79, 1980.

Peden N.R., Taha A., McSorley P.D., Bryden G.T., Murdoch I.B. and Anderson J.M. Industrial exposure to hydrogen cyanide: Implications for treatment. Br. Med. J. **293:** 538, 1986.

Pitt B.R., Radford E.P., Gurtner G.H. and Traystman R.J. Interaction of carbon monoxide and cyanide on cerebral circulation and metabolism. Arch. Environ. Health. **34:** 354-359, 1979.

Rosling H. Cassava associated neurotoxicity in Africa. In *Proceedings of the 5th International Congress of Toxicology*, Volans G.N., Sims J., Sullivan F.M. and Turner P. eds. Taylor and Francis, Brighton, England, 605-614, 1989.

Ryan J.G. Cyanide. In *Emergency Toxicology*, 2nd edn., Viccellio, P., ed., Lippincott-Raven Pub., Philadelphia, 969-978, 1998.

Salkowski A.A. and Penney D.G. Cyanide poisoning in animals and humans: A review. *Vet. Human. Toxicol.* **36:** 455-466, 1994.

Schwartz C., Morgan R.L., Way L.M. and Way J.L. Antagonism of cyanide intoxication with sodium pyruvate. Toxicol. Appl. Pharmacol. **50:** 437-441,1979.

Slater A.F.G., Stefan C., Nobel, I., van den Dobbelsteen D.J. and Orrenius S. Signalling mechanisms and oxidative stress in apoptosis. Toxicol. Lett. **82/83:** 149-153, 1995.

Solomonson L.P. Cyanide as a metabolic inhibitor. In *Cyanide in Biology*, Vennesland B. Conn E.E., Knowles C.J., Westley J. and Wissing F. eds. Academic Press, San Diego, California, U.S.A. 1981, 11-28.

Suchard J.R., Wallace K.L. and Gerkin R.D. Acute cyanide toxicity caused by apricort kernel ingestion, Ann. Emerg. Med, **32:** 742-744, 1998.

Sun P., Borowitz J.L., Kanthasamy A.G., Kane M.D., Gunasekar P.G. and Isom G.E. Antagonism of cyanide toxicity by isosorbide dinitrate: Possible role of nitric oxide. Toxicology. **104:** 105-111, 1995.

van Dijk A., van Heijst A.N.P. and Douze J.M.C. Clinical evaluation of the cyanide antagonist 4-DMAP in a lethal cyanide poisoning case. Vet. Hum. Toxicol. **2:** 38-39, 1987.

van Heijst A.N.P., Douze J.M.C., van Kesteren R.G., van Bergen J. and van Dijk A. Therapeutic problems in cyanide poisoning. Clin. Toxicol. **25:** 383-398, 1987.

van Heijst A.N.P. and Meredith J.J, Antidotes for cyanide poisoning. In *Basic Science in Toxicology*, Volanis G.N., Sims J., Sullivan F. and Turner P. eds. Taylor & Francis, Brighton, England, 1990, 558-566.

Vesey C.J. and Cole P.V. Blood cyanide and thiocyanate concentrations produced by long-term therapy with sodiun nitroprusside. Br. J. Anaes. **57:** 148-155, 1985.

Way J.L. Cyanide antagonism. Fundam. Appl. Toxicol. **3:** 383-386, 1983.

Way J.L. Cyanide intoxication and its mechanism of antagonism. Ann. Rev. Pharmacol. Toxicol. **24:** 451-481, 1984.

Way J.L., Sylvester D., Morgan R.L., Isom G.E., Burrows G.E., Tamulinas C.B. and Way J.L. Recent perspectives on the toxicodynamic basis of cyanide antagonism. Fundam. Appl. Toxicol. **4:** 231-239, 1984.

Weger N.P. Treatment of cyanide poisoning with 4-dimethylaminophenol (DMAP): Experimental and clinical overview. Fundam. Appl. Toxicol. **3:** 387-396,1983.

Yamamoto H-A., Hyperammonemia, increased brain neutral and aromatic amino acid levels, and encephalopathy induced by cyanide in mice. Toxicol. Appl. Pharmacol. **99:** 415-420, 1989.

Yamamoto H.A. Protection against cyanide induced convulsions with α-ketoglutarate. Toxicology, **61:** 221-228, 1990.

Yamamoto H-A. and Tang H.W. Preventive effect of melatonin against cyanide-induced seizures and lipid peroxidation in mice. Neurosci. Letts. **207:** 89-92, 1996.

Pharmacological Perspectives of Toxic Chemicals and Their Antidotes
Editors: S J S Flora, J A Romano, S I Baskin and K Sekhar

Oxidative Stress in Alzheimer's Disease and Cerebral Ischemia: Implications for Antioxidant Treatment

Y.K Gupta, Monisha Sharma, Geeta Chaudhary
Neuropharmacology Laboratory, Department of Pharmacology
All India Institute of Medical Sciences, New Delhi 110 029, India

OXIDATIVE STRESS

The term "oxidative stress" is used to refer to a situation in which the generation of free radicals significantly exceeds the available antioxidant defense and repair capacities (Halliwell, 2001). Oxidative stress can result from (1) the generation of free radicals at an abnormally high rate (2) insufficient antioxidant defense (3) liberation of transition metal ions (4) a combination of any of the above (Simonian and Coyle, 1996). Oxidative stress can raise the levels of damaged biomolecules (DNA, lipids and proteins) in cells and such increases are collectively referred to as oxidative damage.

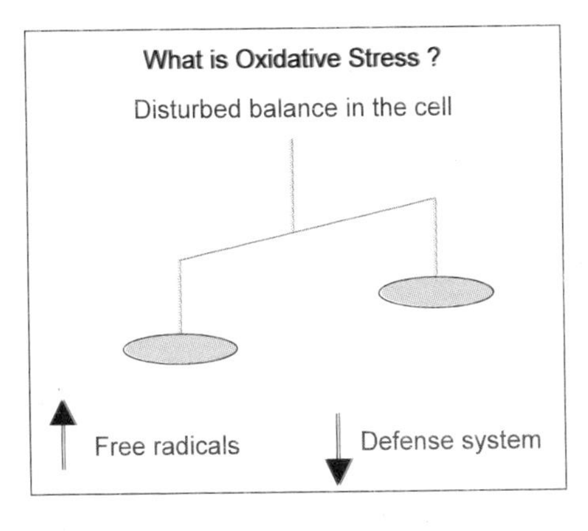

FREE RADICALS

Free radicals are atoms or groups of atoms with an odd (unpaired) number of electrons and can be formed when oxygen interacts with certain molecules. Once formed these highly reactive radicals can start a chain reaction, like dominoes. They are normal products of cellular aerobic metabolism. Superoxide and hydroxyl species are the predominant cellular free radicals. Hydrogen peroxide and peroxynitrite are although not themselves free radicals contribute importantly to the cellular redox state. Together, these molecules are referred to as reactive oxygen species.

ANTIOXIDANTS

An array of cellular defense system exits to counterbalance free radicals. These include enzymatic and non-enzymatic antioxidants that lower the concentration of free radical species and repair oxidative cellular damage. There are three forms of SOD; an extracellular Cu/Zn SOD and a mitochondrial Mn SOD. All three convert oxygen to hydrogen peroxide. Most H_2O_2 in the brain is removed by glutathione peroxidase and catalase, which is found in very low levels in the brain.

The non enzymatic antioxidants include vitamin E, which is the major lipid soluble chain breaking antioxidant, transferrin , ceruloplasmin and ascorbic acid.

HOW OXIDATIVE STRESS CAUSES CELL DEATH?

The oxidative stress has been associated with both necrosis and apoptosis. Morphological and biochemical characteristics distinguish these two from cell death. In necrosis, the selective loss of membrane permeability occurs which results in swelling of organelles, loss of membrane depolarization and rupture of the plasma membrane. In apoptosis, a death promoting signal activates a program of cell death through either new protein and RNA synthesis or depression of the existing pathway.

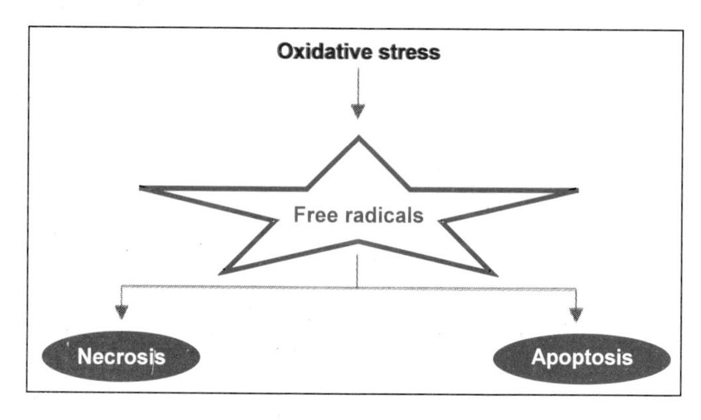

WHY IS THE BRAIN MORE SUSCEPTIBLE TO OXIDATIVE STRESS ?

The brain is highly susceptible to oxidative stress due to many reasons (1) the brain has a very high rate of oxygen consumption (about 20% of the basal oxygen consumption)(Halliwell and Gutteridge, 1999) (2) its extensive use of glutamate as a neurotransmitter and impaired energy metabolism in the brain can lead to excess extracellular glutamate levels whose receptor binding leads to a raising of intracellular calcium to pathological levels, which in the process lead to excess free radical generation (Halliwell , 1999) (3) a high content of lipids (polyunsaturated fatty acids). These lipids are highly susceptible to lipid peroxidation (4) many neurotransmitters are auto-oxidisable (dopamine, noradrenaline) which react with oxygen to form superoxide, hydrogen peroxide and reactive quinines (5) iron is found throughout the brain and important iron containing proteins in brain include cytochromes, ferritin and tyrosine etc (6) a paucity of antioxidant enzymes as compared to the rest of the body. In particular the levels of catalase are low in most brain regions.

Why is the brain more vulnerable to free radicals damage ?

- High metabolic rate and high oxygen consumption
- High lipid content
- Low antioxidant enzymes
- Pressure of Iron
- Auto-oxidisable neurotransmitters
- Glutamate

There is plenty of evidence that free radicals play an important role in, and may even be the causative of the neurological diseases such as Alzheimer's disease, Parkinson's disease, epilepsy and cerebral ischemia.

ALZHEIMER'S DISEASE

Alzheimer's disease (AD) is an irreversible, progressive neurodegenerative disorder that occurs gradually and results in memory loss, unusual behavior, personality changes, and ultimately death. It is the most common form of adult onset dementia (Clegg et al, 2001). It is the 4th leading cause of death in western countries, preceded only by heart disease, cancer and stroke (Bush et al, 1994). Currently there are estimated 2 million AD patients in India and 17-25 million worldwide. Age is by far the main risk factor for AD; its prevalence illustrates an exponential rise with age, approximately doubling each five-year period, from 1% incidence at age 65 years to 16-20% at 85 years (Flynn and Ranno, 1999).

Approximately 5% of all Alzheimer's disease cases have an early onset and are familial, based on mutations of presenilin 1, presenilin 2 or the beta amyloid precursor protein (APP) (Selkoe, 2001). By contrast approximately

95% of all Alzheimer's disease cases represent the sporadic or late onset form of disease showing no mutations of presenilin 1, presenilin 2 or APP (Cummings and Cole, 2002).

Effective treatment and prevention strategies and management of attendant behavioral disturbances are paramount to the management of this illness.

Advances have been made in the past decade in understanding of this neurodegenerative disorder. However, at present there is no definitive treatment or cure. Since AD is a progressive disorder, the efficacy of agents may be dependent upon the stage of the disease. In addition AD appears to be heterogeneous disorder with the potential for patients to respond differently to various therapeutic agents (Sramek and Cuttler, 1999)

Current therapies for AD are based on increasing the amount of acetylcholine to the neurons. However, cholinergic drugs target only 1 of several dysfunctional neurobiological systems in AD and they improve cognition and reduce behavioral symptoms in about 15-40 % of the patients (Parnetti et al., 1997).

EVIDENCE OF OXIDATIVE STRESS IN ALZHEIMER'S DISEASE

Much of the AD research performed over the past decade has focused on the importance of oxidative stress mechanisms in disease pathogenesis (Pratico and Delanty, 2000; Smith et al., 2000; Smith et al., 2000; Christen., 2000).

A great deal of evidence has been collected which implicates oxidative stress and free radical damage as important to neuron degeneration and death in this disorder (Pratico and Delanty, 2000; Smith et al., 2000; Christen, 2000).

The direct evidence supporting increased oxidative stress in AD is an increased brain Fe, Al and Hg in AD, capable of stimulating free radical generation (Markesbery, 1997; Xu et al., 1992; Deibel et al., 1996; Stacey and Kappus, 1982); Increased lipid peroxidation and decreased PUFA in the AD brain (Lovell et al, 1995; Marcus et al, 1998; Palmer and Burns, 1994) , and increased hydroxynonenal, an aldehyde product of lipid peroxidation in AD ventricular fluid (Gutteridge , 1995; Markesbery and Lovell, 1998), increased protein and DNA oxidation in the AD brain, diminished energy metabolism (Beal, 1995) and decreased cytochrome oxidase in the brain in AD (Cottrell et al., 2002), presence of advanced glycation end products (AGE) (Yan et al., 1994), malondialdehyde, SOD-1 in NFT and senile plaques and studies showing that amyloid beta peptide is capable of generating free radicals (Hensley et al., 1994). Supporting indirect evidence comes from a variety of *in vitro* studies showing that free radicals are capable of mediating neuron degeneration and death. Overall, these studies indicate that free radicals are possibly involved in the pathogenesis of neuron death in AD. Even if free radical generation is secondary to other initiating causes they are deleterious and part of a cascade of events that can lead to neuron death, suggesting that therapeutic efforts aimed at removal of reactive oxygen species (ROS) or prevention of these formations may be beneficial in AD (Markesbery, 1997).

SCHEMATIC DIAGRAM REPRESENTING SOURCES OF OXIDATIVE STRESS IN AD

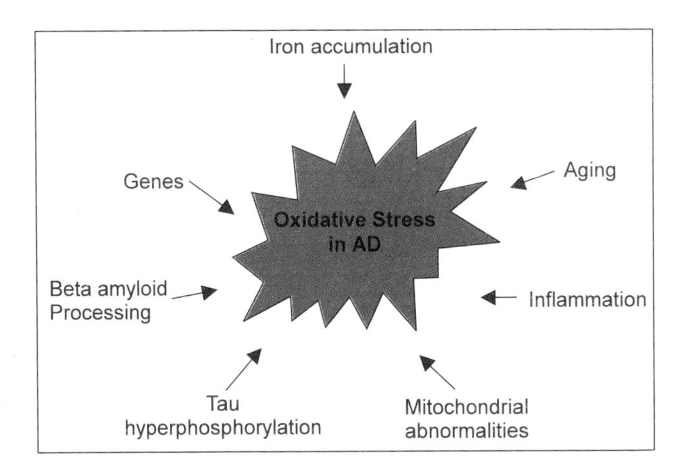

Therapeutic Potential of Antioxidants in Alzheimer's Disease

Oxidative stress has been implicated in AD neurodegeneration. This raises the possibility of the therapeutic use of free radical scavengers and antioxidants (Munch et al., 2002; Moosmann and Behl, 2002).

According to the oxidative stress hypothesis of AD, numerous approaches for an effective antioxidant neuroprotection have been developed. Numerous free radical scavengers have been used in experimental paradigms of neuronal cell death *in vitro* and *in vivo*, such as vitamin E (α-tocopherol), the lazoroids (21-aminosteroids) or mifepristone (RU486) (Moosmann and Behl, 2002; Behl and Moosmann, 2002). All these antioxidants share their high lipophilicity as a consequence of their chemical structure and their free radical scavenging moieties. Although most of these compounds are currently still studied at the preclinical level, employing *in vitro* models, and a possible clinical application is not possible, yet, such molecules are of basic interest and represent lead structures for an improved drug design.

Moreover, many free radical scavengers are known and many (e.g. vitamins E and C, melatonin, flavanoids and carotenoids) have no major side effect (Christen, 2000).

In our laboratory, we have shown that the ICV streptozotocin model of AD is appropriate to evaluate antioxidants as potential neuroprotective agents (Sharma and Gupta, 2000).

We have studied the effect of antioxidants melatonin (a pineal hormone), trans resveratrol (a polyphenolic compound found in grapes and red wine), alpha lipoic acid and alpha tocopherol (endogenous antioxidants) for their neuroprotective effect in the ICV streptozotocin model of AD (Sharma and Gupta 2001; Sharma and Gupta, 2002; Sharma and Gupta (In press).

All the four antioxidants at different doses prevented the cognitive impairment, oxidative and neurochemical deficit induced by ICV STZ suggesting the potential of antioxidants in AD.

EVALUATION OF ANTIOXIDANTS IN AD PATIENTS

Clinical trials are necessary to prove an antioxidant activity *in vivo* before the above-mentioned compounds could be considered as drugs for the prevention and therapy of AD. A first clinical trial employing the; antioxidant vitamin E has successfully been completed just recently. In a multicenter clinical trial on AD patients suffering from a moderately severe impairment, vitamin E effectively slowed down the progression of the disease (Sano et al., 1997). This first success raised high hopes with respect to the concept of an antioxidant therapy and fuelled further intensive research and additional clinical trials on this topic. Of course, novel antioxidants should have improved properties compared to vitamin E, such as an increased lipophilicity, an easy blood-brain barrier crossing and, therefore, enhanced activity.

In a recent Cochrane review (Rosler et al., 1998) the outcomes of all unconfounded, double blind, randomised trials were analysed, in which treatment with vitamin E at any dose was compared with placebo for patients with Alzheimer's disease. Only one study (published by Sano and coworkers (Sano et al, 1997) was identified as meeting the inclusion criteria of the Cochrane analysis. In this study, focusing on patients diagnosed as having mild to moderate Alzheimer's disease, a slower progression of Alzheimer's disease in patients using vitamin E has been shown as compared with patients not using this antioxidant. In the corresponding Cochrane review it has been concluded that there is– albeit insufficient – evidence for an efficacy of vitamin E in the treatment of people with Alzheimer's disease, justifying further studies on this topic.

Since oxidative stress seems to be involved in the earliest phases of Alzheimer's disease, it is an attractive issue to focus therapeutic interventions on the early phases of this disease. Accordingly, very recently three studies on this topic have been published. First, Commenges and coworkers (Commenges et al., 2000) found an inverse relation between intake of flavonoids and the risk of incident dementia of which Alzheimer's disease is one possible cause. Second, Morris and coworkers (Morris et al., 2002) investigated the relationship between Alzheimer's disease incidence and intake of the antioxidant nutrients, vitamin E, vitamin C, and β-carotene The results of this study indicate that only vitamin E from food, but not other antioxidants or vitamin E from supplements, may be associated with a reduced risk for Alzheimer's disease (Morris et al, 2002). Third, Engelhart and coworkers (Engelhart et al., 2002) determined whether dietary intake of β-carotene, flavonoids, vitamin C and vitamin E is related to the risk for Alzheimer's disease in a large population based, prospective cohort study (the so-called Rotterdam study). The results, obtained after a mean follow-up of 6 years, indicate that high dietary intake of vitamins C and E was associated with a significantly lower risk for Alzheimer's disease. This relation was even more pronounced in current smokers, for whom intake of β-carotene and flavonoids was also preventive.

CEREBRAL ISCHEMIA

Cerebral Ischemia (Stroke) is a neurological condition occurring as a result of loss of blood supply to the brain. According to WHO definition stroke is a "rapidly developing clinical sign of focal (at times global) disturbance of cerebral function, lasting more than 24 hours or leading to death with no apparent cause other than that of vascular origin".

It is the second largest cause of mortality world-wide and is surpassed only by that of heart disease (Bonita, 1992). It is estimated to be responsible for 9.5% of all deaths. In the U. S. more than 700,000 new strokes occur annually. Stroke is also the leading cause of disability with an estimated 4 million stroke survivors living with stroke related deficits in US alone. More than 70% of stroke survivors remain vocationally impaired, more than 30 % require help with activities of daily living, and more than 20 % walk only with assistance (Broderick, 1998; Tuhrim, 2002). In India the prevalence of stroke is estimated as 203 per 100,000 population above 20 years amounting to a total of about 1 million cases. The male to female ratio is 1:7. Around 12% of all strokes occurs in population below 40 years. The total number of deaths due to stroke is 102,000 which represents 1.2% of total deaths in the country (Anand et al., 2002).

The principal pathophysiological processes in ischemic stroke are energy failure, loss of cell ion homeostasis, acidosis, increased intracellular calcium, excitotoxicity, and free radical-mediated toxicity. The relative contributions of each processes are believed to vary significantly especially in relation to the level of cerebral blood flow. There is also growing interest in changes in gene expression after ischemia. Although advances have been made, there are still no proven pharmacological therapies to rescue ischemic human neurons. Therefore there is a constant need to develop newer agents for the treatment of stroke (Macdonald and Stoodley , 1998; Lee et al., 1999).

OXIDATIVE STRESS IN CEREBRAL ISCHEMIA

Oxidative stress has been implicated as a potential contributor to the pathogenesis of acute ischemic injury. Free radicals are important pathophysiological mediators of ischemia-induced toxicity. After brain injury by ischemic or hemorrhagic stroke, the production of free radical increases through several different cellular pathways, including calcium activation of phospholipases, nitric oxide synthase, xanthine oxidase, the Fenton and Haber-Weiss reactions by inflammatory cells, leading to tissue damage. Oxidative damage does not occur in isolation but participates in the complex interplay between excitotoxicity, apoptosis, and inflammation in ischemia and reperfusion (Barone and Feuerstein 1999; Chan 1996). Excessive free radical production causes the peroxidation of lipids, proteins, and nucleic acids. Free radical generation also occurs during the time of reperfusion. Reoxygenation during reperfusion causes an increase in oxygen to levels that cannot be utilized by the mitochondria under normal physiological conditions. As a result there is perturbation of the antioxidative defense mechanisms as a result of overproduction of oxygen radicals, inactivation of detoxification systems, consumption of antioxidants and failure to replenish antioxidants in the ischemic brain tissue (Chan, 1996) . If cellular defense systems are weakened, increased production of free radicals will lead to oxidation of lipids, proteins, and nucleic acids, which may alter cellular function in a critical way (Lewen et al., 2000).

Evidence suggests that pharmacological upregulation of free radical scavenging enzymes can reduce infarct volumes.

Schematic Diagram Representing the Role of Free Radicals in Cerebral Ischemia

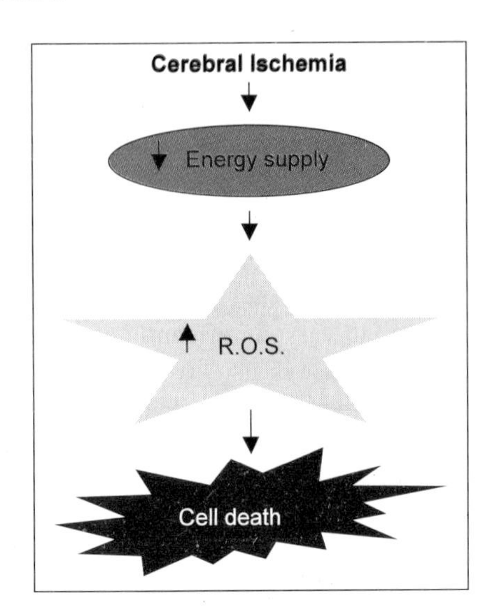

Schematic Diagram Representing the Involvement of Free Radicals in Reperfusion Injury

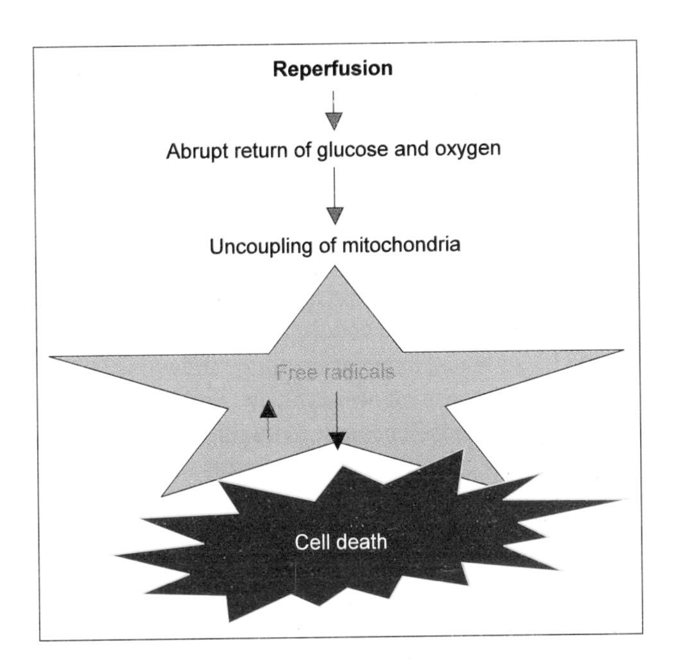

ANTIOXIDANTS IN STROKE

Tirilazad mesylate is a non glucocorticoid 21 – aminosteroid with multiple mechanisms of action including inhibition of iron dependant lipid peroxidation within membranes, effects that are mediated through free radicals scavenging of lipid peroxyl and hydroxyl groups, reduction of the formation of hydroxyl radicals and membranes phospholipid fluidity and maintenance of endogenous antioxidant levels (Bath et al., 2000). Based on the beneficial effects observed with tirilazad in several preclinical experimental models, Pharmacia advanced this compound to phase–III trials for acute stroke. Two randomized trials of Tirilazad mesylate in patients with acute stroke (RANTTAS I & II) were conducted and ended prematurely due to a lack of effect on outcome when evaluated by a preplanned interim analysis (Haley, 1998; Scott et al., 1996).

NXY-059 is a neuroprotectant , under development for the treatment of stroke. It is a free radical trapping agent that stabilizes ischemic reperfusion injury protecting mitochondrial function blocking cytochrome efflux and subsequent caspase activation. NXY-059 has been shown to reduce infarct volume in rodent models of transient (Kuroda et al, 1999), permanent (Zhao et al, 2001) and hemorrhagic stroke (Peeling et al, 2001) in addition to improving neurological outcome in a primate model of stroke (Marshall et al, 2001). In a placebo controlled phase – IIa trial NXY-059, 2-4 mg/kg/ h for 72 h was shown to be safe (Lees et al, 2001). Further phase – II studies designed to define the optimal dose for pivotal efficacy are completed and the drug has entered the phase – IIb/ III trials.

Egb –761 is a free radical scavenger derived from a concentrated extract of *Ginkgo biloba* enriched in flavanoids and terpens (Diamond et al., 2000). It inhibits neuronal death caused by various agents such as beta– amyloid protein, protects against ischemic processes and facilitates decision making. It is currently in phase- II clinical trial.

Ebselen a seleno-organic compound with antioxidant activity through a glutathione peroxidase like action. Ebselen penetrates cells as a result of its binding to the intracellular thiol groups such as glutathione. It also inhibits the peroxidation of the membrane phospholipids, blocks the production of the superoxide anions by activated leucocytes, inhibit inducible NO synthase and exhibit a sustained defense against peroxynitrite. Experimental studies in rats and dogs have revealed that ebselen is able to inhibit both vasospasm and tissue damage in stroke models, which correlates with its inhibitory effects on oxidative processes. Results from randomized, placebo-controlled, double-blind clinical studies on the neurological consequences of acute ischemic stroke, subarachnoid haemorrhage and acute middle cerebral artery occlusion, have revealed that ebselen significantly enhances outcome in patients who have experienced occlusive cerebral ischemia of limited duration. Safety and tolerability are good and no adverse effects have become apparent (Parnham and Sies, 2000; Yamaguchi et al, 1998).

Although the antioxidants so far tried in cerebral ischemia have not shown promising results however use of antioxidants still remains a lucrative approach and lot of work is going on in this direction. In our laboratory we have studied different antioxidants adenosine (an inhibitory neurotransmitter), trans resveratrol (a polyphenolic compound), melatonin in middle cerebral artery occlusion model of stroke in rats. Pretreatment with antioxidants prevented the motor deficit, attenuated the oxidative stress markers and decreased the infract volume as evidenced by diffusion weighted imaging (Sinha et al, 2001; Sinha et al, 2002). At present our laboratory is engaged in evaluating Indian herbal plants with antioxidant property in the middle cerebral artery occlusion model of stroke in rats.

CONCLUSION

Accumulating evidence indicates a major role of free radicals and oxidative stress in the pathogenesis and pathophysiology of Alzheimer's disease and cerebral ischemia. Reports indicate that administration of antioxidants may be useful in prevention and treatment of such diseases. In this respect scavenging of free radicals by non-enzymatic/exogenous antioxidants seems to be the most practical approach. This is due to the fact that many different nonenzymatic/

exogenous antioxidants without major side effects are available and there is currently clinically no therapeutic approach known to increase levels of enzymatic/endogenous antioxidants in humans.

References

Anand K., Chowdhury D., Singh K.B., Pandav C.S. and Kapoor S.K. Estimation of mortality and morbidity due to strokes in India. Neuroepidemiology. 3: 208-11, 2001.

Barone F.C. and Feuerstein G.Z. Inflammatory mediators and stroke: new opportunities for novel therapeutics. J Cereb Blood Flow Metab.**19:** 819-34, 1999.

Bath P.M., Iddenden R., Bath F.J., Orgogozo J.M. Tirilazad for acute ischaemic stroke. Cochrane Database Syst. Rev. **4:** CD002087, 2001.

Beal M.F. Aging, energy, and oxidative stress in neurodegenerative diseases. Ann. Neurol. **38:** 357-66, 1995.

Behl C. and Moosmann B. Antioxidant neuroprotection in Alzheimer's disease as preventive and therapeutic approach. Free Radic Biol Med. 33(2): 182-91, 2002.

Bonita R. Epidemiology of stroke. Lancet. **339**(8789): 342-4, 1992.

Broderick J.P. Recanalization therapies for acute ischemic stroke. Semin Neurol. **18:** 471-84, 1998.

Bush T.L., Miller S.R., Criqui M.N. et al. Risk factors for morbidity and mortality in older population: an epidemiological approach. In: hazard WR, Beirman EL, Blass JP et al., editors Principles of geriatric medicines and gerontology, New York, McGraw hill. 153-166, 1994.

Chan P.H. Role of oxidants in ischemic brain damage. Stroke. **27:** 1124-9, 1996.

Christen Y. Oxidative stress and Alzheimer disease. Am. J. Clin. Nutr. **71:** 621S-629S, 2000.

Clegg A., Bryant J., Nicholson T., McIntyre L., De Broe S., Gerard K. and Waugh N. Clinical and cost-effectiveness of donepezil, rivastigmine and galantamine for Alzheimer's disease: a rapid and systematic review. Health Technol Assess. **5:**1-137, 2001.

Commenges D,. Scotet V, Renaud S., et al. Intake of flavonoids and risk of dementia. Eur. J. Epidemiol. **16:** 357-363, 2000.

Cottrell D.A., Borthwick G.M., Johnson M.A., Ince P.G. and Turnbull D.M. The role of cytochrome c oxidase deficient hippocampal neurones in Alzheimer's disease. Neuropathol Appl Neurobiol. **28:** 390-6, 2002.

Cummings J.L. and Cole G. Alzheimer disease. JAMA. **287:** 2335-8, 2002.

Deibel M.A., Ehmann W.D. and Markesbery W.R.. Copper, iron, and zinc imbalances in severely degenerated brain regions in Alzheimer's disease: possible relation to oxidative stress. J. Neurol. Sci. **43:** 137-42, 1996.

Diamond B.J., Shiflett S.C., Feiwel N., Matheis R.J., Noskin O., Richards J.A. and Schoenberger N.E. Ginkgo biloba extract: mechanisms and clinical indications. Arch Phys Med Rehabil. **81:** 668-78, 2000.

Engelhart M.J., Geerlings M.I., Ruitenberg A., et al. Dietary intake of antioxidants and risk of Alzheimer disease. JAMA. **287:** 3223-3229, 2002.

Flynn B.L. and Ranno A.E. Pharmacologic management of Alzheimer disease, Part II: Antioxidants, antihypertensives, and ergoloid derivatives.Ann Pharmacother. **33:** 188-97, 1999.

Gutteridge J.M.. Lipid peroxidation and antioxidants as biomarkers of tissue damage. Clin. Chem. **41:** 1819-28, 1995.

Haley E.C. Jr. High-dose tirilazad for acute stroke (RANTTAS II). RANTTAS II Investigators.Stroke. **29:** 1256-7, 1998.

Halliwell B. and Gutteridge J.M.C. Free radicals in biology and medicine. 3rd Edition. Oxford :University Press, 1999.

Halliwell B. Role of free radicals in neurodegenerative diseases. Drugs and Aging. **18:** 685-716, 2001.

Hensley K., Carney J.M., Mattson M.P., Aksenova M., Harris M., Wu J.F., Floyd R.A., Butterfield D.A. A model for beta-amyloid aggregation and neurotoxicity based on free radical generation by the peptide: relevance to Alzheimer disease.Proc Natl Acad Sci U S A. **91:** 3270-4, 1994.

Kuroda S., Tsuchidate R., Smith M.L., Maples K.R., Siesjo B.K. Neuroprotective effects of a novel nitrone, NXY-059, after transient focal cerebral ischemia in the rat. J. Cereb Blood Flow Metab. **19:** 778-87, 1999.

Lee J.M., Zipfel G.J. and Choi D.W. The changing landscape of ischaemic brain injury mechanisms. Nature. **399**(6738 Suppl): A7-14, 1999.

Lee J.M., Zipfel G.J. and Choi D.W. The changing landscae of ischaemic brain injury mechanisms. Nature. **399**(6738 Suppl):A7-14, 1999.

Lees K.R., Sharma A.K., Barer D., Ford G.A., Kostulas V., Cheng Y.F. and Odergren T. Tolerability and pharmacokinetics of the nitrone NXY-059 in patients with acute stroke. Stroke. **32:** 675-80, 2001.

Lewen A., Matz P. and Chan P.H. Free radical pathways in CNS injury. J Neurotrauma. **17:** 871-90, 2000.

Lovell M.A., Ehmann W.D. and Butler S.M., Markesbery WR.Elevated thiobarbituric acid-reactive substances and antioxidant enzyme activity in the brain in Alzheimer's disease. Neurology. **45:** 1594-601, 1995.

Macdonald R.L., Stoodley M. Pathophysiology of cerebral ischemia. Neurol Med Chir (Tokyo). **38:** 1-11, 1998.

Macdonald R.L. and Stoodley M.. Pathophysiology of cerebral ischemia. Neurol. Med. Chir. (Tokyo). **38:** 1-11, 1998.

Marcus D.L., Thomas C., Rodriguez C., Simberkoff K., Tsai J.S., Strafaci J.A., Freedman M.L. Increased peroxidation and reduced antioxidant enzyme activity in Alzheimer's disease. Exp. Neurol. **150:** 40-4, 1998.

Markesbery W.R. Oxidative stress hypothesis in Alzheimer's disease. Free Radic Biol Med. **23:** 134-47, 1997.

Markesbery W.R., Lovell M.A. Four-hydroxynonenal, a product of lipid peroxidation, is increased in the brain in Alzheimer's disease. Neurobiol Aging. **19:** 33-6, 1998.

Marshall J.W., Duffin K.J., Green A.R., Ridley R.M. NXY-059, a free radical—trapping agent, substantially lessens the functional disability resulting from cerebral ischemia in a primate species. Stroke. **32:** 190-8, 2001.

Moosmann B., Behl C. Antioxidants as treatment for neurodegenerative disorders. Expert Opin Investig Drugs. **11:** 1407-35, 2002.

Morris M.C., Evans D.A., Bienias J.L. et al. Dietary intake of antioxidant nutrients and the risk of incident Alzheimer disease in a biracial community study. JAMA. **287:** 3230-3237, 2002.

Munch G., Deuther-Conrad W. and Gasic-Milenkovic J. Glycoxidative stress creates a vicious cycle of neurodegeneration in Alzheimer's disease—a target for neuroprotective treatment strategies? J. Neural. Transm. Suppl. **62:** 303-7, 2002.

Palmer A.M. and Burns M.A. Preservation of redox, polyamine, and glycine modulatory domains of the N-methyl-D-aspartate receptor in Alzheimer's disease. J Neurochem. **62:** 187-96, 1994.

Parnetti L., Senin U. and Mecocci P. Cognitive enhancement therapy for Alzheimer's disease. The way forward. Drugs. **53:** 752-68, 1997.

Parnham M. and Sies H. Ebselen: prospective therapy for cerebral ischaemia. Expert Opin Investig Drugs. **9:** 607-19, 2000.

Peeling J., Del Bigio M.R., Corbett D., Green A.R., Jackson D.M. Efficacy of disodium 4-[(tert-butylimino)methyl]benzene-1,3-disulfonate N-oxide (NXY-059), a free radical trapping agent, in a rat model of hemorrhagic stroke. Neuropharmacology. **40:** 433-9, 2001.

Pratico D. and Delanty N. Oxidative injury in diseases of the central nervous system: focus on Alzheimer's disease. Am. J. Med. **109:** 577-85, 2000.

Rosler M., Retz W., Thome J. and Riederer P. Free radicals in Alzheimer's dementia: currently available therapeutic strategies. J. Neural Transm. Suppl. **54:** 211-219, 1998.

Rutten B.P., Steinbusch H.W., Korr H. and Schmitz C. Antioxidants and Alzheimer's disease: from bench to bedside (and back again). Curr Opin Clin Nutr Metab Care. **5:** 645-651, 2002.

Sano M., Emesto C., Thomas R.G., et al. A controlled trial of selegiline, alha-tocopherol, or both as treatment for Alzheimer's disease. The Alzheimer's Disease Cooperative Study. N. Engl. J. Med. **336:** 1216-1222, 1997.

Scott. A. randomized trial of tirilazad mesylate in patients with acute stroke (RANTTAS). The RANTTAS Investigators.Stroke. **27:** 1453-8, 1996.

Selkoe D.J. Physiological production of the β-amyloid protein and the mechanism of Alzheimer's disease. TINS. **16:** 403-409, 1993.

Sharma M. and Gupta Y.K. Effect of chronic treatment of melatonin on learning, memory and oxidative deficiencies induced by intracerebroventricular streptozotocin in rats. Pharmacol. Biochem. Behav. **70:** 325-31, 2001.

Sharma M. and Gupta Y.K. Intracerebroventricular injection of streptozotocin in rats produces both oxidative stress in the brain and cognitive impairment. Life Sci. **68:** 1021-9, 2001.

Sharma M. and Gupta Y.K. Chronic treatment with trans resveratrol prevents intracerebroventricular streptozotocin induced cognitive impairment and oxidative stress in rats. Life Sci. **71:** 2489-98, 2002.

Sharma M. and Gupta Y.K. Effect of alpha lipoic acid on intracerebroventricular streptozotocin model of cognitive impairment in rats. Eur Neuropsychopharmacol 2003 *In press.*

Simonian N.A. and Coyle J.T. Oxidative stress in neurodegenerative diseases. Annu. Rev. Pharmacol. Toxicol. **36:** 83-106, 1996.

Sinha K., Chaudhary G. and Gupta Y.K. Protective effect of resveratrol against oxidative stress in middle cerebral artery occlusion model of stroke in rats. Life Sci. **71:** 655-65, 2002.

Sinha K., Degaonkar M.N., Jagannathan N.R. and Gupta Y.K. Effect of melatonin on ischemia reperfusion injury induced by middle cerebral artery occlusion in rats. Eur. J. Pharmacol. **428:** 185-92, 2001.

Smith M.A., Nunomura A., Zhu X., Takeda A. and Perry G. Metabolic, metallic, and mitotic sources of oxidative stress in Alzheimer disease. Antioxid Redox Signal **2:** 413-20, 2000.

Smith M.A., Rottkamp C.A., Nunomura A., Raina A.K. and Perry G. Oxidative stress in Alzheimer's disease. Biochim Biophys Acta. **1502:** 139-44, 2000.

Sramek J.J. and Cutler N.R. Recent developments in the drug treatment of Alzheimer's disease. Drugs Aging. **14:** 359-73, 1999.

Stacey N.H. and Kappus H. Cellular toxicity and lipid peroxidation in response to mercury. Toxicol. Appl. Pharmacol. **63:** 29-35, 1982.

Tuhrim S. Management of stroke and transient ischemic attack. Mt Sinai J. Med. **69:** 121-30, 2002.

Xu N., Majidi V., Markesbery W.R. and Ehmann W.D. Brain aluminum in Alzheimer's disease using an improved GFAAS method. Neurotoxicology. **13:** 735-43, 1992.

Yamaguchi T., Sano K., Takakura K., Saito I., Shinohara Y., Asano T. and Yasuhara H. Ebselen in acute ischemic stroke: a placebo-controlled, double-blind clinical trial. Ebselen Study Group. Stroke. **29:** 12-7, 1998.

Yan S.D., Chen X., Schmidt A.M., Brett J., Godman G., Zou Y.S., Scott C.W., Caputo C., Frappier T., Smith M.A., et al. Glycated tau protein in Alzheimer disease: a mechanism for induction of oxidant stress. Proc Natl Acad Sci USA. **91:** 7787-91, 1994.

Zhao Z., Cheng M., Maples K.R, Ma J.Y. and Buchan A.M. NXY-059, a novel free radical trapping compound, reduces cortical infarction after permanent focal cerebral ischemia in the rat. Brain Res. **909:** 46-50, 2001.

Pharmacological Perspectives of Toxic Chemicals and Their Antidotes
Editors: S J S Flora, J A Romano, S I Baskin and K Sekhar

Succimer and its Analogues: Antidotes for Metal Poisoning

Ashish Mehta, Govinder Flora[a], Shashi N. Dube and Swaran J.S. Flora[*]
Department of Pharmacology and Toxicology, Defence Research and Development Establishment, Jhansi Road, Gwalior 474 002, India
[a]Present address: Department of Surgery/Neurosurgery, University of Kentucky Medical Center, Lexington, KY 40536, USA
[*]Corrosponding author

INTRODUCTION

The use of chelators for decreasing metal and metalloid toxicity started nearly 100 years back with the collaboration between Alfred Werner in Zurich and Paul Ehrlich in Frankfurt, to find less toxic arsenic compounds for syphilis treatment. During 1920-1940, similar attempts were made to reduce the toxicity of arsenical drugs for trypanosomiasis by Voegtlin and antimony drugs for schistosomiasis by Schmidt (Anderson, 1999).

During the World War II, British Anti-Lewisite (BAL) was developed as an antidote for war gas, dichlorivinyl arsine (Lewisite) (Peter et al., 1945). Lewisite was, however, never used, so the first clinical use of BAL was in intoxication due to treatment of syphilis with organic arsenical drugs. The next chelator to come into clinical use was ethylene diamine tetra acetic acid (EDTA), initially to combat lead intoxication and for decorporation of radionucleotides, the latter role was played more efficiently by diethylene triamine pentaacetic acid (DTPA). The value of EDTA as a clinical chelating agent was reduced by the need of slow intravenous administration, low intestinal uptake, exclusive extra-cellular action and high stability constants with essential metals (Anderson, 1999).

The use of chelating agents was of immense importance for the afflicted individual in the treatment of metal storage diseases like Wilson's disease, with D-pencillamine (DPA), chemically a 3,3-dimethyl-substitute of cysteine

(Walshe, 1956). Walshe later used triethylenetetramine (TETA) to treat patients developing pencillamine intolerance. Today, TETA is considered the drug of first choice in copper storage disease.

CHELATING AGENTS AND CHELATION THERAPY

Chelating agents are organic compounds capable of linking together metal ions to form complex ring-like structure called chelates. 'Chelate' is a Greek word meaning the claws of a lobster. Chelators act according to a general principle: the chelator forms a complex with the respective (toxic) ion, and these complexes reveal a lower toxicity and are more easily eliminated from the body. It could be represented in a chemical equation as -

Metal-Tissue + Chelating Agent $\rightarrow$ Tissue + Metal-Chelating agent complex

It is essential that the chemical affinity of the complexing agent for the metal ion is higher than the affinity of the metal for the sensitive biological molecules. Thus, chemical measurement of the stability constants of the metal-complexes formed may give a first indication of the effectiveness of a particular chelating agent. Chelation can thus be defined as the incorporation of a metal ion into a heterocyclic ring structure. An ideal chelating agent should possess the following characteristic:

- Greater affinity for the toxic metal that has to be chelated
- Ability to chelate with natural chelating groups found in biological system
- Low toxicity
- Ability to penetrate cell membrane
- Minimal metabolism
- Rapid elimination of metal
- High water solubility

CHEMISTRY AND MECHANISM OF CHELATE ACTION

The design of chelating agents for a toxic metal ion must take into consideration the coordination number, net ionic charge and stereochemistry of that metal ion. The use of a chelating agent that will occupy more of the coordination position of a metal ion will generally (not always) give a complex of greater stability than is found for those complexes with chelating agents, which occupy fewer positions.

The metal chelate complexes have a reduced tendency to undergo exchange reactions once they are formed. However, it is frequently advantageous to use a preferred donor atom in a chelating agent of lower dentisity. Thus, 2,3 dimercaptosuccinic acid (DMSA) is preferable to EDTA in treatment of lead intoxication (Cory Slechta et al., 1988; Jones, 1994). It is also necessary to keep in mind that the introduction of the chelating agent into any intracellular space requires its passage through the cell membrane. This passage can be accomplished either (a) by passing through the lipid part of the membrane as an uncharged molecule or (b) via utilizing one of the anion/ cation transport systems present in the membrane (Jones, 1994). There is a hypothesis that large ion complex with a positive charge will pass out of a cell very slowly because of their inability to pass through either the lipid portion of the cellular membrane or the cation transport system designed to move ions with +1 or a +2 charge across the membrane (Kontoghiorghes et al., 1987; Jones, 1994). Another important property of metal complexes is the stereochemistry of the toxic metal ion. Chelating agents tie up all the coordination position of a metal ion (Clevette and Orivig, 1990; Hancock and Martell, 1989; Hancock, 1990; Busch and Stephenson, 1990, Lindoy, 1989; Izatt et al., 1991). It should be noted that metal chelating agents usually contain more than one functional group, in order to provide a chemical 'claw' to chelate the toxic metal. The link formed between the metal and chelating agent is of coordinate type, which is generally similar to covalent type but the major difference is that both the electrons forming the link are supplied by the binding atom (the resulting compound is called 'metal complex' or 'coordinate compound'). A simplest example in this case can be a link formed by proton (hydrogen atom with a positive charge). Beside hydrogen atom there are number of other atoms which can take part to coordinate complex formation like sodium, magnesium, copper, zinc and various other transition elements such as manganese, iron and cobalt. Chelation treatment can now be more completely defined as an equilibrium reaction between a metal ion and a complexing agent, characterized by the formation of more than one bond between the metal and a molecule of the complexing agent resulting in the formation of a heterocyclic ring structure incorporating the metal ion.

Other determining factor for complex formation is the hardness/ softness characteristics of electron donor and acceptor. This characteristic not only determines the stability of the formed complex but also the chelating agent's degree of metal selectivity in relation to competing with essential metals present in the biological fluid. Softness character can be explained as the ability of the empty frontier orbital of metal ions for accepting electrons and to the deformability of the outer most occupied electron orbital of donor groups. So-called soft metal cations have large atomic radius and a high number of electron pair in the outer shell in contrast to the hard ions. Formation of a metal complex ML, involves that the metal cation, M, coordinates or accept free electron pairs that are furnished by electron-donor groups from the ligands, L. The interaction between a chelating agent and a toxic metal can be expressed in term of stability constants. Assuming the formation of the simple mononuclear complex only, the equilibrium concentration can be calculated from the law of mass action:

$$\text{Stability Constant, K} = \frac{[ML]}{[M][L]}$$

Values within the [ ] denotes concentration in mol/l (Jones, 1994)

A metal with a higher stability constant competes for the chelating agent with a lower stability value and sooner or later removes the metal with a lower constant from the complex, even if this has already been formed. Relative concentration also influences this release and this is the major reason that calcium which is readily available in the body, binds to EDTA in large quantities if disodium EDTA is administered, even though it has a stability constant lower than that of lead.

The chelating agents have been divided into various groups depending on their structural features as given below :

Group	*Chelating Agents*	*Structural Features*
A	EDTA, DTPA, etc	Polyaminocarboxylic acids
B	BAL, DMPS, DMSA	Vicinal –SH groups
C	DPA, L-cysteine	β- mercapto-a-amino acid
D	Salicylic acid, Triton	ortho hydroxycarboxylic acid or ortho diphenol
E	Desferrioxamine	hydroxamic acid
F	Diethylthiocarbamate	$N\text{-}CS_2^-$
	Thiosulfate	$S_2O_3^-$

(from Jones and Basinger 1982)

CONVENTIONAL CHELATING AGENTS AND THEIR DRAWBACKS

In general terms the most commonly used chelating agents that have been the forerunners in chelation therapy have belonged to the polyaminocarboxylic groups. As the name indicates these chelators utilize the amino and the carboxylic groups to scavenge the toxic metal from the system.

In this category, calcium disodium ethylene diamine tetra acetic acid ($CaNa_2EDTA$) is a derivative of ethylene diamine tetra acetic acid (EDTA), a synthetic polyamino-polycarboxylic acid was used for the treatment of metal poisoning had been the mainstay of chelation therapy for many years. It is a white crystalline solid with a molecular weight of 374.28 and empirical formula is $C_{10}H_{12}CaN_2Na_2O_8$. It is a weak tetra basic acid. Another member belonging to this family is diethylene triamine pentaacetic acid DTPA is a synthetic polyaminocarboxylic acid with properties similar to EDTA (Hammound, 1971) having a molecular weight 497.4 and empirical formula is $C_{14}H_{23}N_2O_{10}$. DTPA can bind atoms of plutonium and other actinides thus forming a complex that is quickly excreted from the body. Only a small amount of calcium disodium EDTA is absorbed in the gastro-intestinal tract following oral administration in humans and laboratory animals hence cannot be administered orally. Hammond and co-workers (1967) demonstrated that the apparent distribution volume of EDTA is 26% of the total body weight and therefore its distribution is limited to intravascular and interstitial fluids. It can be affirmed that EDTA does not penetrate cell membranes and has a biological half-life of 50-60 minutes; 90% is excreted within 6-8 hours after administration. Renal clearance is mainly through active tubular secretion without any significant re-absorption. Variations in the pH and diuresis do not affect the excretion rate (Forland et al., 1966). $CaNa_2EDTA$ has the LD_{50} value of 16.4 mmol/kg in mouse (Cantilena and Klaassen, 1981). Intravenous administration of this drug results in good absorption but very painful at the injection site. Hence intravenous injection could be given either by diluting in 5% dextrose or saline (Klaassen, 1990). Hypocalcemia is reported with the administration of Na_2EDTA. CaEDTA has the major toxic effects on the renal system causing the necrosis of tubular cells. Severe, hydropic degeneration of proximal tubule cells has also been reported. These lesions along-with some alterations in the urine like hematuria, proteinuria and elevated BUN are generally reversible when the treatment ceases. Another side effect of EDTA is its ability to chelate various essential metals endogenous to the body, zinc in particular (Flora and Tandon 1990, Powell et al., 1999). Zinc administration during EDTA administration is generally recommended to reduce toxicity (Flora and Tandon, 1990).

It has been well established that administration of EDTA during pregnancy can result in teratogenic effects especially when administered between days 11 to 14 at doses comparable to humans (Brownie at al., 1986). Tuchmann-Duplessis and Mercier-Parort (1956) were the first to report teratogenic effect of EDTA. It was observed that when 2-3% disodium EDTA was given in feed along with 100 ppm of zinc to pregnant rats from the gestation day 6 through 21, all the full term young had gross congenital malformations which included clubbed legs, micro or anophthalimia, micro or agnathia, cleft palate, fused or missing digits, brain malformation and curly, short or missing tail. EDTA readily crossed the placenta and equilibrates in the maternal fetal circulation and protects against fetotoxicity of lead in rats (Flora and Tandon, 1987). The use of EDTA to remove endogenous zinc appeared to offer a mechanism for studying the effects of short-term zinc supplementation at critical periods in pregnant zinc deficient rats. Kimmel (1977) showed that effects on teratogenicity varied with the route of administration of EDTA. Subcutaneous administration of the disodium salt at a dose of 375 mg/kg was only maternally toxic, but did not cause any malformation, while the gavage administration exhibited more signs of toxicity. Absorption into the circulation, potential interaction with essential trace elements, and the stress associated with the administration of the compound were suggested to be the possible factors involved in the differences in EDTA-induced maternal and developmental toxicity (Kimmel, 1977). Brownie et al. (1986) also reported teratogenic effects. Another reported disadvantage of $CaNa_2EDTA$ is that it redistributes lead to the brain. Cory Slechta et al. (1987) and Flora et al. (1995) in separate studies provided evidence that rats given lead as lead acetate in their drinking water and then treated with $CaNa_2EDTA$ mobilized lead from their tissues and redistributed to brain and liver on the first day of treatment. The large number of side effects due to the administration of these chelating agents prompted in the commercialization of chelators containing thiol or sulfhydryl groups.

D-Penicillamine (DPA) is 3,3 dimethylcysteine, a sulfhydryl containing amino acid, first introduced in clinical practice by Walshe (1956) but was tried by Ohlsson (1962) as an antidote for low or mild lead poisoning. It has an empirical formula is $C_5H_{11}NO_2S$ and molecular weight is 149.21. It is available as a hydrochloride salt in capsule, which is slightly hygroscopic crystalline powder, soluble in water and ethanol. It can penetrate cell membranes and then get metabolized. It can be absorbed through the gastro intestinal tract and thus can be administered orally. Its absorption from the gastrointestinal tract is between 40 to 70% (Netter et al., 1987). It is fairly stable as its SH group is very resistant to oxidation *in vivo*, attack from enzymes such as cysteine desulfhydrase and L-amino acid oxidase, compared to other monothiols. Excretion of DPA through urine is very fast. Small amount is also reported to cross hepatocyte membrane and excreted through bile. However, the major toxic effect of DPA is antagonizing pyridoxine and inhibiting pyridoxine dependent enzyme such as transaminases. It is thus advisable to supply a dietary supplement of pyridoxine in prolonged administration and also an additional supply of essential metals like iron. Other toxic effects include hypersensitive allergic reactions like fever, skin rashes, leukopenia and thrombocytopenia (Shannon et al., 1988). In few reports nephrotoxic effects too have been observed along with penicillin allergic reaction in sensitive individual due to cross reactivity. Prolonged treatment may also lead to anorexia, nausea, vomiting in human. Apart from this DPA is also a well recognized teratogen and lathyrogen that causes skeletal, palatal, cutaneous and pulmonary abnormalities (Kilburn and Hess, 1982; Myint, 1984; Roussaeux and MacNabb, 1992). As compared to other chelators the developmental toxicity of DPA is abundant in both human and experimental animals. First report on human embryopathy associated with DPA was published by Mjolnerod et al. (1971). Author described the effect of DPA on the infant with generalized connective tissue defects including lax-skin, hyperflexibility of joints, vein fragility, vericosities and impaired wound healing, the child died at a age of 7 weeks. Since DPA chelated copper, it was hypothesis that the drug might be teratogenic (Keen et al., 1982a). Various investigations were performed in the early eighties to test the hypothesis (Keen et al., 1982a, 1982b; Keen et al., 1983; Mark-Savage et al., 1983) and it was observed that when pregnant rats were given DPA along with their diet, there was a high incidence of malformations. The frequency of reabsorption and the frequency and severity of malformations increased in the rats in a dose dependent manner (Keen et al., 1983). However, literature also suggests that the administration of DPA during pregnancy protects the mother from the relapse of Wilson's disease, while it would carry few risks to the fetus (Hartard and Kunze, 1994). DPA have been tried safely throughout pregnancy in women with Wilson's disease, suggesting that the excessive copper stores improve tolerance (Soong et al., 1991). The American Academy of Pediatrics (1995) recommends pencillamine use only when unacceptable adverse reactions to both DMSA and EDTA have occurred. However, Kreppel et al. (1989) reported that pencillamine was ineffective in reducing arsenic burden in rats.

NEWER THIOL CHELATING AGENTS: SODIUM 2,3 DIMERCAPTOPROPANE 1-SUPHONATE (DMPS) AND MESO 2,3 DIMERCAPTOSUCCINIC ACID (DMSA)

In the early eighties it was shown that some newer complexing agents like DMPS and DMSA were effective against mercury, arsenic and lead poisoning. When compared to BAL these newer chelating agents were of significant lower toxicity and moreover they could be administered orally or intravenously (Aaseth, 1983). In addition to their heavy metal chelating properties, these agents have a dithiol group that may act as an oxygen radical scavenger and thus inhibit lipid peroxidation (Gersl et al., 1997; Benov et al., 1990, 1992). Chemical structures and some of the other biochemical and pharmacological properties of some of the newer thiol chelators are summarized in Figure 1 and Table 1.

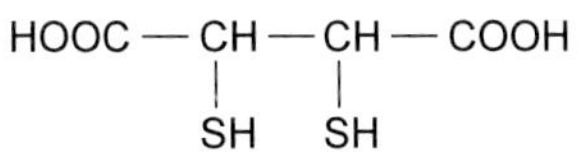

$HOOC-CH(SH)-CH(SH)-COOH$

Meso 2,3-dimercaptosuccinic acid (DMSA)

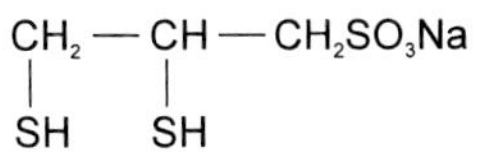

$CH_2(SH)-CH(SH)-CH_2SO_3Na$

Sodium 2,3-dimercaptopropane sulfonate (DMPS)

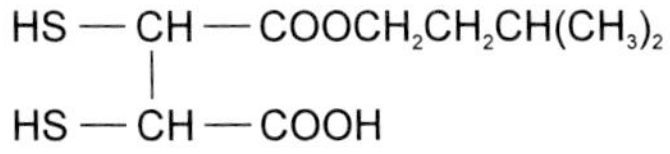

$HS-CH-COOCH_2CH_2CH(CH_3)_2$ / $HS-CH-COOH$

Monoisoamyl meso 2,3-dimercaptosuccinic acid (MiADMSA)

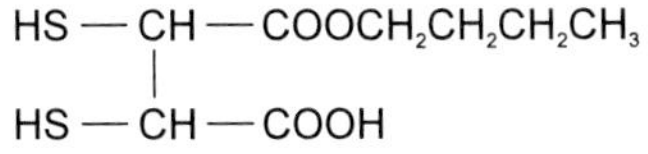

$HS-CH-COOCH_2CH_2CH_2CH_3$ / $HS-CH-COOH$

Mono-n-amyl meso 2,3-dimercaptosuccinic acid (Mn-ADMSA)

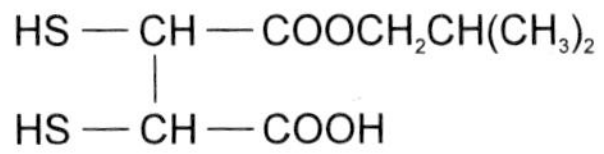

$HS-CH-COOCH_2CH(CH_3)_2$ / $HS-CH-COOH$

Mono-I-butyl meso 2,3-dimercaptosuccinic (Mi-BDMS)

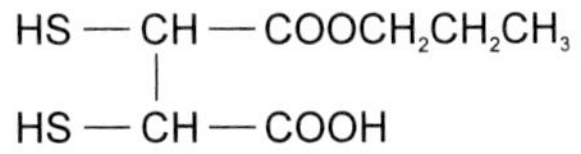

$HS-CH-COOCH_2CH_2CH_3$ / $HS-CH-COOH$

Mono n-butyl meso 2,3-dimercaptosuccinic (Mn-BDMS)

Table 1 Comparative pharmacological and toxicological properties of DMPS, DMSA and MiADMSA

Chelating Agents	*DMPS*	*DMSA*	*MiADMSA*	*References*
Chemical formula	$C_3H_7O_3S_3Na$	$C_4H_6O_4S_2$	$C_9H_{16}O_4$	—
Molecular weight	210	182	252	—
LD_{50} (i.p) mmol/kg	6.53 (mouse)	13.73 (i.p.) 23.8 (oral)	3.0	Aposhian et al. 1995; Jones 1991
Antidote for toxic metals	Mercury, lead, arsenic	Lead, arsenic, antimony	Cadmium, copper, arsenic, gallium arsenide, mercury	Aposhian et al. 1995, Jone 1991, Flora 2000, Flora et al. 2003
Possible effects	Allergic reaction,	GI discomfort, skin reaction, mild neutropenia,	Not known	Jones et al. 1991
Toxic effects	Low systemic and local toxicity,	mild liver disorders and neutropenia	Copper loss, mild hepatotoxicity, mild maternal toxicity	Bosque et al. 1994, Mehta et al. 2002

SODIUM 2,3 DIMERCAPTOPROPANE 1-SUPHONATE (DMPS)

DMPS was first introduced in Soviet Union in the 1950s as 'Unithiol'. DMPS is mainly distributed in the extracellular space; it may enter cells by specific transport mechanism. After i.p. injection of lethal doses the animals were highly irritable for some minutes before they became apathetic and breathing ceased (Planas Bohme et al., 1980). DMPS is rapidly eliminated from the body through the kidneys. The serum half-life is about 20 to 60 minutes. Following oral administration, about 60 to 30% of the administered dose are absorbed in dogs (Wiedelmann et al., 1982) and 30 % in rats (Gabard, 1980), and plasma peak levels are reached after 30 to 45 minute (Wieldenauer et al., 1982). Rapid oxidation of DMPS after intravenous administration to disulfide forms is well reported in blood (Klimmek et al 1993, Maiorino et al., 1988). Fifteen minutes after iv administration of DMPS(3 mg/kg) to humans only 12% of the total DMPS was oxidized to disulfides (Hurlbut et al 1994). DMPS is not involved in important metabolic pathway and parts of administered substance are excreted in an unchanged form. Maiorino et al. (1988) showed several acyclic and cyclic oxidized metabolites in the urine of rabbits. Both urinary and biliary excretion of DMPS occurs (Zheng et al., 1990). By the parenteral route LD_{50} for various species is about 1 g/kg to 2 g/kg. No major adverse effects following DMPS administration in humans or animals have been reported (Hruby and

Donner, 1987). However, a dose dependent decrease in the copper contents was found in the serum, liver, kidneys and spleen. Szincicz et al. (1983) observed that after 10 weeks of DMPS administration there was a decrease in copper concentration in serum and in various organs, an increase in iron contents of liver and spleen and a decrease in hemoglobin, hematocrit, red blood cells, activity of alkaline phosphatase and zinc contents in the blood. Gersl et al. (1997) performed toxicity evaluation of DMPS in rabbits through the oral (0.5g/kg/10weeks) and intravenous (single dose 50mg/kg) routes and observed that there was a low toxicity of the chelator based on the biochemical and hematological variables. However, they also reported some altered biochemical variables like Na, urea, ALT, GSH-Px during the initial phases of experimentation. They also reported a significant loss of serum calcium due to the chelating activities of DMPS. Information regarding the developmental toxicity of DMPS is rather scarce. No abnormalities in the offspring with chronic oral DMPS treatment are reported. Oral administration of DMPS did not adversely affect late gestation, parturition, or lactation in mature mice and fetal and neonatal development does not appear to be adversely affected (Domingo, 1988).

DMPS although known for its antidotal efficacy against mercury, it has been reported to be an effective drug for treating arsenic poisoning. This drug too can be administered both orally and intravenously. An oral dose of 100 mg/kg thrice a day for 10-12 days is effective against mild arsenic poisoning while no recommendation for treating chronic arsenic poisoning is available (Angle, 1996). In experimental animals i.p. administration of DMPS increased the lethal dose of sodium arsenite in mice by four folds. A quantitative evaluation of three drugs reveals that DMPS is 28 times more effective than BAL in arsenic therapy in mice (Hauser and Weger, 1989), while DMSA and DMPS are equally effective. DMPS also appeared to be effective at least in reducing the body lead and gold burden (Gabard, 1980, Twarog and Cherian 1984). By the parenteral route, treatment of adults may be started with 250 mg DMPS and continued with 250 mg every 4 hours on the first day. On the second day 250 mg may be given every 6 hours. On the following days dosages should be adjusted to the clinical status and the results of the toxicological analysis. In children, 5 mg/kg per single dose should be given (Kemper et al., 1990). Oral DMPS treatment in adults may be given with an initial dose of 100-300 mg and continued with 100 mg every 6 or 8 hours. In children, the oral dosage is 5 mg/kg/day.

SUCCIMER OR MESO 2,3-DIMERCAPTOSUCCINIC ACID (DMSA)

The one chemical derivative of dimercaprol, which has gained more and more attention these days, is DMSA also known as Succimer. Succimer is an orally active chelating agent, much less toxic than BAL and its therapeutic index is about 30 times higher (Kuntzelman et al., 1990). US FDA has approved this compound in 1991 for the treatment of children whose blood lead concentration was above 45 µg/dL (FDA, 1991). The empirical formula of DMSA is $C_4H_6O_4S_2$ and its molecular weight is 182.21. It's a weak acid soluble in water.

DMSA distribution is predominantly extracellular since it is unable to cross hepatic cell membrane and excreted by the kidney with a half-life of about two days (Aaseth 1989). Over 95% of blood DMSA is bound mainly to albumin (Maiorino et al., 1990). DMSA appears to be transported by plasma albumin. It has been reported that 2-4 hours after DMSA administration only 12% of meso DMSA excreted in urine was unaltered whereas about 88% oxidized to form disulfates (DMSA attached to one or 2 cysteine molecules). No mixed disulfates are found in the blood (Miller 1998, Aposhian et al., 1992, Maiorino et al., 1989, 1990). The absorption of DMSA after oral administration is about 60%. Studies addressing the possibility that DMSA may chelate metal stored in the gut, because a significant percentage of an oral dose is not absorbed, have yet to be elucidated (Miller, 1998).

The LD_{50} value of sodium salts of DMSA in mice are: iv 2.4, im 3.8, ip 4.4 and po 8.5 g/kg respectively. Using a percutaneous route the acute LD_{50} for rats and mice is about 2 g/kg. Graziano et al. (1978) reported that i.p. administration of 200 mg/kg DMSA could produce only a marginal change in growth but did not elicit any appreciable change in histopathological alterations in tissue or cause hematological or biochemical change in blood. No significant loss of essential metals like zinc, iron, calcium or magnesium was observed. A slight increase in transaminase activities in serum of human and animals has been reported after DMSA treatment (Graziano et al., 1985; Flora and Kumar, 1993). Adverse reaction to DMSA includes gastrointestinal discomfort, skin reaction, mild neutropenia and elevated liver enzymes. No redistribution of lead also occurred on DMSA administration in rats (Cory Slechta, 1987).

Orally administered DMSA caused no marked adverse reactions but some sulphurous odor in the mouth, weakness, abdominal distension and anorexia. These reactions were mild and disappeared quickly after withdrawal of DMSA. No pathological findings were observed in the blood, urine, ECG or ultrasonography of liver and spleen.

Maternal and developmental toxicity was evaluated in pregnant mice injected subcutaneous with 410, 820 and 1640 mg/kg/day on gestation days 6 through 15. Effects on fetal development were seen in the 1640 mg/kg/day group, with the drug resulting in a significant increase in prenatal death, fetal retardation and incidence of stunning, fetal body weight loss and soft tissue abnormalities (Domingo et al. 1988). When DMSA was given orally, the drug was embryofetotoxic at lower doses also. The embryo/fetal toxicity of DMSA was suggested to be the result of direct contact of the chelator and/or its metabolites with the embryo or fetal tissue more than the indirect result of maternal toxicity. There is no reported teratogenic effect of DMSA reported till date either by sc or oral route (Domingo et al., 1990a).

It has also been found that DMSA resulted in low maternal liver copper and calcium concentration whereas high iron levels, while the fetal copper, calcium and zinc levels decreased (Paternain et al., 1990). Although the results suggest that DMSA induced developmental toxicity was due to an induced zinc deficiency, additional investigations showed that the embryo/fetal toxicity of DMSA might be mediated, at least in part, through altered fetal copper metabolism (Taubeneck et al., 1992). In contrast, the oral route did not cause any adverse affects on the offspring survival and development (Domingo et al., 1990b).

Although these doses are very high as compared to the usual drug dose given to human metal intoxication (10 to 30 mg/kg/day, po) (Aposhian and Aposhian, 1990). If the drug has to be used clinically, its adverse effects on pregnancy, maternal and fetal essential metal status should be evaluated. As to date there is no available data on the effects of DMSA on pregnant women (Domingo, 1998).

DMSA has been tried successfully in animal as well as in few cases of human arsenic poisoning (Flora and Tripathi, 1998). DMSA has been shown to protect mice due to lethal effects of arsenic. A subcutaneous injection of DMSA provided 80-100% survival of mice injected with sc sodium arsenite (Ding and Liang, 1991). Flora and Tripathi (1998) also reported a significant depletion of arsenic and a significant recovery in the altered biochemical variables of chronically arsenic exposed rats. This drug can be effective if given either oral or i.p. route. Patients treated with 30 mg/kg DMSA per day for 5 days showed significant increase in arsenic excretion and a marked clinical improvement. In a case of attempted suicide by ingesting 2g of arsenic the patient was given a course of 300 mg DMSA orally every 6 hours for 3 days with good results (Aposhian, 1983). It has been recommended that for treating mild arsenic poisoning an oral dose of 10 mg/kg DMSA thrice a day for 5-7 days may be given followed by two daily doses of 10 mg/kg for another 10-14 days. While for severe arsenic poisoning an oral dose of 18 mg/kg thrice a day for first 5-7 days followed by 2 doses of same strength for next 10-14 days are recommended. Number of other studies appeared in the recent past have recommended that DMSA could be safe and effective for treating arsenic poisoning. However, in a double blind, randomized controlled trial study conducted on few selected patients from arsenic affected West Bengal (India) regions with oral administration of DMSA suggested that DMSA was not effective in producing any clinical or biochemical benefits or any histopathological improvements of skin lesions (Guha Mazumder et al., 1998). In an experimental study recently conducted, provided an *in vivo* evidence of arsenic induced oxidative stress in number of major organs of arsenic exposed rats and that these effects can be mitigated by pharmacological intervention that encompasses combined treatment with N-acetylcysteine and DMSA (Flora, 1999).

US FDA has recently licensed the drug DMSA for reduction of blood lead levels. It was reported that EDTA increases the lead content in the brain due to redistribution (Cory-Slechta, 1987). DMSA when administered either alone or in combination with EDTA decreases the lead concentration in the brain (Cory-Slechta, 1987; Flora et al., 1995). Besunder et al. (1997) too recently confirmed these findings in rats and recommended the administration of DMSA and EDTA to children hospitalized for combined chelation therapy.

ESTERS OF SUCCIMER (DMSA)

A large number of esters of DMSA have been synthesized for achieving optimal chelation as compared to DMSA. These esters are mainly the mono and dimethyl esters of DMSA that have been studied experimentally with the aim of enhancing tissue uptake of chelating agents (Aposhian et al., 1992). In order to make the compounds more lipophillic the carbon chain length of the parent DMSA was increased by controlled esterification with the corresponding alcohol (methyl, ethyl, propyl, isopropyl, butyl, isobutyl, pentyl, isopentyl and hexyl). A large number of esters have been synthesized and are being tried for the treatment of metal poisoning. It has also been reported that these mono and diesters have a better potential in mobilizing cadmium and lead from the tissues in mice (Jones et al., 1992, Walker et al., 1992). Rivera et al. (1989) reported that that the dimethyl ester of DMSA (meso-DiMeDMSA) increased the excretion of cadmium. They also reported that when rabbit liver metallothionein was incubated with the diester, 32% of the cadmium and 87% of zinc bound metallothionein was removed from the system (Rivera et al., 1991). Although, the diester entered the cell but it caused severe zinc depletion (Rivera et al., 1991). Singh et al. (1989) examined the efficacies of three diesters of DMSA and found that these diesters were effective in reducing the soft organ lead concentrations when compared to BAL. Kreppel et al. (1993) reported the therapeutic efficacy of six analogues of DMSA in mice. They administered mice with a single LD_{80} dose of arsenic trioxide followed by a single dose of these six analogues of DMSA. They found that meso 2,3-di(acetylthio) succinic acid (DATSA) and 2,3-di(benzoylthio) succinic acid (DBTSA) increased the survival rates by 29% and 43% respectively when administered via gastric tube (i.g) and 89% when administered intraperitoneally (i.p). Administration of dimethyl DMSA (DMDMSA) through i.g and i.p and diethyl DMSA (DEDMSA), di-n-propyl DMSA (DnPDMSA) and diisopropyl DMSA (DiPDMSA) through i.g route did not reduce the lethality. While the i.p. administration of DnPDMSA increased the survival rate by 72% whereas DEDMSA and DiPDMSA increased it by 86% (Kreepel et al., 1993). Kreepel et al. (1995) also reported that effects of 4 monoesters of DMSA in increasing the survival and arsenic elimination in various organs in mice. It was observed that all the monoesters, MiADMSA (mono- isoamyl), MnDMSA (mono n-amyl), MnBDMSA (mono n-butyl) and MiBDMSA (mono i-butyl) markedly decreased the arsenic content in most of the organs as soon as 1.5 hrs after administration. They found that MiADMSA and MnADMSA were the most effective in increasing the survival of mice (Kreepel et al., 1995). Similar studies were also performed by Flora et al. (1997) where they investigated the effect of DMDMSA, DEDMSA DiPDMSA and diidoamyl DMSA (DiADMSA) on sub chronically arsenic treated rats. The results suggested that the diesters reduced the arsenic burden in blood and soft tissue but were only moderately effective in reversing the biochemical recoveries when compared to DMSA (Flora et al., 1997).

Gale et al. (1993) reported therapeutic effects of monoesters of DMSA against mercury intoxication. They evaluated 7 monoesters of DMSA in mobilizing and promoting the excretion of mercury in mice and observed that after the first dose of DMSA and DMPS corporal mercury burden reduced by 16% and 24% respectively while in animals treated with the monoesters, the reduction varied from 35% with MEDMSA to 49% with MiADMSA. After the second dose, DMSA and DMPS caused a decrease in mercury burden by 24% and 38% respectively while with the monoesters it varied from 52% MEDMSA to 61% with MnBDMSA (n-butyl DMSA).

Walker et al. (1992) studied the effects of seven different monoalkyl esters of DMSA on the mobilization of lead in mice and observed that after a single parenteral dose of the chelator DMSA there was a 52% reduction in the lead concentrations while with the monoesters the reduction varied from 54% to 75%. Jones et al. (1992) reported the efficacy of ten different monoesters through oral and i.p. route on cadmium mobilization in mouse. Out of the ten monoesters studied they found MiADMSA to be the most effective in reducing the cadmium concentrations from the liver and kidneys.

In all of the reported literature, it was observed that the analogues of DMSA are capable of crossing the membranes and was more effective in reducing the metal burden in acute and sub-chronic metal intoxication. Most of the studies have also suggested that the monoesters are more effective in treatment of experimentally induced metal intoxication.

MONOISOAMYL DMSA (MiADMSA)

Among these new chelators, monoisoamyl ester of DMSA (MiADMSA; a C_5 branched chain alkyl monoester of DMSA) has been found to be the more effective than DMSA in reducing cadmium and mercury burden (Xu et al., 1995; Gale et al., 1993). It is reported that the toxicity of DMSA with LD_{50} of 16 mmol/kg is much lower than the toxicity of MiADMSA with LD_{50} of 3 mmol/kg but lesser than BAL (1.1 mmole/kg). The interaction of MiADMSA and DMSA with essential metals is same. Mehta and Flora (2001) reported for the first time the comparison of different chelating agents (3 amino and 4 thiol chelators) on their role on metal redistribution, hepatotoxicity and oxidative stress in chelating agents induced metallothionein in rats. We suggested that out of all the 7 chelators, MiADMSA and DMSA produced the least oxidative stress and toxicity as compared to all other 5 chelators (Mehta and Flora, 2001). However, no reports are available about the toxicity of this metal complexing agent except for its developmental toxicity. No observed adverse effect levels (NOAELs) for maternal and developmental toxicity of MiADMSA were 47.5 mg/kg and 95 mg/kg/day respectively indicating that MiADMSA would not produce developmental toxicity in mice in the absence of maternal toxicity (Blanusa et al., 1997). Bosque et al. (1994) reported that administration of MiADMSA through the parenteral route to pregnant mice during organogenesis produced maternal toxicity at a dose of 95 and 195 mg/kg with a significant decrease in the body weight and an increase in the liver weights. They also reported that MiADMSA caused embryo/fetotoxicity at a dose of 190 mg/kg by significantly increasing the embryolethality and non-significant increase in the skeletal defects. Taubeneck et al. (1992) showed that the developmental toxicity of DMSA is mediated mainly through disturbed copper metabolism and this may also be true for MiADMSA. Recently, our group was the first to report the toxicological data of MiADMSA when administered in male and female rats (Mehta et al., 2002; Flora and Mehta, 2003) through the oral as well as the intraperitoneal route (25, 50 and 100 mg/kg /3 weeks). We observed that there was no major alteration in the heme biosynthesis pathway except for a slight rise in the zinc protoporphyrin levels suggesting mild anemia at the highest dose. The oral route of administration was also seen to be better when compared to the ip route based on the histopathological studies of the liver and kidney tissues. MiADMSA was seen to be slightly more toxic in terms of copper loss and some biochemical variable in the hepatic tissue in females as compared to male rats. The studies concluded that the administration of MiADMSA in female rats is confounded with side effects and may require caution during its use (Mehta et al., 2002; Flora and Mehta, 2003). Since administration of a chelating agents during pregnancy is always with caution, we studied the effects of MiADMSA administration from day 14 of gestation to day 21 of lactation at different doses through oral and ip routes to examine the maternal and developmental toxicity in the pups (Mehta and Flora, 2003a unpublished report). Results suggested that MiADMSA had no effect on length of gestation, litter-size, sex ratio, and viability and lactation. No skeletal defects too were observed following the administration of the chelator. However, MiADMSA administration produced some marginal maternal oxidative stress at the higher doses (100mg/kg and 200 mg/kg) based on thiobarbituric acid reactive substances (TBARS) in RBCs and decrease in the δ-aminolevulinic acid dehydratase (ALAD) activity. MiADMSA administration too caused some changes in the essential metal concentration in the soft tissues especially the copper loss in lactating mothers and pups, which would be of some concern. Apart from copper, changes too were observed in the zinc concentrations in mothers and pups following administration of MiADMSA. The study further suggested that the chelator could be administered during pregnancy as it does not cause any major alteration in the mothers and the developing pups (Mehta and Flora, 2003a unpublished report). Since chelating agents are administrable to individuals of all ages, we investigated the effect of MiADSMA administration in different age groups of male rats (young, adult and old rats) based on the fact that whether MiADMSA, a dithiol agent was a prooxidant or an antioxidant (Mehta and Flora, 2003b unpublished report). Results suggested that MiADMSA administration increased in activity of ALAD in all the age groups and increased blood GSH levels in young rats. MiADMSA also potentiated the synthesis of MT in liver and kidneys and GSH levels in liver and brain. Apart from this it also significantly reduced the GSSG levels in tissues. MiADMSA was found to be safe on adult rats followed by young and old rats (Mehta and Flora, 2003b unpublished report).

A large number of reports are available on the therapeutic efficacy of the MiADMSA. Pande et al. (2001) found that MiADMSA was effective in prevention and treatment of acute lead intoxication. Walker et al. (1992) reported that MiADMSA administration reduced the brain lead concentrations by 75% when compared to 35% with DMSA whereas the ip administration reduced kidney lead levels by 93% while oral administration reduced the kidney lead by 94% (Walker et al., 1992). MiADMSA completely prevented the testicular damage after intraperitoneal administration of cadmium chloride at a dose of 0.03 mmol/kg (Xu et al., 1995). Jones et al. (1992) reported that MiADMSA enhanced the cadmium elimination through urine by 3.6% compared to 0.02% of the controls and 24% in faeces compared to 0.11% in controls.

Therapeutic effects of MiADMSA against mercury burden have shown that MiADMSA is capable of decreasing mercury concentration by 59% and 80% after two doses when compared to DMSA (25% and 54% respectively). The total corporal mercury burden 29.25mg was reduced to 21.06mg with DMSA after a single injection of 0.5 mmol/kg. The same dose of MiADMSA effected a reduction to 12.09 mg (Gale et al., 1993). Belles et al. (1996) assessed the protective activity of MiADMSA against methyl mercury-induced maternal and embryo/fetal toxicity in mice. Oral methyl mercury administration increased the number of resorptions, decreased fetal weights and increased skeletal abnormalities. MiADMSA administration could not reverse the embryolethality but fetotoxicty was significantly reduced by the administration of these agents at different doses.

Recently, Flora et al. (2002a) reported the effect of MiADMSA on the reversal of gallium arsenide (GaAs) induced changes in the hepatic tissue. Rats were exposed for 24 weeks with 10 mg/kg GaAs, orally, once daily and treated with 0.3 mmol/kg of MiADMSA or DMSA for two courses. They observed that MiADMSA was better than DMSA in mobilizing arsenic and in the turnover of the GaAs sensitive biochemical variables. Histopathological lesions, also responded more favorably to chelation therapy with MiADMSA. In another study, does dependent therapeutic potential of MiADMSA was compared with monomethyl ester and DMSA in sub-chronically GaAs treated rats and it was found that MiADMSA was highly effective in the reversal of altered biochemical variables and in the mobilization of arsenic (Flora et al., 2002b).

Dose and route dependent efficacy of MiADMSA against chronic arsenic poisoning has also suggested that the chelator is highly effective through oral route in reversing the arsenic induced changes in the variables indicative of oxidative stress in major organs as well as in mobilization of arsenic (Flora et al., 2002c). Kreppel et al. (1995) reported that MiADMSA was effective in increasing the survival of arsenic exposed mice when compared to its parent DMSA.

Despite a few drawbacks/side effects associated with MiADMSA, the above results suggests that MiADMSA may be a future drug of choice owing to its lipophilic character and the absence of any metal redistribution. However, significant copper loss requires further studies. Moderate toxicity after repeated administration of MiADMSA may be reversible after the withdrawl of the chelating agent.

FUTURE CHALLENGES

Much of the current interest in chelating agents comes from the concerns about the possible beneficial effects of removal of toxic metals in patients with low-level exposure without any signs of toxicity. During the last 15 years, several important developments have occurred in the possibility and clinical practice of chelation therapy of acute and chronic metal intoxication. During this period DMSA and DMPS have gained more general acceptance among clinicians, undoubtly improving the management of many human metal intoxications. The other aspect that makes this field interesting is the toxicity caused by the administration of chelating agents. This toxicity can be in the form of nephrotoxicity (as in EDTA), essential metal imbalance (mostly all chelating agents) or developmental and post-natal defects (as in EDTA, DPA). These challenges make the administration of most of the chelating agents restricted. However, efforts have been made to synthesis better chelating agents (like analogue of DMSA) that are devoid of such complications but it is very important to evaluate these compounds for their toxicity when administered individually without administration of any toxic metal through different route in different animal models. Once the toxicity data is generated, more precise dosage of administration could be achieved without any major side effects in patients suffering from low levels of metal intoxication.

Apart from this, still further knowledge is needed in several basic research areas within the field of *in vivo* chelation of metals and call for studies on (a) Molecular mechanism of action of clinically important chelators, (b) Intracellular and extracellular chelation in relation to mobilization of aged metal deposits and the possible redistribution of toxic metal to sensitive organs as the brain, (c) Effect of metal chelators on biokinetics during continued exposure to metal, especially possible enhancement or reduction of intestinal metal uptake, (d) Combined chelation with lipophilic and hydrophilic chelators, which presently has a minimal clinical role, (e) Minimization of the mobilization of essential trace elements during long-term chelation, (f) Fetotoxic and teratogenic effects of chelators.

References

Aaseth J.O. Presentation of preliminary review on DMSA for heavy metal poisoning. Annual Metal European Associ. of Poison Control, 1989.

Aaseth J. Recent advances in the therapy of metal poisoning with chelating agents. Hum. Toxicol. **2:** 257-272, 1983.

Anderson O. Principle and recent development in chelation treatment of metal intoxication. Chem. Rev. **99:** 2683-2710, 1999.

Angle C.R. Chelation therapies for metal intoxication. In Toxicology of Metals ed. Chang L.W. CRC Press, 487-504, 1996.

Aposhian H.V. DMSA and DMPS- water soluble antidotes for heavy metal poisoning. Ann. Rev. Pharmacol. Toxicol. **23:** 193-215, 1983.

Aposhian H.V. Aposhian MM. Meso-2, 3-dimercapto-succinic acid: chemical, pharmacological and toxicological properties of an orally effective metal chelating agent. Ann. Rev. Pharmacol. Toxicol. **30:** 279-306, 1990.

Aposhian H.V., Maiorino R.M. and Reviera M. Human studies with the chelating agents, DMPS and DMSA. J. Toxicol. Clin. Toxicol. **30:** 505-528, 1992.

Belles M., Sanchez D.J., Gomez M., Domingo J.L., Jones M.M., Singh P.K. Assessment of the protective activity of monoisoamyl meso-2,3-dimercaptosuccinate against methylmercury-induced maternal and embryo/fetal toxicity in mice. Toxicol. **106:** 93-97, 1996.

Benov L.C., Benchev I.C., Monovich O.H. Thiol antidotes affect on lipid peroxidation in mercury-poisoned rats. Chem Biol Interact. **76:** 321-332, 1990.

Benov L.C., Ribarov S.R. and Monovich O.H. Studies of activated oxygen productionby some thiols using chemiluminiscence. Gen Physiol Biophys. **11:** 195-202, 1992.

Besunder J.B., Super D.M., Anderson R.L. Comparison of dimercaptosuccinic acid and calcium disodium ethylenediaminetetraacetic acid versus dimercaptopropanol and ethylenediaminetetraacetic acid in children with lead poisoning. J. Pediatr. **130:** 966-971, 1997.

Blanusa M., Prester L., Piasek M., Kostial K., Jones M.M. and Singh P.K. Monoisoamyl ester of DMSA reduces $^{203}Hg(NO_3)_2$ retention in rats: 1. Chelation therapy during pregnancy. J. Trace Elem. Exp. Med., **10:** 173-181, 1997.

Bosque M.A., Domingo J.L., Cobella J., Jones M.M., Singh P.K. Developmental toxicity evaluation of monoisoamyl meso 2,3- dimercaptosuccinate inmice. J. Toxicol. Environ. Health. **42:** 443-448, 1994.

Brownie C.F., Brownie C., Noden D., Krook L., Haluska M., Aronson A.L. Teratogrnic effect of calcium edetate (CaEDTA) in rats and the protective effect of zinc. Toxicol. Appl. Pharmacol., **82:** 426-443, 1986.

Buscha D.H. and Stephenson N.A. Molecular organization, portal to supramolecular chemistry. Structural analysis of the factors associated with molecular organization in co-ordination and inclusion chemistry including the cordination template effects. Coord. Chem. Rev. **100:** 119, 1990.

Cantilena Jr L.R. and Klaassen C.D. Comparison of effectiveness of several chelators after single administration on the toxicity, excretion, and disribution of cadmium. Toxicol. Appl. Pharmacol. **58:** 452-460, 1981

Clevette D.J. and Orivig C. Comparison of ligand of different denticity and basicity for the in vivo chelation of aluminum and gallium. Polyhedron. **9:** 151-161, 1990.

Cory- Slechta D.A. Mobilization of lead over the course of DMSA chelation therapy and long term efficacy. J. Pharmacol. Exp. Ther. **246:** 84-91,1988

Cory-Slechta D.A., Weiss B. and Cox C. Mobilization and redistribution of lead over the course of calcium disodium ethylenediamine tetraacetate chelation therapy, J. Pharmacol. Exp. Ther. **243:** 804-813, 1987.

Ding G.S. and Liang Y.Y. Antidotal effects of dimercaptosuccinic acid. J. Appl. Toxicol. **11:** 7-14, 1991.

Domingo J.L. Developmental toxicity of metal chelating agents. Reprod. Toxicol. **12:** 499-510, 1998.

Domingo J.L., Ortrga A., Paternain J.L., Llobet J.M. and Corbella J. Oral meso 2,3-dimercaptosuccinic acid in pregnant Spraque-Dawley rats: Teratogenicity and alterations in mineral metabolism. I. Teratological evaluation. J. Toxicol. Environ. Health. **30:** 181-190, 1990a.

Domingo J.L., Bosque M.A. and Corbella J. Effects of oral meso 2,3-dimercaptosuccinic acid (DMSA) administration on late gestation and post natal developments in mouse. Life Sci. **47:** 1745-1450, 1990b.

Domingo J.L., Paternain J.L., Llobet J.M., Corbella J. Developmental toxicity of subcutaneously meso 2,3-dimercaptosuccinic acid in mice. Fund. Appl. Toxicol. **11:** 715-722, 1988.

FDA (Food and Drug Administration), Succimer (DMSA) approved for severe lead poisoning. JAMA, **265:** 1802, 1991.

Flora S.J.S. Arsenic induced oxidative stress and its reversibility following combined administration of N-acetyl cyseine and meso 2,3-dimercaptosuccinic acid in rats. Clin. Exp. Pharmacol. Physiol. **26:** 865-869, 1999.

Flora S.J.S. and Kumar P. Biochemical and Immunological evaluation of metal chelating drugs in rats. Drug Invest. **5:** 269-273, 1993.

Flora S.J.S. and Mehta A. Haematological, hepatic and renal alterations after repeated oral and intraperitoneal administration of monoisoamyl DMSA. II. Changes in female rats. J. Appl. Toxicol. **23:** 97-102, 2003.

Flora S.J.S. and Tandon S.K. Beneficial effects of zinc supplementation during chelation treatment of lead intoxication in rats. Toxicol. **64:** 129-139, 1990.

Flora S.J.S., Tandon S.K. Influence of calcium disodium edetate on the toxic effects of lead administration in pregnant rats. Ind. J. Physiol. Pharmacol. **31:** 267-272, 1987.

Flora S.J.S. and Tripathi N. Treatment of arsenic poisoning: an update. Ind. J. Pharmacol. **30:** 209-217, 1998.

Flora .J.S., Dubey R., Kannan G.M., Chauhan S., Pant, BP, Jaiswal, DK. meso 2,3-dimercaptosuccinic acid (DMSA) and monoisoamyl DMSA effect on gallium arsenide induced pathological liver injury in rats. Toxicol. Lett., 132, 9-17, 2002a

Flora S.J.S., Mehta, A., Kannan G.M., Dube S.N. and Pant B.P. Therapeutic Potential of Few Monoesters of Meso 2,3-dimercaptosucccinic acid in Experimental Gallium Arsenide Intoxication in Rats. Toxicology, In Press, 2004.

Flora S.J.S., Dube S.N., Tandon S.K. Chelating agents and their use in metal poisoning. In Modern Trends in Environmental Biology (G. Tripathi, Ed.), CBS Publisher, New Delhi, India, 209-227, 2002c.

Flora S.J.S., Bhattacharya R, Vijayaraghavan R. Combined therapeutic potential of meso 2,3- dimercaptosuccinic acid and calcium disodium edetate in the mobilization and distribution of lead in experimental lead intoxication in rats. Fund. Appl. Toxicol. **25:** 233-240, 1995.

Flora S.J.S., Pant B.P., Tripathi, N. Kannan, GM, Jaiswal, DK. Therapeutic efficacy of a few diesters of meso 2,3- dimercaposuccinic acid during sub-chronic arsenic intoxication in rats. J. Occup. Health **39:** 119-123, 1997

Forland M., Pullman T.N., Lavender A.R. and Aho I. The renal excretion of ethylenediamine tetraacetic acid in the dog. J. Pharmacol. Exp. Ther. **153:** 142-147, 1966.

Gabard B. Removal of internally deposited gold by 2,3-dimercaptopropane sodium sulfonate (Dimaval) Brit. J. Pharmacol. **68:** 607-610, 1980.

Gale G.R., Smith A.B., Jones M.M. and Singh P.K. Meso-2,3-dimercaptosuccinic acid monoalkyl esters: effects on mercury levels in mice. Toxicol. **81:** 49-56, 1993.

Gersl V., Hrdina R., Vavrova J., Holeckova M., Palicka V., Vogkova J., Mazurova Y. and Bajgar J. Effects of repeated administration of dithiol chelating agent- sodium 2,3-dimercapto 1-propanesulphonate (DMPS)- on biochemical and heamatological parameters in rabbits. Acta Medica. **40:** 3-8, 1997.

Graziano J.H., Siris E.S., LoIacono N., Silverberg S.J., Turgeon L. 2, 3-dimercaptosuccinic acid as an antidote for lead intoxication. Clin. Pharmacol. Ther., **37:** 431-438, 1985

Graziano J.H., Cuccia D. and Friedheim E. The pharmacology of 2,3-dimercaptosuccinic acid and its potential use in arsenic poisoning. J. Pharmacol. Exp. Ther., **207:** 1051-1055, 1978.

Guha Mazumder D.N., Ghoshal U.C., Saha J., Santra A., De B.K., Chatterjee A., Dutta S., Angle C.R. and Centeno J.A. Randomized placebo-controlled trial of 2,3-dimercapto succinic acid in therapy of chronic arsenicosis due to drinking arsenic contaminated subsoil water. Clin. Toxicol. **36:** 683-690, 1998.

Hammound P.B. Effect of chelating agents on the tissue distribution and excretion of lead. Toxicol. Appl. Pharmacol. **18:** 296-310, 1971

Hammond P.B., Aronson A.L. and Olson W.C. The mechanism of mobilization of lead by ethylenediaminetetraaceticacid. J. Pharmacol. Exp. Ther. **157:** 196-206, 1967.

Hancock R.D. Molecular mechanics calculation and metal ion recognition. Acc. Chem. Res. **23:** 253 - 257, 1990

Hancock R.D. and Martell A.E. Ligand design for selective complexation of metal ion in aqueous solution. Chem. Rev. 89: 1875- 1914,1989.

Hartard C. and Kunze K. Pregnancy in a patient with Wilson's disease treated with D-pencillamine and zinc sulfate. A case report and review of literature. Eur. Neurol. **34:** 337-340, 1994.

Hauser W. and Weger N. Treatment of arsenic poisoning in mice with sodium dimercaptopropane 1-sulfonate. In Proc. Inter. Cong. Pharmacol., Paris, (abst), 1989.

Hruby K. and Donner A. 2,3-dimercapto 1-propane sulfonate in heavy metal poisoning. Med. Toxicol. **2:** 317-323, 1989

Hurlbut K.M., Maiorino R.M., Mayersohn M., Dart R.C., Bruce D.C. and Aposhian H.V. Determination and metabolism of dithiol chelating agents. XVI. Pharmacokinetics of 2,3 dimercapto 1-propanesulphonate after intravenous administration to humanvolunteers. J Pharmacol Exp Ther. **268:** 662-668, 1994.

Izatt R.M., Pawlak K., Bradshaw J.S. and Brevning RL. Thermodynamic and kinetic data for macrocycle interaction with cations and anions. Chem. Rev. **91:** 1721-2085, 1991.

Jones M.M. Design of new chelating agents for removal of intracellular toxic metals, In coordination chemistry: A century of progress, Eds. Kauffman, GB, American Chemical Society, 427-438, 1994.

Jones M.M.. New developments in therapeutic chelating agents as an antidote for metal poisoning. Crit. Rev. Toxicol.. **21:** 209-233, 1991.

Jones M.M., Basinger M.A. A hypothesis for the selection of chelate antidotes for metal poisoning. Med. Hypotheses **9:** 445-453, 1982.

Keen C.L., Mark-Savage P., Lonnerdal B., Hurley L.S. Teratogenic effects of D-pencillamine in rats. Relation to copper deficiency. Drug-Nutrient Interact. **2:** 17-34, 1983.

Keen C.L., Lonnerdal B. and Hurley L.S. Teratogenic effects of copper deficiency and exces. In: Sorenson, JRJ ed, Inflammatory diseases and copper. Clifton, NJ: The human press, pp 109-121, 1982a.

Keen C.L., Mark-Savage P., Lonnerdal B. and Hurley L.S. Teratogenesis and low copper status resulting from D-pencilliamine in rats. Teratol. **26:** 163-165, 1982b.

Kemper F.H., Jekat F.W., Bertram H.P. and Eckard R. New Chelating Agents. In Basic Science in Toxicology. Volnasis GN, Srinis J, Sullivan FW, Turner P. Taylor & Francis Brighton, England, 523-546, 1990

Kilburn R.H. and Hess R.A. Neonatal deaths and pulmonary dysplasia due to D-pencillamine in the rat. Teratology. **26:** 1-9, 1982.

Kimmel C.A. Effect of route of administration on the toxicity of EDTA in the rat. Toxicol. Appl. Pharmacol. **40:** 299-306, 1977.

Klaassen C.D. Heavy metals and heavy metal antagonist in Goodman and Gilman‘s. The Pharmacological Basis of Therapeutics, Pergamon Press. USA. 1592-1614, 1990.

Klimmek R., Krettek C. and Werner H.W. Acute effects of the heavy metal antidotes DMPS and DMSA on circulation, respiration and blood homeostasis in dogs. Arch. Toxicol., **67:** 428-434, 1993.

Kontoghiorghes G.J., Sheppard L. and Chambers S. New synthetic approach and iron chelating studies on 1-alkyl-2-methyl 3-hydroxyl pyrid-4-ones. Arzneim Forsch/Drug Res. **37:** 1099-, 1987

Kreppel H., Reichl F.X., Kleine A., Szinicz L., Singh P.K. and Jones M.M. Antidotal efficacy of newly synthesized dimercaptosuccinic acid (DMSA) monoesters in experimental arsenic poisoning in mice. Toxicol. Appl. Pharmacol. **26:** 239-245, 1995

Kreppel H., Paepcke U., Thiermann H., Szinicz L., Reichl F.X., Singh P.K., Jones, MM. Therapeutic efficacy of new dimercaptosuccinic acid (DMSA) analogues in acute arsenic trioxide poisoning in mice. Arch. Toxicol. **67:** 580-585, 1993.

Kreppel H., Reichl F.X., Forth W. and Fichtl B. Lack of effectiveness of D-pencillamine in experimental arsenic poisoning. Vet. Hum. Toxicol. **31:** 1-5, 1989.

Kuntzelman D.R., England K.E. and Angle C.R. Urine lead (UPb) in outpatient treatment of lead poisoning with dimercaptosuccinic acid (DMSA). Vet. Human Toxicol. **4:** 364-371, 1990.

Lindoy L.F. The chemistry of macrocyclic ligand complexes. Cambridge University press, Cambridge, 1989.

Maiorino R.M., Akins J.M. and Blaha K. Determination and metabolism of dithiol chelating agents X: In humans meso 2,3-dimercaptosuccinic acid is bound to plasma proteins via mixed disulfide formation. J. Pharmacol. Exp. Ther. **254:** 570-577, 1990.

Maiorino R.M., Bruce D.C. and Aposhian H.V. Determination and metabolism of dithiol chelating agents VI: Isolation and identification of the mixed disulfides of meso 2,3-dimercaptosuccinic acid with L-cysteine in human urine. Toxicol. Appl. Pharmacol. **97:** 338-349, 1989.

Maiorino R.M., Weber G.L. and Aposhian H.V. Determination and metabolism of dithiol chelating agents. III. Formation of oxidized metabolites o 2,3- dimercaptopropane 1-sulfonic acid in rabbit. Drug Metabol. Disposit., **16:** 455-463, 1988.

Mark-Savage P., Keen C.L. and Hurley L.S. Reduction of copper supplementation of teratogenic effects of D-pencillamine. J. Nutr., 113, 501-510, 1983

Mehta A. and Flora S.J.S. Possible role of metal redistribution, hepatotoxicity and oxidative stress in chelating agents induced hepatic and renal metallothionein in rats. Fd Chem Toxicol., **39,** 1039-1043, 2001.

Mehta A. and Flora S.J.S. MiADMSA and DMSA induced changes in pregnant female rats during late gestation and lactation. 2003a unpublished report.

Mehta A. and Flora S.J.S. Prooxidant / Antioxidant effects of MiADMSA in male rats: age related effects. 2003b, unpublished report.

Mehta A., Kannan G.M., Dube S.N., Pant B.P., Pant S.C., Flora S.J.S. Haematological, hepatic and renal alterations after repeated oral and intraperitoneal administration of monoisoamyl DMSA. Part I. Changes in male rats. J Appl. Toxicol. **22:** 359-369, 2002.

Miller A.L. Dimercaptosuccinic acid (DMSA) a non-toxic, water-soluble treatment for heavy metal toxicity, Altern. Med. Res. **3:** 199-207, 1998.

Mjolnerod O.K., Rasmussen K., Dommerund S.A. and Gjerulsen S.T. Congenital connrctive-tissue defect probably due to D-pencillamine treatment in pregnancy. Lancent, **1,** 673-675, 1971.

Myint B. D-pencillamine-induced cleft palate in mice. Teratology, **30,** 333-340, 1984.

Netter P., Bannwaith B., Pera P. and Nicolas A. Clinical pharmacokinetics of D-pencillamine. Clin. Pharmacokinet. **13:** 317-333, 1987.

Ohlsson W.T.L. Penicillamine as lead-chelating substance in man. BMJ, ii, 1454-1456, 1962.

Pande M., Mehta A., Pant B.P. and Flora S.J.S. Combined administration of a chelating agent and an antioxidant in the prevention and treatment of acute lead intoxication in rats. Environ. Toxicol. Pharmacol. **9:** 173-184, 2001.

Paternain J.L., Ortrga A., Domingo J.L., Llobet J.M. and Corbella J. Oral meso 2,3-dimercaptosuccinic acid in pregnant Spraque-Dawley rats: Teratogenicity and alterations in mineral metabolism. II. Effects on mineral metabolism. J. Toxicol. Environ. Health. **30:** 191-197, 1990.

Peter R.A., Stocken L.A. and Thompson R.H.S. British anti-lewsite (BAL). Nature. **156:** 616-619, 1945.

Planas Bohne F., Gabard B. and Schaffer E.H. Toxicological studies on sodium 2,3-dimercaptopropane 1-sulfonate in the rat. Arzneim Forch/ Drug Res., **30:** 1291-1294, 1980.

Powell J.J., Burden T.J., Greenfield S.M., Taylor P.D. and Thompson R.P.H. Urinary excretion of essential metal following intravenous calcium disodium edetate: an estimate of free zinc and zinc status in man. J. Inorgan. Biochem. **75:** 159-165, 1999.

Rivera M., Levine D.J., Aposhian H.V. and Fernando Q. Synthesis and properties of monomethyl ester of meso dimercaptosuccinic acid and its chelators of lead (II), cadmium (II), mercury (II). Chem. Res. Toxicol. **4:** 107-114, 1991.

Rivera M., Zheng W., Aposhian H.V. and Fernando Q. Determination and metabolism of dithiol chelating agents VIII: metal complexes of meso dimercaptosuccinic acid. Toxicol. Appl. Pharmacol. **100:** 96-106, 1989.

Roussaeux C.G. and MacNabb L.G. Oral administration of D-pencillamince causes neonatal mortality without morphological defects in CD-1 mice. J. Appl. Toxicol. **12:** 35-38, 1992.

Shannon, M, Graef, J, Lovejoy FH. Efficacy and toxicity of D-penicillamine in low level lead poisoning. J. Pediatr., **112,** 799-804, 1988.

Singh P.K., Jones M.M., Gale G.R., Atkins L.M. and Smith A.B. The mobilization of intracellular cadmium by butyl and amyl esters of meso 2,3-dimercaptosuccinic acid. Toxicol. Appl. Pharmacol., **97,** 572-579, 1989.

Soong Y.K., Huang H.Y., Huang C.C. and Chu N.S. Successful pregnancy after D-pencillamine therapy in patients with Wilson's disease. J. Formos. Med. Assoc. **90:** 693-696, 1991.

Szincicz L., Wiedeman P., Haring H. and Weger N. Effects of repeated treatment with sodium 2,3-dimercaptopropane-1-sulfonate in beagle dogs. Drug Res. **33:** 818-821, 1983.

Taubeneck M.W., Domingo J.L., Llobet J.M. and Keen C.L. Meso 2,3-dimeracaptosuccinic acid (DMSA) affects maternal and fetal copper metabolism in swiss mice. Toxicol. **72:** 27-40, 1992.

Tuchmann-Duplessis H. and Mercier-Parort L. Influence d'um corps de chelation l'acide ethylene diaminetetra acetique sur la gestation et le developmental doetal du rat. C.R. Acad. Sci. Paris. **243:** 1064, 1956.

Twarog T. and Cherian M.G. Chelation of lead by dimercaptopropane sulfonate and possible diagnostic uses. Toxicol. Appl. Pharmacol. **72:** 550-556, 1984.

Walker E.M., Stone A., Milligan L.B., Gale G.R., Atkins L.M., Smith A.B., Jones M.M., Singh P.K. and Basinger M.A. Mobilization of lead in mice by administration of monoalkyl esters of meso 2,3-dimercaptosuccinic acid. Toxicol. **76:** 79-87, 1992.

Walshe J.M. Pencillamine, a new oral therapy for Wilson's disease. Am. J. Med. **21:** 487, 1956.

Wieldemann P., Fichtl B. and Sizinicz L. Pharmacokinetic of ^{14}C-DMPS (^{14}C - 2.3-dimdercaptopropane 1-sulphonate) in beagle dogs. Biopharm. Drug Dispos. **3:** 267-274, 1982.

Xu C., Holscher M.A., Jones M.M. and Singh P.K. Effect of monoisoamyl meso-2, 3-dimercaptosuccinate on the pathology of acute cadmium intoxication. J. Toxicol. Environ. Health. **45:** 261-277, 1995.

Zheng W., Maiorino R.M. and Brendel K. Determination and metabolism of dithiol chelating agents. VII. Biliary excretion of dithiols and their interactions with cadmium and metallothionein. Fundam. Appl. Toxicol. **14:** 598-607, 1990.

Pharmacological Perspectives of Toxic Chemicals and Their Antidotes
Editors: S J S Flora, J A Romano, S I Baskin and K Sekhar

ECGs and Metabolic Networks: An *in silico* Exploration of Cyanide-caused Cardiac Toxicity

C.K. Zoltani[a], S.I. Baskin[b], G.E. Platoff[b]
[a]U.S. Army Research Lab
Computational and Information Sciences Directorate
Aberdeen Proving Ground, MD 21005-5066, USA
[b]U.S. Army Medical Research Institute of Chemical Defense
Pharmacology Division
Aberdeen Proving Ground, MD 21010-5400, USA

INTRODUCTION

The primary effect of cyanide (CN) on cardiac tissue is the interference with the cellular energy balance, that is, inhibition of the mitochondrial enzyme cytochrome oxidase preventing the transfer of electrons from cytochrome c to molecular oxygen. The impacted aerobic metabolism results in lactic acid acidosis and reduced adenosine triphosphate (ATP) concentration leading to anoxia. In tandem, with a reduction in cellular energy, a change in the calcium ion concentrations in several cellular compartments, calcium current modulation and an inotropic effect develops, Way (1984). Simultaneously, a gradual decrease in action potential duration is noted. The subsidiary cellular processes are still being elucidated, but the sine-non-quo for the development of effective therapeutic regimes is management of these effects. An *in silico* approach is presented in this paper showing how available data and models can be used in the development of new therapies.

Toward these ends, an ECG of an individual under CN intoxication is taken as the starting point, Wexler et al. (1947). The deviation from homeostasis, that is, from normal conditions, is interpreted in terms of the modulation of the membrane ionic currents that comprise the action potential. The nature of the deviation suggests the pharmacological intervention required to stop and possibly reverse the toxicology.

The need for a therapy for CN poisoning is especially urgent in situations where a dosage of one LD_{50} of CN is encountered, that is where the dosage of the toxin is 0.2 mg per kg of bodyweight or greater. In these cases, as shown by Baskin et al. (1987), (1991) and Ansell and Lewis (1970), the poison primarily affects the heart, expressed through effect on the ventricles. Reminiscent of severe ischemia, lesions in the heart tissue were noted by Suzuki (1968). Initially, upon exposure to the poison, the heart slows down into the bradycardia regime, then in time transits to arrhythmia. The ventricles are severely affected, expressed by symptoms of repolarization disorders culminating in Torsade de Pointes (TdP) and ventricular fibrillation (VF).

An overview of cardiac cell energy metabolism is addressed in the next section, setting the stage for a quantification of the changes brought about by cyanide perturbation of the metabolism of cardiac tissue. The following section delves into the issue of metabolic networks, and sketches the opportunities, in conjunction with high performance computing requirements, to deepen our understanding of the cellular processes. This is followed by an interpretation of the salient changes in the ECG of a subject under severe CN intoxication. The next section shows how *in silico* techniques can shed light on the modulations in the action potential caused by CN and point to the required characteristics of an antidote. The paper concludes with a discussion, acknowledgement and an extensive bibliography.

CARDIAC ENERGY METABOLISM

The cardiac myocyte derives its energy primarily from fatty acids, making up to 90% of the total energy required. The energy is derived from glucose and lactate metabolism, Opie (1998). The mitochondrion is the source of the ATP produced from adenosine diphosphate (ADP) and inorganic phosphate. Glucose oxidation, though only a minor part of the total energy budget of the cell, plays a crucial role in ion homeostasis. An elaborate system of enzymes regulates the glycolytic pathway. Glycolysis converts glucose to pyruvate which in turn is transformed to acetyl-CoA and enters the Krebs cycle. In a diseased state breakdown of glucose or glycogen to lactate replaces the oxidative metabolism for the production of anaerobic energy. In anoxia, ATP levels decrease with the cessation of oxidative metabolism and glycolysis provides limited amounts of energy.

The three stages of energy metabolism are the input to the Krebs cycle, the Krebs cycle where nicotinamide adenine dinucleotide (NADH) is produced, and the oxidative phosphorylation (OxPhos). Oxidization takes place by the electron transport chain, passing electrons to molecular oxygen by means of cytochrome oxidase to form water.

The Krebs cycle and glycolysis are the depositories of the energy obtained from catabolism of substrates. The products of the cycle are oxidized by the electron transport chain. ATP production is controlled at the substrate level by F_1F_0-ATPase by ADP through a negative feedback mechanism. That is, an increase in ATP/ADP ratio inhibits OxPhos. Also, homeostasis requires control of Mg^{2+} and Ca^{2+} levels, Wang et al. (2003).

Energy metabolism may be subject to a number of pathologies. Reduction of oxygen uptake due to CN leads to slowing of oxidative phosphorylation, and increase of NADH levels. In addition, acidification of the cell, increased ADP and creatine and decreased creatine phosphate is observed. Mitochondrial myopathies and cytosolic pathologies have been studied, shedding light on defects in the respiratory chain in electron transport and substrate defects that can affect fatty acid utilization. Cytochrome oxidase deficiency and defect in the respiratory chain, and decrease of enzyme activity affect the cellular energy metabolism.[1]

Calcium balance controls mitochondria respiration. Mitochondrial calcium acts as a buffer for the cytosol. Its importance is underlined by the fact that excess calcium under cytosolic calcium overload conditions limits the proton gradient thereby limiting the synthesis of ATP.

This short caricature of energy metabolism has not touched upon many of the subsidiary issues including the kinetics of key regulatory enzymes, substrates and products, see Balaban (2002), Jafri et al. (2001). Rather, our aim was to make the reader aware of the necessity to consider the transformations, components of metabolic networks, for understanding the processes accompanying CN intoxication. Interference with respiration is not the only process that may play a determining role.

METABOLIC NETWORKS: WORK IN PROGRESS

Metabolic and protein networks of toxin intoxicated tissue contain a wealth of information needed for the development of antidotes and therapy. The networks are best pictured as consisting of nodes and links. The former may be a substrate or a protein, and the link the interaction between two chemical nodes, a chemical transformation. The interconnected nodes form a structure.

The last few years has witnessed an explosion in understanding the structure of live and inanimate networks and attempts to relate them to performance. The research of Barabasi and his group, Jeong et al. (2000), Albert and Barabasi (2001), Barabasi (2002) showed that a network's topology is the key to its dynamic behavior and robustness when under attack.

[1] Letellier et al. (1994, 1998) have modeled the block of cytochrome oxidase by cyanide using metabolic control analysis.

A notable insight gained from tests on forty-three organism is that their structures are scale-free, that is, no single node represents all the nodes. In fact the distribution of nodes in the organism follows a power law, a few nodes linked with many neighbors, forming hubs, but the vast majority having only few neighbors. The hubs also lend heterogeneity to the topology. To test the robustness of the system to perturbations, links are removed at random. Most deletions do not affect the overall functioning of the system, but removal of hubs does. Both in a biological network as well as the internet, removal of hubs is catastrophic in the sense that the network no longer is able to function as intended. Thus for example catalytic enzymes may be removed from an organism without adverse effect on the overall performance. Removal of any of the hubs, on the other hand, could have disastrous consequences.

Questions of the effect of threat agents on an organism can be reformulated in terms of networks. The identification of the affected link gives clues on the cellular problem and possibly points to alternative or compensatory links that could be established to overcome the network fault, a fault that may adversely affect the functioning of the organism.

The study of the topology of metabolic networks, the identification of hubs, and the kinetics regulating the links coupled with flux analysis offers a further means to the understanding of the mechanisms involved in the action of toxins, Bhalla et al. (1999). It can shed light on the necessity for survival of certain node-link combinations and whether, under injury, how would it be possible to offer alternate routes or actions that would enhance the chance for survival. Ancillary work along these lines to determine the effect of the deletion (or turn-off) of certain genes on red blood cell function and e-coli have been performed. For the cardiac myocyte network, ATP is a hub. Its deletion or isolation leads to apoptosis, cell death. This is explored further in this paper.

CYANIDE-CAUSED CHANGES IN THE ECG

Lead I ECG from a sodium cyanide gas intoxicated adult male, at a lethal dose level of 0.2 mg/kg of body weight is shown in Figure 1. The ECG traces, interpreted in terms of myocyte membrane currents give a time-resolved record of the effect of cyanide on the heart. Some of the significant anomalies and deviations from the norm in this ECG include the following:

(1) The heart rate (HR) initially is 150 beats per minute (BM) that changes to a rate of 37.5 BM at the 3 minute mark, signaling bradycardia. At five minutes the rate increases to 54.5 BM and at 10 minutes, before severe arrhythmia develops, to 85.7 BM.

(2) Morphological changes in the T-wave, the repolarization of the ventricles, are apparent from the time t = 3 minutes on. There is a notable widening of the base of the wave, and fusing of the T-wave with QRS.

(3) The ST segment represents the early part of the ventricular repolarization. Under normal conditions, its duration is 0.2 sec or less dependent on HR. An ST segment changed by more than 1.0 mm from the horizontal, homeostatic position signals widespread myocardial injury, usually of the transmural kind, Li et al. (2000), Yan et al. (1999). Such a change is evident in Figure 1.

(4) Missing P wave. By t = 3 min, the P wave has disappeared, indicating no atrial activity. Absence of the P wave persists from that time on.

AN *IN SILICO* EXPLORATION OF CYANIDE-AFFECTED CARDIAC TISSUE

The ECG modulation due to the myocyte's membrane current changes is used as the basis of probing the effect of CN on the electrophysiology of the cell. Effects that diminish or block membrane currents dictate the requirements on the therapeutic agent. The computer simulations allow identifying those cellular parameters that can be most effective in controlling these changes.

Available models consider only indirectly if at all, the effect of metabolic changes on the electrophysiology. Myocardial ischemia causes anoxia,

Table 1 Summary some of the deviations from the norm in the ECG and its relationship to the myocyte membrane currents of a CN intoxicated individual.

Symptom	*Possible Cause*	*Primary Membrane current affected*
Disappearance of P waves	Auricular arrest	Depolarization of left and right atria by I_{Na}
T-wave modulation		Abnormal repolarization
Peaked upright	Subendocardial ischemia	I_K
Inverted	Subepicardial ischemia	
Polarity, width	Gap junction	
ST-segment	Ischemia, hyperkalemia	I_{KATP} activation
Elevation, shortening disappearance		Inactivation of I_{Na}, Brugada syndrome
QRS right axis deviation	Right bundle branch left posterior fascular block	Depolarization currents of right and left ventricles
Wenckebach Phenomenon	Defective conduction through AV node	
(Progressive lengthening of PR, until absence of QRS)		
Bradycardia, HR < 50	AV block, prolonged conduction between atria and ventricles	Decreased L-calcium channel activity
Tachycardia HR > 100	Increased activity of adrenergic system	I_f in sinus node
Torsade de Pointes, Ventricular Fibrillation	Various, including QT interval lengthening	

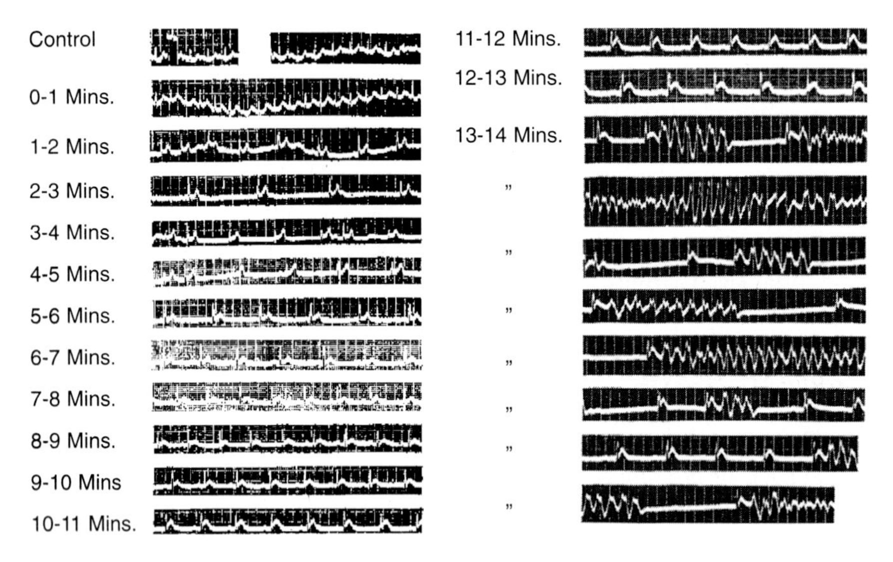

Fig. 1 Significant cyanide-induced changes in the ECG of a male adult. Reprinted from American Heart Journal, Vol. 34, Wexler et al. The effect of cyanide on the electrocardiogram of man, pp. 163-173, copyright (1947), with permission from Elsevier. Additional discussion of ECG modulation due to CN is given in Katzman and Penney (1993) and Salkowsky and Penney (1995).

acidosis and increase in the potassium concentration. In some respects it is characterized by symptoms analogous to CN intoxication. A number of models of ischemia have been proposed, as discussed by Ferrero et al. (1996), Shaw and Rudy (1997), and snapshots of the action potential calculated for specific levels of ATP, based on the experimental observations that in anoxia, the ancillary metabolic changes lead to a decline in the concentration of ATP. Below an ATP concentration threshold, the I_{KATP} current is activated which in turn affects the action potential duration and shape, Elliott et al. (1989), Nichols et al. (1991) and Light et al. (2001). These concepts can be extended for an explanation of the ECG changes in CN intoxication.

Cyanide inhibits cytochrome c oxidase activity that in turn causes a decrease in the mitochondrial respiration and exhibits a threshold behavior akin to some of the mitochondrial diseases. The reduction of ATP concentration opens K_{ATP} channels, Borowitz et al. (2001). In addition, increases in catecholamine levels, thought to contribute to the genesis of arrhythmias, Baskin et al. (1987), intracellular Na^+ accumulation, Kupriyanov et al. (1996), increases in inorganic phosphate, lactate production and elevated calcium levels, Kondo et al. (1998), are noted. This paper addresses the issue of ATP concentration change and its effect on I_{KATP} in atrial tissue.

Atrial K_{ATP} currents are incompletely understood at the present time. Baron et al. (2001) reported that pituitary adenylate cyclase-activating polypeptide (PACAP) activates the K_{ATP} current in rat atrial myocytes through the PKA and PKC pathways. 1 nmol/L of PACAP activates a current of 85 pA/pF compared to 26 pA/pF with 5 mmol/L of ATP. A number of peptides present in the heart, including VIP (vasoactive intestinal peptide), also activate I_{KATP} though not as effectively as PACAP. Thus the ultimate model will have to take several subsidiary, and heretofore neglected events, into account.

Metabolic impairment that activates I_{KATP} is thought to have cardioprotective function in ischemia, but also shortens the action potential duration [APD] explained on basis of increased $[K^+]_0$. Several models of I_{KATP} exist. Ferrero et al. (1996), Shaw and Rudy (1997) have included it in their model of the electrophysiology of the ventricles. These models propose a relationship between the current density and ATP concentration and give snapshots of the state of the tissue as I_{KATP} is activated. Since CN intoxication shares a number of characteristics of ischemia, including anoxia and acidosis, models of the latter are convenient starting points for CN affected tissue simulation.

The effect of the CN loading on the atrial tissue is modeled by including the I_{KATP} current, adjusting the potassium and sodium concentrations and $I_{Ca(L)}$.

A modified atrial model, Nygren et al. (1998), is used to demonstrate that modulation of the potassium currents, especially when I_{KATP} is present, can explain the AV blockage that is expressed by disappearance of the P wave on the ECG in CN poisoning. For this purpose, an ATP-dependent current is added to the membrane model and expressed as follows:

$I_{KATP} = fac1 * fac2 * fac3 * f_{ATP} * (V_m - E_K)$.

$fac1$ = channel density ~4.0 channels/μm^2,

$fac2$ = channel conductance ~ 40.0×10^{-3} nS/cm^2,

$fac3$ = effect of the presence of Mg^{2+} ions

f_{ATP} = a function of the ATP concentration,

with E_K the Nernst potential and V_m is the voltage in the membrane.

In the resting state, the ATP concentration is taken as 5 mmol/L. Three cases are calculated, the resting, the 50% reduction in ATP concentration and the case when the ATP has been used up.

For the calculations reported, the monodomain model of cardiac tissue was assumed to be valid. The properties of the cells are averaged over the whole domain without taking the anisotropy of the cardiac tissue into account. The action potential propagation is determined from the solution of the following equation:

$$\frac{\partial v}{\partial t}(\bar{x},t)=\frac{1}{C_m}\left[-i_{ion}(\bar{x},t)-i_{stim}(\bar{x},t)+\frac{1}{\beta}\left(\frac{\kappa}{\kappa+1}\right)\nabla\circ(D_i(\bar{x})\nabla v(\bar{x},t))\right]$$

where v is the voltage, *i* the current and β and κ characterize the tissue with D_i the diffusion coefficient, and C_m the cell membrane capacitance.

The calculations were carried out using CardioWave on the ARL Major Shared Resource Center's IBM SP3 and SP4 assets. To obtain a typical solution (i.e., the time-resolved electrophysiological state of the atria) for 300 ms real time in a tissue of 3 cm x 3 cm required 7 hr of nondedicated time using 16 processors. Data was stored at 5 ms intervals, and TIFF files were constructed of the voltage as a function of position in the two-dimensional plane of the tissue. The panels displayed were constructed using Tecplot, ver. 9.01.

RESULTS AND DISCUSSION

CN interference with the energy production of atrial tissue was simulated by studying the effect of the changes in magnitude of the I_{KATP} current on the action potential in order to shed more light on the processes leading to the disappearance of the P-wave in the CN-affected cardium. In addition, ischemic conditions were used as the hypothesized state of the tissue with $I_{Ca(L)}$ partially blocked.

Fig. 2 shows the atrial action potential at several concentrations of ATP, starting with the normal, homeostatic and corresponding to [ATP] concentrations in excess of 6.0, at 2.0, and 1.0 mmol/L respectively. Notable shortening of the action potential is observed, from 300 ms down to 10 ms at [ATP] = 1.0 mmol/L. In tandem, the slope of the repolarization changes. The maximum value of the membrane voltage drops from around 20 mV to the single digit values under conditions of increased Na_i^+ and K_0^+ that is accumulation of potassium ions outside of the cell. When the potassium and sodium is held at homeostatic values, the action potential does not reach the threshold voltage, indicating impending shutdown of the cell.

Under conditions of elevated potassium and sodium concentration values, Fig. 3, with a slightly elevated Mg^{2+} concentration, 3.5 mmol/L, approximating CN-toxicity conditions, with a 75 % block of the $I_{Ca(L)}$ current, drastic shortening of the action potential cycle is observed. Again the homeostatic cycle length (CL) of around 300 ms is shortened to 50 ms at [ATP] = 2.0 mmol/L, with CL =~10 ms at ATP = 1.0 mmol/L. The action potential maximum voltage value stays positive, though considerably lower, ~8.0 mV, than the normal value. In addition, the voltage never returns to the baseline value of –74.0 mV, but approaches –68.0 mV.

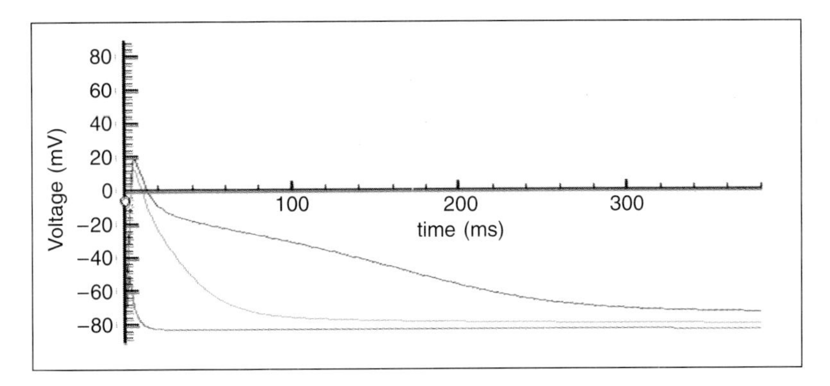

Fig. 2 The effect of the change in ATP concentration on the atrial action potential. The top curve denotes the homeostatic condition. The second curve from the top has [ATP] = 2.0 mmol/L and the bottom curve gives the condition for [ATP] = 1.0 mmol/L.

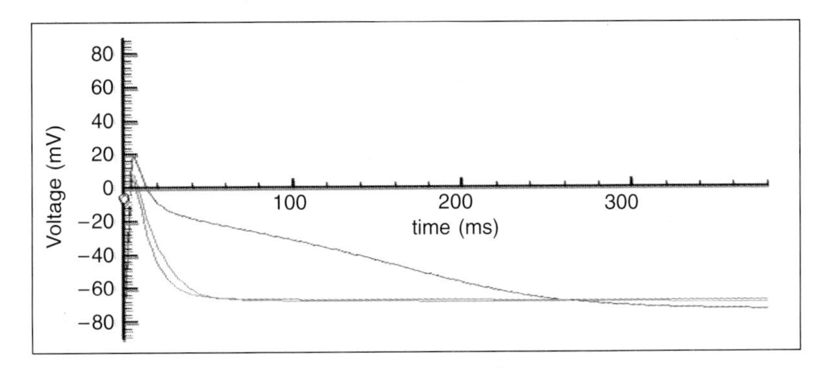

Fig. 3 CN toxicity causes ischemia-like conditions that increases the $[Na^+]_i$ and $[K^+]_o$ levels in the tissue. The action potential in the atria comparing the homeostatic (top) with conditions of [ATP] = 1.5 (middle) and 1.0 (bottom) at $[Na^+]_i$ and $[K^+]_o$ 17.0 and 12.0 with slightly elevated $[Mg^{2+}]$ of 3.5 mmol/L.

At elevated $[Na^+]_i$ value of 17.0 mmol/L the action potential hyperpolarizes. The hyperpolarization disappears as the $[K^+]_0$ is elevated, indeed under these conditions, the normal resting voltage is never reached.

Fig. 5 contrasts the normal action potential with the case when [ATP] = 2.0 mmol/L and $[Na^+]_i = 4.5$ mmol/L, subnormal condition when $[K^+]_0$ is held at 4.5 mmol/L. Again the CL is shortened, the maximum amplitude is lowered and the slope of the repolarization changes drastically.

The reduction in time available for the depolarization to take place, Fig. 4, as well as the voltage amplitude observed under these conditions, makes a strong case for the diminution and eventual complete disappearance of the P-wave of the ECG. In addition, blockage of the depolarization wave is a plausible scenario.

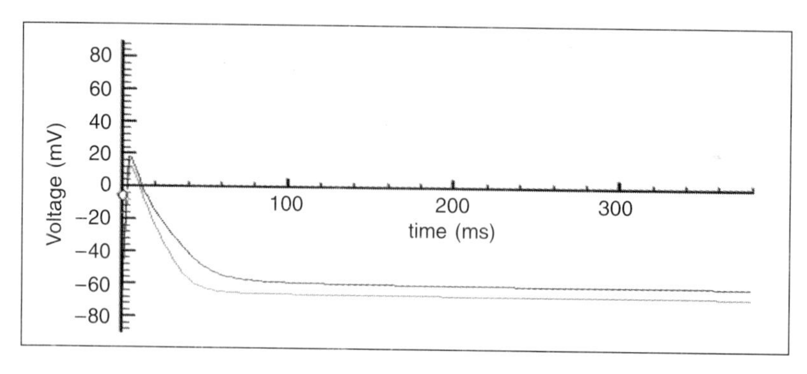

Fig. 4 In an atrial tissue with elevated $[K^+]_0$ = 12.0, the initial resting voltage is no longer reached. An elevation of $[Na^+]_i$ to 17.0 mmol/L, lower curve, depresses the resting voltage, but the trace remains above the original value.

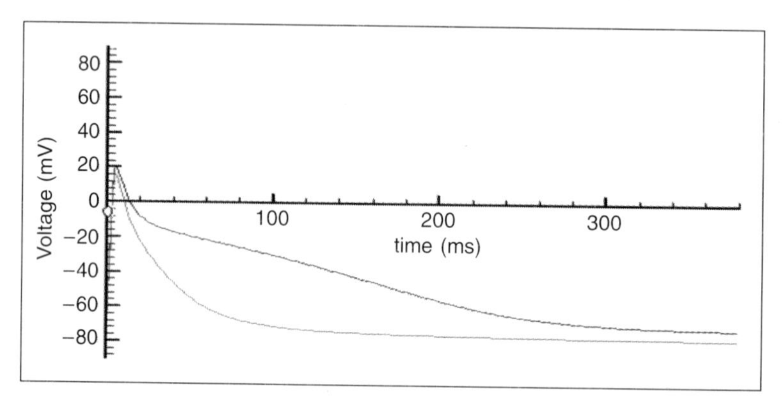

Fig. 5 Contrast of a normal atrial action potential (top) with one when the ATP concentration is depleted to 2.0 mmol/L and both potassium and sodium are at subnormal levels of 4.5 mmol/L.

This study addressed the effect of the flow and concentration changes of calcium ions only indirectly, by including a partial block of the $I_{Ca(L)}$. The rundown of the ATP energy supply for the SR (sarcoplasmic reticulum) pump that transfers calcium ions into the extracellular space against a concentration gradient leads to calcium overload. This may lead to abnormal calcium-induced currents, precursors to arrhythmia. The change in the morphology of the action potential is one of the expressions and possibly a contributor to the modulation of the calcium current, $I_{Ca(L)}$. The threshold for the activation of this current for myocytes is around −25 mV. Half-maximum deactivation is observed at or below −20 mV. One of the effects of CN on the action potential is narrowing of the area under the curve with decline in the ATP availability. This is signaled by the increase of the angle, measured from the horizontal, from the maximum of the voltage versus time plot. The narrowing indicates shortening of the time that the calcium ion channels are open and operative, and thus the amount of Ca^{2+} ions that are transferred. Consequently, the total ion flow is reduced.

The electrophysiological changes under CN intoxication of the atria lead to profound changes in the membrane currents needed for the maintenance of the P-wave of the ECG. Indeed, the effect of the CN is to render the membrane current levels insufficient to maintain the required wave shape and duration. These results suggest that pharmacological intervention could seek the restoration of the ion concentration levels as a possible approach to the restoration of the homeostasis.

Acknowledgments

It is a pleasure to thank Dr. John Pormann of Duke University for giving expert advice in the simulations. This work has benefited from computer time made available by the Department of Defense High Performance Computing Modernization Program. The computer calculations were made at the Major Shared Resource Center at ARL, Aberdeen Proving Ground, MD on the IBM SP3 and SP4.

References

Albert, R., Barabasi, A-L. Statistical mechanics of complex networks. Phys. Rev. 2001, pp.1-54.

Ansell, M., Lewis, F. A. S. A review of cyanide concentrations found in human organs. J. Forensic Med. 1970; **17:** 148-154.

Balaban, R. S. Cardiac energy metabolism homeostasis: role of cytosolic calcium. J. Mol. Cell Cardiol. 2002; **34:** 1259-71.

Barabasi, A-L. *Linked. The new science of networks*. Perseus Publishing, Cambridge, MA 2002.

Baron, A., Monnier, D., Roatti, A., Baertschi, A.J. Pituitary adenylate cyclase-activating polypeptide activates K_{ATP} current in rat atrial myocytes. Am. J. Physiol. Heart Circ. Physiol. 2001; **280:** H1058-H1065.

Baskin, S. I. The cardiac effects of cyanide. In *Cardiac Toxicology*, S. I. Baskin (editor), Boca Raton, FL: CRC Press, 1991, pp. 419-430.

Baskin, S. I., Wilkerson, G., Alexander, K., Blitstein, A. G. Cardiac effects of cyanide. In *Clinical and Experimental Toxicology of Cyanides*. B. Ballantyne and T. C. Marrs (editors) Bristol, England: Wright, 1987, pp. 138-155.

Bhalla, U. S., Iyengar, R. Emergent properties of networks of biological signaling pathways. Science 1999; **283:** 381-387.

Borowitz, J. L., Isom, G. E., Baskin, S. I. Acute and chronic cyanide toxicity. In *Chemical Warfare Agents: Low Level Toxicity.*, S. Somani and J.A. Romano, Jr. (editors), Boca Raton, FL: CRC Press, 2001, pp. 301-319.

Elliott, A. C., Smith, G. L, Allen, D. G. Simultaneous measurements of action potential duration and intracellular ATP in isolated ferret hearts exposed to cyanide. Circ. Res. 1989; **64:** 583-591.

Ferrero, J. M. Jr., Saiz, J., Ferrero, J. M., Thakor, N. V. Simulation of action potentials from metabolically impaired cardiac myocytes. Circ. Res. 1996; **79:** 208-221.

Jafri, M. S, Dudycha, S. J., O'Rourke, B. Cardiac energy metabolism: models of cellular respiration. Annu. Rev. Biomed. Eng. 2001; **3:** 57-81.

Jeong, H., Tombor, B., Albert, R., Oltvai, Z. N., Barabasi, A.-L. The large-scale organization of metabolic networks. Nature 2000; **407:** 651-654.

Katzman, G. M., Penney, D. G. Electrocardiographic responses to carbon monoxide and cyanide in the conscious rat. Toxicol. Lett. 1993; **69:** 139-153.

Kondo, R. P., Apstein, C. S., Eberli, F. R., Tillotson, D. L., Suter, T. M. Increased calcium loading and inotropy without greater cell death in hypoxic rat cardiomyocytes. Am. J. Physiol. 1998; **275:** H2272-H2282.

Kupriyanov, V. V., Yang, L., Deslauriers, R. Cytoplasmic phosphates in Na^+-K^+ balance in KCN-poisoned rat heart: a ^{87}Rb-, ^{23}Na-, and ^{31}P-NMR study. Am. J. Physiol. 1996; **270:** H1303-H1311.

Letellier, T., Heinrich, R., Malgat, M., Mazat, J. P. The kinetic basis of threshold effects observed in mitochondrial diseases: a systematic approach. Biochem. J. 1994; **302:** 171-174.

Letellier, T., Malgat, M., Rossignol, R., Mazat, J. P. Metabolic control analysis and mitochondrial pathologies. Mol. Cell Biochem. 1998; **184:** 409-17.

Light, P. E., Kanji, H. D., Fox, J. E. M., French, R. J. Distinct myoprotective roles of cardiac sarcolemmal and mitochondrial K_{ATP} channels during metabolic inhibition and recovery. FASEB J. 2001; **15:** 2586-2594.

Li, R. A., Leppo, M., Miki, T., Seino, S., Marban, E. Molecular basis of electrocardiographic ST-segment elevation. Circ. Res. 2000; **87:** 837-839.

Nichols, C. G., Ripoll, C., Lederer, W. J. ATP-sensitive potassium channel modulation of the guinea pig ventricular action potential and contraction. Circ. Res. 1991; **68:** 280-287.

Nygren, A., Fiset, L., Firek, J.W., Clark, D.S., Lindblad, R.B., Giles, W.R. Mathematical model of an adult human atrial cell. The role of K^+ currents in repolarization. Circ. Res. 1998; **82:** 63-81.

Opie, L. H. *The heart: physiology, from cell to circulation.* Lippincott Williams & Wilkins, Philadelphia, 1998.

Shaw, R. M., Rudy, Y. Electrophysiologic effects of acute myocardial ischemia. Circ. Res. 1997; **80:** 124-138.

Suzuki, T. Ultrastructural changes of heart muscle in cyanide poisoning. Tohoku J. Exp. Med. 1968; **95:** 271-87.

Salkowsky, A. A, Penney, D. G. Metabolic, cardiovascular, and neurologic aspects of acute cyanide poisoning in the rat. Toxicol. Lett. 1995; **75:** 19-27.

Wang, Y. X, Zheng, Y. M, Abdullaev, I., Kotlikoff, M. I. Metabolic inhibition with cyanide induces calcium release in pulmonary artery myocytes and *Xenopus* oocytes. Am. J. Physiol. Cell Physiol. 2003; **284:** C378-C388.

Way, J. L. Cyanide intoxication and its mechanism of antagonism. Annu. Rev. Pharmacol. Toxicol. 1984; **24:** 451-481.

Wexler, J., Whittenberger, J. L, Dumke, P. R. The effect of cyanide on the electrocardiogram of man. Am. Heart J. 1947; **34:** 163-173.

Yan, G. X., Antzelevitch, C. Cellular basis for the Brugada syndrome and other mechanisms of arrhythmogenesis associated with ST-segment elevation. Circulation. 1999; **100:** 1660-1666.

Index

About the Editors

DR. S.J.S. FLORA

Dr. S.J.S. Flora received his Ph.D. from Industrial Toxicology Research Center, Lucknow, India in 1985. He was a post-doctoral fellow at Utah State University and presently serves as the Deputy Director at Defence Research and Development Establishment (DRDE), Gwalior, India. Dr. Flora has made significant contribution to the clinical therapeutic measures for the treatment of metal poisoning. He has published over 125 original research papers/ review articles. He was awarded the prestigious Dr Kailash Nath Katju award for scientific excellence by the Government of Madhya Pradesh, India in the year 2003, Shakuntala Amir Chand award of Indian Council of Medical Research and Archna Gold Medal of Academy of Environmental Biology (India) in the year 2003. He is a member of the Royal Society of Chemistry (U.K.), fellow of Academy of Environmental Biology and is on the editorial board of seven international journals.

COLONEL JAMES A. ROMANO

Colonel James Romano received his Ph.D. from Fordham University in 1975. In 1994 he was certified as a Diplomate in General Toxicology by the American Board of Toxicology. He has been a university professor at Manhattan College in New York City and has served since 1978 as a researcher and scientific administrator in the United States Army. Colonel Romano was the Commander of the US Army Medical Research Institute of Chemical Defense (USAMRICD) and presently serves as the Deputy Commander of the US Army Medical Research and Material Command at Fort Detrick, MD. He is the author of more than 90 papers in the area of medical chemical defense and the editor of a successful textbook, "Chemical Warfare Agents: Toxicity at Low Levels."

DR. STEVEN I. BASKIN

Dr. Steven I. Baskin, Pharm.D, PhD, D.A.C.T., is a pharmacologist/toxicologist with a speciality in cardiovascular, neuro-, endocrine, and chemical terrorism. His areas of research interest are drug discovery, antidotes, analysis and biochemistry. Dr. Baskin has proposed, new, unique antidotes and models for the treatment of cyanide, mustard and organophosphates. He is an author or co-author of more than 100 scientific papers and technical reports. Dr. Baskin is currently Team Leader of the Biochemical Toxicology and Pharmacology Division of the USAMRCID at Aberdeen Proving Ground, MD, USA.

KRISHNAMURTHY SEKHAR

Mr. Krishnamurthy Sekhar, Director, DRDE, Gwalior, obtained his B.Tech and M.Tech in Chemical Engineering from IIT, Kharagpur and IIT, Madras in 1971 and 1973 respectively with specialization in Process Dynamics and Control. Under his able guidance first liquid propellant stage for *Prithvi* and *Agni* materialized. As Director, Reliability and Quality Assurance, he ensured a high quality of system and equipments development for the missile program leading to the successful induction of *Prithvi* and *Agni* systems. He is the recipient of DRDO award for excellene for his contribution to liquid propulsion technology.